III-Nitride Semiconductors: Optical Properties I

Optoelectronic Properties of Semiconductors and SuperLattices

A Series edited by M.O. Manasreh, Department of Electrical and Computer Engineering University of New Mexico, Albuquerque, USA

Volume 1 **Long Wavelength Infrared Detectors**
Edited by Manijeh Razeghi

Volume 2 **GaN and Related Materials**
Edited by Stephen J. Pearton

Volume 3 **Antimonide-Related Strained-Layer Heterostructures**
Edited by M.O. Manasreh

Volume 4 **Strained-Layer Quantum Wells and Their Applications**
Edited by M.O. Manasreh

Volume 5 **Structural and Optical Properties of Porous Silicon Nanostructures**
Edited by G. Amato, C. Delerue and H.-J. von Bardeleben

Volume 6 **Long Wavelength Infrared Emitters Based on Quantum Wells and Superlattices**
Edited by Manfred Helm

Volume 7 **GaN and Related Materials II**
Edited by Stephen J. Pearton

Volume 8 **Semiconductor Quantum Wells Intermixing**
Edited by E. Herbert Li

Volume 9 **InP and Related Compounds: Materials, Applications and Devices**
Edited by M.O. Manasreh

Volume 10 **Vertical-Cavity Surface-Emitting Lasers: Technology and Applications**
Edited by J. Cheng and N.K. Dutta

Volume 11 **Defects in Optoelectronic Materials**
Edited by K. Wada and S.W. Pang

Volume 12 **II-VI Semiconductor Materials**
Edited by Maria C. Tamargo

Volume 13 **III-Nitride Semiconductors: Optical Properties I**
Edited by M.O. Manasreh and H.X. Jiang

III-Nitride Semiconductors: Optical Properties I

Edited by

M. O. Manasreh and H. X. Jiang

Taylor & Francis Books, Inc.

New York • London

Denise T. Schanck, *Vice President*
Robert H. Bedford, *Editor*
Catherine M. Caputo, *Assistant Editor*
Thomas Hastings, *Marketing Manager*
Mariluz Segarra, *Marketing Assistant*

Dennis P. Teston, *Production Director*
Anthony Mancini Jr., *Production Manager*
Brandy Mui, *STM Production Editor*

Published in 2002 by
Taylor & Francis
29 West 35th Street
New York, NY 10001

Published in Great Britain by
Taylor & Francis
11 New Fetter Lane
London EC4P 4EE

Printed in the United States of America on acid-free paper.

10 9 8 7 6 5 4 3 2 1

Library of Congress Cataloging-in-Publication Data

III-nitride semiconductors: optical properties/edited by Omar Manasreh and H.X. Jiang,
p. cm.—(Optoelectronic properties of semiconductors and superlattices; v. 14–15)
Includes index.
ISBN 1-56032-972-6 (v. 1: alk. paper)—ISBN 1-56032-973-4 (v. 2: alk. paper)
1. Semiconductors—Optical properties. 2. Nitrides—Optical properties. 1. Title: 3-nitride semiconductors. II. Manasreh, Mahmoud Omar. III. Jiang, H.X. IV. Series.

QC611.6.O6 A 18 2002
537.6′226—dc21

2002018101

CONTENTS

ABOUT THE SERIES

The series *Optoelectronic Properties of Semiconductors and Superlattices* provides a forum for the latest research in optoelectronic properties of semiconductor quantum wells, superlattices, and related materials. It features a balance between original theoretical and experimental research in basic physics, device physics, novel materials and quantum structures, processing, and systems—bearing in mind the transformation of research into products and services related to dual-use applications. The following sub-fields, as well as others at the cutting edge of research in this field, will be addressed: long wavelength infrared detectors, photodetectors (MWIR–visible–UV), infrared sources, vertical cavity surface-emitting lasers, wide-band gap materials (including blue-green lasers and LEDs), narrow-band gap materials and structures, low-dimensional systems in semiconductors, strained quantum wells and superlattices, ultrahigh-speed optoelectronics, and novel materials and devices.

The main objective of this book series is to provide readers with a basic understanding of new developments in recent research on optoelectronic properties of semiconductor quantum wells and superlattices. The volumes in this series are written for advanced graduate students majoring in solid state physics, electrical engineering, and materials science and engineering, as well as researchers involved in the field of semiconductor materials, growth, processing, and devices.

PREFACE

This is the first part of a two part volume focusing on the optical properties of III-nitride semiconductors. The companion volume, #14 in the *Optoelectronic Properties of Semiconductors and Superlattices* is also available.

Research advances in III-nitride semiconductor materials and device have led to an exponential increase in activity directed towards electronic and optoelectronic applications. There is also great scientific interest in this class of materials because they appear to form the first semiconductor system in which extended defects do not severely affect the optical properties of devices. This volume consists of chapters written by a number of leading researchers in nitride materials and device technology with the emphasis on the time-resolved photoluminescence and Raman spectroscopies, optical properties of InGaN based III-nitride heterostructures, optical properties of homoepitaxial GaN, physics and optical properties of GaN/AlGaN quantum wells, characterization of GaN and related alloys by Raman scattering, Raman studies of wurtzite GaN and related compounds, and light emission from rare earth doped GaN. This unique volume provides a comprehensive review and introduction of defects and structural properties of GaN and related compounds for newcomers to the field and stimulus to further advances for experienced researchers.

Given the current level of interest and research activity directed towards nitride materials and devices, the publication of this volume is particularly timely. Early pioneering work by Pankove and co-workers in the 1970s yielded a metal-insulator-semiconductor GaN light-emitting diode (LED), but the difficulty of producing p-type GaN precluded much further effort. The current level of activity in nitride semiconductors was inspired largely by the results of Akasaki and co-workers and of Nakamura and co-workers in the late 1980s and early 1990s in the development of p-type doping in GaN and the demonstration of nitride-based LED's at visible wavelengths. These advances were followed by the successful fabrication and commercialization of nitride blue laser diodes by Nakamura et al. at Nichia. The chapters contained in this volume constitutes a mere sampling of the broad range of research on nitride semiconductor materials and defect issues currently being pursued in academic, government, and industrial laboratories worldwide.

We would like to thank all authors of the chapters, whose excellent efforts have made this volume possible.

M. O. Manasreh
University of New Mexico

and

H. X. Jiang
Kansas State University

June 2002

CHAPTER 1

Introduction

M.O. MANASREH[1] and H.X. JIANG[2]

[1]*Department of Electrical & Computer Engineering, University of New Mexico, Albuquerque, NM 87131, USA*
[2]*Department of Physics, Kansas State University, Manhattan, KS 66506-2601, USA*

III-Nitride wide bandgap semiconductors have been recognized as technologically very important materials. They have recently attracted considerable interest due to their applications in optical devices that are active in the blue/green and ultraviolet (UV) wavelength regions and in electronic devices capable of operation at high temperatures, at high power levels, and in harsh environments. Silicon- and GaAs-based devices operate only in the red and near-infrared spectral regions and cannot tolerate elevated temperatures or chemically hostile environments due to their low bandgap, uncontrolled generation of intrinsic carriers, and low resistance to caustic chemicals. Bright blue light-emitting diodes (LEDs) based on III-nitrides paved the way for full-color displays as well as for the mixing of the three primary colors to obtain white light for illumination by semiconductor LEDs. If used in place of incandescent light bulbs, these LEDs would provide compactness and longer lifetimes (>10,000 hours for LEDs compared with 2000 hours for incandescent light bulbs) while consuming only 15–20% of the power for the same luminous flux, resulting in significant energy savings. Short-wavelength laser diodes (LDs) based on III-nitrides are essential for high-density optical storage applications because the diffraction-limited optical storage density increases quadratically as the probe laser wavelength is reduced. Due to their large dielectric strengths, GaN-based devices can operate at much higher voltages for any dimensional configuration. Furthermore, they are virtually immune from environmental attack.

The foregoing are just a few examples of III-nitride device applications. The further development of III-nitride technologies has become a worldwide high priority. Devices based on III-nitride materials will lead to many technological improvements in a variety of systems and in many areas, including communications, transportation, energy, indoor and outdoor full-color displays, lighting and signs, medical instrumentation, flat-panel displays, and space and solar monitoring.

All optoelectronic applications based on III-nitrides are a direct consequence of their unique optical properties. In optoelectronic devices based on III-nitrides, such as UV/blue/green light emitters and UV detectors, the fundamental optical properties, such as optical transitions, carrier dynamic processes, and carrier–photon and carrier–phonon interactions predominantly determine device performance. Thus, knowledge of optical constants and their variations with external conditions are crucial for optimization of device performance.

The history of conventional III–V semiconductor optoelectronic device development has shown that understanding fundamental optical properties is crucial for the development of suitable material qualities and device structures; however, in the past, growth of group III nitrides suffered from poor crystalline quality and high n-type background carrier concentrations, and it was not possible to study the properties of many of the fundamental optical transitions in the system. The recent rapid progress in the III-nitride field has allowed the fundamental optical properties and parameters of these materials to be measured and their optical processes to be studied. Much information that is critical to the design and improvement of optoelectronic devices based on group III nitrides has been obtained by the efforts of the whole semiconductor community, this information includes the detailed band structure near the band edge, the effective masses of electrons and holes, the dielectric constants, the exciton binding energies, and the recombination lifetimes in epilayers and quantum wells (QWs). Knowledge concerning the band alignment among InGaN/GaN and GaN/AlGaN heterostructures, as well as the mechanisms of spontaneous, stimulated, and lasing emissions in III-nitride materials and LED/LD structures has also advanced significantly.

Until now, however, no text has been dedicated to the optical processes in III-nitrides. Many excellent books and review articles concerned with the growth and the general physical properties and prospects of III-nitrides have been published recently and have served as important references as well as guides for research for the nitride community; however, optical properties are only summarized within one or two chapters, and much important progress in the field has been left out. The tremendous research effort in the last decade has advanced knowledge concerning the optical properties of III-nitrides. Thus, to provide a more complete overview of the fundamental optical properties and processes in III-nitrides and related device applications, a text that combines contributions from active experts in the field with diverse backgrounds would be useful.

Rapid advances in III-nitride technologies build on the knowledge accumulated over the past few decades of research in III–V and II–VI semiconductors; however, due to the unique properties of III-nitrides, they being with them many particular problems in growth, characterization, and device fabrications. Due to the lack of GaN substrates, III-nitride epilayers are most commonly grown on sapphire or SiC substrates, which contain a high dislocation density of the order of $10^{10}/cm^2$. This causes complications in measuring and understanding many fundamental optical properties and parameters. In many cases, optical transitions with the same physical origins can behave quite differently in samples grown by different methods (e.g., MOCVD vs. MBE). One example is the exciton recombination lifetimes in unintentionally doped GaN epilayers. Values differing by more than one order of magnitude have been reported by different groups due to the fact that such quantities depend on the excitation intensity used in the experiment, the sample quality and purity, as well as the strain presented in the material under investigation, all of which can vary in different samples.

InGaN and AlGaN are direct bandgap semiconductor alloys with energy bandgaps varying from 1.9 to 3.4 eV (InGaN) and 3.4 to 6.2 eV (AlGaN), with a predominant wurtzite structure for both alloys at all compositions. It is well known that localized exciton recombination is the dominant optical process in many semiconductor alloys at low temperatures, including CdSSe, GaAsP, and CdMnTe alloys. It has also been shown theoretically that the amplitude of the fluctuating potential at the band edges caused by alloy fluctuation is strongly correlated to the energy gap difference between the two semiconductors, e.g., between GaN and AlN for AlGaN alloys, in which the energy gap difference is 2.8 eV. This is the largest energy gap offset ever studied for semiconductor alloys and is much larger than the typical value of a few hundred meV in II–VI semiconductor alloys, in which a strong exciton localization effect is known to be present. It is thus expected that alloy fluctuation will play an important role in determining the optical properties of nitride alloys.

Another important phenomenon in III-nitride alloys is the segregation of InN, or InN quantum dot formation, in InGaN alloys, particularly in materials, with high indium content due to the low solubility of indium in GaN. There is an evidence that the alloy fluctuation or InN segregation in InGaN alloys induces carrier and exciton localization, and reduces nonradiative recombination processes, i.e., enhances the radiative recombination rate or quantum efficiency. In spite of the recent intensive studies of optical properties of InGaN and AlGaN alloys, there are still many unknowns concerning the indium-rich InGaN alloys and aluminum-rich AlGaN alloys. For InGaN,

InN segregation makes it difficult to grow indium-rich InGaN alloys. For aluminum-rich AlGaN, in addition to the difficulties in material growth, it is very difficult to measure the optical properties in the deep UV region ($< 240\,\text{nm}$). Though AlN has been widely used in buffer layers as well as to form alloys with GaN, until now photoluminescence (PL) data near the band edge ($\sim 200\,\text{nm}$) of AlN are still absent.

Optical properties of III-nitride QWs are more complicated in comparison with other better understood III–V QWs due to the lack of GaN substrates and the large energy bandgap offset and lattice mismatch between the well and the barrier materials. Although some of the properties of III-nitride QWs are similar to those of more conventional semiconductor QWs, the combination of various effects in III-nitride QWs make them unique. For example, $In_xGa_{1-x}N/In_yGa_{1-y}N$ ($x \neq y$) or InGaN/GaN QWs have been used as active media in UV/blue light emitters. In addition to the variations in the QW structural parameters, such as well and barrier widths, and InN composition, the properties of the optical transitions in the well regions are affected significantly by many other effects such as (a) piezoelectric field, (b) polarization field, (c) well width fluctuation, (d) alloy fluctuation, and (e) InN segregation or InN quantum dot formation. Only effect (c) is known to affect the optical transitions in the well regions in more conventional QWs such as GaAs/AlGaAs. All the effects mentioned are not necessarily detrimental to device performance. For example, the presence of the polarization and piezoelectric fields in AlGaN/GaN heterostructures is known to be beneficial to the performance of AlGaN/GaN heterojunction field-effect transistors (HFET), because they enhance the 2D electron density and hence improve the overall performance of HFETs. On the other hand, these built-in fields are also known to decrease the radiative recombination rate in III-nitride QWs. Thus, understanding these effects on the optical properties of III-nitride QWs is important in terms of both fundamental physics as well as device applications.

Due to the wide bandgaps of III-nitrides, the experimental setups for optical studies of them are also more complicated and expensive than those for conventional III–V semiconductors. The optical characterization system typically consists of a laser and a detection system with either UV or deep-UV capability. Due to the various effects present in III-nitride materials previously mentioned, time-resolved PL spectroscopy has become an important technique for characterizing material quality and purity as well as for understanding the mechanisms of optical transitions in III-nitrides because such a technique can provide much information concerning the temporal behaviors of optical transitions and carrier recombination dynamics. Nonetheless, a time-resolved spectroscopy system for studying III-nitrides must be equipped with ultrafast excitation laser and detection systems with UV capabilities. It is interesting to note that the research and development of

III-nitrides has also promoted advances in various technologies and improvements in experimental measurement methods. For example, prior to 1980, time-resolved PL spectroscopy was employed to study impurity-related transitions such as donor–acceptor pair recombination in GaN semiconductors in the wavelength region limited to above 400 nm with a nanosecond or even microsecond time resolution. Now, many groups have the capability of performing time-resolved PL measurements down to 300 nm with a picosecond or femtosecond time resolution. Further advancements in this technique that will eventually allow measurements to be performed on pure AlN down to 200 nm with a picosecond time resolution are in progress. We believe that the availability of such a measurement system will accelerate the development of deep-UV optoelectronic devices based $Al_xGa_{1-x}N$ alloys with high Al contents. III-Nitride research has surely accelerated the development of various other technologies worldwide, including those for MOCVD growth and plasma dry etching processes.

The progress in group III nitride material growth and device fabrication is truly breathtaking, but many problems remain to be solved, primarily with material quality. As demonstrated by III-nitride bright blue LEDs, LDs, and HFETs, all III-nitride–based devices must take advantage of heterojunctions and QWs and the tunability of the bandgap in the alloys, from InN (1.9 eV) to GaN (3.4 eV) and to AlN (6.2 eV). Thus, another important issue is QW or heterostructure device structural perfection, which requires further improvements in the material quality of the alloys and in the interface properties of low-dimensional systems. The aim of this text is to review in-depth the most recent experimental and theoretical results on the optical properties of III-nitrides including epilayers, heterostructures, and QWs, which are crucial to the design of optoelectronic devices based on them. We hope that this text will aid in the further development and improvement of material quality and device structures and that it will provide insights regarding future research trends. We hope also that this book will also be of use to incoming new graduate students and researchers in physics, materials sciences, and engineering.

Since we are dealing with a large and rapidly evolving field with extensive current research activity, we are unable to include some of the most recent developments, such as the optical properties of InAlGaN quaternary alloys. These alloys have recently been recognized as having the potential to overcome some of the shortcomings of GaN as well as InGaN and AlGaN alloys. By varying In and Al compositions in InAlGaN, one can tailor the energy bandgap while keeping the lattice matched with GaN. The potential applications of InAlGaN quaternary alloys in InGaN/InAlGaN QW light emitters, GaN/InAlGaN HFETs, and UV detectors have recently been demonstrated, and optical studies of InAlGaN quaternary alloys are burgeoning. Although the growth of high-quality and high-purity InAlGaN alloys presents a challenge, because more parameters must be controlled and optimized, the advantages of lattice match with GaN and bandgap engineering of these alloys will eventually be utilized in many existing and future III-nitride devices. This research trend will be similar to that with the more conventional semiconductor systems, since many of the current high-performance optoelectronics devices utilize quaternary alloys. The foregoing is just one example; many more applications may result from the ongoing extensive research activities.

In order to provide a coherent picture of the fundamental optical properties and processes in III-nitrides, each chapter in this text reviews only one specific topic.

Chapter 2 reviews time-resolved PL studies of the III-nitrides including GaN epilayers, InGaN and AlGaN alloys, $In_xGa_{1-x}N/GaN$, $In_xGa_{1-x}N/In_yGa_{1-y}N$ $(x\gamma y)$, and $GaN/Al_xGa_{1-x}N$ QWs. The recombination dynamic processes of fundamental optical transitions in GaN involving free exciton, impurity-bound exciton, band-to-impurity, and band-to-band recombination are reviewed. Exciton localization dynamics in InGaN and AlGaN alloys are discussed. For III-nitride QWs, the recombination dynamics of optical transitions in both $In_xGa_{1-x}N/GaN$ and $GaN/Al_xGa_{1-x}N$ QWs grown by different methods (MOCVD and MBE) are summarized. The use of time-resolved PL to probe the effects of the internal built-in fields on the optical properties and carrier recombination dynamics and to identify the mechanisms of stimulated emission and optical gain is described, and the optimal growth conditions and optimal QW structural parameters, are discussed. The chapter also provides information on the required instrumentation for carrying out time-resolved PL studies. Despite the fact that the investment in such a measurement system is large, the number of time-resolved PL systems designated exclusively for III-nitride studies has significantly increased worldwide due to their unique capabilities to provide insights into improvements in material quality and device design as well as to elucidate the dynamic processes of optical transitions in this material system.

Carrier dynamics, thermalization time, and distributions are important parameters for designing faster and better devices, since carrier injections are involved in many optoelectronic devices. Carrier relaxation also provides information about different scattering processes and interactions, such as electron–electron, hole–hole, and electron–hole scattering as well as electron–phonon and hole–phonon, and phonon–phonon interactions. Chapter 3 summarizes the results of ultrafast carrier relaxation dynamics and carrier distributions in GaN obtained through the use of time-resolved Raman spectroscopy. The experimental setup and approach are described. The results of dynamical properties of GaN epilayers including electron–phonon scattering rates, phonon–phonon interaction times, decay channels of optical phonons, and nonequilibrium electron distributions are summarized.

Chapter 4 summarizes the optical properties of InGaN-based III-nitride heterostructures studied by PL, absorption, temporal evolution, and stimulated emission. As active media for commercial UV/blue/green LEDs and LDs, InGaN alloys and InGaN/GaN QWs have attracted extensive research efforts in areas such as the effects of alloy fluctuations, and quantum dot–like indium phase separations. The effects of piezoelectric and polarization fields on the optical properties of InGaN-based heterostructures studied by a wide range of techniques are surveyed. Nonlinear optical properties as well as possible mechanisms for stimulated emission in InGaN-related structures are also discussed.

Since GaN substrates are not yet commercially available, most III-nitride epilayers are grown on foreign substrates, most commonly on sapphire or SiC. Most optical data accumulated over the last decade are for III-nitride materials heteroepitaxially grown on sapphire. The lack of good-quality lattice-matched substrates for epitaxial growth caused many problems in the development of III-nitride related knowledge and technology. With recent progresses in GaN bulk growth it is now possible to grow homoepitaxial nitride materials and to study their optical properties. For example, line widths as narrow as 0.1 meV have been observed in the PL emission spectra of homoepitaxially grown GaN epilayers. Optical studies of homoepitaxial materials, which are summarized in Chapter 5, provide much information about the fundamental properties of this system. The potential of III-nitrides cannot be fully utilized until high-quality nitride substrates become available. With worldwide efforts in research and development, nitride substrates could become available commercially in a few years. Thus, understanding the optical properties of homoepitaxial nitride epilayers and other structures will be very useful for future technology based on III-nitride materials.

Chapter 6 covers optical properties as well as carrier dynamics in GaN/AlGaN QWs, which have applications in UV emitters and detectors. Elucidating the properties of GaN/AlGaN heterostructures can also provide useful information for AlGaN/GaN HFETs. The major differences between III-nitride QWs and conventional III–V semiconductor QWs, say GaAs/AlGaAs, are the lattice mismatch between the barrier and well layers, and the wurtzite structure, which induce piezoelectric and polarization electric fields along the growth direction. These fields significantly affect the overall material quality, such as interface quality and critical thickness. They also make the analysis of optical properties of III-nitride QWs more complex compared with that of other conventional III–V semiconductors. Among InGaN/GaN and GaN/AlGaN QWs, the optical processes in GaN/AlGaN QWs are less complex, since the wells are formed by binary compound semiconductors, and the effects of alloy fluctuations on the well transitions can be eliminated. Another advantage of the GaN/AlGaN QW system is that high-quality samples can be produced by either MOCVD or MBE. Thus, differences in optical properties due to different growth methods can be distinguished.

In addition to PL, Raman spectroscopy is also very commonly used for the optical characterization of III-nitrides. It provides information regarding material quality and purity, surface and interface properties, and strain of heterostructures, and so forth. Fundamental parameters of phonons are also very important for many device applications, since phonon–carrier interactions can affect the speed and efficiency of the devices. Characterization of phonon–carrier interactions is also important for understanding the many optical processes involved. Chapters 7 and 8 summarize phonon properties in III-nitrides as well as the use of Raman spectroscopy to characterize III-nitrides, including GaN, AlN, and InN epilayers, and AlGaN and InGaN alloys. The effects of impurity doping on phonon properties are also discussed, and the evaluation of strain, composition fluctuation, and other parameters probed by Raman scattering is reviewed. Phonon properties in QWs with effects of quantum confinements and strain-induced electric field as well as in quantum dots are also summarized.

Chapter 9 summarizes the optical properties of rare earth–doped GaN epilayers. It has been realized recently that III-nitrides serve as a good host for rare earth elements and that they have many potential applications in optical communications such as temperature-stable optical amplifiers and LEDs operating at wavelengths from visible to infrared. Studies of rare earth–doped semiconductors have shown that thermal quenching of rare earth element–related emission intensity decreases with increasing bandgap of the host materials. This phenomenon has naturally led recent research efforts toward rare earth doping within wide bandgap III-nitride semiconductors. PL spectra of rare earth–doped GaN epilayers at both IR and visible regions for the materials prepared by ion implantation as well as by in situ doping are surveyed. Potential applications of rare earth–doped III-nitrides for visible LEDs are discussed.

CHAPTER 2

Time-Resolved Photoluminescence Studies of III-Nitrides

H.X. JIANG[1], J.Y. LIN[1], and W.W. CHOW[2]

[1]*Department of Physics, Kansas State University, Manhattan, KS 66506-2601*
[2]*Sandia National Laboratories, Albuquerque, NM 85718-0601*

2.1 INTRODUCTION

III-Nitride based devices have great potential in applications such as UV/blue laser diodes (LDs) and light-emitting diodes (LEDs), solar-blind UV detectors, and high-power/high-temperature electronic devices. Researchers in this field have made extremely rapid progress toward material growth as well as device fabrication [1]. The commercial availability of superbright blue and green LEDs and blue/UV LDs based on III-nitrides is a clear indication of the great potential of this material system [2–4]. Recently, there has been much research concerning the fundamental optical transitions in III-nitrides, including GaN epilayers, InGaN and AlGaN alloys, and multiple quantum wells (MQWs) comprising alternating layers of $In_xGa_{1-x}N/GaN$, $In_xGa_{1-x}N/In_yGa_{1-y}N$ $(x \neq y)$ and $GaN/Al_xGa_{1-x}N$. In particular, studies of optical transitions in these materials are extremely important for understating their fundamental properties as well as for their practical applications. Fundamentally important information regarding exciton binding energy and Bohr radius, carrier– and exciton–phonon interactions, decay lifetimes, and quantum efficiencies can be obtained from these studies [6].

For practical applications, optical transitions provide a simple and effective way for calibrating sample quality. It is well known that free excitonic transitions can be observed only in high-quality samples, and the line widths of these transitions are directly correlated with parameters such as sample uniformity and stoichiometry [6]. More importantly, the dynamics of various optical transitions can provide important information regarding excitation and energy transformation processes and recombination lifetimes of injected carriers, which are strongly correlated with quantities such as quantum efficiency and optical gain in III-nitrides. These basic quantities are crucial to the design of optoelectronic devices. For example, if the radiative channel has a characteristic recombination time constant τ, and the nonradiative channel has a recombination time constant τ', the effective recombination time constant τ_{eff} is given by [Ref. 5 and refs. therein]:

$$\frac{1}{\tau_{eff}} = \frac{1}{\tau} + \frac{1}{\tau'}, \tag{2.1}$$

where τ_{eff} is also the experimentally measured recombination lifetime. The magnitude of τ_{eff} is the length of time a photoexcited carrier can remain in the conduction (or valence) band and is thus directly correlated to material quality, purity, and doping level. For an intrinsic material, the total recombination rate is

$$R_T = \left(\frac{1}{\tau_{eff}}\right)\left(\frac{np}{2n_i}\right) \tag{2.2}$$

where n_i is the intrinsic carrier concentration, and n and p are the injected electron and hole concentrations, respectively. The radiative recombination rate is

$$R = \left(\frac{1}{\tau}\right)\left(\frac{np}{2n_i}\right). \tag{2.3}$$

The radiative efficiency is then

$$\eta = \frac{R}{R_T} = \frac{\tau'}{(\tau + \tau')}, \tag{2.4}$$

which is directly correlated with recombination lifetimes. Furthermore, one of the most important parameters determining the performance and design of many semiconductor devices is the lifetime of excess minority carriers, which can be obtained by measuring the dynamical behaviors of optical emissions involved using time-resolved photoluminescence (PL). With recent advancements in epitaxial growth techniques, remarkable improvement in III-nitride crystal quality has been achieved. Investigations of the fundamental optical transitions in III-nitrides and their dynamical processes have become increasingly important for understanding the physical properties of this important new class of materials.

In this chapter, time-resolved PL studies of III-nitrides including GaN epilayers, InGaN and AlGaN alloys, $In_xGa_{1-x}N/GaN$, $In_xGa_{1-x}N/In_yGa_{1-y}N$ ($x \neq y$) and $GaN/Al_xGa_{1-x}N$ MQWs are reviewed. For GaN epilayers, the results for n- and p-type (Mg doped) and semi-insulating GaN epilayers are summarized, which include the optical transitions involving free exciton, impurity-bound exciton, band-to-impurity, and band-to-band recombination. For MQWs, the recombination dynamics of optical transitions in both $In_xGa_{1-x}N/GaN$ and $GaN/Al_xGa_{1-x}N$ MQWs grown by different methods (MOCVD and MBE) are summarized. The results of MQWs are compared with one another as well as with GaN epilayers and InGaN and AlGaN alloys to extrapolate the mechanisms and quantum efficiencies of the optical emissions in these structures. The implications of these results on device applications, particularly for the UV/blue/green LEDs and LDs as well as on the lasing mechanisms in III-nitride lasers, are also discussed.

2.2 INSTRUMENTATION

A time-resolved PL spectroscopy system can be used to study the time evolution of the PL emission or the dynamical processes of photoexcited carriers, which are beyond the reach of continuous-wave (cw) PL spectroscopy techniques. A cw PL system provides only time-integrated (or average) information, whereas a time-resolved PL study can reveal both time-resolved and time-integrated information. Temporal responses of PL emissions from a sample cannot be measured through the use of a cw PL system but can be obtained by using a time-resolved PL system. In particular, time-resolved PL spectroscopy is very productive for obtaining the recombination rates (or lifetimes) of various transitions. Thus, determining the dynamical processes including emission line energies and the associated recombination rates as well as quantum efficiencies is one of the basic purposes of time-resolved PL studies. As we described in the previous section, a thorough knowledge of dynamical processes of optical transitions is of both fundamental and technological importance for III-nitride semiconductor materials. In order to study the carrier dynamical processes experimentally; however, one needs a sufficiently good time resolution, which means that the resolvable minimum time resolution Δt must be at least a factor of 5 to 10 shorter than the time scale t of the optical process under investigation. In this section, methods and instruments for measuring PL decay lifetimes and time-resolved spectra are presented.

A time-resolved PL measurement system includes a fast-pulsed laser system for excitation, a monochromator for dispersing the emitted photons, and a fast detector for signal detection. The emission wavelength of GaN epilayers is around 360 nm, which leaves only a few choices for the laser system for the photoexcitation of nitrides. The two most commonly used laser systems in time-resolved PL spectroscopy are the frequency-tripled femtosecond (fs) or picosecond (ps) Ti:sapphire lasers and the frequency-quadrupled ps neodymium-doped yttrium–aluminum–garnet (YAG) lasers. Coherent Inc. and Spectra Physics are the two primary vendors for these types of laser systems. The frequency-tripled Ti:sapphire laser can provide excitation laser wavelengths as short as 260 nm. The frequency-quadrupled YAG lasers can provide excitation wavelengths as short as 266 nm, but with a pulse width of approximately 100 ps and a fixed repetition rate around 76 MHz. It is thus more advantageous to insert a cavity-dumped dye laser between the frequency-doubled YAG laser and the second frequency doubler [7]. The cavity dumper and the cavity-dumped dye laser can control the repetition rate, increase the pulse energy, and compress the pulse width from 100 ps to approximately 8 ps. In this way the excitation wavelength can be tuned approximately from 285 to 310 nm if Rhoadamine-6G dye is used in the dye laser.

Two techniques are commonly used for detecting the time-resolved PL signals. One involves the use of synchroscan streak cameras, which are provided either by Hamamatsu or Handland Photonics, Inc. At present the only detector that can attain a time resolution of about 1 ps is the streak camera system. The other technique utilizes a time-correlated single-photon-counting detection system together with a microplate-channel photomultiplier tube (MPC-PMT from Hamamatsu), which can provide a time resolution of about 25 ps if a deconvolution technique is employed. One advantage of using a time-correlated single-photon-counting detection system together with a MPC-PMT is that the cost is relatively low. The disadvantages are that such a technique gives a lower system time resolution and requires more effort in data gathering than the streak camera system, particularly for obtaining the time-resolved PL spectra; however, recent developments (by PicoQuant) in personal computer–based time-correlated single-photon-counting measurement electronics have made the technology much more accessible to routine use and have increased its performance and functionality.

As an illustration, Figure 2.1 shows a time-resolved PL image of a GaN epilayer taken by the streak camera detection system in the authors' laboratory at Kansas State University, in which the horizontal and vertical axes represent the PL emission wavelength and delay time, respectively. The density of the white dots of the image indicates the PL emission intensity at different delay times and wavelengths. Three horizontal and two vertical boxes are also included in the image. The data from the two vertical boxes provide the PL transient behaviors (or temporal responses) at two representative wavelengths, which are plotted in the top half of Figure 2.2, which

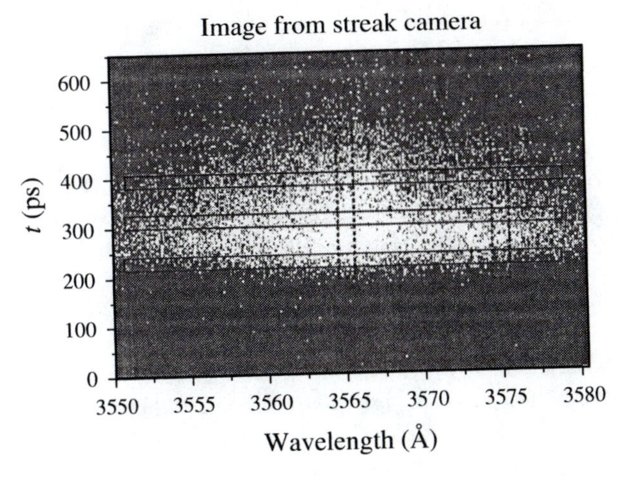

Figure 2.1 PL emission image of a GaN sample recorded by a streak camera.

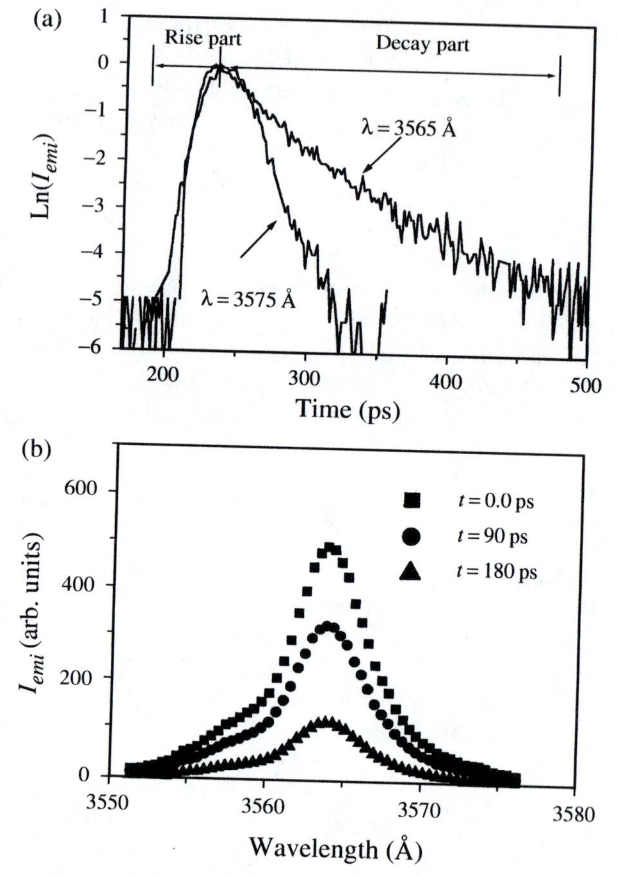

Figure 2.2 Top: Temporal responses of PL emission from a GaN sample measured at two different wavelengths obtained by a streak camera (from data in Figure 2.1). Bottom: An example of time-resolved PL spectra recorded at three different delay times obtained by a streak camera (from data in Figure 2.1).

is a semilogarithmic plot of the PL transient data and contains two parts, the rise and decay parts of the signal. It is clearly seen that in this case the PL decays are exponential, and the decay lifetimes can thus be determined. Each of the three horizontal boxes provides an emission spectrum at a fixed delay time. Time-resolved PL spectra (i.e., emission spectra at three different delay times) can thus be obtained if the three spectra are plotted together, as shown in the bottom half of Figure 2.2. These are the two most important data analysis techniques employed in time-resolved PL studies. When the dependences on other experimental parameters such as sample temperature, excitation intensity, or pumping power are measured, time-resolved PL is a very powerful technique for revealing optical recombination processes in semiconductors.

2.3 RECOMBINATION DYNAMICS IN GaN EPILAYERS

2.3.1 Free Excitons

If the quality and purity of a semiconductor material are sufficiently high (low densities of defects and impurities), the photoexcited electrons and holes pair off into free excitons (FX) by Coulomb interaction, which recombine, emitting a narrow spectral line. In a direct-gap semiconductor like GaN, where momentum is conserved in a simple radiative transition, the energy of the emitted photon by free exciton recombination is simply

$$h\nu = E_g - E_x, \tag{2.5}$$

where E_x is the binding energy of the free exciton.

In pure hexagonal GaN epilayers, there exist three types of free excitons, including free A-, B-, and C-excitons (with the holes of excitons from the A-, B-, and C-valence bands, respectively) [8–14]. At the cost of a lower transition probability, a direct transition can also occur with the emission of one or more optical phonons, with the energy of the emitted photon being [Ref. 5 and refs. therein]

$$h\nu = E_g - E_x - mE_p, \quad (m = 0, 1, 2, 3, \ldots) \tag{2.6}$$

where m is the number of optical phonons emitted per transition. In general, the larger m is, the lower the transition probability and the weaker the corresponding emission line.

Figure 2.3 shows four low-temperature (10 K) PL spectra obtained for three GaN epilayers grown on sapphire substrates by MOCVD, samples A ($n \sim 5 \times 10^{16}$ cm^{-3}), B ($n \sim 3 \times 10^{17}$ cm^{-3}), C ($p \sim 1 \times 10^{17}$ cm^{-3}), and an MBE-grown GaN epilayer on a sapphire substrate (sample D, insulating) [10–13]. For high-purity epilayers grown by MOCVD [e.g., sample A], two emission lines with energy peak positions at about 3.485 eV and 3.491 eV were identified as due to the recombination of the ground state of free A and B excitons [A(n = 1) and B(n = 1)], FX_A and FX_B. For an MBE-grown high-quality and -purity epilayer (sample D), three emission lines at 3.483 eV, 3.489 eV, and 3.498 eV were observable, which coincided very well with the corresponding transitions of the ground state of FX_A, FX_B, and the first excited state of FX_A seen in the MOCVD-grown layers [9–12]. The results revealed that the A-exciton binding energy is around 20 meV, and the

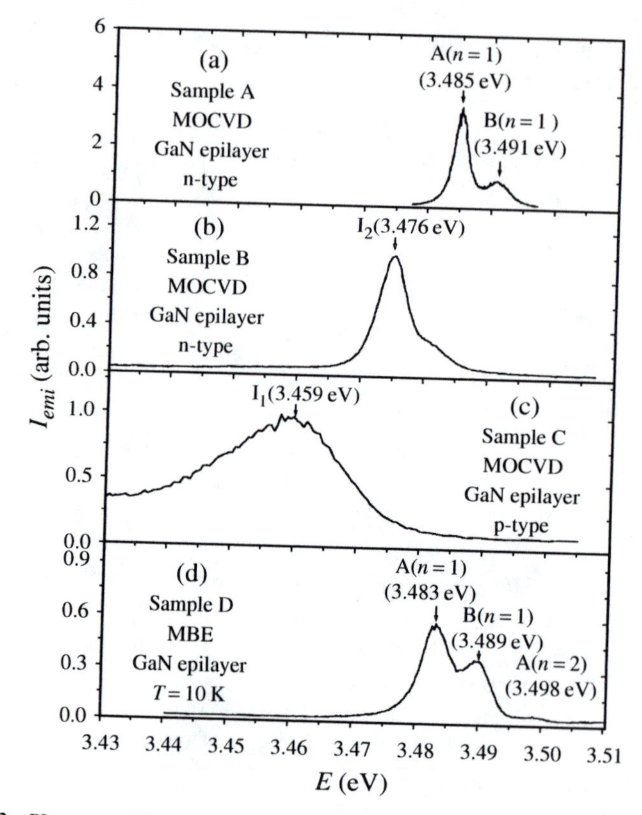

Figure 2.3 PL spectra of four different GaN epilayers measured at 10 K [after Refs. 10 and 11].

energy separation between the A and B valance bands is about 6 meV with the assumption that A and B exciton binding energies are comparable [11]. It is generally known that the spectral peak position as well as the binding energy of excitons in GaN epilayers may vary from one sample to another depending on the nature of substrate as well as the layer thickness.

The decay lifetimes of the A- and B-excitons have also been measured in high-purity and -quality GaN epilayers at low temperatures [10,11]. The exciton luminescence decays exponentially, $I(t) = I_0 e^{-t/\tau}$, where τ defines the recombination lifetimes. Figure 2.4a shows the recombination lifetimes of the A- and B-excitons measured at their respective emission peak positions at 40 K as functions of relative excitation intensity, I_{exc}. The B-exciton recombination lifetimes were about 15% shorter than those of the A-excitons. The lifetime dropped from 0.31 to 0.20 ns for the A-exciton, while it decreased from 0.27 to 0.17 ns for the B-exciton as the relative excitation intensity was increased from 0.1 to $1.0 I_0$. The observed behavior was attributed to exciton–exciton interaction, which caused a reduction in the

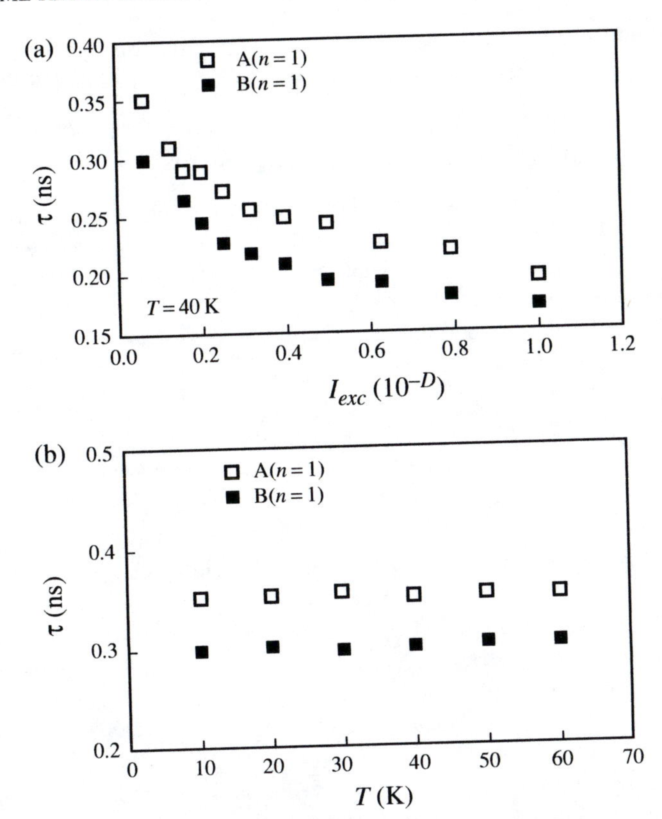

Figure 2.4 The decay lifetime of the A-exciton (□) and B-exciton (■) as a function of (a) excitation intensity, I_{exc} ($\sim 10^{-D}$) measured at 40 K, where I_{exc} was controlled by a set of neutral density filters with different optical densities D, and (b) sample temperature [after Ref. 11].

free-exciton recombination lifetime; however, such an interaction is absent at low free-exciton densities.

The temperature dependence of the recombination lifetime of $A(n = 1)$ and $B(n = 1)$ measured at low excitation intensity is shown in Figure 2.4b. The lifetimes of both A- and B-excitons were nearly temperature independent at $T < 60$ K, which may imply that radiative recombination was the dominant process at $T < 60$ K. Additionally, because of the temperature-independent behavior of the recombination lifetimes, the dependence of excitation intensity on temperature is expected to be as shown in Figure 2.4a. Thus, the radiative recombination lifetimes of single excitons can be obtained by extrapolating the curves in Figure 2.4a to the lowest excitation intensity. The results thus suggest that Figure 2.4a the radiative recombination lifetimes for single excitons are about 0.35 and 0.3 ns for the A- and

B-excitons, respectively, and Figure 2.4b the radiative decay rate increases with an increase of excitation pumping power for both A- and B-excitons.

2.3.2 Excitons Bound to Neutral Impurities

In the presence of impurities, bound excitons may be obtained. These bound excitons have much smaller kinetic energies than free excitons. Thus, when these excitons recombine in materials of high crystalline qualities (low defect density), their emission is characterized by a spectral line at a lower photon energy and narrower linewidth than that of the free excitons. The energy of the photon emitted is

$$h\nu = E_g - E_x - E_{bx}, \tag{2.7}$$

where E_x is the free-exciton binding energy, and E_{bx} is the additional energy binding the free exciton to the impurity center. The radiative decay lifetime of bound excitons is predicted to increase with binding energy E_{bx} [15,16]. In p-type GaN materials, the free exciton can be bound to a neutral acceptor (A^0), and the corresponding PL emission line is called A^0X transition (or I_1). In n-type GaN materials, a free exciton can be bound to a neutral donor (D^0) or to an ionized donor (D^+), and the corresponding PL emission lines are called D^0X and D^+X transitions (or I_2 and I_3), respectively, by adapting the notations for hexagonal II–VI semiconductors [17–19].

As temperature increases, the bound excitons become thermally ionized. It is possible to statistically predict the number of bound excitons at a given temperature. The ionization energy (or the binding energy E_{bx}) of the bound excitons can be determined from

$$\frac{I(T)}{I(0)} = \frac{1}{1 + CT^{3/2}e^{-(E_{bx}/kT)}}, \tag{2.8}$$

where $I(T)$ and $I(0)$ are the emission intensities at temperatures T and 0 K, respectively.

Excitons bound to neutral donors [I_2 or (D^0X)], to neutral acceptors [I_1 or (A^0X)], and to ionized donors [I_3 or (D^-X)] have been reported in GaN epilayers [20–37]. Since unintentionally doped GaN epilayers are most often n-type, the most extensively studied impurity-bound excitons are I_2 transitions. Optical properties of I_1 and I_2 have been summarized in several review articles. Here we concentrate on the recombination dynamics and time-resolved PL studies of these impurity-bound excitons.

In samples with relatively high background donor concentrations (e.g., sample B in Figure 2.3) the dominant PL emission line is due to the recombination of the excitons bound to neutral donors (I_2) [20–23]. Figure 2.3b shows the PL spectrum for such an epilayer grown on sapphire (sample B in Figure 2.3) [23], in which the I_2 transition occurs at about 3.476 eV. The shoulder at about 3.484 eV in sample B was due to the free A-exciton ($n = 1$) recombination. Thus, the binding energy of the neutral-donor-bound exciton was determined to be about 8–9 meV. In Mg-doped p-type epilayers (e.g., sample C in Figure 2.3), very often the dominant emission line is due to the recombination of neutral-acceptor-bound excitons (I_1) at around 3.46 eV [24–26]. A value of about 26 meV was thus obtained for the binding energy of the neutral-acceptor-bound exciton in GaN. These values correlated very well with the ionization energies of I_2 and I_1 deduced from Eq. (2.8) when the temperature dependences of the PL emission intensities of the I_2 and I_1 transitions were measured [23,26], which further confirmed the assignments of these emission lines.

The temperature dependence of the I_2 recombination decay lifetime measured at the spectral peak position is plotted in Figure 2.5a. It shows that the decay lifetime decreased with temperature, which was due to an increased nonradiative decay rate as well as the dissociation rate of the neutral-donor-bound excitons to neutral donors and free excitons at higher temperatures, described by

$$(\mathrm{D}^0, \mathrm{X}) \rightarrow \mathrm{D}^0 + \mathrm{X}. \tag{2.9}$$

By extrapolating the temperature dependence of τ shown in Figure 2.5a to $T = 0$ K, a value of about 0.13 ns was obtained for the I_2 recombination.

Figure 2.5b shows the emission energy dependence of the I_2 decay lifetime measured at $T = 10$ K, which shows τ decreasing with increasing emission energy. It has been predicted and confirmed in many semiconductors that the radiative recombination lifetime of a bound exciton increases with an increase in its binding energy, E_{bx} [15,16]. The emission energy dependence of the I_2 decay lifetime was thus a consequence of the existence of a distribution of E_{bx}. Since higher emission energies correspond to lower values of E_{bx}, τ thus decreases with increasing emission energy.

From the well-resolved free exciton and I_2 emission peaks of an MOCVD-grown GaN epilayer, Shan et al. [27] have also measured decay lifetimes of FX and I_2 transitions at low temperatures. The time evolution for both FX and I_2 emission was dominated by exponential decay with decay lifetimes of about 35 and 55 ps for FX and I_2, respectively. The radiative lifetime of FX and I_2 transition have also been estimated from the following equation [28,29]:

$$\tau_R = \frac{2\pi \varepsilon_0 m_0 c^3}{n e^2 \omega^2 f}, \tag{2.10}$$

where f is the oscillator strength of the transition, n is the refractive index, and the other symbols have their usual meanings. The radiative lifetime of

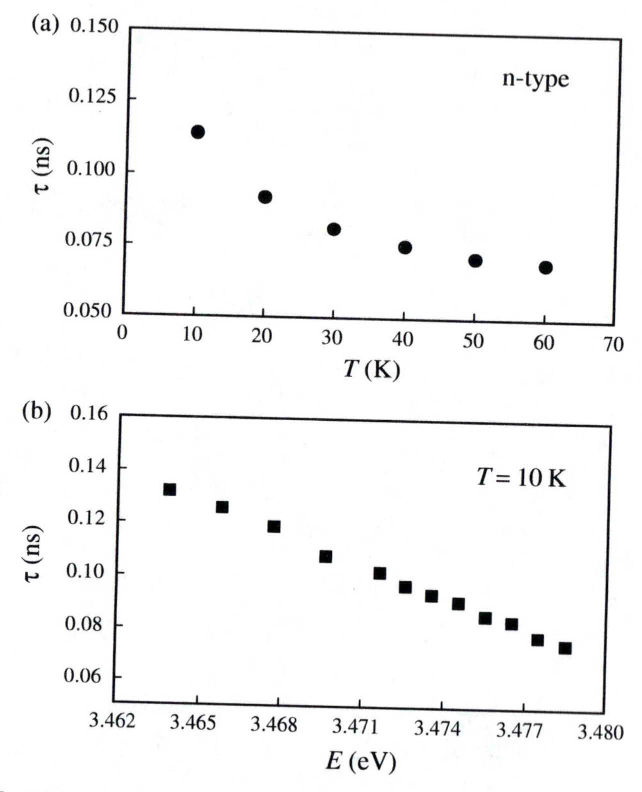

Figure 2.5 The decay lifetime of a neutral-donor-bound exciton, or the I_2 transition line in an n-type GaN epilayer (sample B of Figure 2.3), as a function of (a) temperature and (b) emission energy measured at 10 K [after Ref. 23].

bound excitons in GaN was thus estimated on the order of several hundreds of picoseconds by taking the upper limit of its oscillator strength as unity ($f \sim 1$), while that of FX was on the order of several tens of nanoseconds.

The discrepancy between the measured and the estimated radiative decay lifetimes of FX and bound excitons was discussed in terms of nonradiative relaxation processes in competition with the radiative channels. It was found that the measured PL decay lifetimes were directly correlated with the defect-related yellow line (YL) at about 2.2 eV. It was found that the stronger the YL emission the shorter the PL decay lifetime in a GaN sample. Figure 2.6 shows a comparison of FX PL decay between two samples with strong and weak YL emissions [27]. The deduced decay lifetime of FX emission in the sample with stronger YL emission was less than 15 ps, which reflected the fast trapping of excitons and carriers at defects and impurities. Thus, the measured decay lifetime of FX emission was governed by nonradiative recombination. This may also be one of the reasons for the discrepancies

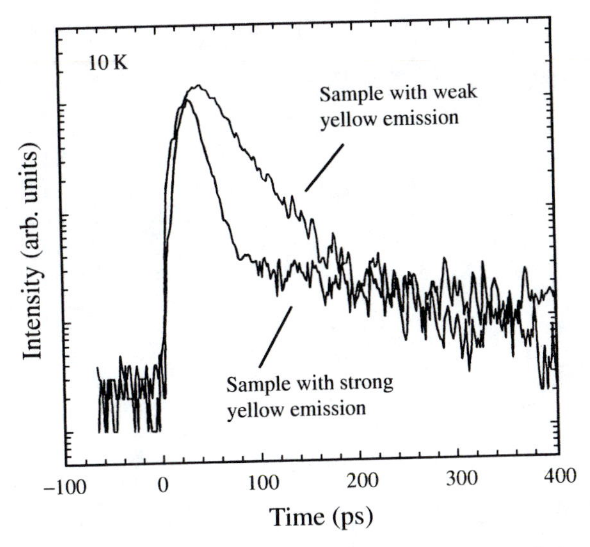

Figure 2.6 Comparison of the temporal response of the free-exciton emission between two GaN epilayers with a relative intensity ratio of 100 : 1 for the yellow emission under the same excitation conditions [after Ref. 27].

between measured exciton lifetimes in different samples obtained in various laboratories. Additionally, as illustrated in Figure 2.4a, the measured exciton lifetimes in many cases are also dependent on the excitation intensities used in different laboratories.

The PL decay behaviors of a Mg-doped p-type GaN epilayer (sample C in Figure 2.3) have also been studied under different conditions [26]. Figure 2.7a plots emission energy dependence of the recombination lifetime of the I_1 transition measured around the I_1 spectral peak at $T = 10\,\text{K}$. At 10 K, the recombination lifetime was observed to decrease monotonically from 0.56 to 0.30 ns with increasing emission energy, while a larger change in τ was observed in the energy region between 3.445 and 3.465 eV. This emission energy dependence of τ was similar to that of the I_2 transition [23] and can be explained by the existence of a distribution of E_{bx} of neutral-acceptor-bound excitons.

The temperature dependence of the recombination lifetime of the I_1 transition measured at its spectral peak positions is plotted in Figure 2.7b. As can be seen, τ was independent of temperature below 20 K and decreased with increasing temperature above 20 K, similar to the behavior seen for the I_2 transition shown in Figure 2.5a and was due to the increased nonradiative recombination rate as well as the dissociation rate of I_1 to A^0 and free excitons at higher temperatures described by

$$(\text{A}^0, \text{X}) \rightarrow \text{A}^0 + \text{X}. \tag{2.11}$$

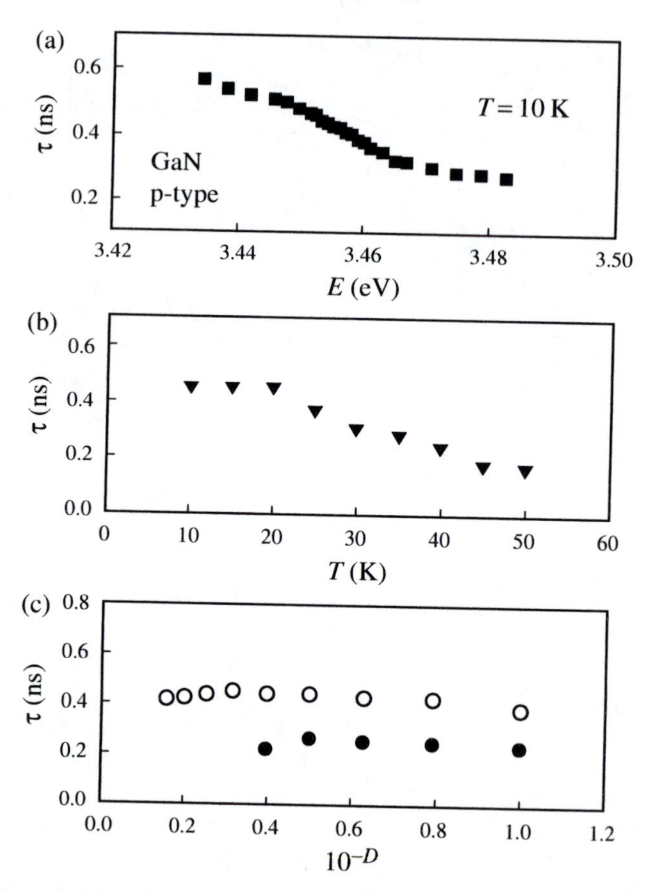

Figure 2.7 (a) Emission energy, (b) temperature, and (c) excitation intensity dependence of the recombination lifetime of the I_1 transition in a p-type GaN epilayer (sample C of Figure 2.3) measured (a) around the I_1 spectral peak at $T = 10$ K, (b) at the I_1 spectral peak positions, and (c) at two representative temperatures $T = 10$ K (○) and 35 K (•) [after Ref. 26].

The temperature dependence of the recombination lifetime τ shown in Figure 2.7b could also explain the rapid luminescence intensity quenching of the I_1 emission line at higher temperatures.

Figure 2.7c plots the excitation intensity dependence of the recombination lifetime of the I_1 transition measured at its spectral peak position for two representative temperatures, $T = 10$ K (○) and 35 K (•). Weak excitation intensity dependence has been observed at both of these temperatures. Figure 2.7c indicates that even with very high Mg doping concentration (typically in the range of 10^{19} cm^{-3}), the excitons bound to neutral Mg acceptors were isolated under the highest excitation intensity used in the experiment. Therefore, the measured recombination lifetimes

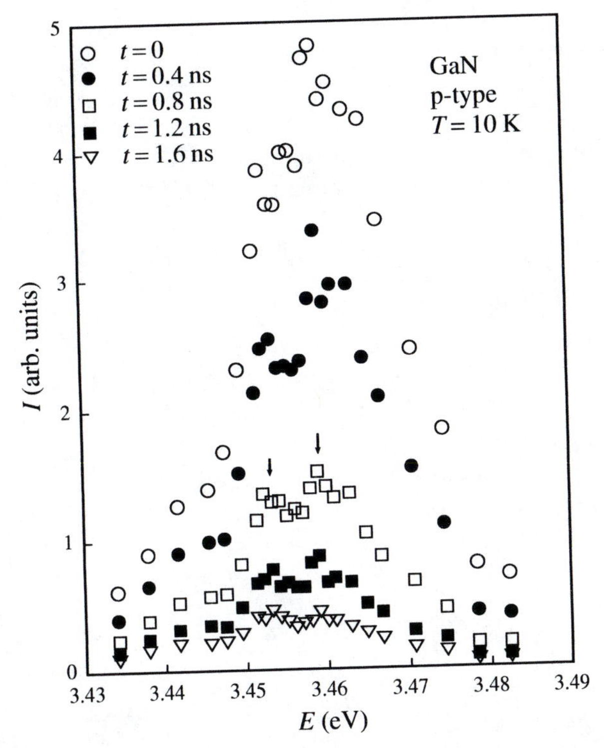

Figure 2.8 Time-resolved PL spectra in the energy region of the I_1 transition line in a p-type GaN epilayer (sample C of Figure 2.3) at several representative delay times at $T = 10$ K [after Ref. 26].

should represent the recombination lifetimes of isolated acceptor-bound excitons.

Other important information that is obtainable only from time-resolved PL measurements is the time-resolved emission spectra, which are the PL emission spectra at different delay times. Time-resolved PL spectra may reveal more detailed or extra features that cannot be resolved by cw PL measurements. Figure 2.8 shows the time-resolved PL spectra obtained in the energy region of the I_1 transition line at several representative delay times at $T = 10$ K [26]. At delay time $t = 0$, an emission spectrum that was almost identical with the cw spectrum shown in Figure 2.3c was observed; however, an additional emission line that was absent in the cw spectrum was resolved in the time-resolved spectra at later delay times. As indicated by the two arrows in the $t = 0.8$ ns spectrum, two emission lines at 3.453 and 3.459 eV were seen clearly. In fact, the intensity ratio of these two emission lines evolved with delay time. At delay time $t = 1.6$ ns, the intensity of the 3.453 eV emission line was even higher than that of the dominant emission line at

3.459 eV, although the former one was hardly observable in the spectrum of $t = 0$ or in the cw spectrum. It was speculated that the 3.459 eV emission line was due to the recombination of the acceptor-bound excitons associated with the activated Mg impurities, while the 3.453 eV line might be due to the recombination of the acceptor-bound excitons associated with inactivated Mg impurities. This speculation was supported by the fact that the 3.453 eV line observed in the time-resolved PL emission spectra was the peak position of the I_1 emission line observed in early high-resistive Mg-doped samples [24], considering the temperature difference involved.

Time-resolved PL was also employed to study the decay dynamics of free excitons (FX), neutral-donor-bound excitons (D^0X) at 3.478 eV, and two neutral-acceptor-bound excitons at 3.473 eV (A^0X_1) and 3.461 eV (A^0X_2) at low temperatures on thick (about 80 μm) GaN layers grown on sapphire substrates by hydride vapor-phase epitaxy (HVPE) [30,31]. A value of 3.6 ns was observed for the radiative lifetime of the acceptor-bound exciton transition at 3.461 eV. The dominant mechanism responsible for the nonradiative recombination of the bound excitons was shown to relate to dissociation of the bound excitons into free excitons as observed in GaN epilayers grown by MOCVD.

The low-temperature (4 K) decay lifetimes τ in HVPE-grown samples were found to be about 45, 70, and 480 ps, respectively, for the FX, D^0X, and A^0X_1 transitions in one report [30], while a more recent work reported the radiative recombination lifetimes of about 295 and 530 ps at 4 K for FX and I_2 transitions, respectively [31]. A surprisingly long lifetime (3.6 ns) was observed for the transition labeled A^0X_2 at 3.461 eV [30]. The temperature dependence of the recombination lifetime τ for the FX, D^0X, and A^0X_1 and A^0X_2 transitions measured at their spectral peak positions is shown in Figure 2.9 [30]. For transitions involving the impurity-bound excitons, D^0X, A^0X_1 and A^0X_2, τ decreased with temperature above 30 K due to an increased nonradiative recombination rate. It was assumed that the radiative decay rate (W_r) or lifetime (τ_r) for the bound excitons is temperature independent and can be found by extrapolating the temperature dependence of τ to $T = 0$ and that the nonradiative recombination rate (W_{nr}) is thermally activated according to

$$W_{nr} = A \exp(-E_{bx}/kT), \tag{2.12}$$

where E_{bx} is again the activation energy (or binding energy) of the bound excitons, and A is a temperature-independent constant. This assumption gives the following expression for the experimentally measured PL decay lifetimes τ:

$$\tau^{-1} = W_r + W_{nr} = \tau_r^{-1} + A \exp(-E_{bx}/kT). \tag{2.13}$$

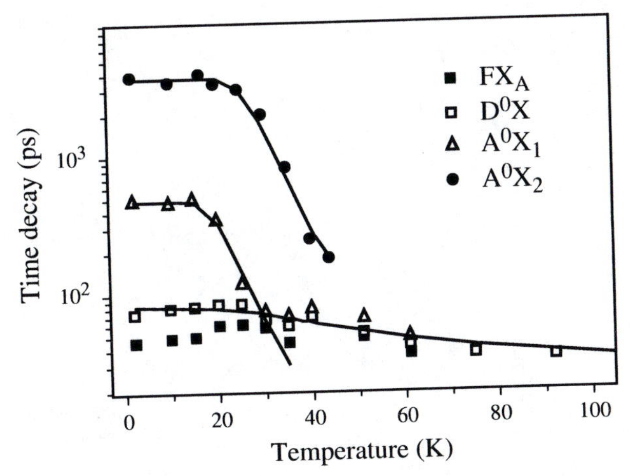

Figure 2.9 Temperature dependence of the recombination lifetimes of the optical transitions, in an 80-μm GaN layer grown by HVPE, involving the free exciton FX_A, neutral-donor-bound exciton D_0X, and two different neutral-acceptor-bound excitons A_0X_1 and A_0X_2 measured at their spectral peak positions. Dots are experimental PL decay times. Solid lines represent the fit using Eq. (2.13) [after Ref. 30].

The solid lines in Figure 2.9 were the results of the best fit using Eq. (2.13). The activation energy E_{bx} was determined to be 8.2, 13.9, and 23.3 meV for the D^0X, A^0X_1, and A^0X_2 transitions, respectively. These values correlated rather well with the binding energies of the bound excitons, that is, the separation between the PL peak position of the FX and that of the bound excitons, i.e., 6.2, 12.2, and 23.4 meV for the D^0X, A^0X_1 and A^0X_2 transitions, respectively. Thus, the main nonradiative process for the bound excitons is related to the dissociation of the neutral donor- (or acceptor-) bound excitons into free excitons and neutral impurities described by Eqs. (2.9) and (2.11).

The decay dynamics of free and bound excitons at low temperatures (below 70 K) in homoepitaxial GaN epilayers grown by MOCVD have also been studied [32]. The advantage of using homoepitaxial epilayers (in comparison with more common heteroepitaxial layers) was the superior optical quality, which resulted in narrower emission linewidths (<1 meV), due to the reduced strain and hence dislocation density. Figure 2.10 shows a low-temperature PL spectrum of a Mg-doped GaN homoepitaxial layer [32]. The intrinsic A-exciton (FE or FX) structure was weakly observed at about 3.78 eV, the neutral-donor-bound exciton (DBE or D^0X) was observed at around 3.472 eV, and the neutral-acceptor-bound exciton (ABE or A^0X) was seen at about 3.4664 eV. FX decay lifetime was measured at different temperatures. The A-exciton recombination typically showed a fast initial decay (with a time constant τ_{FX} of about 40 ps) followed by a much slower secondary decay at longer times (after about 200 ps). The deduced decay lifetimes, assuming exponential decays, were evaluated and plotted as a

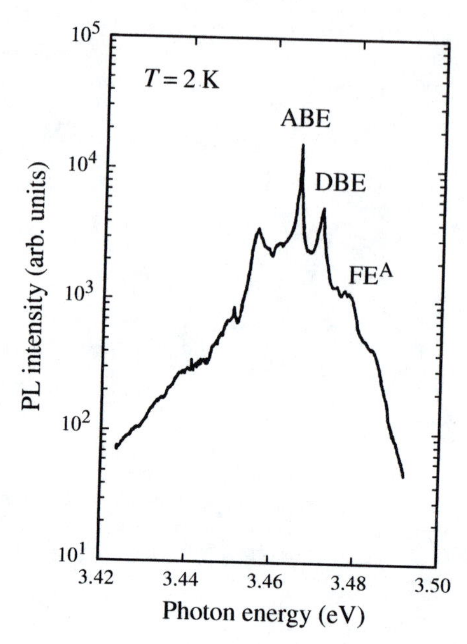

Figure 2.10 PL spectrum at 2 K of a lightly Mg-doped GaN homoepitaxial layer [after Ref. 32].

function of temperature in Figure 2.11. In Mg-doped homoepitaxial layers, the FX decay lifetime τ_{FX} was shortest, with a minimum value of about 25 ps at low T, mainly due to the capture of FX to the shallow impurities as well as to the nonradiative recombination at defects. The value τ_{FX} increased to a maximum value of about 90 ps at 35 K, then decreased with temperature again at higher T. The donor-bound exciton decay lifetime in Mg-doped samples was about 100 ps at low temperatures (below 30 K). The neutral-acceptor-bound exciton decay lifetime was the longest, around 800 ps, and constant over the temperature range 2–15 K, but decreased quickly with increasing of temperature due to the dissociation process described by Eq. (2.11).

Free and bound exciton decay dynamics have also been studied for GaN epilayers grown by a sublimation technique on 6H–SiC substrates at low temperatures [33]. The measured decay lifetimes were $\tau = 20 \pm 5$ ps for the FX at 3.4805 eV, very close to the limit of the system time resolution; $\tau = 34 \pm 5$ ps for the donor-bound exciton, I_2, at 3.4722 eV; and $\tau = 160 \pm 5$ and 370 ± 40 ps for two different acceptor-bound excitons with energies at 3.459 eV (I_1) and 3.4672 eV (I_1'). The measured decay lifetimes of bound excitons had an approximate relationship with their binding energies according to the theory of Rashba and Gurgenishvili [15], $\tau \backsim E_{bx}^{3/2}$. An upper limit of the FX oscillator strength of 0.0046 ± 0.0005 for the GaN

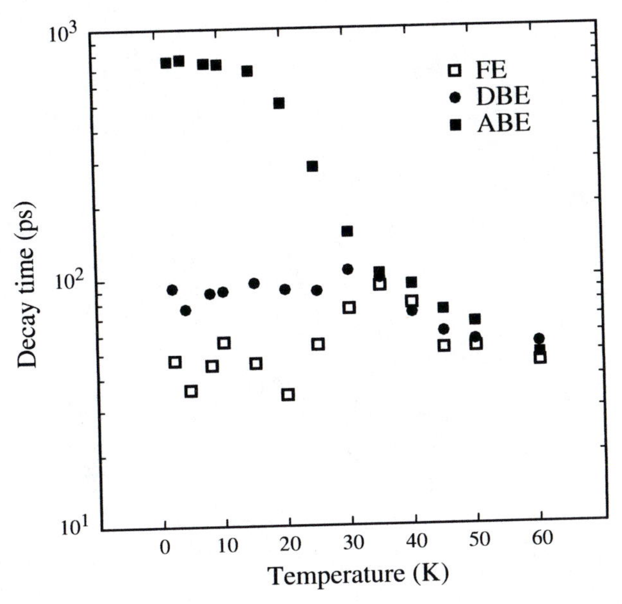

Figure 2.11 Temperature dependence of the measured decay times for the three major excitonic transition lines shown in Figure 2.10 [after Ref. 32].

epilayer was deduced [34], which was in the range of those reported for other direct-wide-gap semiconductors such as ZnSe and ZnO [34,35].

2.3.3 Excitons Bound to Ionized Impurities

The ionized-donor-bound exciton (D^+X) was also recently identified [25,36,37]. Time-resolved PL spectroscopy has also been used to study the radiative recombination of excitons bound to ionized donors in MOCVD-grown GaN epilayer co-doped with both Mg and Si at concentrations of $5 \times 10^{18}\,\text{cm}^{-3}$ and $1.5 \times 10^{17}\,\text{cm}^{-3}$, respectively [37]. The temporal character of the PL features exhibited by a co-doped GaN epilayer was recorded with a streak camera. Figure 2.12 shows time-resolved PL spectra recorded at three representative delay times relative to the excitation laser pulse. The spectra recorded at times $t = 0$, 50, and 1250 ps were arbitrarily scaled for presentation. The very short lived peak at 3.4845 eV was identified as the free exciton associated with the A-valence band (FX_A) [9–14]. The FX_A position implied a binding energy of 20.5 meV for the 3.464 eV peak, which identified it as A^0X (or I_1) [24–26]. The remaining peak at 3.473 eV had a binding energy of 11.5 meV, which was in very good agreement with the reported D^+X (or I_3) binding energy as determined from cw PL spectroscopy measurements [25,33]. The (D^+X) assignment was further supported by the

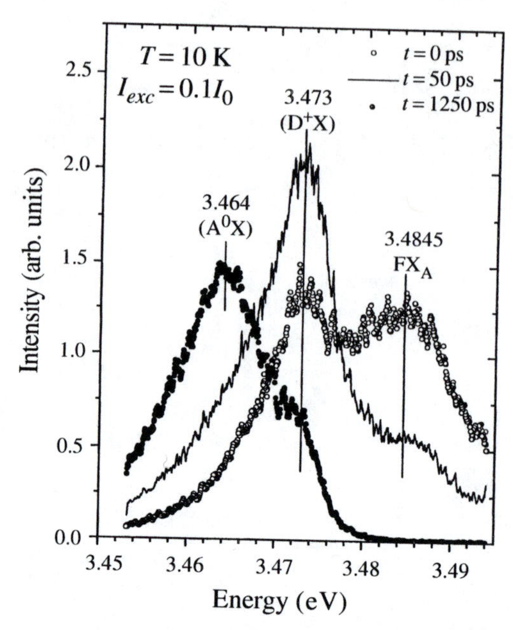

Figure 2.12 Time-resolved PL spectra recorded at three representative delay times relative to the incident laser pulse for a co-doped (with Si and Mg) GaN epilayer. The spectra recorded at times $t = 0$, 50, and 1250 ps were arbitrarily scaled for presentation [after Ref. 37].

expectation that the Si donors in this sample were fully compensated by the higher concentration of Mg acceptors.

The FX_A emission lifetime in these co-doped samples was very short ($\leq$20 ps), which implied that excitons were either rapidly captured by ionized donors or neutral acceptors or quenched by other nonradiative processes. Furthermore, the time-resolved data revealed that in co-doped GaN epilayers the FX_A emission decays before the onset of PL decay at the (D^+X) or (A^0X) emission energies. These facts supported a model for (D^+X) formation whereby ionized donors capture free excitons. It may also be possible that (D^+X) states form in a two-step process involving free electrons and holes ($D^+ + e^- \rightarrow D^0$ followed by $D^0 + h^+ \rightarrow (D^+X)$). Such a formation process was not precluded by any of these experimental observations; however, because the (A^0X) state most likely involved the capture of a free exciton, it was believed that sufficient exciton formation must occur in this sample so that a two-step capture process need not be invoked for (D^+X) formation.

Figure 2.13 shows the PL decay at the (D^+X) peak energy of 3.473 eV for two excitation intensities. Emission lifetimes were determined by least-squares fits of one- or two-exponential decay form. The decays exhibited by these co-doped epilayers shown in Figure 2.13 were not single

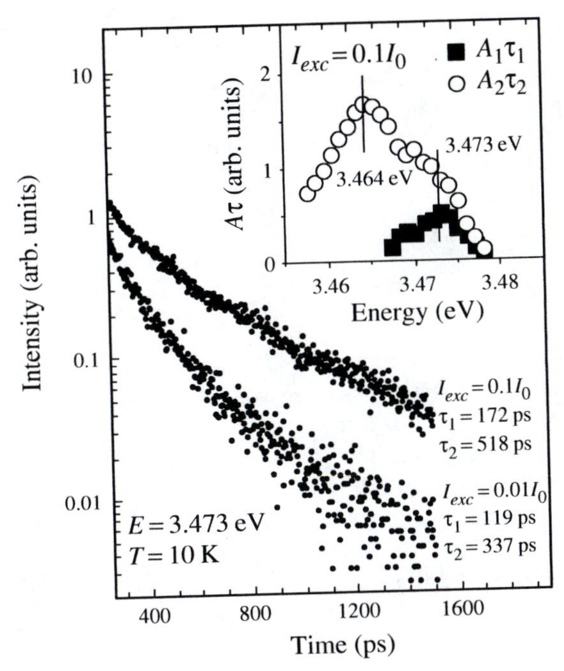

Figure 2.13 PL decay at the (D^+X) peak energy of 3.473 eV for two excitation intensities. Emission lifetimes were determined by least-squares fits of the one- or two-exponential decay form [after Ref. 37].

exponential because they represented PL contributions at 3.473 eV from two distinct but overlapping transitions, as shown in Figure 2.12. The long-lifetime component of $\tau_2 = 518$ ps ($I_{exc} = 0.1I_0$) was comparable with a previously reported measurement of the A^0X (or I_1) emission lifetime for a GaN:Mg epilayer [26]. The short-lifetime component, $\tau_1 = 172$ ps, was related to (D^+X) emission. The two-exponential fit had the form $I(t) = A_1 \exp(-t/\tau_1) + A_2 \exp(-t/\tau_2)$. In this way, the products $A_1\tau_1$ and $A_2\tau_2$ represented the time-integrated intensities for each of two distinct emissions. The inset in Figure 2.13 shows the emission energy dependence of the two decay components ($A_1\tau_1$, $A_2\tau_2$) extracted from a series of decay curves. The resulting profiles resembled very closely the time-resolved emission peaks shown in Figure 2.12 and clearly demonstrated that the decay components τ_1 and τ_2 represented the recombination lifetimes of the (D^+X) and (A^0X) emission, respectively.

The expected lifetime for the (D^+X) emission may be estimated by using a theoretical relationship between the bound (F_b) and free exciton (F_{FX}) oscillator strength of [15,16]:

$$F_b = (E_0/E_{bx})^{3/2} F_{FX}. \tag{2.14}$$

Here, E_b is the binding energy of the bound exciton, and $E_0 = (2^2_\hbar/m) \times (\pi/\Omega_0)^{2/3}$, where m is the effective mass of the free exciton, and Ω_0 is unit cell volume. This relationship states that the oscillator strength of a bound exciton transition decreases with increased binding energy. From Eq. (2.14), the ratio of radiative lifetimes for two distinct bound exciton transitions is expressed as

$$\frac{\tau_1}{\tau_2} = \frac{F_2}{F_1} = \left[\frac{(E_{bx})_1}{(E_{bx})_2}\right]^{3/2}. \tag{2.15}$$

Taking 130 ps for the lifetime of (D^0X) emission with a 8.5 meV binding energy [11,23], together with the measured (D^+X) binding energy of 11.5 meV, yielded from Eq. (2.15) an expected lifetime of 205 ps for (D^+X) emission. Although the expected lifetime was somewhat longer than the measured (D^+X) emission lifetime of $\tau = 160$ ps, the trend of increased lifetime for increased binding energy was evident. It should be noted that Eq. (2.15) is meant to compare radiative lifetimes. It is likely that both the (D^0X) lifetime of 130 ps and the (D^+X) lifetime of 160 ps are shorter than the radiative lifetimes for those states because of possible nonradiative contributions to the total decay rate.

Decay lifetimes were also measured as a function of PL emission energy for different excitation intensities. Both the (A^0X) and (D^+X) emissions exhibited a trend of decreasing lifetime with increasing emission energy. For example, at $I_{exc} = 0.1 I_0$ the (D^+X) lifetime decreased monotonically from $\tau = 180$ ps at $E = 3.468$ eV to $\tau = 120$ ps at $E = 3.478$ eV, similar to the behaviors observed for excitons bound to neutral acceptors and donors in GaN [23,26], which may be understood as the result of a distribution of exciton binding energies due to local variations of the crystal quality at the impurity sites. According to Eq. (2.15), lower energy emission (larger binding energy) will exhibit a longer radiative lifetime than higher energy emission. The decay lifetimes were measured for three excitation intensities. The average (D^+X) lifetime increased slightly with increasing I_{exc} from approximately $\tau = 120$ ps at $I_{exc} = 0.01 I_0$ to $\tau = 160$ ps at $I_{exc} = I_0$. The (A^0X) lifetime, on the other hand, was maximum for $I_{exc} = 0.1 I_0$ ($\tau = 660$ ps) and decreased somewhat for the highest intensity, $I_{exc} = I_0$ ($\tau = 510$ ps).

Table 2.1 displays a summary of parameters for the fundamental bound exciton states in hexagonal GaN [37]. The quantities listed include bound exciton binding energy (E_{bx}), emission lifetime (τ_L), and thermal dissociation energy. No parameters are indicated for the ionized acceptor bound exciton (A^-X) because, to our knowledge, it has not been observed. In fact, effective mass arguments preclude the existence of this state in hexagonal GaN [25,36,38]. The values given in Table 2.1 represent a survey of

Table 2.1 Comparison of measured parameters from low-temperature PL emissions associated with bound exciton states in hexagonal GaN.

Bound state	E_{bx} *(meV)*	τ_L *(ps)*	*Thermal dissociation (meV)*	*Relevant material/structure*
(A^0X)	20–26	370–3600	23	p-type
(D^0X)	6–9	35–530	6–8	n-type, intentional and unintentional
(A^-X)	—	—	—	Existence is precluded
(D^+X)	11.5	160	No data	Bipolar structures, intentional and unintentional co-doping

Note: Listed quantities include bound exciton binding energy (E_{bx}), emission lifetime (τ_L), and thermal dissociation energy.

references 9–42. Although this survey is not exhaustive, it is representative of the majority of published data for the listed parameters. It can be seen that large variations in bound exciton parameters, especially in lifetimes, have been observed in GaN, due to variations in crystal quality, nature of strain associated with different types of substrates, and layer thickness, as well as in the growth method and growth parameters.

2.3.4 Band-to-Impurity Transitions

Time-resolved PL has been employed to study the mechanisms of band-edge emissions in relatively heavily doped Mg-doped p-type GaN epilayers [43]. Two emission lines at about 3.21 and 2.95 eV have been observed. The 2.95-eV emission band has been observed in early MOCVD-grown and post-growth thermal-annealed p-GaN films and has been utilized for violet-blue emission in p-n junction GaN LEDs [44–47]. These emission lines exhibited lifetimes on the order of subnanoseconds, comparable to that of a band-to-impurity transition in a highly n-type GaN epilayer involving a donor and the valence band [12,48]. Figure 2.14 presents the time-resolved emission spectra of the 3.21-eV band measured at $T = 10\,\mathrm{K}$. Figure 2.14 shows that the spectral line shape of the 3.21-eV emission band was quite similar for different delay times due to the broadness of the emission band as well as because of its subnanosecond recombination lifetimes. In contrast, the emission spectral line shape of a donor–acceptor pair (DAP) transition is expected to be delay time dependent with a microsecond time scale for delay times [49,50]. Furthermore, based on a theoretical model developed by Avouris and Morgan, the decay kinetics of both the band-to-impurity and DAP recombination are not necessarily exponential. The asymptotic decay at long times is t^{-1} for the DAP recombination [51]. On the other hand, the decay of the

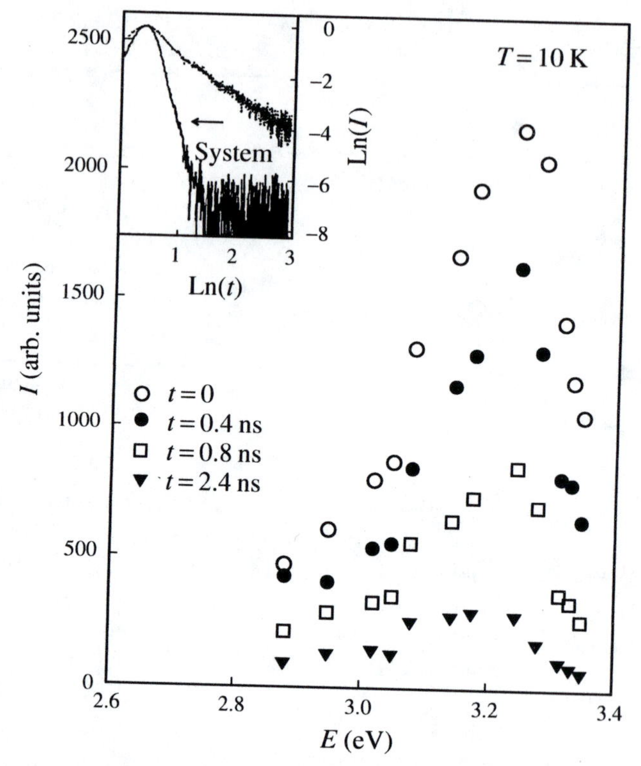

Figure 2.14 Time-resolved emission spectra for the 3.21-eV band in a relatively heavily Mg-doped GaN epilayer at $T = 10\,\mathrm{K}$. Inset: the 10 K PL temporal response obtained at the spectral peak position plotted on a double logarithmic scale, which clearly shows the decay at the longer delay times following a power law, $I \backsim t^{-\alpha}$, with an exponent α greater than 1.0 [after Ref. 43].

band-to-impurity recombination at longer times should follow $t^{-\alpha}$, with α being greater than 1.0, because the product of the time variations of the free carriers and the DAP recombination gives the decay of the band-to-impurity recombination. In the inset of Figure 2.14, the 10 K PL temporal response obtained at the spectral peak position is plotted in a double logarithmic scale, which clearly shows that the decay at longer delay times followed a power law with an exponent greater than 1.0. These results supported the hypothesis that the dominant emission lines in highly Mg-doped GaN samples were of a band-to-impurity nature, although a weak contribution from DAP recombination could not be completely precluded, especially at low temperatures and low excitation intensities. More specifically, based on the excitation intensity and sample temperature dependences of the PL lineshape, the 3.21-eV line was attributed to the conduction band–to-impurity transition involving shallow Mg impurities, while the line at about 2.95 eV was attributed to the conduction band–to-impurity transition involving Mg-related deep-level centers (or complex). The results also indicated a possibility for deep-level impurity band formation in heavily doped p-type GaN.

2.3.5 Band-to-Band Recombination

Most of the optical measurements on GaN were carried out for n-type (doped or undoped) and p-type epilayers. Very little work has been reported for undoped high-quality insulating GaN epilayers due to the difficulties in producing these materials; however, understanding the properties of insulating GaN epilayers may provide basic information regarding how to control the properties of GaN epilayers, such as from n-type to insulating, as well as on how to optimize the doping process and the sample quality. Insulating GaN layers are needed for the design of many III-nitride optoelectronic and electronic devices.

It was reported that the conductivity of undoped GaN epilayers was a strong function of the ammonia (NH_3) flow rate during MOCVD growth [52]. For layers grown by low-pressure (76 Torr) MOCVD at a temperature of 1050 °C on a sapphire substrate with a 50-nm low-temperature AlN buffer layer, the as-grown layers exhibited n-type conductivity when the ammonia (NH_3) flow rate was 2 standard liters per minutes (slm); however, the conductivity of the as-grown GaN epilayers decreased with the reduction of the NH_3 flow rate. When the NH_3 flow rate was lowered to 1.5–1.8 slm, insulating GaN epilayers were obtained. PL spectra of a high-quality insulating GaN epilayer measured at a low excitation intensity ($I_{exc} \sim 11\,W/cm^2$) are shown in Figure 2.15 for temperatures from 9 to 50 K. These insulating GaN epilayers exhibited three emission lines with peak positions at 3.491 eV, 3.503 eV, and 3.512 eV, which were quite different from those of as-grown n-type GaN epilayers produced by MOCVD or those of as-grown insulating epilayers produced by reactive MBE. A typical PL spectrum for a relatively high quality MOCVD-grown n-type or reactive MBE–grown insulating GaN epilayer on sapphire shows either I_2 transition near 3.478 eV or free A- and B-exciton transitions around 3.485 eV and 3.491 eV, respectively [9–35]. It has been pointed out that the free A-exciton transition energy can be as high as 3.494 eV due to the presence of a higher biaxial strain in certain GaN epilayers [53,54]; however, a value as high as 3.503 eV at low temperatures was never observed in other GaN epilayers. The peak energy at 3.503 eV seen in insulating GaN epilayers was exactly the expected value of the fundamental bandgap energy (involving the A-valence band) of the wurtzite GaN epilayer. The transition lines at 3.503 eV and 3.512 eV for the insulating GaN epilayer were thus assigned to the band-to-band transitions involving the A- and B-valence bands, respectively. The 9-meV difference between the A- and B-valence bands observed in these layers was slightly larger than

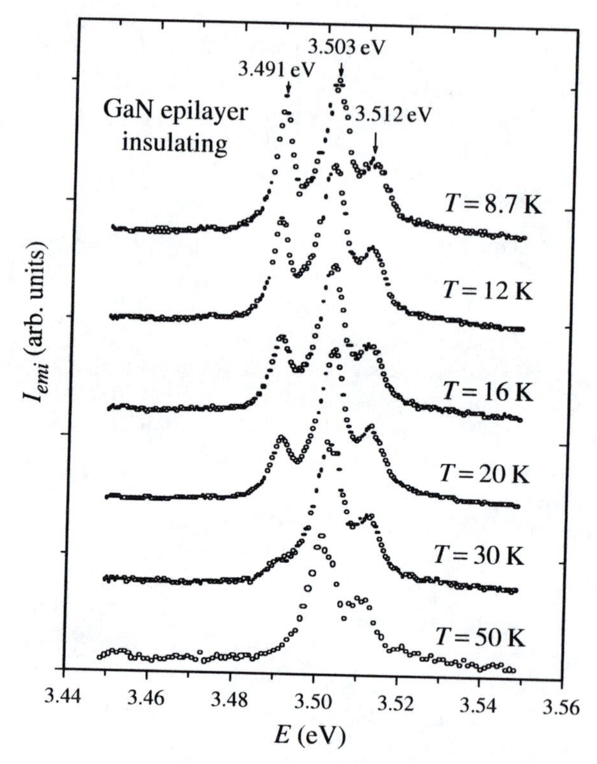

Figure 2.15 PL spectra of a high-quality insulating MOCVD-grown GaN epilayer measured at different temperatures. The spectra have been shifted vertically for clearer presentation [after Ref. 52].

the 6-meV difference observed in undoped n-type epilayers produced by MOCVD or insulating epilayers produced by reactive MBE [10–13].

The assignment of the band-to-band transition lines at 3.503 eV and 3.512 eV was further supported by the excitation intensity (I_{exc}) dependence of the PL peak positions. It was observed that the dominant emission line redshifted from 3.503 to 3.496 eV as I_{exc} increased from 60 to 6000 W/cm^2. More specifically, as shown in Figure 2.16, the spectral peak position of the dominant emission line in these layers (at 3.503 eV at the lowest I_{exc}), E_p, as a function of I_{exc} followed

$$E_p = E_g(n \to 0) - \gamma I_{exc}^{1/3}, \tag{2.16}$$

where $E_g(n \to 0)$ is the energy gap of the GaN epilayer as free carrier concentration approaches zero (the fundamental energy band gap), and γ is a proportionality constant. The relationship described in Eq. (2.16) is a

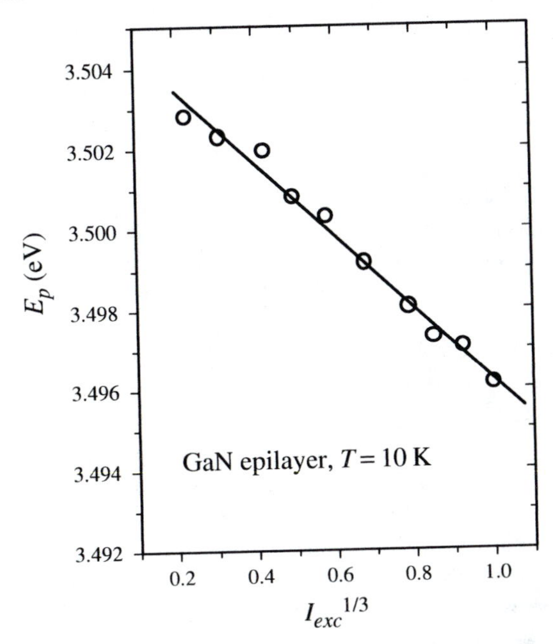

Figure 2.16 The emission peak position of the 3.503-eV line in a high-quality insulating GaN epilayer grown by MOCVD, E_p, as a function of $I_{exc}^{1/3}$ measured at 10 K. The solid line is the least-squares fit of the experimental data with Eq. (2.16) [after Ref. 52].

direct consequence of enhanced many-body and free carrier screening effects [55]. This behavior can be taken as a direct evidence for the band-to-band transition [56].

The decay kinetics of the band-to-band transitions in these layers were not one or two exponential, nor can they be described by a bimolecular decay, which is expected for pure band-to-band transition. By considering the fact that the band-to-band transitions were under the influence of the band-to-impurity transition at 3.491 eV and taking into account the contributions of the nonradiative recombination, one can model the decay by [57]

$$\frac{dp}{dt} = -\beta(N_D p + N_A n + pn) - \frac{p}{\tau_{nr}}, \tag{2.17}$$

where $p(n)$ and $N_D(N_A)$ are the photoexcited hole (electron) and unintentionally doped donor (acceptor) concentrations, respectively, τ_{nr} represents the total nonradiative recombination decay lifetime (including both the bulk and interface nonradiative recombination), and β is the radiative recombination coefficient. From Eq. (2.17), one obtains an equation describing the decay of the band-to-band recombination under the influence of a band-to-impurity transition,

$$p(t) = \frac{p_0 e^{-t/\tau}}{[1 + \alpha(1 - e^{-t/\tau})]}, \tag{2.18}$$

where p_0 is the photoexcited hole concentration at delay time $t = 0$, $\alpha = p_0/[(N_D + N_A) + 1/(\beta\tau_{nr})]$, and $\tau\{= 1/[1/\tau_{nr} + \beta(N_D + N_A)]\}$ is the total decay lifetime determined by the total impurity concentration $(N_D + N_A)$, β, and τ_{nr}.

The PL decay measured at the band-to-band transition lines in insulating GaN layers, 3.503 eV or 3.512 eV, was well described by Eq. (2.18), as shown by the solid fitting curves in Figure 2.17 for two representative excitation intensities, $I_{exc} = 23\ \mathrm{W/cm^2}$ and $73\ \mathrm{W/cm^2}$. The fitted values of α under different I_{exc} are plotted in the inset of Figure 2.17, which shows that α increased linearly with I_{exc}, as expected, since $\alpha \propto p_0$, and $p_0 \propto I_{exc}$. Here, p_0 was estimated to range from 1.76 to $5.55 \times 10^{17}/\mathrm{cm^3}$ as I_{exc} varied from 23 to 73 $\mathrm{W/cm^2}$. The fitted value of α varied from 13.8 to 38.8, which gave $p_0/\alpha = (N_D + N_A) + 1/(\beta\tau_{nr}) \approx 1.27 \times 10^{16}/\mathrm{cm^3}$. The upper-limit value of the total impurity concentration $(N_D + N_A)$ was thus estimated to be smaller than $1.27 \times 10^{16}/\mathrm{cm^3}$ (if $1/\tau_{nr} = 0$, then $N_D + N_A \approx 1.27 \times 10^{16}/\mathrm{cm^3}$), which was consistent with the fact that these GaN epilayers were insulating. The fitted value of τ was about 3.7 ns for both the 3.503-eV and 3.512-eV emission lines independent of I_{exc}, which was expected from the relation $\tau = 1/[1/\tau_{nr} + \beta(N_D + N_A)]$. From α and τ, a value of $\beta = \alpha/(p_0\tau) \approx 1.96 \times 10^{-8}\mathrm{cm^3/s}$ for the radiative recombination coefficient in these insulating GaN epilayers was also obtained.

Many important issues remain to be understood concerning the optical properties of insulating GaN epilayers, including (a) why the GaN epilayer became insulating under the growth conditions described and (b) why the band-to-band transitions were dominant in these MOCVD-grown insulating GaN epilayers. The concentration of nitrogen vacancies was expected to increase with a decrease in the ammonia flow rate. Thus, one would expect the GaN epilayers to be more heavily n-type; however, the results were contrary to this expectation, i.e., the GaN epilayers became more resistive and eventually became insulating when the ammonia flow rate was further reduced. One possibility was that more Ga atoms may have occupied N sites as the ammonia flow rate was reduced, which resulted in more impurities. These impurities behaved as compensation centers and made the GaN epilayers more resistive. These impurities may also act as scattering centers that prevent the formation of excitons, even at low temperatures. As a consequence, the band-to-band transitions were the dominant optical transitions in these insulating layers. A full understanding of the mechanism as well as the procedure for obtaining high-quality insulating GaN epilayers is very

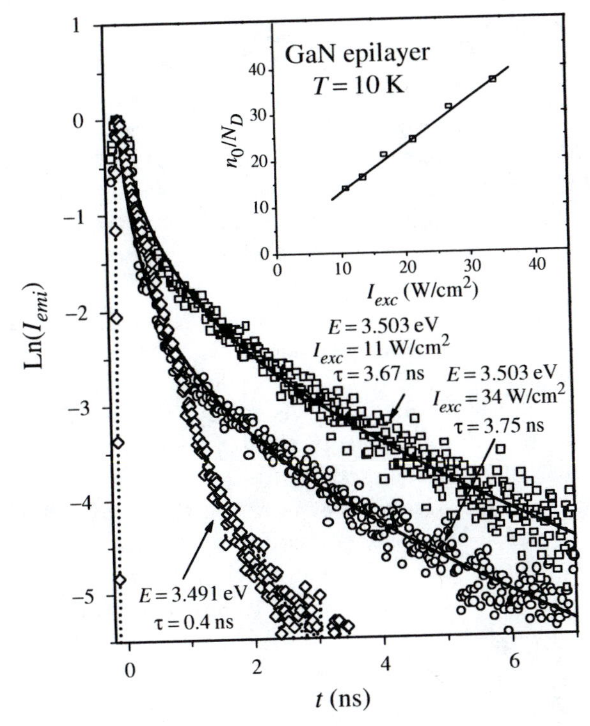

Figure 2.17 The PL temporal responses in a high-quality insulating GaN epilayer: the 3.491-eV emission line measured at $I_{exc} = 23$ W/cm^2 (open diamonds), and the 3.503-eV emission line measured at $I_{exc} = 11$ W/cm^2 (open squares) and 34 W/cm^2 (open circles) at 10 K. The solid lines are the least-squares fits of the experimental data with Eq. (2.18). The fitted values of α (open squares) for the 3.503-eV emission line under different I_{exc}, and a linear fit (solid line) of α versus I_{exc} are plotted in the inset [after Ref. 52].

important for many III-nitride device applications such as heterostructure field-effect transistors (HFETs) and p-i-n detectors.

2.4 RECOMBINATION DYNAMICS OF LOCALIZED EXCITONS IN III-NITRIDE ALLOYS

It is well known that the localized exciton transition is the dominant optical process in many semiconductor alloys at low temperatures, including CdS_xSe_{1-x}, $GaAs_xP_{1-x}$, and $Zn_{1-x}Cd_xTe$ [58–63]. A previous theoretical calculation indicated that the amplitude of the fluctuating potential at the band edges caused by alloy disorder is strongly correlated to the energy gap difference between the two semiconductors, e.g., between InN and GaN for $In_xGa_{1-x}N$ or between GaN and AlN for $Al_xGa_{1-x}N$ [63]. InN and GaN (GaN and AlN) form a continuous alloy system whose direct bandgap ranges from 1.9 to 3.4 eV (3.4 to 6.2 eV), giving an energy gap difference ΔE_g of 1.5 eV (2.8 eV). This is much larger than the typical value of a few hundred meV in II–VI semiconductor alloys, in which a strong exciton localization effect is known to present. Two distinctive phenomena, indium segregation and composition fluctuation, have also been studied in both InGaN alloy and InGaN/GaN MQWs and have many important consequences. For indium segregation, although it is still a controversial issue, many groups report that indium tends to be segregated when the In content is above a certain value.

In semiconductor alloys, a critical energy (called the mobility edge E_m), which separates the localized and delocalized exciton states in random potentials induced by compositional fluctuations, is expected [59]. An exciton created at a point where its energy is greater than E_m will be quickly transferred to lower-energy sites by phonon emission, while an exciton created below E_m is expected to decay predominantly radiatively at the same sites [61]. Much recent work has shown that PL emission in InGaN and AlGaN alloys results primarily from localized exciton recombination [64–75]. Time-resolved PL studies have proved to be very fruitful for elucidating properties of localized excitons in III-nitride alloys.

2.4.1 InGaN Alloys

The carrier decay dynamics of spontaneous and stimulated emission in $In_xGa_{1-x}N$ alloys at low temperatures have been investigated by time-resolved PL spectroscopy [69]. PL spectra of $In_xGa_{1-x}N$ alloys ($x = 0.08$ and 0.14) were measured by varying the excitation photon energy, E_{exc}, at $T = 2$ K to determine the mobility edge, E_m, for the localized excitons. It was found that the PL emission peak position depends on E_{exc}. Above a certain excitation energy (3.175 eV for $x = 0.08$ and 3.07 eV for $x = 0.14$), the emission peak position did not vary with E_{exc}. Below that energy value, the emission peak position showed a red shift with decreasing E_{exc}. The mobility edges E_m were thus determined to be at 3.175 and 3.07 eV for $In_xGa_{1-x}N$ alloys with $x = 0.08$ and $x = 0.14$, respectively.

The stimulated emission in $In_xGa_{1-x}N$ alloys ($x = 0.08$ and $x = 0.14$) was observed to occur just below the mobility edges in surface-emitting configuration. It was thus suggested that the localized excitons were responsible for the stimulated emission at low temperatures, a case that is similar to that in II–VI semiconductors [70–72]. The stimulated emission showed a dependence of decay lifetime on emission energy similar to that of the spontaneous emission, i.e., the decay lifetimes increased with decreasing emission energies; however, the decay time of the stimulated emission was much faster and was less than 30 ps compared with values between 0.5 and 1 ns observed for the spontaneous emission.

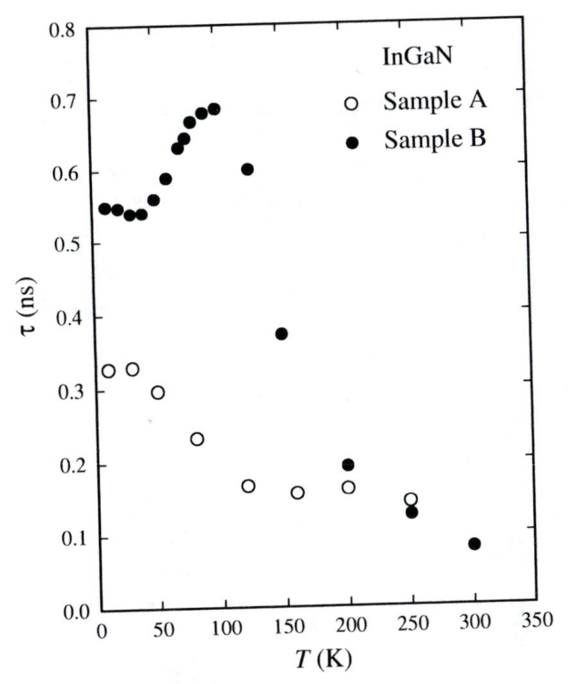

Figure 2.18 Temperature dependence of the PL recombination lifetimes measured at the spectral peak positions of two different InGaN epilayers, samples A and B [after Ref. 73].

It was shown that the recombination lifetime, as well as its temperature dependence, of the localized excitons in InGaN alloys depended strongly on sample quality [73]. Figure 2.18 shows the PL recombination lifetimes measured at the spectral peak positions from 10 K to room temperature for two different InGaN epilayers, sample A and sample B. The emission linewidth of sample A (110 meV) was much larger than that of sample B, implying sample B was of higher quality than sample A. At room temperature the lifetimes in both samples were comparable (about 0.1 ns); however, their lifetime temperature dependencies were quite different. In sample A, the recombination lifetime was about 0.33 ns at low temperatures ($T < 30$ K) and decreased gradually with temperature. In contrast, in sample B, the PL lifetime was nearly temperature independent (about 0.53 ns) at $T < 40$ K, a typical characteristic of radiative recombination of localized excitons. Above 40 K, the PL lifetime increased almost linearly with temperature up to 100 K, a signature of radiative recombination involving partially localized excitons. The lifetime reached its longest value of about 0.7 ns at 100 K and then decreased as temperature further increased. These results demonstrated that time-resolved PL studies uniquely provide opportunities for understanding basic optical processes as well as for identifying high-quality materials.

2.4.2 AlGaN Alloys

Time-resolved PL has also been employed to investigate the compositional dependence of the optical properties of the $Al_xGa_{1-x}N$ alloy system [74]. Room- (300 K) and low-temperature (10 K) cw PL spectra for several MOCVD-grown $Al_xGa_{1-x}N$ alloys with $0 \leq x \leq 0.35$ are presented in Figure 2.19a and b. Apart from seeing the shift in the peak position with increasing Al content, one also notices a considerable decrease in the PL intensity and increase in the full width at half maximum (FWHM), which was caused by alloy broadening. For clarity, PL intensity and FWHM as functions of Al content at room and low temperatures are depicted in the insets of each figure. Specifically, the PL intensity at room temperature decreased exponentially with increasing Al content. This behavior was very similar to that of the room-temperature electron Hall mobility in $Al_xGa_{1-x}N$ alloys, which also decreased exponentially with Al content [75].

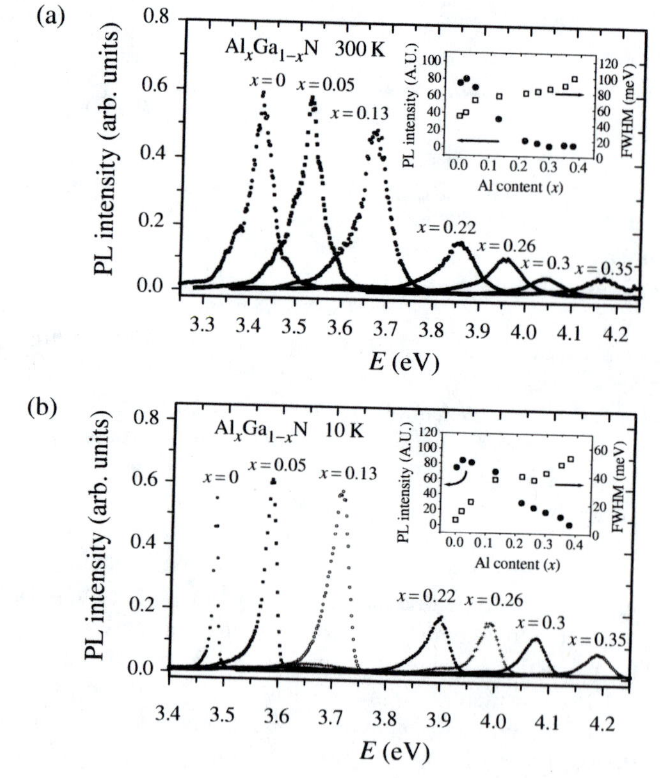

Figure 2.19 (a) Room- (300 K) and (b) low-temperature (10 K) cw PL spectra for $Al_xGa_{1-x}N$ alloys with $0 \leqslant x \leqslant 0.35$. The insets show the Al content x, dependence of the full width at half maximum (FWHM), and the PL intensity in $Al_xGa_{1-x}N$ alloys [after Ref. 74].

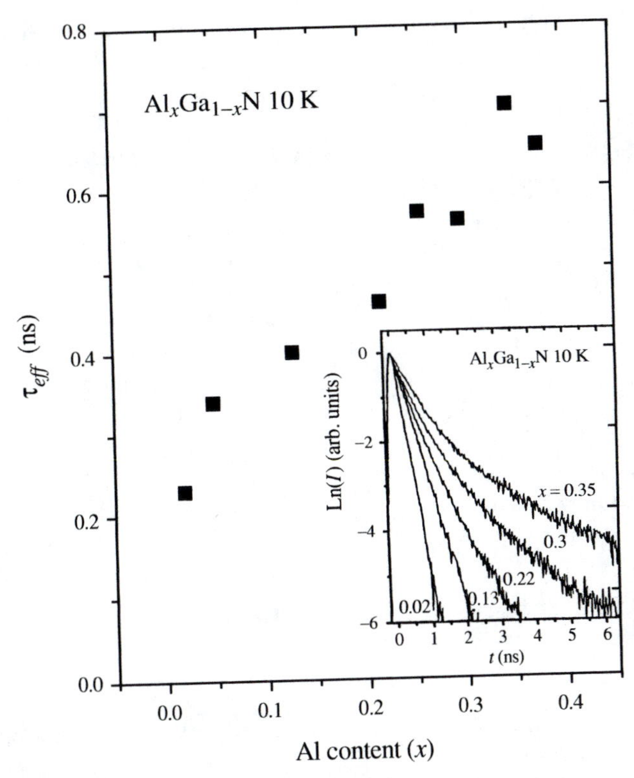

Figure 2.20 Effective PL decay time τ_{eff} as a function of Al content in $Al_xGa_{1-x}N$ alloys. Decay profiles were measured at the PL emission peak energies at $T = 10$ K. The inset shows a semilogarithmic plot of the temporal response for several representative Al contents [after Ref. 74].

PL decay behavior at 10 K was investigated for various AlGaN samples. Figure 2.20 shows the effective decay time (τ_{eff}) measured at the PL peak energy plotted as a function of Al content, x. Here, τ_{eff} was defined as the time at which the PL intensity decayed to $1/e$ of the maximum intensity. This definition for decay time was used because, as seen in the inset of Figure 2.20, the PL decay was only single exponential for the alloy samples with the lowest Al content. It is clear from Figure 2.20 that the effective lifetime increased with increasing Al content. Specifically, τ_{eff} increased approximately linearly from 0.2 to 0.7 ns as x varied from 0.02 to 0.35. This behavior was consistent with previous theoretical arguments that predicted that the radiative lifetime of bound excitons increases with binding energy [15,16]. Within the AlGaN samples, excitons are energetically and spatially localized due to compositional fluctuations. This localization is analogous to the binding of an exciton to an impurity, so that a larger characteristic localization energy (E_0) should result in a longer radiative lifetime. It is expected that E_0 will increase with increasing Al content. The measured decay time (τ) is related to the radiative lifetime (τ_r) and the nonradiative lifetime (τ_{nr}) by $\tau = \tau_r \tau_{nr}/(\tau_{nr} + \tau_r)$. Therefore, we expect a localization-induced increase in the radiative lifetime, as evidenced by an increase in the measured PL decay time (τ), but the magnitude of the increase in τ is not necessarily as great as the increase in τ_r.

The emission energy dependence of the PL decay was also studied at 10 K for the AlGaN samples. As shown in Figure 2.21, τ_{eff} increased with decreasing emission energy for all the represented samples. Such a dependence of decay lifetime on emission energy is a well-known manifestation of a localized exciton distribution within a semiconductor alloy [76,77]. Within the localization model, highly localized (lower-energy) excitons decay primarily via radiative recombination, while less localized (higher-energy) excitons exhibit a decreased decay time due to the additional channel of transfer to lower-energy sites. The data of Figure 2.21 were qualitatively in agreement with this model.

2.5 CARRIER DYNAMICS IN GaN/InGaN QUANTUM WELLS

The past several years have witnessed rapid advances in the development of III–V nitride-based devices such as blue and UV LEDs and cw operation of InGaN MQW LDs. With continued progress in fabrication and optimization of InGaN/GaN or $In_xGa_{1-x}N/In_yGa_{1-y}N$ ($x \neq y$) QW LDs, it becomes increasingly important to understand the fundamental properties of this new class of LDs. One of the key issues is, of course, the lasing mechanism in InGaN/GaN QW LDs—the nature of the optical recombination that supplies the optical gain within the active medium, namely, within the InGaN/GaN or $In_xGa_{1-x}N/In_yGa_{1-y}N$ ($x \neq y$) MQWs. There are several important differences between InGaN/GaN MQWs and, for example, conventional III–V MQWs such as GaAs/AlGaAs MQWs. The well in InGaN/GaN MQWs is an alloy, whereas that in GaAs/AlGaAs MQWs is a binary compound semiconductor. Since the important and dominant optical transitions are from the well regions, the properties of InGaN alloy in the wells are thus very important. Optical transitions in InGaN/GaN and $In_xGa_{1-x}N/In_yGa_{1-y}N$ MQWs have been investigated extensively by time-resolved PL and pump-and-probe measurements [78–99]. Although many controversial issues concerning several phenomena remain to be resolved, including mechanisms of stimulated emissions and lasing in InGaN/GaN MQWs and LDs, many important optical properties have been revealed.

Several lasing mechanisms for the III–V-nitride lasers have been proposed, which include (a) localized exciton recombination in the In-rich

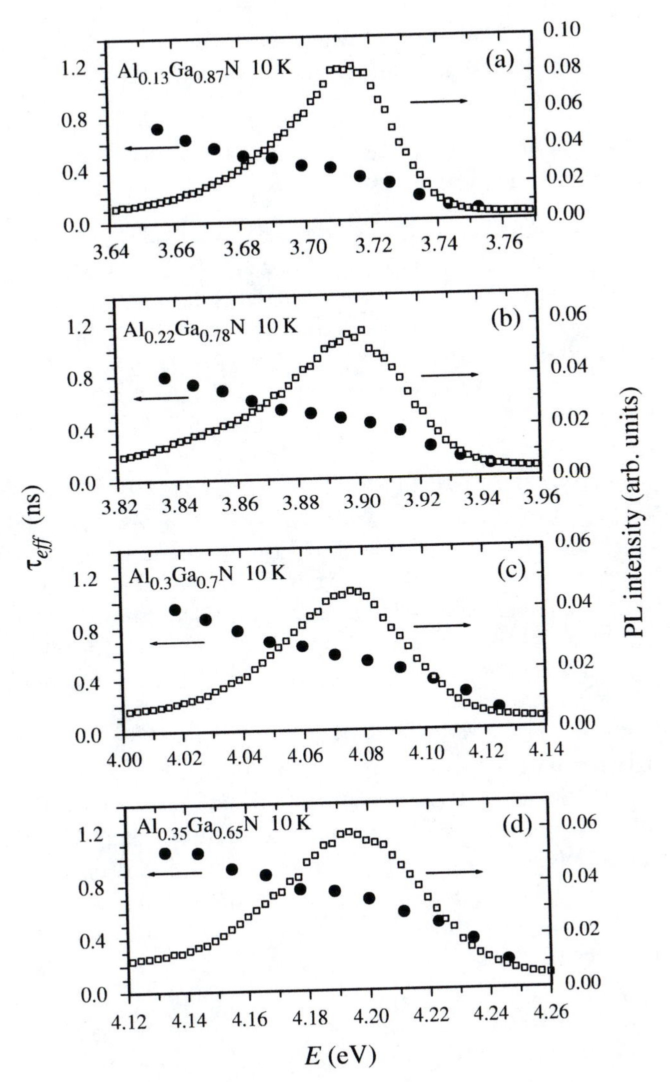

Figure 2.21 The emission energy dependence of the effective PL decay lifetime τ_{eff} for the main emission lines of $Al_xGa_{1-x}N$ alloys for (a) $x = 0.13$, (b) 0.22, (c) 0.3, and (d) 0.35 measured at 10 K [after Ref. 74].

regions (or the so-called quantum dot regions) [64,65] and (b) conventional electron–hole plasma (EHP) recombination [100–102]. The origin of optical gain in III–V nitride lasers has also been theoretically studied [67,103]. Calculated results have suggested that an interacting EHP recombination provides the optical gain in InGaN LDs. The optical gain spectra of a Nichia's InGaN MQW blue LD were measured and it was suggested that EHP recombination is responsible for the gain mechanism [68].

2.5.1 Recombination Processes in InGaN/GaN QW LED and LD Structures

Stimulated emissions from optically pumped $In_xGa_{1-x}N$ (2 nm)/$In_yGa_{1-y}N$ (4 nm) ($x = 0.11$, $y = 0.03$) MQWs at 2 K and room temperature have been observed, and the excitation intensity dependence of the exciton decay dynamics has been measured [99]. The dynamics of the optical gain of the stimulated emission in this MQW sample at low temperatures was very similar to that in the InGaN alloys [69]. Under low excitation density, the temporal change of the spontaneous emission indicated slow dynamical features of the 2D exciton localization, with decay lifetimes in the range of hundreds of picoseconds to a few nanoseconds [69,99]. Figure 2.22 shows temporal responses of the low-temperature [Figure 2.22a and b] and room-temperature [Figure 2.22c and d] PL emissions measured at different emission energies [99]. The decay kinetics of the spontaneous emission showed a nearly single exponential decay, and the decay time increased with decreasing detection photon energy, a typical behavior of localized excitons. Under an excitation energy density of 25 $\mu J/cm^2$, a sharp change in the PL decay behavior took place at emission energies higher than 3.16 eV, and a stimulated emission band at the higher-energy side of the spontaneous emission was observed at $T = 2$ K. The emission line width was reduced from 185 meV for the spontaneous emission to about 20 meV for the stimulated emission. The stimulated emission threshold was found to be around 17 $\mu J/cm^2$ (corresponding to an average carrier area density of $n = 1.1 \times 10^{12}/cm^2$) at $T = 2$ K, obtained from the excitation intensity dependence of the PL intensity. This value was below the Mott density, the density necessary for the formation of an EHP, which is given approximately by $n_p = 1/\pi(a_B^{2D})^2 = 1 \times 10^{13}\,cm^{-2}$, where $a_B^{2D} = 1.8$ nm is the two-dimensional (2D) Bohr radius of the exciton estimated from the 3D Bohr radius of the exciton by the relation $a_B^{2D} = a_B^{3D}/2$. The decay lifetime of the stimulated emission in InGaN/GaN was found to be very fast—less than 30 ps. Above the stimulated emission threshold the localized states were saturated, and many electron–hole pairs existed in the delocalized states, from which the optical gain originated in terms of the stimulated emission process. A carrier transformation from stimulated to spontaneous emission was observed at low temperatures. The stimulated emission was thus related to the filling of the localized states.

Furthermore, above the stimulated emission threshold, the stimulated emission could be observed even at room temperature. The decay of the spontaneous emission was nearly single exponential, and the decay time

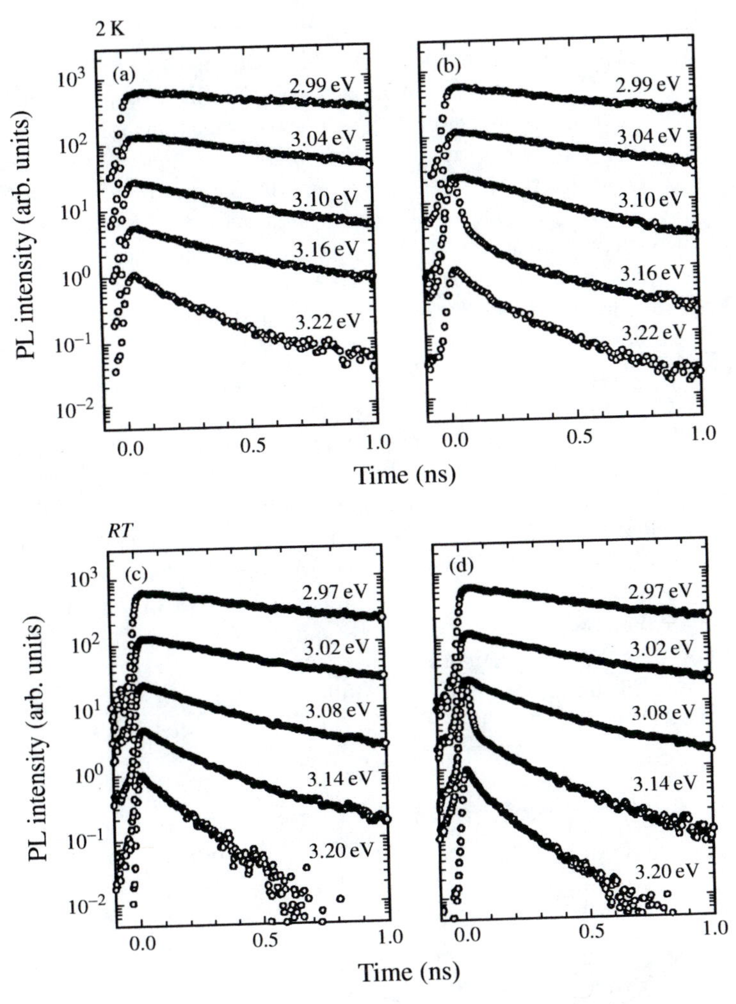

Figure 2.22 Temporal responses of low-temperature [(a) and (b)] and room-temperature [(c) and (d)] PL emissions measured at different emission energies for $In_xGa_{1-x}N$ (2 nm)/$In_yGa_{1-y}N$ (4 nm) ($x = 0.11,\ y = 0.03$) MQWs [after Ref. 99].

increased with decreasing emission photon energy, similar to what happened at 2 K. The experimental features at room temperature were so similar to those at 2 K that the physical mechanisms of the emission at the two temperatures were considered to be the same.

A related important experimental observation in InGaN laser diodes was that the measured mode spacing of the lasing spectra could be one order of magnitude larger than that "calculated" from the known cavity length [2,3]. A formula that accurately determines the mode spacing in InGaN LDs was derived [100]. Analysis has shown that the discrepancy between the "expected" and observed mode spacing was due to the effect of carrier-induced reduction of the refractive index under lasing conditions, and this discrepancy decreases and naturally disappears as the threshold carrier density required for lasing decreases. Since the carrier-induced reduction of the refractive index is expected only from an electron–hole plasma state, the results naturally implied that electron-hole plasma recombination provides the optical gain in InGaN LDs, like in all other conventional III–V semiconductor lasers.

Recently, Mukai et al. reported that the addition of a small percentage of indium to GaN active layers resulted in a significant improvement (about 10 times) of external quantum efficiency (η_{ext}) of UV LEDs [104]. The emission mechanism of these UV $In_{0.02}Ga_{0.98}N$ LEDs was studied by time-resolved PL spectroscopy together with photoinduced voltage and electroreflectance spectroscopy [105]. Radiative and nonradiative recombination lifetimes were derived from the measured temperature-dependent PL integrated intensity and decay lifetimes. It was concluded that the improvement of the quantum efficiency by the addition of a small amount of In to the active layer can be attributed to the suppression of the density of the nonradiative recombination centers (NRC) as well as possibly to where the origin of the NRC was changed.

Figure 2.23 shows a time-integrated PL spectrum and the PL decay lifetime of a UV LED [105]. The structure of UV LEDs studied as a function

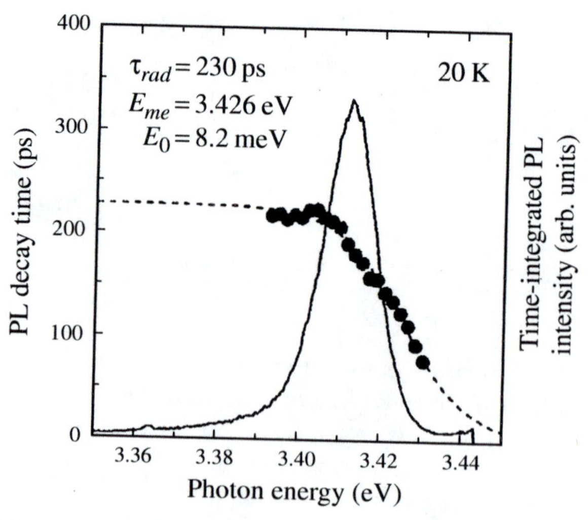

Figure 2.23 Time-integrated PL spectrum and the PL decay lifetime as a function of emission photon energy at 20 K of an $In_{0.02}Ga_{0.98}N$/GaN UV LED. The dotted curve is the theoretical fit with Eq. (2.19) [after Ref. 105].

of emission photon energy at 20 K comprised an n-GaN:Si layer (4 μm), an n-$Al_{0.10}Ga_{0.90}N$:Si cladding layer (30 nm), an undoped $In_{0.02}Ga_{0.98}N$ active layer (40 nm), a p-$Al_{0.15}Ga_{0.85}N$:Mg cladding layer (60 nm), and a p-GaN:Mg layer (120 nm). The PL decay time decreased with increasing emission photon energy, which is typical for localized exciton transition in semiconductor alloys. By assuming that the density of the tail state was an exponential function, approximated as $\exp(-E/E_0)$, and that the radiative recombination lifetime (τ_{rad}) did not change with emission energy, the authors then expressed the observed PL decay lifetime $\tau(E)$ as [77]

$$\tau(E) = \tau_{rad}\left\{1 + \exp\left[\frac{(E - E_{me})}{E_0}\right]\right\}^{-1}, \qquad (2.19)$$

where E_0 describes the density of the tail state distribution, and E_{me} is the characteristic energy at which the transfer and decay rates are equal. The result of the least-squares fit of data with Eq. (2.19) is shown as a dotted line in Figure 2.23, and the fitted results were $\tau_{rad} = 230$ ps, $E_0 = 8.2$ meV, and $E_{me} = 3.426$ eV.

Figure 2.24 shows the temperature dependence of the radiative and nonradiative recombination lifetimes (τ_{rad} and $\tau_{non\text{-}rad}$) in the UV LED obtained from the temperature dependence of the time-integrated PL intensity and PL decay lifetime [105]. It can be seen from Figure 2.24 that the τ_{rad} value increased gradually with temperature because of the localization of excitons.

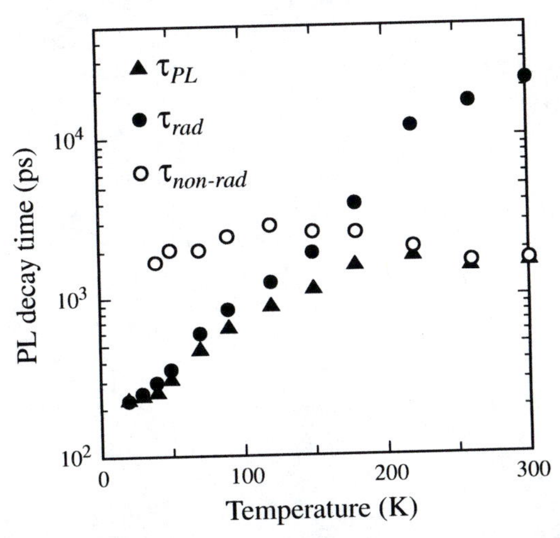

Figure 2.24 Temperature dependence of the radiative and nonradiative recombination lifetimes (τ_{rad} and $\tau_{non\text{-}rad}$) in an $In_{0.02}Ga_{0.98}N$/GaN UV LED obtained from the temperature dependence of the time-integrated PL intensity and PL decay lifetime [after Ref. 105].

The value increased proportionally to $T^{3/2}$ between 50 and 200 K, which was suggested to be a characteristic behavior of a nearly free 3D exciton above 50 K, as predicted by calculations for 3D excitons [107,108]. On the other hand, the derived $\tau_{non\text{-}rad}$ was almost constant at 2 ns. This behavior was different from the temperature dependence of GaN epilayers, in which the $\tau_{non\text{-}rad}$ decreased with increasing temperature, reaching 100 ps at room temperature. Since the nonradiative decay rate was proportional to the density of nonradiative recombination centers, N_t, it was suggested that N_t was greatly suppressed by the addition of a small amount of In to the active layer.

The radiative recombination process of the blue emission band in an InGaN single QW (SQW) LED has been investigated by measuring the dependence of the PL and time-resolved PL spectra on an external electric field [109]. It was found that the PL intensity decreased dramatically with increasing reverse-bias voltage, V_{RB}, at room temperature. The model based on field ionization of excitons cannot explain the experimental results. It was suggested that in these LED structures the free-carrier recombination process was dominant at room temperature [110].

Figure 2.25 shows the time-resolved PL spectra at 77 K of an InGaN SQW blue LED under a reverse-bias voltage of $V_{RB} = -6\,V$ [109]. Two emission components with a separation of about 50 meV were clearly observed after a delay time of 2.5 ns. Neither the higher- nor the lower-emission component exhibited a simple exponential decay profile. If the dominant effect under the reverse bias was the ionization of excitons or an increased mean separation distance between the exciton's electron and hole pair along the growth direction, an increased recombination lifetime would have been observed with increasing V_{RB}, as a consequence of the decreased electron–hole wavefunction overlap; however, the decay times of the higher- and lower-emission components decreased with increasing V_{RB} from 13 to 1.7 ns and from 19 to 2.1 ns, respectively. It was thus suggested that this decrease in decay times was due to the increased transport of carriers, which were separated by the applied reverse-bias voltage, and hence an enhanced capture process by nonradiative recombination centers.

2.5.2 Free-Carrier Recombination in InGaN/GaN QWs

Time-resolved PL was also employed to study the carrier recombination kinetics in an $In_{0.13}Ga_{0.87}N$/InGaN SQW with a 5-nm well width and InGaN:Si barrier layer of graded In composition from 4% to 13% [78]. The low-temperature radiative lifetime of PL was measured to be on the order of 250 ps at a generated carrier density of $10^{12}/cm^2$, while a lifetime of 130 ps was observed at room temperature. It was found that the PL recombination kinetics cannot be described by a single exponential decay, as expected from

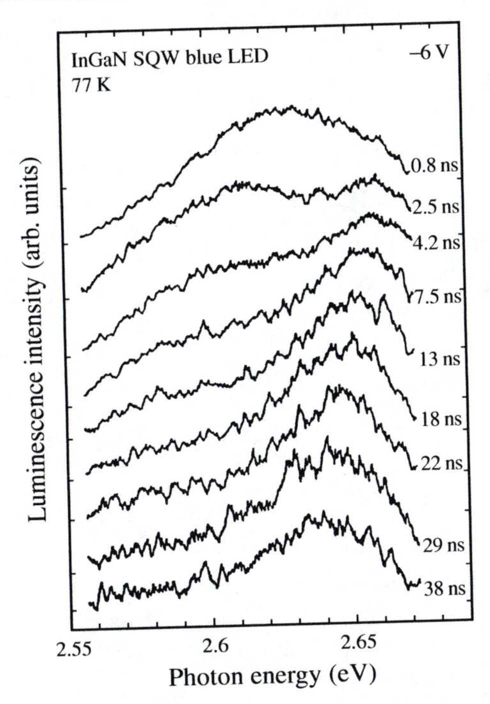

Figure 2.25 Time-resolved PL spectra at 77 K from an InGaN active layer of an InGaN SQW blue LED under a reverse-bias voltage of $V_{RB} = -6$ V [after Ref. 109].

an exciton recombination, but can be well fitted to a hyperbolic decay, which is a characteristic of a bimolecular recombination of free carriers. A convolution fit using a decay function of bimolecular recombination resulted in a recombination coefficient $B = 1.5 \times 10^{-9}$ cm^3/s at 7 K. The observation of free-carrier recombination in this MQW sample instead of the localized exciton recombination seen by most groups may be related to the special structures of the SQW used, namely, Si doping as well as graded In composition in the barriers. It was speculated that this special structure together with the existing strong scattering due to alloy fluctuation and well width roughness in MQWs may enhance the ionization of excitons and result in a dominant free-carrier population at a carrier density of 10^{12}/cm^2, which is three times less than the Mott density.

Figure 2.26 shows the measured decay time constant, τ, obtained from a single exponential fit with the PL decay profile, as a function of temperature measured at the emission spectral peak for this barrier-graded $In_{0.13}Ga_{0.87}N$/InGaN SQW sample [78]. It was seen that τ increased from 230 to 260 ps as temperature increased from 7 to 21 K, then decreased with further increasing temperature, which indicated that carrier recombination was no longer dominated by radiative processes at temperatures

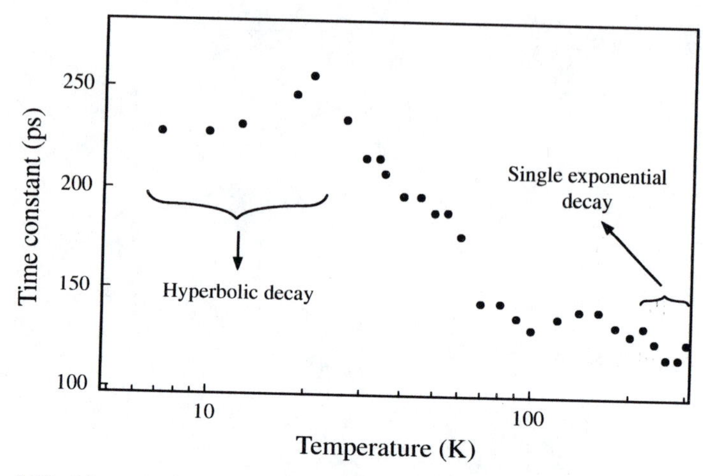

Figure 2.26 Measured PL lifetime as a function of temperature for a graded InGaN/GaN MQW sample. Measurements were performed at the PL peak wavelength. A single-exponential fit was used to determine the PL lifetime [after Ref. 78].

higher than 21 K. Nonradiative recombination became more important at higher temperatures. At room temperature, the carrier recombination was found to be dominated by nonradiative processes with a measured decay lifetime of about 130 ps, and the measured time-resolved PL could no longer be fitted to a bimolecular recombination mechanism.

2.5.3 Doping Effects on Recombination Processes

Effects of impurity doping on the radiative recombination of $In_xGa_{1-x}N$/GaN MQWs have been investigated between 2 to 300 K for Si doping in the wells [111] as well as in the barriers [112]. For Si doping in the wells, the screening effects from donor electrons and excited electron–hole pairs were observed to be important. A strong blue shift of the PL emission photon energy was observed with donor doping, which could be explained by screening of both the piezoelectric field and the localization potential. A 2D model for electron- and donor-screening was proposed and was in reasonable agreement with the experimental data, if rather strong localization potentials of short range (of the order of 100 Å) were present [111].

Figure 2.27 compares the time-resolved PL spectra of Figure 2.27a an undoped and Figure 2.27b a Si-doped (in wells $n = 2 \times 10^{18}/\text{cm}^3$) sample [111]. The two sets of spectra show behavior similar to that in the cw spectra regarding the blue shift with Si doping. The spectra for the doped samples were considerably narrower, giving the impression that the lower-energy part of the spectrum in Figure 2.27a was simply missing in spectrum Figure 2.27b.

For InGaN/GaN with Si doping in the barriers, the interface properties of the MQWs were improved with increasing Si doping density N_{Si}.

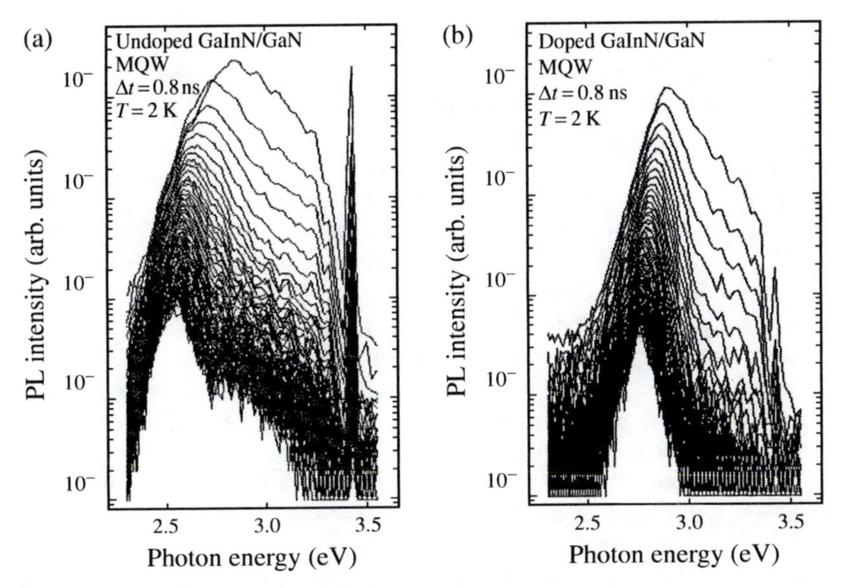

Figure 2.27 Time-resolved PL spectra of (a) an undoped and (b) a Si-doped (in wells $n = 2 \times 10^{18}/\mathrm{cm}^3$) InGaN/GaN MQW sample. The time interval between spectra was 0.8 ns. Note the strong spectral shift between the two samples [after Ref. 111].

The radiative recombination lifetime at 10 K was observed to decrease from about 30 ns (for $N_{\mathrm{Si}} < 1 \times 10^{17}/\mathrm{cm}^3$) to about 4 ns (for $N_{\mathrm{Si}} = 3 \times 10^{19}/\mathrm{cm}^3$) with increasing Si doping concentration [112]. In order to determine whether this behavior was due to the predominance of nonradiative recombination processes, the temperature dependence of the decay lifetime was also investigated. It was found that the decrease in decay lifetime was due to a decrease in the radiative recombination lifetime itself rather than to an increased influence of nonradiative recombination processes. It was thus concluded that the decrease in lifetime with increasing N_{Si} was due to decreasing potential fluctuation (or exciton localization) in the MQWs active layers.

2.5.4 Well Width Dependence

Well width dependence of the carrier decay dynamics in InGaN/GaN MQWs has also been studied by time-resolved PL [80,113]. At room temperature, the carrier recombination was found to be dominated by interface-related nonradiative processes. The dominant radiative recombination was thought to be through the band-to-band free carriers in one report [80] and through localized excitons in another [113]. In combination with a quantum efficiency measurement, the room-temperature radiative carrier lifetime and the room-temperature bimolecular radiative recombination coefficient B were obtained. For samples grown at a higher growth rate, a longer PL decay lifetime was observed, which was attributed to an improved QW interface. A decrease of the radiative lifetime with decreasing well width was observed [80].

It was found that the Stokes shift, the energy difference between the absorption edge and the PL peak, and the PL decay time drastically increased with $In_xGa_{1-x}N$ layer thickness [113]. The decay time increased with sample temperature. The radiative and nonradiative decay times of excitons in these structures were determined by measuring the temperature dependence of the decay times, the integrated PL intensities, and the PL intensities immediately after the picosecond excitation pulse. The fitted single-exponential characteristic decay times for the MQW samples are displayed in Figure 2.28 as a function of temperature [113]. Three curves were obtained for each sample. Solid circles represent lifetimes deduced from the transients of the PL peaks, and solid triangles represent lifetimes measured from transients at the lower- (higher-) energy half-maximum-intensity side of the PL peaks. It was found that the wider the well widths the longer the PL decay times [114–116]. It was also found that the measured lifetimes at temperatures below 130 K were longer for lower emission energies. Above 130 K the entire PL line had one characteristic decay time. The observation that the PL lifetime increased with temperature led the authors to conclude that at

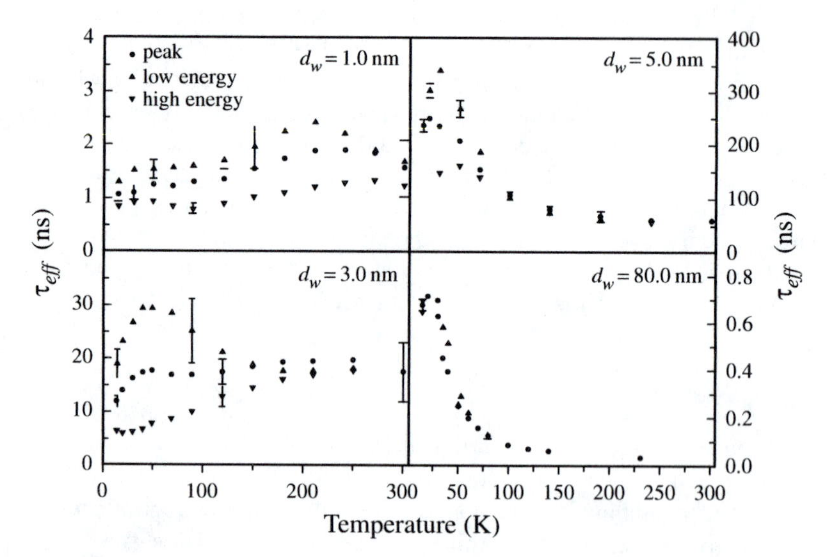

Figure 2.28 Measured PL decay times versus temperature at various emission energies for four different InGaN/GaN QW samples of different well widths, d_w = 1.0, 3.0, 5.0, and 80.0 nm. Solid circles (•) represent lifetimes deduced from the transients of the PL peaks, and solid triangles (▲) (▼) represent lifetimes measured from transients at the lower (higher) energy half-maximum intensity side of the PL spectrum [after Ref. 113].

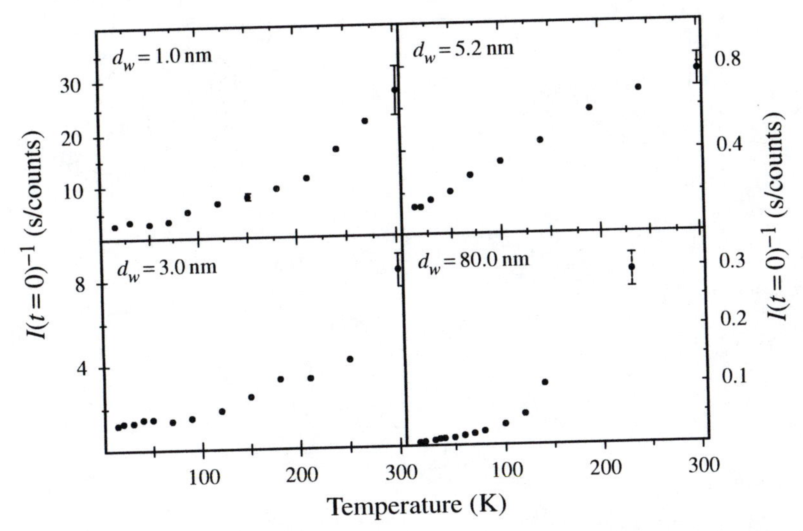

Figure 2.29 The reciprocal of the measured PL peak intensity during a time window of 200 ps immediately after the pulsed excitation, as a function of the sample temperature for four different InGaN/GaN QW samples of different well widths, d_W = 1.0, 3.0, 5.2, and 80.0 nm [after Ref. 113].

low temperatures the nonradiative decay rates were slower than the radiative ones. It was then deduced that the radiative recombination lifetime (τ_R) was inversely proportional to the PL intensity at a given exciton density. From the temperature dependence of the measured PL intensity during a short temporal window (which was significantly shorter than the effective lifetime) around the excitation time, the temperature dependence of the radiative lifetime was obtained. Figure 2.29 plots the reciprocal of the measured PL peak intensity during a time window of 200 ps immediately after the pulsed laser excitation as a function of the sample temperature [113]. From Figure 2.29, the radiative lifetime was roughly temperature independent below 50 K and increased linearly with temperature between 50 and 220 K.

The deduced radiative and nonradiative lifetimes as functions of temperature for the MQWs are shown in Figure 2.30 [113]. The nonradiative decay time decreased rapidly with temperature, whereas the radiative decay time, consistent with the results shown in Figure 2.29, increased linearly with temperature. Both radiative and nonradiative decay times increased rapidly with well width. The radiative decay times were temperature independent at low temperatures, which was expected for fully localized excitons [107,117]. Above 50–70 K, the radiative decay times increased linearly with temperature up to about 250 K. This behavior was an unambiguous signature of the 2D excitonic system [107]. In contrast, the increase with temperature of the radiative decay time of the reference bulk sample was faster than linear, as expected for a 3D system [107]. The effective PL lifetimes of the bulk

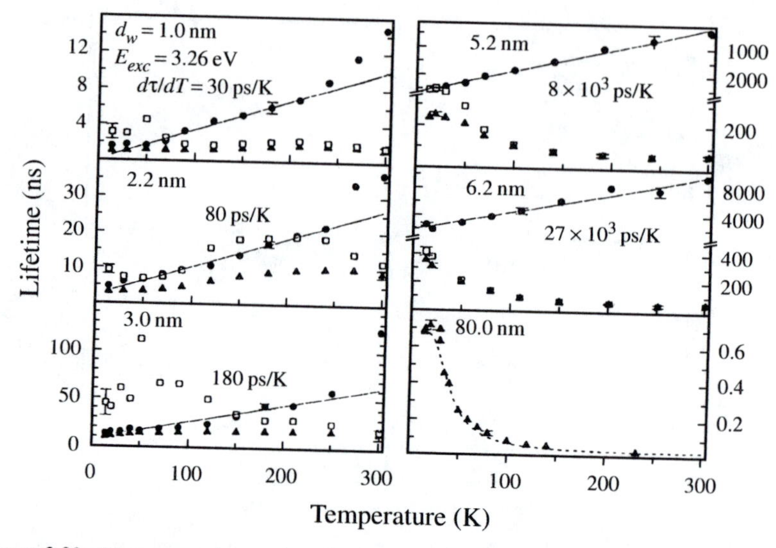

Figure 2.30 The measured τ_{eff} (▲) and the deduced radiative τ_R (•) and nonradiative τ_{NR} (□) lifetime versus temperature for InGaN/GaN QWs of different well widths. Solid lines represent the fitted linear temperature dependence of τ_R [after Ref. 113].

sample were much shorter than the lifetimes of the MQW samples. They also decreased more rapidly with sample temperature at high temperatures, $T > 250$ K.

These authors have also deduced the intrinsic radiative lifetime as a function of the QW well width from their experimental results shown in Figure 2.30, and the result is shown in Figure 2.31 [113]. A strong increase in the radiative lifetime with QW width was observed in these InGaN/GaN MQWs. Furthermore, the increase in the radiative lifetime was much larger than that expected for a symmetric QW structure. Such a strong dependence of the radiative lifetime on the well width was a consequence of the built-in piezoelectric field, which separated the electron–hole pairs and reduced their overlap integral. As a consequence, the oscillator strength for their optical recombination decreased, whereas the radiative recombination lifetime increased. This effect was estimated by an eight-band $\vec{k} \cdot \vec{p}$ model to calculate the oscillator strength, which is inversely proportional to the radiative lifetime, for the optical transitions across the bandgap in the presence of a 900 kV/cm piezoelectric field in these InGaN/GaN MQWs. The solid line in Figure 2.31 represents the reciprocal of the calculated oscillator strength versus the QW width. A relatively good agreement with the experimentally determined dependence of the intrinsic radiative lifetimes on the QW width was obtained. These results clearly demonstrated that the strong dependence of the radiative lifetime on the QW well width was due to a large lattice-mismatch strain-induced piezoelectric field along the

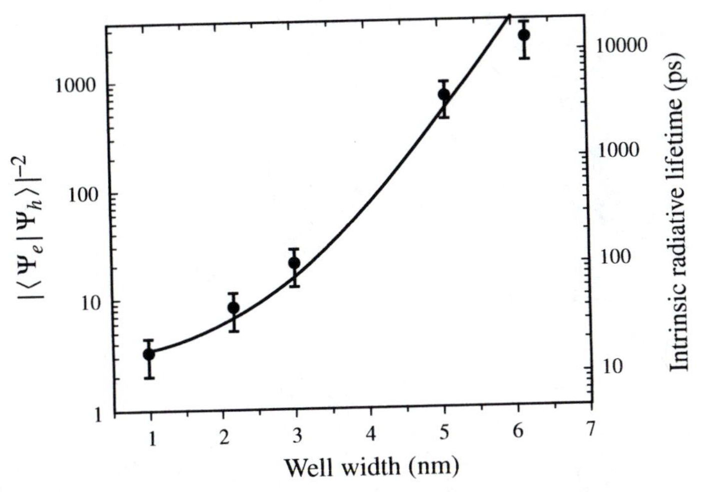

Figure 2.31 Intrinsic radiative lifetime versus QW width. The solid line represents the reciprocal of the calculated electron–hole wavefunction overlap integral [after Ref. 113].

growth axis. It was shown that the piezoelectric field was also responsible for the large Stokes shift in these MQW samples.

2.6 CARRIER DYNAMICS IN GaN/AlGaN QUANTUM WELLS

2.6.1 Piezoelectric Effect

It has been demonstrated by several groups that the piezoelectric and polarization fields can have a strong influence on the optical properties of InGaN/GaN and GaN/AlGaN QWs. Furthermore, these fields can also exert a substantial influence on carrier and electric field distributions in III-nitride heterostructures [117–121]. Redshifts associated with optical transitions and reduced radiative recombination rate in the well regions of GaN/AlGaN and InGaN/GaN MQWs due to the piezoelectric field have been observed by many groups [122–126].

Time-resolved PL studies have been carried out for MBE-grown GaN/AlGaN MQW samples with varying well thicknesses [126,128]. Low-temperature (10 K) cw PL spectra for two representative GaN/$Al_xGa_{1-x}N$ MQW ($x \backsim 0.1$) samples with well thicknesses $L_w = 25$ Å and $L_w = 50$ Å are presented in Figures 2.32a and 2.32b, respectively [127]. For comparison, the PL spectrum of a GaN epilayer deposited under similar conditions is shown in Figure 2.32c. In the GaN epilayer, the dominant transition at 3.485 eV at 10 K was due to the recombination of the ground state of A-excitons. In the 25 Å well MQW sample, the excitonic transition peak position at 10 K was blueshifted with respect to the epilayer

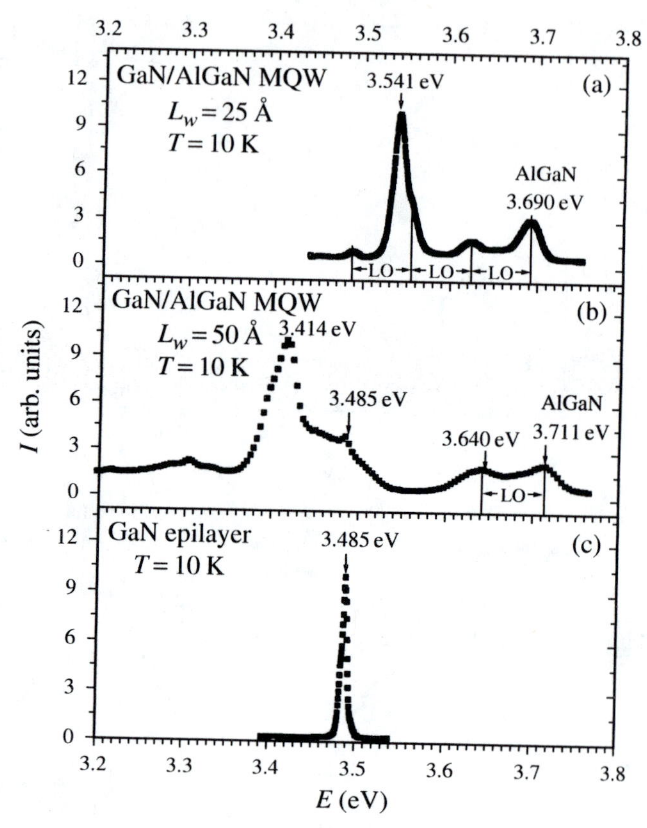

Figure 2.32 Low-temperature (10 K) PL spectra of a nominally undoped (a) GaN/AlGaN MQW sample with well thickness L_w = 25 Å; (b) GaN/AlGaN MQW sample with well thickness L_w = 50 Å; and (c) GaN epilayer grown under identical conditions as the MQW samples [after Ref. 127].

by 56 meV, to 3.541 eV. This 56-meV blueshift was predominantly due to the well-known effect of quantum confinement of electrons and holes. The transition peaks at higher emission energies in MQWs were attributed to an exciton transition and its LO phonon replicas in the AlGaN barrier regions. The observed LO phonon energy was modified due to the symmetry properties of the MQWs. In contrast, the blue shift associated with the exciton transition at 3.485 eV was not observed in the 50 Å well MQW sample at 10 K (Figure 2.32b), indicating a negligible quantum confinement in this QW sample. Moreover, the dominant PL emission line at low temperatures in the 50 Å MQW sample appeared at 3.414 eV, which was 71 meV below the exciton transition line. MQW samples with 40 Å and 60 Å well widths behaved similar to the 50 Å well MQW sample. On the other hand, a 20 Å well MQW behaved similar to the 25 Å well MQW sample; however, with

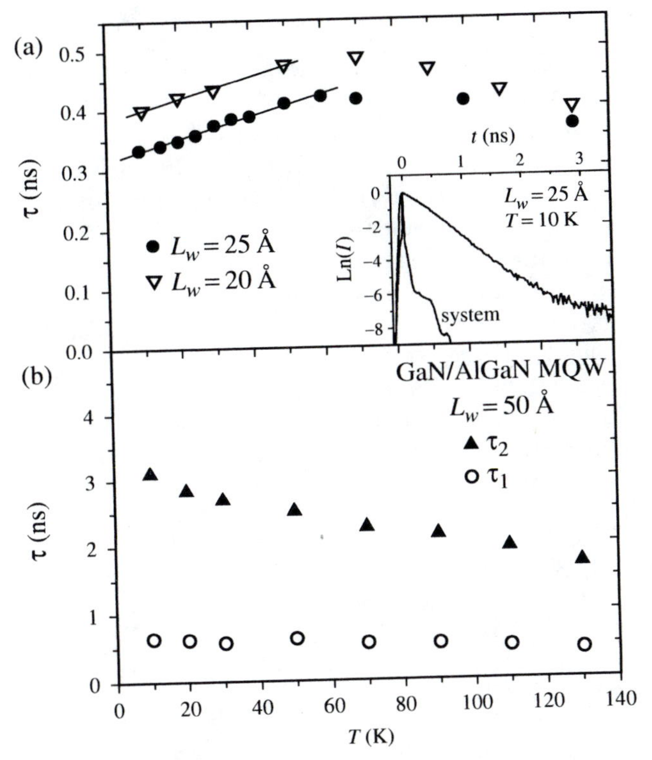

Figure 2.33 (a) The recombination lifetime of the dominant PL emission line resulting from the GaN wells as a function of temperature for GaN/AlGaN MQW samples with well thicknesses $L_w = 20$ Å and $L_w = 25$ Å. The inset shows an example of the PL temporal response at 10 K, which illustrates that the low-temperature ($T < 150$ K) PL decay in these MQWs was exponential, $I(t) = A\exp(-t/\tau)$. The detection system response to the laser pulse (7 ps) is indicated as "system", which was about 20 ps. (b) The decay time constants of the dominant PL emission line resulting from the GaN wells as functions of temperature for a GaN/AlGaN MQW sample with well thickness $L_w = 50$ Å, where τ_1 and τ_2 were obtained by fitting the PL decay to a two-exponential function, $I(t) = A_1\exp(-t/\tau_1) + A_2\exp(-t/\tau_2)$ [after Ref. 127].

a larger spectral blueshift and a broader exciton emission linewidth due to the enhanced quantum confinement and interface roughness effects.

The decay of the exciton recombination resulting from the well regions of the 20 Å and 25 Å MQWs was exponential at temperatures below 150 K. An example of the PL temporal response is shown in the inset of Figure 2.33a. The main figure of Figure 2.33a shows that the exciton recombination lifetime in both 20 Å and 25 Å MQWs increased linearly with temperature up to 60 K, similar to the behavior seen in the GaAs/AlGaAs QWs. This is a well-known property of exciton radiative recombination in QWs, which is observable only in high-quality samples [107,108]. In contrast, the decay of the dominant transition in the 50 Å MQW (at 3.414 eV at 10 K) was nonexponential and could be fitted with a two-exponential function, $I(t) = A_1 \exp(-t/\tau_1) + A_2 \exp(-t/\tau_2)$. The decay time constants τ_1 and τ_2 as functions of temperature ($T < 140$ K) are shown in Figure 2.30b. The faster decay time constant τ_1 was almost independent of temperature, whereas the slower decay time constant τ_2 decreased monotonically with temperature. These results indicated that GaN/AlGaN MQWs of high optical quality can be achieved when the well width is narrow, around 25 Å or below.

Furthermore, time-resolved PL studies revealed that the PL emission peak positions in GaN/AlGaN MQWs exhibiting strong piezoelectric effect are in fact blueshifted at early delay times due to quantum confinement and carrier screening, but they shift toward lower energies as the delay time increases, and become redshifted at longer delay times [126]. Figure 2.34 illustrates this and shows time-resolved emission spectra of the main emission line of an MBE-grown 40 Å well GaN/$Al_xGa_{1-x}N$ ($x = 0.15$) MQW sample measured at $T = 10$ K at several representative delay times [126].

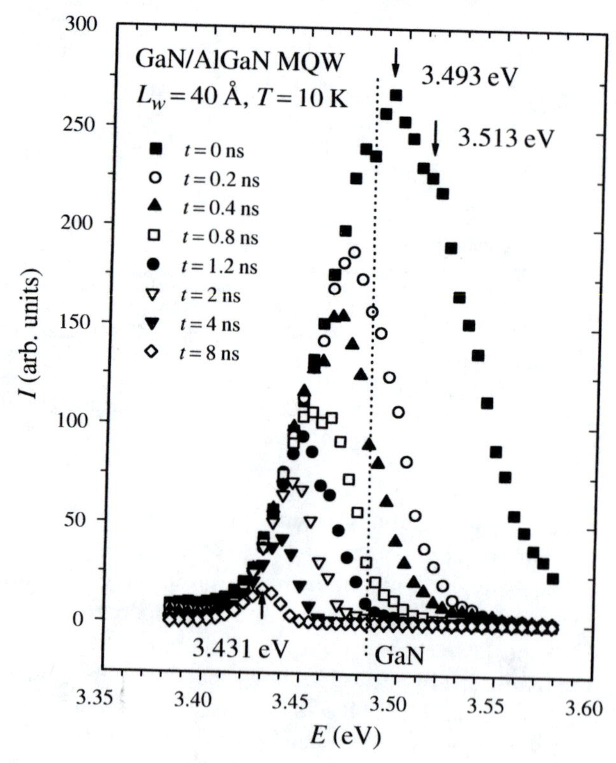

Figure 2.34 Low-temperature (10 K) time-resolved PL spectra of the main emission line in a 40 Å well GaN/$Al_xGa_{1-x}N$ MQW sample. The dotted line indicates the position of the excitonic transition peak in GaN epilayers grown under similar conditions [after Ref. 126].

The arrows in Figure 2.34 indicate the spectral peak positions at delay times $t_d = 0$ and $t_d = 8$ ns. Several features can be observed in Figure 2.34. First, the spectral peak at delay time $t_d = 0$ was blueshifted with respect to the peak position of the emission in the GaN epilayer line (dashed vertical line) grown under similar conditions (3.485 eV). Second, the peak positions of the emission line shifted toward lower energies with increasing delay time, with a total shift of 62 meV from $t_d = 0$ to $t_d = 10$ ns. Third, the linewidth of the emission line decreased with delay time. The time-resolved PL results were a direct consequence of the piezoelectric effects. As illustrated in Figure 2.35, under the influence of the piezoelectric field, injected (either optically or electrically) electrons and holes in the well regions drift apart. At the same time, these spatially separated electrons and holes induce an electric field that screens the piezoelectric field; however, the screening field strength decreases with decreasing carrier concentration. Thus, after a pulsed excitation the strength of the screening field decreases with delay time due to the recombination of carriers. As shown in Figure 2.35, the mean distance between the ground-state wavefunctions of electrons and holes in GaN/AlGaN MQWs increases and the transition energy decreases with a gradually decreasing screening field (or increasing delay time) due to the recombination of electrons and holes. The magnitude of the spectral redshift with delay time is directly correlated with the piezoelectric field strength within the wells. From the measurement results in Figure 2.35 together with

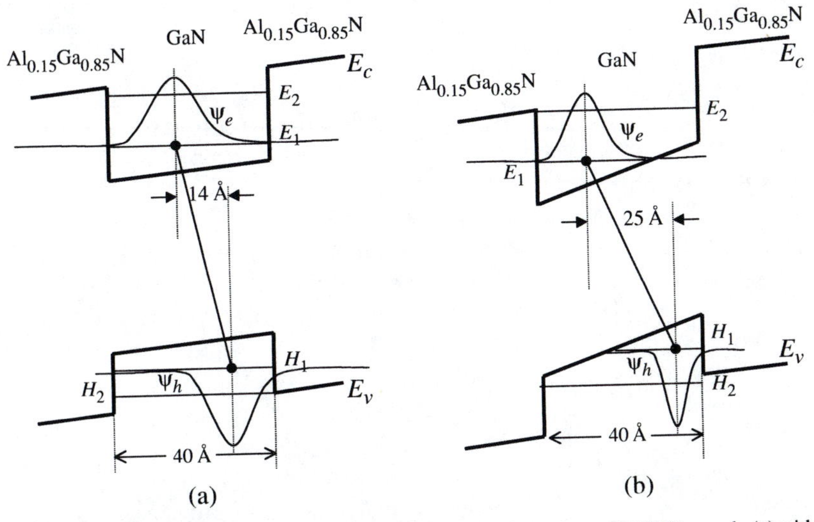

Figure 2.35 Schematic energy band diagrams of the GaN/$Al_xGa_{1-x}N$ MQW sample (a) with the piezoelectric field under the influence of screening by photoexcited carriers (or $t_d = 0$) and (b) the original piezoelectric field in the absence of carrier screening (or after long delay times) [after Ref. 126].

a theoretical calculation to be discussed in the theory section (Section 2.7), a low limit value of about 0.6×10^6 V/cm was obtained for the piezoelectric field strength in a GaN/$Al_{0.15}Ga_{0.85}N$ MQW sample [126].

It was shown that the measured decays in GaN/AlGaN MQWs are controlled not only by radiative lifetimes, which depend on the fields inside GaN wells, but also on the nonradiative escape of carriers through AlGaN barriers, which depends on their widths and on the electric field in these layers [128]. Thus, built-in electric fields present in barrier materials should be considered when dealing with III-nitride MQW systems. Time-resolved PL studies have also shown that electric fields in these MQWs are not solely of piezoelectric origin and are not a simple function of lattice-mismatch strains but depend also on respective thicknesses of wells and barriers [128].

2.6.2 Optimizing Growth Conditions by Time-Resolved PL

The unique features of time-resolved PL results, such as the temperature dependence of the exciton lifetime and time-resolved spectra, were shown to be extremely useful for optimizing the growth conditions of GaN/$Al_xGa_{1-x}N$ QWs for light emitter applications. To optimize LEDs and LDs structures it is crucial to maximize the optical emission or quantum efficiencies (QE) in the quantum confined states in the well regions. It is expected that the QE of MQWs will depend strongly on the growth conditions. For the growth of GaN/$Al_xGa_{1-x}N$ MQWs, one can choose the growth conditions to be optimal for either GaN epilayers (GaN-like) or $Al_xGa_{1-x}N$ epilayers ($Al_xGa_{1-x}N$-like). It was demonstrated by time-resolved PL results that the optimal growth conditions for GaN/AlGaN MQWs by MOCVD were GaN-like [129].

The cw PL spectra measured at 10 K for a set of GaN/$Al_xGa_{1-x}N$ MQW samples A, B, C, D, and E are shown in Figure 2.36a, b, c, d, and e, respectively. Samples A and E represent those grown under the optimal GaN growth conditions (or GaN-like), at a temperature and pressure of 1050 °C and 300 torr, respectively. The well widths of samples A and E were 30 Å and 18 Å, respectively, based on the growth rate of GaN epilayers under the same growth conditions. Sample D represents those grown under the optimal $Al_xGa_{1-x}N$ growth conditions (or $Al_xGa_{1-x}N$-like), at a temperature and pressure of 1060 °C and 150 torr, respectively. The barrier width was targeted at 50 Å for sample D based on the $Al_xGa_{1-x}N$ growth rate. Samples B and C represent those grown under growth conditions between the optimal GaN and $Al_xGa_{1-x}N$ growth conditions, with either the pressure or temperature being varied systematically.

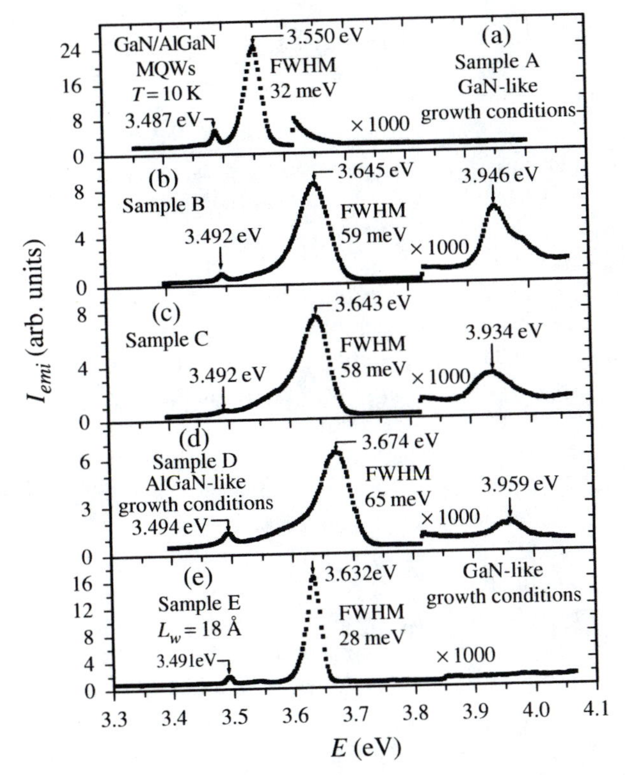

Figure 2.36 Low-temperature (10 K) cw PL spectra of GaN/$Al_xGa_{1-x}N$ MQWs. (a) Sample A grown under optimal GaN growth conditions (GaN-like) with well thickness $L_w = 30$ Å, (b) and (c) samples B and C grown under conditions in between optimal GaN and $Al_xGa_{1-x}N$ growth conditions, (d) sample D grown under optimal $Al_xGa_{1-x}N$ growth conditions ($Al_xGa_{1-x}N$-like) with barrier thickness of 50 Å, and (e) sample E grown under GaN-like growth conditions with well thickness $L_w = 18$ Å. The growth time for the well and barrier layers was fixed for samples A, B, C, and D. Note that samples A and E exhibited the highest PL intensities and narrowest linewidths from the well regions as well as no PL emission from the barrier regions [after Ref. 129].

The main emission peaks in Figure 2.36 were due to excitonic recombination in the GaN well regions, which were all blueshifted with respect to an emission line at about 3.49 eV from the underlying GaN epilayer. The most important feature shown in Figure 2.36 is that PL efficiencies of samples A and E (grown under the GaN-like conditions) were about two to four times higher than those of other samples. The linewidths of the well transitions of samples A and E were about 32 and 28 meV, respectively; however, the linewidths of samples B, C, and D were between 54 and 65 meV. The broadening of the linewidth and the lower quantum efficiency were attributed to the relatively poor qualities of samples B, C, and D (grown under other than the GaN-like conditions). Thus, GaN/$Al_xGa_{1-x}N$ MQWs grown under the GaN-like conditions had the highest quantum efficiency and narrowest linewidth, i.e., the highest optical quality.

In designing MQW laser structures, another important consideration is to minimize the optical transitions from the barrier regions, since emissions from the barrier regions represent losses in quantum efficiency or optical gain [130]. For samples A and E grown under the GaN-like growth conditions, there were hardly any observable emission intensities from the barrier regions [Figure 2.36a and e]. No barrier transitions were observed for samples A and E even at room temperature. Samples B, C, and D, grown under other conditions, exhibited clear barrier emission peaks around 3.95 eV, as shown in Figure 2.36b, c, and d. These results showed that for GaN/$Al_xGa_{1-x}N$ MQW structures carrier leakage from well to barrier regions can be minimized by employing GaN-like growth conditions. The temperature dependence of the PL emission intensity of the five MQW samples A, B, C, D, and E were also measured. Samples A and E exhibited the highest emission efficiency at all temperatures.

Time-resolved PL measurements provided crucial information regarding the quality and quantum efficiency of these MQWs grown under different conditions. Figure 2.37 shows the temperature dependence of the PL decay lifetime τ measured at the peak positions of the well transitions in these MQWs. For sample A grown under GaN-like growth conditions (Figure 2.37a), the PL decay lifetime increased with temperature up to 70 K. As illustrated in Figure 2.37b, c, and d, a linear increase of τ with T was absent in samples grown under conditions other than GaN-like conditions. The time-resolved PL results thus suggested that the radiative recombination was dominant at low temperatures probably only in MQW samples grown under GaN-like conditions. Thus, time-resolved PL data further corroborate that GaN/$Al_xGa_{1-x}N$ MQWs grown under GaN-like growth conditions have the best optical qualities as well as the highest quantum efficiencies.

Time-resolved PL studies showed that when GaN/$Al_xGa_{1-x}N$ MQWs were grown by MOCVD under optimal GaN-like growth conditions, these GaN/AlGaN MQW structures exhibited weak piezoelectric effects and hence enhanced QE. The time-resolved spectra of a 48 Å GaN/$Al_xGa_{1-x}N$ MQW sample grown under such conditions are presented in Figure 2.38, where the well emission peak position demonstrated a total redshift with a delay time of only about 5 meV [131]. Similar behaviors have been observed for GaN/$Al_xGa_{1-x}N$ MQW samples of different well widths, implying the contribution from the piezoelectric field is quite small in GaN/$Al_xGa_{1-x}N$ MQWs grown under GaN-like conditions by MOCVD. These results are in contrast with the time-resolved spectra of the well emission seen in the MBE-grown GaN/$Al_xGa_{1-x}N$ MQWs shown in Figure 2.34.

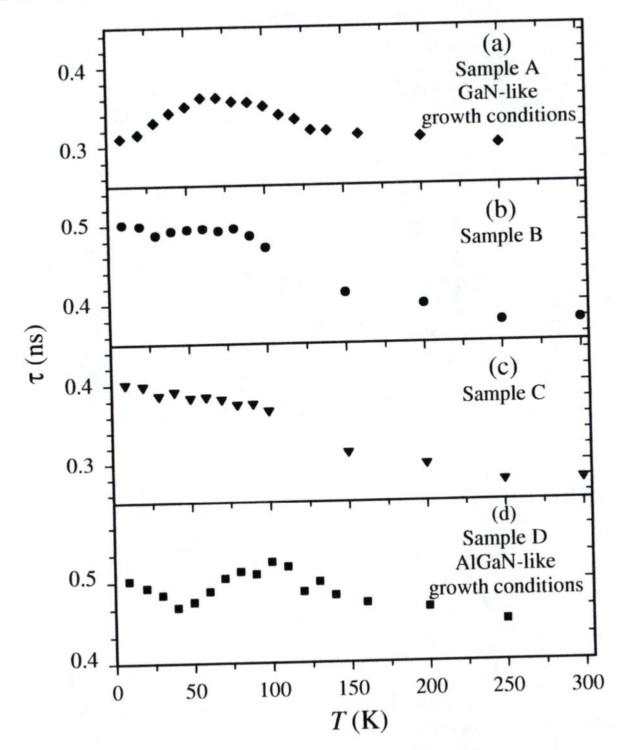

Figure 2.37 Temperature dependence of the well-transition PL decay lifetime τ for samples A, B, C, and D of Figure 2.36. A linear increase of τ with temperature was observed for sample A up to 70 K (a) [after Ref. 129].

2.6.3 Optimizing QW Structures by Time-Resolved PL

MQW structural parameters such as well and barrier widths are expected to affect the QE as well. Time-resolved PL measurements have been employed to identify the optimal structures of GaN/AlGaN MQWs [131,132]. The low-temperature ($T = 10$ K) PL spectra of a set of 30 Å well GaN/$Al_xGa_{1-x}N$ ($x \approx 0.2$) MQW samples grown under GaN-like conditions, with barrier widths $L_B = 30$, 40, 50, 80, and 100 Å, are presented in Figure 2.39. The excitonic transition peak resulting from the well regions was significantly blueshifted with respect to the underlying GaN epilayers by about 70 meV (10 K). Another important feature exhibited by these 30 Å well MQW structures was that the ratio of their well to barrier emission intensity was about 10^4 at 10 K.

The barrier width dependence of the integrated PL intensity of the well transition for these 30 Å well MQW structures can also be obtained from Figure 2.39 and is plotted in Figure 2.40a. The total integrated PL intensities

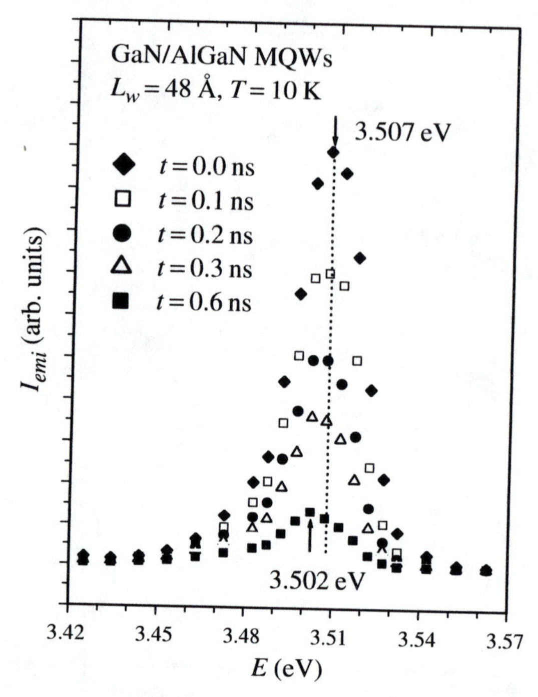

Figure 2.38 Time-resolved PL spectra of a 48 Å well GaN/$Al_xGa_{1-x}N$ MQW sample grown under optimized GaN-like conditions by MOCVD, showing a total redshift with a delay time of only about 5 meV [after Ref. 131].

of these MQW samples were reduced by only one order of magnitude as the temperature rose from 10 to 300 K, indicating high PL efficiencies even at room temperature. The most important result shown in Figure 2.40 was that the integrated PL intensities (or QE) of these MQW samples increased monotonously with increasing barrier width up to 80 Å.

The decay lifetime for the well transitions increased almost linearly from 0.2 to 0.4 ns when the barrier width varied from 30 to 80 Å. This is clearly illustrated in Figure 2.40b, where the PL temporal responses as well as the barrier width dependence of the PL decay lifetime are shown for the well transitions. Furthermore, a linear increase of the well transition lifetime with temperature was observed in these MQW structures when the barrier width was below 100 Å, which reflected that radiative exciton recombination dominated in these MQW samples. The enhanced decay lifetime with barrier width up to 80 Å shown in Figure 2.40b was consistent with the QE enhancement with barrier width shown in Figure 2.40a. For the 100 Å barrier MQW sample studied here, well transition lifetime decreased with temperature, which was an indication of increased nonradiative recombination rates at higher temperatures in this MQW sample.

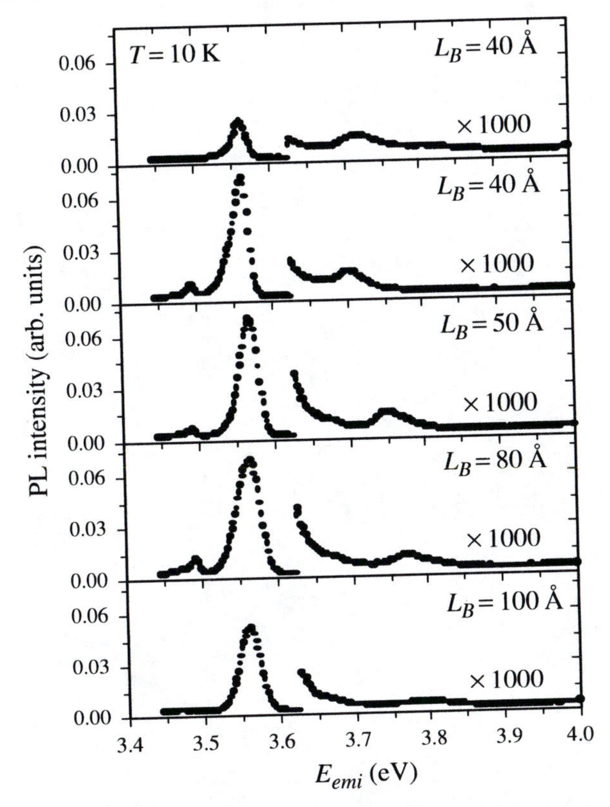

Figure 2.39 PL spectra of a set of 30 Å well GaN/$Al_xGa_{1-x}N$ MQW samples with varying barrier widths, $L_B = 30, 40, 50, 80$, and 100 Å, measured at 10 K. The MQW structures were grown under identical GaN-like conditions by MOCVD [after Ref. 132].

The PL spectra of a set of 50 Å barrier GaN/$Al_xGa_{1-x}N$ ($x \approx 0.2$) MQWs with well widths varying from 6 to 48 Å grown under GaN-like conditions by MOCVD, measured at 10 K, are shown in Figure 2.41. The main emission peaks (varying from 3.507 to 3.693 eV) in these spectra were again due to excitonic recombinations in the GaN well regions. The low-energy emission peaks around 3.490 eV were from the underlying undoped GaN epilayers. Again, no transition peaks from the $Al_xGa_{1-x}N$-barrier regions were observed, indicating that the PL emission and carrier confinement in the well regions were highly efficient. The integrated well emission intensity versus L_w measured at 10 K for these GaN/$Al_xGa_{1-x}N$ MQWs is plotted in Figure 2.41, which shows that highest QE can be achieved when L_w is between 12 and 42 Å. The uncertainty in the integrated emission intensity was due mainly to the slight variations in crystal growth between different runs; however, the general trend shown in Figure 2.42 was still quite clear

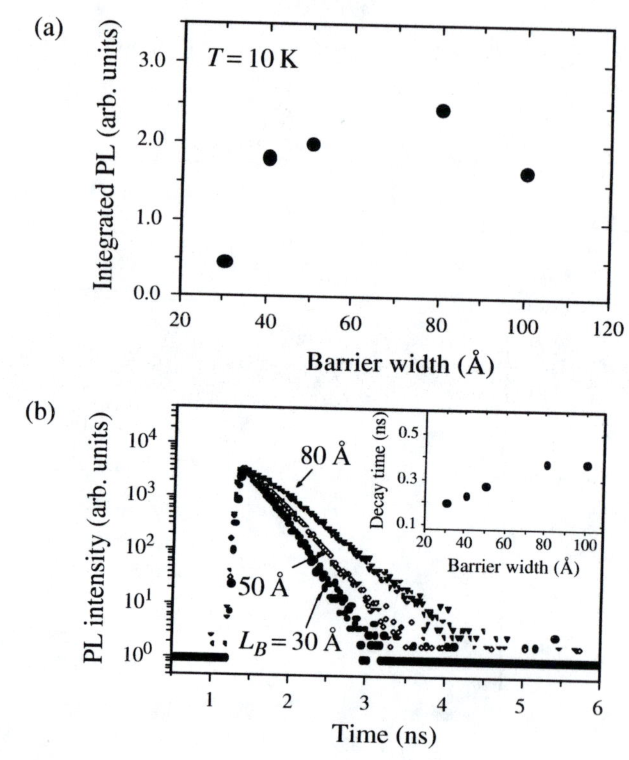

Figure 2.40 (a) Integrated PL intensity of the well transition for the GaN/$Al_xGa_{1-x}N$ MQW samples as a function of L_B measured at 10 K. (b) The temporal responses of the well transitions in the GaN/$Al_xGa_{1-x}N$ MQW samples for various barrier widths at 10 K. The PL decay in all MQW samples studied here can be well described by a single exponential decay function. The inset shows the barrier width dependence of the decay lifetime for the well transition. The MQW structures were grown under identical GaN-like conditions by MOCVD [after Ref. 132].

despite the large experimental uncertainties. The high QE resulted from an improved quantum well quality, a reduced nonradiative recombination rate, and a decreased piezoelectric effect in these MQWs.

In addition to the reduced piezoelectric effect, the high QE achieved in the GaN/$Al_xGa_{1-x}N$ MQWs with L_w between 12 and 42 Å was attributed to the reduced nonradiative recombination rate as well as to the improved quantum well interface quality. The decreased QE seen in GaN/$Al_xGa_{1-x}N$ MQWs with larger L_w(>42 Å) suggested that the nonradiative recombination channels started to play an important role, which was consistent with the observation that the recombination lifetime of the well transition was shortest in the 48 Å GaN/$Al_xGa_{1-x}N$ MQW sample. In Figure 2.43, the PL temporal responses of the well transitions for three representative MQW samples are shown. One sees that the recombination lifetime

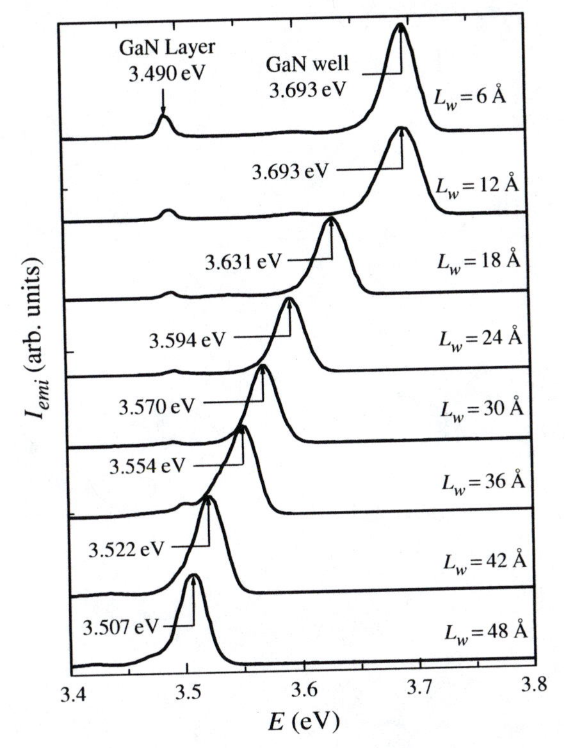

Figure 2.41 PL spectra of a set of GaN/$Al_xGa_{1-x}N$ MQW samples with well width varying from 6 to 48 Å and a fixed barrier width of 50 Å measured at 10 K. The MQW structures were grown under identical GaN-like conditions by MOCVD [after Ref. 131].

of the well transition in the 48 Å GaN/$Al_xGa_{1-x}N$ MQWs was 0.25 ns, which was shorter than the value in the 6 Å (0.35 ns) or in the 12 Å (0.42 ns) GaN/$Al_xGa_{1-x}N$ MQW sample. The low-temperature lifetimes in all other GaN/$Al_xGa_{1-x}N$ MQWs ranged between 0.30 and 0.40 ns (not shown). The reduced recombination lifetime in the 48 Å well width GaN/$Al_xGa_{1-x}N$ MQW sample was due to an increased nonradiative recombination rate. It was expected, since the misfit dislocation density in the GaN-well regions increased sharply as the well width approached the critical thickness of MQWs, which resulted in an enhanced nonradiative interface recombination rate and thus lower QE.

On the other hand, the decreased QE in GaN/$Al_xGa_{1-x}N$ MQWs with small L_w (less than 12 Å) was due to the enhanced carrier leakage to the underlying GaN epilayer. For MQWs with narrow L_w, the electron and hole wave functions extended farther into the barrier regions, as illustrated schematically in the inset of Figure 2.43, which led to an increased (decreased) carrier recombination outside (inside) the well regions.

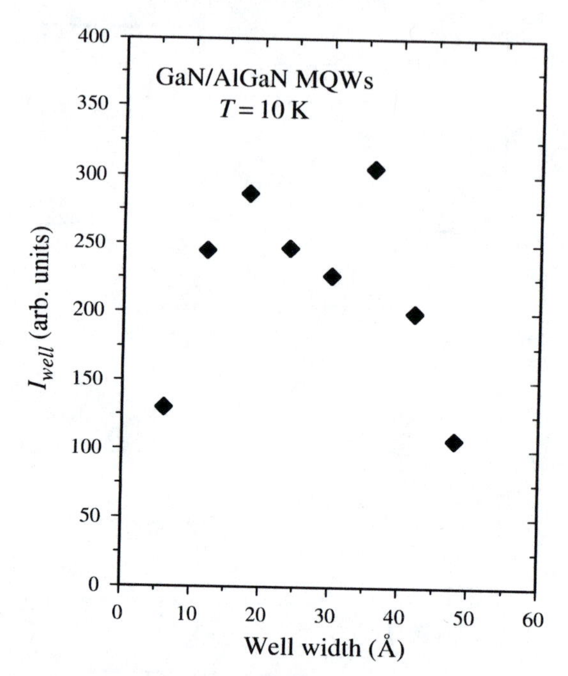

Figure 2.42 The integrated well emission intensity versus well width for GaN/Al_xGa_{1-x}N MQWs measured at 10 K. The MQW structures were grown under identical GaN-like conditions by MOCVD [after Ref. 131].

As shown in Figure 2.41, PL emission intensity at 3.490 eV from the underlying GaN epilayer was largest in the 6 Å well GaN/Al_xGa_{1-x}N MQW sample and decreased with increasing L_w. This clearly indicated that carrier leakage from the well regions to the underlying GaN epilayer increased with decreasing L_w.

These studies showed that the optimal GaN/AlGaN ($x \sim 0.2$) MQW structures for UV light emitter applications are those with barrier widths ranging from 40 to 80 Å and well widths ranging from 12 and 42 Å. The decreased quantum efficiency in GaN/Al_xGa_{1-x}N MQWs with $L_w < 12$ Å is due to enhanced carrier leakage to the underlying GaN epilayers, whereas the decreased quantum efficiency in MQWs with $L_w > 42$ Å is associated with an increased nonradiative recombination rate as L_w approaches the critical thickness of MQWs. For the barrier width dependence, when the barrier width is below the critical thickness, the nonradiative recombination rate increases with decreasing barrier width due to enhanced probabilities of electron and hole wavefunctions at the interfaces as well as in the AlGaN barriers. On the other hand, the misfit dislocation density increases as the barrier width approaches the critical thickness, which can result in an enhanced nonradiative interface recombination rate.

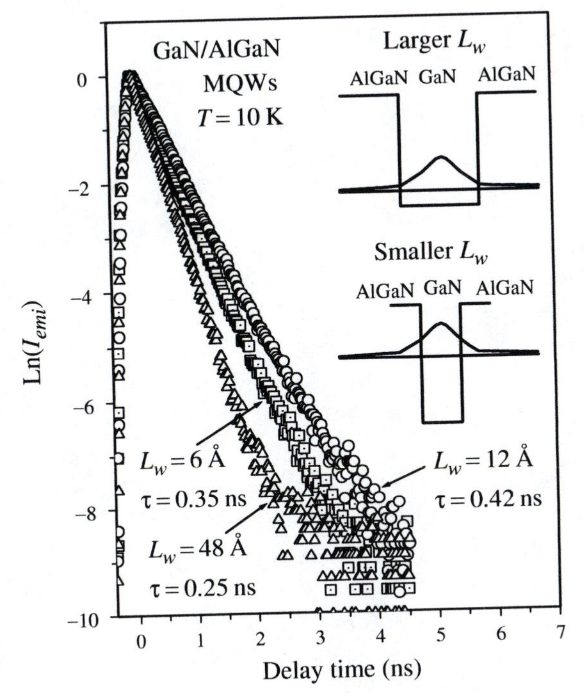

Figure 2.43 PL temporal responses of the well transitions (10 K) of three representative $GaN/Al_xGa_{1-x}N$ MQWs with L_w = 6, 12, and 48 Å of Figure 2.41. The inset shows a schematic of electron wavefunction distribution in $GaN/Al_xGa_{1-x}N$ MQWs with large and small well widths. The MQW structures were grown under identical GaN-like conditions by MOCVD [after Ref. 131].

2.7 THEORY OF SPONTANEOUS EMISSION IN III-NITRIDE QWs

When analyzing experimental data, it is helpful to have available a theory that describes the underlying physics at a microscopic level with a minimum of ad hoc parameterization. This section describes one such theory for an electron–hole plasma in a wurtzite III-nitride quantum well structure. The starting point is a Hamiltonian that contains the effects of band structure, carrier–laser field interaction, and carrier–carrier Coulomb interactions. Following a derivation that ensures a consistent treatment of all Coulomb effects to a given order, a set of coupled differential equations is obtained that describes the dynamics of the microscopic polarization $p_{\vec{k}}^{\upsilon_e,\upsilon_h}$, and the electron and hole occupation probabilities $n_{\vec{k}}^{\upsilon_e}$ and $n_{\vec{k}}^{\upsilon_h}$, respectively [133,134], where $\upsilon_{e(h)}$ is the electron (hole) subband, and $\vec{k}$ is the carrier momentum. These equations are often referred to as the *semiconductor Bloch equations*, and they are widely used in investigations involving the optical response of semiconductors. The equation for the microscopic polarization is

$$\frac{dp_{\vec{k}}^{v_e,v_h}}{dt} = -i\omega_{\vec{k}}^{v_e,v_h} p_{\vec{k}}^{v_e,v_h} - i\Omega_{\vec{k}}^{v_e,v_h}\left(n_{\vec{k}}^{v_e} + n_{\vec{k}}^{v_h} - 1\right) + \left.\frac{\partial p_{\vec{k}}^{v_e,v_h}}{\partial t}\right|_{cor},$$

which describes oscillation at the transition frequency (first term, right-hand side), and stimulated emission and absorption processes (second term). These two terms contain the Hartree–Fock many-body contributions resulting in the transition energy and field (Rabi energy) renormalizations. The former gives a carrier density dependence to the transition energy $\omega_{\vec{k}}^{v_e,v_h}$, and the latter, due to the attractive Coulomb interaction between an electron and a hole, renormalizes the Rabi frequency, $\Omega_{\vec{k}}^{v_e,v_h}$. This renormalization of $\Omega_{\vec{k}}^{v_e,v_h}$ is responsible for the Coulomb or Sommerfeld enhancement of the interband transitions at high carrier densities and for the formation of excitons at low carrier densities. The term $\partial p_{\vec{k}}^{v_e,v_h}/\partial t\,|_{cor}$ denotes Coulomb correlation contributions that are beyond the Hartree–Fock level. They include carrier collision effects and plasma screening. To derive an approximate expression for these contributions, we continue with the derivation that brought us the Hartree–Fock contributions. The actual derivation requires aspects of many-body theory, and the details may be found elsewhere [135]. The resulting expression in its simplest form gives

$$\left.\frac{\partial p_{\vec{k}}^{v_e,v_h}}{\partial t}\right|_{cor} = -\left(\Gamma_{\vec{k}}^{v_e} + \Gamma_{\vec{k}}^{v_h}\right) p_{\vec{k}}^{v_e,v_h} + \sum_{\vec{q}}\left(\Gamma_{\vec{k},\vec{q}}^{v_e} + \Gamma_{\vec{k}+\vec{q}}^{v_h}\right) p_{\vec{k}+\vec{q}}^{v_e,v_h}, \quad (2.20)$$

which describes the Coulomb correlation contributions at the level of quantum kinetic theory in the Markovain limit. Equation (2.20) contains diagonal ($p_{\vec{k}}^{v_e,v_h}$ term) and nondiagonal ($p_{\vec{k}+\vec{q}}^{v_e,v_h}$ term) contributions. The imaginary parts of the Γ's give the leading contributions to plasma screening.

The solution of Eq. (2.20) gives microscopic polarization, which is related to the macroscopic polarization P via

$$P = \sum_{v_e,v_h,\vec{k}} \left(\mu_{\vec{k}}^{v_e,v_h}\right)^* p_{\vec{k}}^{v_e,v_h} + c.c. \quad (2.21)$$

where $\mu_{\vec{k}}^{v_e,v_h}$ is the dipole matrix element. From semiclassical theory [136], we obtain for the optical response of the medium (in mks units),

$$\frac{dE}{dz} = -\frac{\alpha}{2}E = -\frac{\omega}{2\varepsilon_0 n_b c V}\mathrm{Im}P, \quad (2.22)$$

which shows that the absorption coefficient α depends on the imaginary part of the active medium polarization amplitude. In Eq. (2.22), ε_0 and c are the permittivity and speed of light in vacuum, n_b is the background refractive index, V is the active region volume, z is the propagation direction, and E is the slowly varying (real) electric field amplitude. To calculate the spontaneous emission spectrum $S(\omega)$, which is often measured in experiments, we make use of the following relationship derived from detailed balance arguments [137]:

$$S(\omega) = -\frac{1}{\hbar}\left(\frac{n\omega}{\pi c}\right)^2 \alpha(\omega)\left[\exp\left(\frac{\hbar\omega - \mu_{eh}}{k_B T}\right) - 1\right]^{-1}, \qquad (2.23)$$

where k_b is the Boltzmann constant, T is the temperature, and μ_{eh} is the electron–hole quasi-chemical potential energy separation. Such an approach is of course not rigorous. It has the advantage of circumventing the complexities associated with quantizing the electromagnetic field. At high carrier densities, it has been shown that this phenomenological approach agrees relatively well with the results from fully quantized semiconductor quantum luminescence theory [138].

The solution of Eq. (2.21) requires as input the band structure properties, specifically, the electron and hole dispersions, $\varepsilon_{e\vec{k}}^{v_e}$ and and $\varepsilon_{h\vec{k}}^{v_h}$, as well as the dipole matrix elements, $\mu_{\vec{k}}^{v_e v_h}$. These quantities may be calculated using $\vec{k}\cdot\vec{p}$ theory and the envelope approximation. In a typical wurtzite group III nitride quantum well structure, there is an electric field E_p due to strain or spontaneous polarization [139,140]. To account for its effects, one can add the off-diagonal contribution to the electron and hole Hamiltonian:

$$H_{v_\alpha v'_\alpha}^{piezo} = eE_p(N)\int_{-\infty}^{\infty} dz\, u_{v_\alpha}(z) z\, u_{v'_\alpha}(z), \qquad (2.24)$$

where v_α and v'_α are the subbands coupled by the Stark effect, and $u_{v_\alpha}(z)$ and $u_{v'_\alpha}(z)$ are the corresponding subband envelope functions. The two-dimensional carrier density N dependence in the piezoelectric field comes from the screening caused by the resulting spatial separation in the electron and hole distributions. The net electric field is

$$E_p(N) = E_p(0) + E_{scr}(N) \qquad (2.25)$$

where

$$E_p(0) = \frac{1}{\varepsilon_b}\left[P_{sp}^{qw} - P_{sp}^{b} + 2d_{31}e\left(C_{11} + C_{12} - \frac{2C_{13}^2}{C_{33}}\right)\right] \qquad (2.26)$$

is the sum of the spontaneous polarization and piezoelectric fields in the absence of carriers. Here, $P_{sp}^{qw(b)}$ is the spontaneous polarization in the quantum well (barrier) region, C_{ij} is an elastic constant, and ε_b is the background permittivity. We estimate the effective screening field by spatially averaging over the quantum well the electric field due to the combined electron and hole distributions:

$$E_{scr}(N) = \frac{1}{d}\int_{-d/2}^{d/2} dz_0 \frac{eN}{2\varepsilon_b}\int_{-\infty}^{\infty} dz\big[|u_e(z)|^2 - |u_h(z)|^2\big]\frac{z - z_0}{|z - z_0|} \quad (2.27)$$

where d is the quantum well width. We use for the spatial carrier distributions $u_e(z)$ and $u_h(z)$ are the lowest-order electron and hole subband envelope functions at zone center. The screening field $E_{scr}(N)$ is determined by iterating (2.25) and the solution to band structure calculation until convergence is reached.

The input to the quantum well band structure calculation is the bulk material properties such as the effective masses for the electron and holes parallel and perpendicular to the c-axis, and the two zone center energy splittings Δ_1 and Δ_2 between the holes states. In addition to the elastic constants, the parameters relating to strain effects are the lattice constants a (|| to the quantum well plane) and c (perpendicular to the quantum well plane), the conduction band deformation potentials a_{cz} and $a_{c\perp}$, and the valence band deformation potentials D_1, D_2, D_3, and D_4. The remaining parameters are the bulk material dipole matrix element μ_{bulk}, the background refractive index n, and the permittivity of the host material ε_b. The values for the preceding bulk wurtzite material parameters for GaN, InN, and AlN are given in Ref. [134]. For the alloys, we use the weighed averages.

A comparison between theoretical and experimental results sometimes has to account for inhomogeneous broadening in the experimental samples because of spatial variations in quantum well thicknesses or compositions. If the dimensions of these spatial variations are large, so that the quantum confinement remains essentially in the epitaxial direction, the effects of inhomogeneous broadening may be approximated by a statistical average of the homogeneous gain spectra [141]:

$$G_{inh}(\omega, N, T) = \int dx\, P(x) G(x, \omega, N, T), \quad (2.28)$$

where $P(x) = (\sqrt{2\pi}\sigma)^{-1}\exp\{-[(x - x_0)/(\sqrt{2}\sigma)]^2\}$ is the normal distribution representing the variation in x, which can either be the quantum well thickness or indium concentration. The distribution is characterized by an average x_0, and a standard deviation σ.

Figures 2.44 and 2.45 show the spontaneous emission spectra computed using Eq. (2.23). Two $In_{0.2}Ga_{0.8}N$/GaN QW structures, with well widths

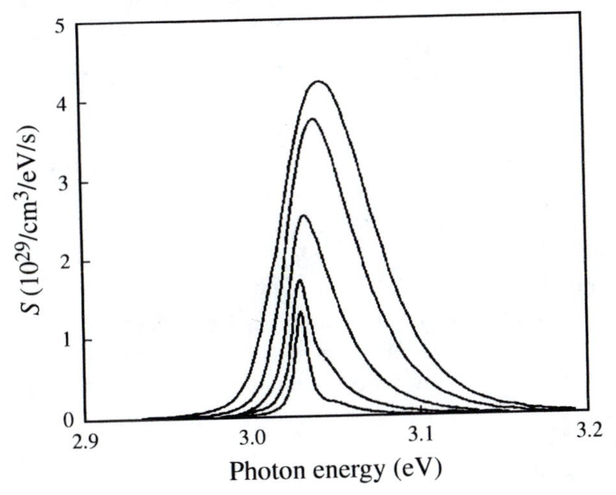

Figure 2.44 Calculated spontaneous emission spectra of a 2-nm-well $In_{0.2}Ga_{0.8}N$/GaN QW at $T = 300$ K, and for carrier density $N = 10^{12}/cm^2$ to $5 \times 10^{12}/cm^2$, in $10^{12}/cm^2$ increments.

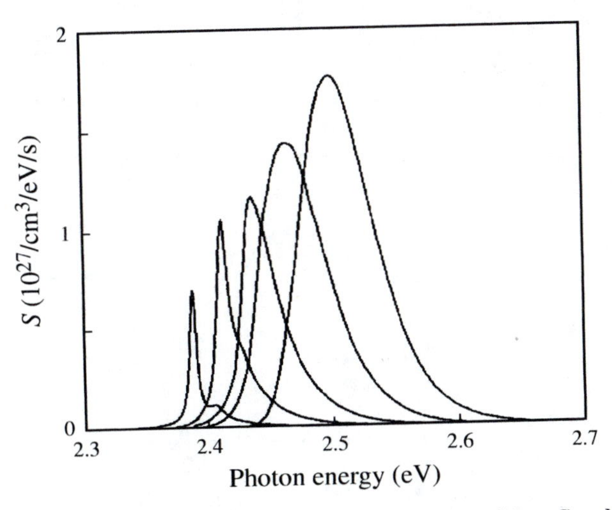

Figure 2.45 Calculated spontaneous emission spectra of a 4-nm-well $In_{0.2}Ga_{0.8}N$/GaN QW at $T = 300$ K, and for carrier density $N = 10^{12}/cm^2$ to $5 \times 10^{12}/cm^2$, in $10^{12}/cm^2$ increments.

of 2 nm and 4 nm, were considered to illustrate the interplay of quantum-confined Stark effect and many-body interactions. The figures clearly depict differences in carrier density dependences for the two quantum well widths. The 2-nm narrow well shows very little shift of the emission peak with carrier density. This behavior mirrors that of excitonic absorption resonance, which is also excitation independent for intrinsic QW systems. The situation is quite different in the 4-nm QW. Here, because of the weaker quantum confinement in this relatively wide QW, the piezoelectric and spontaneous polarization fields are able to significantly redshift the exciton emission and absorption peaks relative to the flat band situation. As the plasma density increases, the screening of these fields weakens the redshift, leading to a net blueshift in the spontaneous emission spectra with increasing plasma density, as shown in Figure 2.45.

Integrating the spontaneous emission spectra in Figures 2.44 and 2.45 gives the spontaneous emission rates, which are the inverse of the radiative carrier lifetimes, plotted in Figure 2.46. The results show that whereas the 2-nm well has lifetimes typical of conventional laser materials, the 4-nm QW has radiative lifetimes at low carrier densities and room temperature that are two orders of magnitude larger. This long lifetime is a direct consequence of

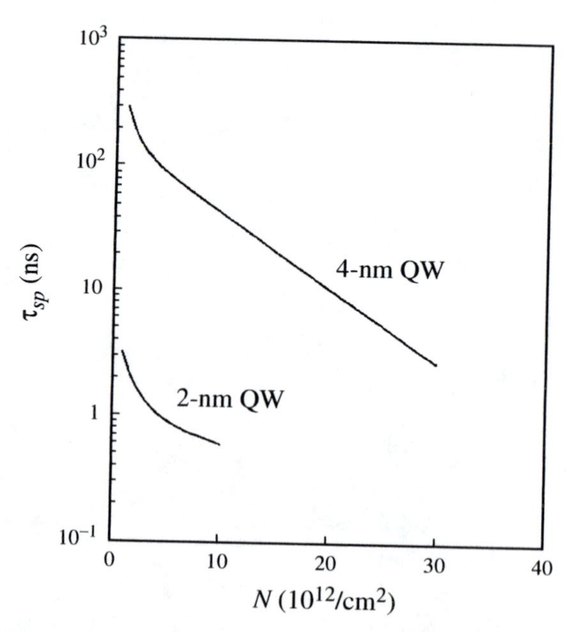

Figure 2.46 Calculated radiative lifetime versus carrier density for 2-nm- and 4-nm-well $In_{0.2}Ga_{0.8}N/GaN$ QW at $T = 300$ K.

the electron–hole wavefunction separation, which is substantial in the 4-nm QW. The results in Figure 2.46 also shows the effects of plasma screening on the radiative lifetime.

The effects of plasma screening on spontaneous emission dynamics have also recently been demonstrated experimentally in a 4-nm/5-nm $In_{0.12}Ga_{0.88}N/In_{0.03}Ga_{0.97}N$ MQW sample, in which the effects of piezoelectric and spontaneous polarization fields are known to be strong, by directly measuring the excitation carrier density dependences of the carrier lifetime and emission energy [142]. Figure 2.47 shows that the PL emission energy decreased with decreasing carrier density nonlinearly for carrier density below $1.5 \times 10^{18}/cm^3$, although the PL energy shift appeared to have a linear characteristic above this value. The authors pointed out that in the case of the low carrier density, the band-filling effect was not the main factor

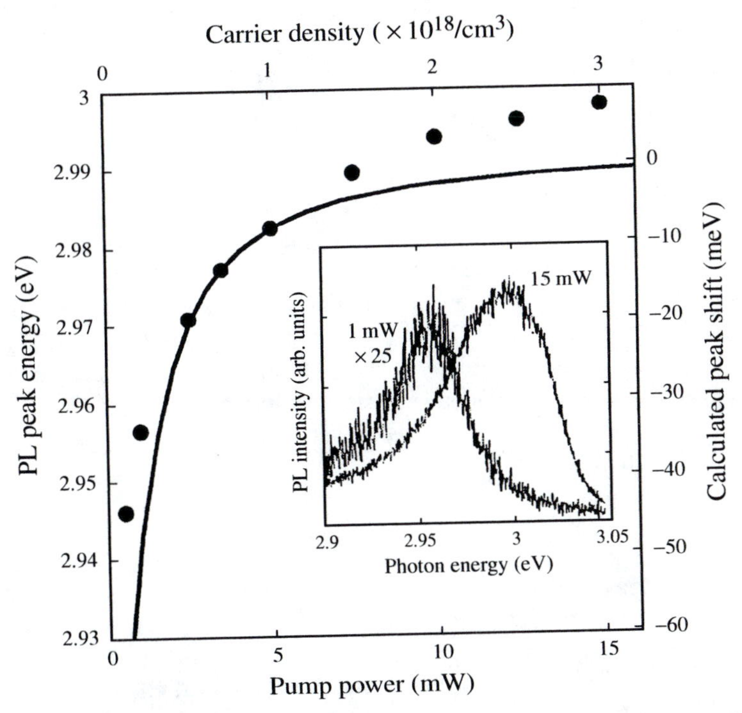

Figure 2.47 Pumping power dependence of the PL peak energy at 10 K in a 4-nm/5-nm $In_{0.12}Ga_{0.88}N/In_{0.03}Ga_{0.97}N$ MQW sample. Solid circles represent the PL peak energies measured at a delay time of 0.5 ns after optical excitation. The solid curve represents the calculated PL energy shift as a function of carrier density assuming an internal electric field of 650 kV/cm. The inset shows the measured PL spectra obtained under average excitation powers of 15 and 1 mW. Each spectrum was time integrated between the time delays of 0.2 and 0.8 ns [after Ref. 142].

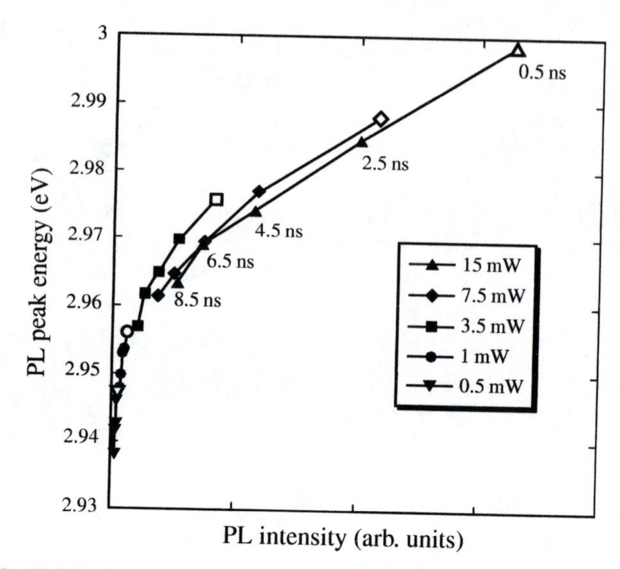

Figure 2.48 Trajectories of PL intensity and PL peak energy for various pump powers at a delay time interval of 2 ns for the same MQW sample as Figure 2.47. The open icons show the results at a time delay of 0.5 ns [after Ref. 142].

responsible for this energy shift. Since the QW has a step-function-like density of states, the PL energy shift should be linear at low temperatures in the frame of the band-filling effect.

The time evolution of the PL emission energy and intensity observed in this QW sample under different pumping intensities is plotted in Figure 2.48 at a 2-ns delay time interval [142]. Immediately after photoexcitation, the carrier density gradually decreased due to carrier recombination. As a consequence, both the emission spectral peak position and intensity decreased with increasing delay time, similar to the behaviors seen in GaN/AlGaN MQWs [127,128]; however, an interesting behavior seen in Figure 2.48 was that the emission peak traced almost a single trajectory. For example, the emission peak trajectory produced under an excitation intensity of 15 mW overlapped perfectly with that produced under an excitation intensity of 7.5 mW after a delay time of 2 ns. Since the PL intensity is governed by the product of the carrier density and the oscillator strength, this behavior implied that the dynamical change in the PL peak energy strongly depended on the carrier density. Figure 2.49 depicts the plasma screening effects on the carrier lifetime, which clearly illustrates the decrease in carrier lifetime with carrier density in this 4-nm/5-nm $In_{0.12}Ga_{0.88}N/In_{0.03}Ga_{0.97}N$ MQW sample, due to the screening (or weakening) of the internal built-in field as well as a reduced electron–hole wavefunction separation.

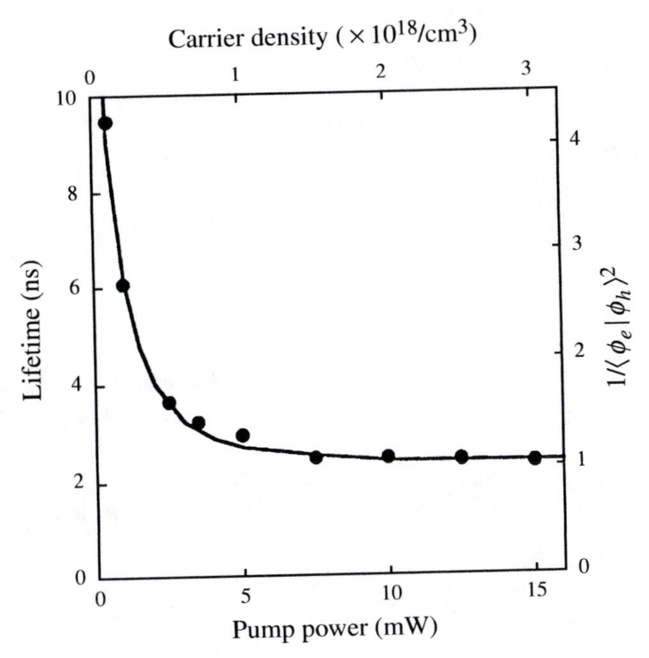

Figure 2.49 Pumping power dependence of the carrier lifetime at 10 K for the same MQW sample as in Figure 2.47. Solid circles represent the measured effective PL decay time determined at the delay time of 0.5 ns. The solid curve represents the calculated recombination rate as a function of carrier density assuming an internal electric field of 650 kV/cm [after Ref. 142].

2.8 CONCLUDING REMARKS

The aim of this chapter was to review time-resolved PL studies on III-nitrides, including GaN epilayers, InGaN and AlGaN alloys, InGaN/GaN, and GaN/AlGaN MQW structures, and to provide an overview on contributions of such studies to the understanding of the many aspects of this material system. It was our intention to cover articles written prior to May 2000, but undoubtedly there were many related articles unintentionally left out.

It can be seen that time-resolved PL, which provides the temporal characters of PL emissions, together with spectral information is indeed a powerful method for determining the dynamics of injected carriers involved in optical processes as well as sample crystalline quality, purity, alloy composition, quantum well interface properties, and quantum efficiencies in different materials and device structures.

Recent optical studies in III-nitrides were driven primarily by technological developments and needs. It is our belief that this trend will be continued. The problems and questions that still stand in the way of practical device implementation in many applications based on III-nitrides are issues of further improvement in material quality and perfection of QW or heterostructure device structures. In particular, research activities on $Al_xGa_{1-x}N$ (3.4 eV $< E_g <$ 6.2 eV) with high AlN mole fractions, and devices operating in the UV spectral region are still in their embryonic state. Several strong effects in AlGaN alloys (e.g., $Al_xGa_{1-x}N$ cannot be made conductive for $x > 0.5$) still need to be elucidated. For deep-UV optoelectronic device applications based on III-nitrides, AlGaN with high AlN fractions must be utilized. Although time-resolved PL studies have contributed significantly to the understanding of mechanisms of optical transitions and impurity properties in III-nitrides, no time-resolved PL work has been done so far on $Al_xGa_{1-x}N$ alloys with $x > 0.5$, because of the difficulty in growing high-quality $Al_xGa_{1-x}N$ alloys with high Al contents, and because there is a lack of suitable ultrafast laser systems that emit photons that can provide the necessary band-to-band excitation. On the other hand, because of the lack of material characterization techniques for $Al_xGa_{1-x}N$ alloys with high Al contents, it has been very difficult to improve overall material quality. Thus, it is imperative to develop time-resolved PL measurement systems with deep-UV capabilities. We believe that the availability of such a measurement system will accelerate the development of deep-UV optoelectronic devices based on $Al_xGa_{1-x}N$ alloys with high Al contents.

Further applications of AlGaN in areas of deep-UV optoelectronic and high-temperature/high-power electronic devices are feasible with increased Al content; however, a drawback to the use of AlGaN alloys is the lattice mismatch with GaN in the AlGaN/GaN heterostructure, in particular at high Al content. The result of this is local strain relaxation at the heterointerface through generation of cracks and/or misfit dislocations. A material that is both lattice-matched with GaN and whose bandgap energy can be adjusted for UV applications is therefore desirable. Such a material would be more versatile than AlGaN for many applications. The growth and properties of InAlGaN quaternary alloys have been reported recently [143–145]. It is expected that the use of this quaternary material will allow control of the lattice mismatch with GaN as well as energy bandgap engineering. Also, since the thermal expansion coefficient of GaN is between those of InN and AlN, we expect the thermal expansion coefficient of the quaternary InAlGaN compared with AlGaN to be better matched with that of GaN. We also expect that time-resolved PL studies will be very fruitful for guiding the material growth technologies for this system. Moreover, the lattice-matched GaN/InAlGaN heterojunction and MQW structures represent an ideal system for studying carrier dynamics in III-nitrides device structures without the influence of the lattice-mismatch-strain-induced piezoelectric fields.

Finally, the successes of conventional III-nitride devices such as edge-emitting lasers and LEDs and detectors are encouraging for the study of microsize devices. Microsize light-emitting devices offer benefits over edge emitters, including the ability to fabricate arrays of individually controllable pixels on a single chip. New physical phenomena and properties begin to dominate as the device size scale approaches that of the wavelength of light or smaller, including modified spontaneous emission, enhanced quantum efficiency, and reduced lasing threshold in microcavities, all of which warrant fundamental investigations [146]. III-Nitride QW microdisk and microring cavities have been fabricated, and an enhanced quantum efficiency and optical resonant modes have been observed in these microcavities [147–149]. Resonant optical modes in microsize GaN pyramids prepared by lateral epitaxial overgrowth (LEO) have also been observed [150]. Optically pumped lasing actions have also been observed in GaN pyramids prepared by LEO [151], in GaN microdisks prepared by reactive-ion etching [152], and in microsize GaN vertical cavity surface emitting laser (VCSEL) structures [153]. Electrically pumped UV/blue microsize LEDs have also been fabricated [154]. It is expected that carrier dynamical behaviors in microsize and nanosize structures and devices are very different from those in conventional broad-area ones. It will be fundamentally as well as practically important to investigate carrier dynamic processes in III-nitride micro- and nanocavities and devices by time-resolved PL studies. These include information regarding processes of carrier injection, transformation, recombination, and lasing, which can be utilized to guide material growth as well as device design.

Acknowledgments

We are indebted to many of the pioneers as well as our respected friends in the field, Professor G. F. Neumark, Professor D. C. Reynold, Professor K. K. Bajaj, Professor J. K. Furdyna, Professor R. K. Chang, Professor D. C. Look, Professor J. J. Song, and Professor B. Monemar, whose work on optical properties of semiconductors has inspired us greatly. We are grateful to Professor H. Morkoc, Professor M. Asif Khan, Prof. S. M. Bedair, Professor N. El-Masry, Professor S. S. Lau, Professor G. Y. Zhang, Professor B. Zhang, Professor J. Edgar, Dr. S. H. Wei, Dr. Eric Jones, and Dr. Sarah R. Kurtz for their long-term collaboration and support. We would like to acknowledge assistance from the following present and former members of our group at Kansas State University: Dr. K. C. Zeng, Dr. R. Mair, Dr. M. Smith, Professor G. D. Chen, Dr. J. Z. Li, J. Li, Professor H. S. Kim, Dr. E. J. Shin, C. Ellis, Professor S. X. Jin, Dr. L. Dai, Dr. T. N. Oder, K. Kim, K. B. Nam, and J. Shakya. We would like to take this opportunity to express our sincere appreciation to Dr. John Zavada, Dr. Kepi Wu, Dr. Yoon Soo Park, Dr. Vern Hess, Dr. Jerry Smith, and Dr. Michael Gerhold for their insights and constant support.

References

1. H. Morkoç, S. Strite, G. B. Gao, M. E. Lin, B. Sverdlov, and M. Burns, *J. Appl. Phys.*, **76**, 1363 (1994).
2. S. Nakamura and G. Fasol, *The Blue Laser Diode*, Chapt. 11 (Springer, New York, 1997).
3. S. Nakamura, M. Senoh, N. Iwasa, and S. Nahahama, *Jpn. J. Appl. Phys. Part 2*, **34**, L797 (1995); S. Nakamura, M. Senoh, S. Nahahama, N. Iwasa, T. Yamada, T. Matsushita, Y. Sugimoto, and H. Kiyoku, *Appl. Phys. Lett.*, **69**, 1568 (1996); *Appl. Phys. Lett.*, **70**, 616 (1997).
4. S. Nakamura, M. Senoh, S. Nahahama, N. Iwasa, T. Yamada, T. Matsushita, H. Kiyoku, Y. Sugimoto, T. Kozaki, H. Umenoto, M. Sano, and K. Chocho, *Appl. Phys. Lett.*, **72**, 2014 (1998).
5. J. I. Pankove, *Optical Processes in Semiconductors* (Dover, New York, 1971).
6. D. C. Reynolds and T. C. Collins, *Excitons—Their Properties and Uses* (Academic, New York, 1981).
7. http://www.phys.ksu.edu/area/GaNgroup/
8. Landolt-Börnstien, *Numerical Data and Functional Relationships in Science and Technology*, edited by P. Eckerlin and H. Kandler (Springer-Verlag, Berlin, 1971) Vol. III.
9. W. Shan, T. J. Schmidt, X. H. Yang, S. J. Hwang, J. J. Song, and B. Goldenberg, *Appl. Phys. Lett.*, **66**, 985 (1995).
10. M. Smith, G. D. Chen, J. Z. Li, J. Y. Lin, H. X. Jiang, A. Salvador, W. K. Kim, O. Aktas, A. Botchkarev, and H. Morkoc, *Appl. Phys. Lett.*, **67**, 3387 (1995).
11. G. D. Chen, M. Smith, J. Y. Lin, H. X. Jiang, S. H. Wei, M. Asif Khan, and C. J. Sun, *Appl. Phys. Lett.*, **68**, 2784 (1996); M. Smith, G. D. Chen, J. Y. Lin, H. X. Jiang, M. Asif Khan, and C. J. Sun, *J. Appl. Phys.*, **79**, 7001 (1996).
12. G. D. Chen, M. Smith, J. Y. Lin, H. X. Jiang, A. Salvador, B. N. Sverdlov, A. Botchkarev, and H. Morkoc, *J. Appl. Phys.*, **79**, 2675 (1996).
13. D. C. Reynolds, D. C. Look, W. Kim, Ö. Aktas, A. Botchkarev, A. Salvador, H. Morkoc, and D. N. Talwar, *J. Appl. Phys.*, **80**, 594 (1996).
14. C. I. Harris, B. Monemar, H. Amano, and I. Akasaki, *Appl. Phys. Lett.*, **67**, 840 (1995).
15. E. I. Rashba and G. E. Gurgenishvili, *Sov. Phys. Solid State*, **4**, 759 (1962).
16. S. Permogorov, A. N. Reznitskii, S. Yu. Verbin, and V. A. Bonch-Bruevich, *Zh. Eksper. Teor. Fiz., Pisma*, **38**, 22 (1983).
17. D. G. Thomas and J. J. Hopfield, *Phys. Rev.*, **116**, 573 (1959).
18. D. G. Thomas and J. J. Hopfield, *Phys. Rev.*, **128**, 2135 (1962).
19. J. J. Hopfield and D. G. Thomas, *Phys. Rev.*, **122**, 35 (1961).
20. M. Ilegems, R. Dingle, and R. A. Logan, *J. Appl. Phys.*, **43**, 3797 (1972).
21. O. Lagerstedt and B. Monemar, *J. Appl. Phys.*, **45**, 2266 (1973).
22. K. Naniwae, S. Itoh, H. Amano, K. Itoh, K. Hiramatsu, and I. Akasaki, *J. Cryst. Growth*, **99**, 381 (1990).
23. G. D. Chen, M. Smith, J. Y. Lin, H. X. Jiang, M. Asif Khan, and C. J. Sun, *Appl. Phys. Lett.*, **67**, 1653 (1995).
24. M. Ilegems and R. Dingle, *J. Appl. Phys.*, **44**, 4234 (1973).
25. B. Santic, C. Merz, U. Kaufmann, R. Niebuhr, H. Obloh, and K. Bachem, *Appl. Phys. Lett.*, **71**, 1837 (1997).
26. M. Smith, G. D. Chen, J. Y. Lin, H. X. Jiang, M. Asif Khan, and C. J. Sun, *Appl. Phys. Lett.*, **67**, 3295 (1995).
27. W. Shan, X. C. Xie, J. J. Song, and B. Goldenberg, *Appl. Phys. Lett.*, **67**, 2512 (1995).
28. D. L. Dexter, *Solid State Physics*, edited by F. Seitz and D. Turnbull (Academic, New York, 1958) Vol. 6.
29. G. W. 't Hooft, W. A. J. A. van der Poel, and L. W. Molenkamp, *Phys. Rev.*, **B35**, 8281 (1987).
30. G. Pozina, J. P. Bergman, T. Paskova, and B. Monemar, *Appl. Phys. Lett.*, **75**, 4124 (1999).
31. G. E. Bunea, W. D. Herzog, M. S. Ünlü, B. B. Goldberg, and R. J. Molnar, *Appl. Phys. Lett.*, **75**, 838 (1999).
32. B. Monemar, J. P. Bergman, I. G. Ivanov, J. M. Baranowski, K. Pakula, I. Grzegory, and S. Porowski, *Solid State Commun.*, **104**, 205 (1997).

33. L. Eckey, J.-Ch. Holst, P. Maxim, R. Heitz, A. Hoffmann, I. Broser, B. K. Meyer, C. Wetzel, E. N. Mokhov, and P. G. Baranov, *Appl. Phys. Lett.*, **68**, 415 (1996).
34. R. Heitz, Ch. Fricke, A. Hoffmann, and I. Broser, *Mater. Sci. Forum*, **83–87**, 1241 (1992).
35. G. H. Kudlek, U. W. Pohl, Ch. Fricke, R. Heitz, A. Hoffmann, J. Gutowski, and I. Broser, *Physica B*, **185**, 325 (1993).
36. D. C. Reynolds, D. C. Look, B. Jogai, V. M. Phanse, and R. P. Vaudo, *Solid State Commun.*, **103**, 533 (1997).
37. R. Mair, J. Li, S. K. Duan, J. Y. Lin, and H. X. Jiang, *Appl. Phys. Lett.*, **74**, 513 (1999).
38. J. J. Hopfield, *Proceedings of the 7th International Conference on Physics of Semiconductors*, Paris, 1964 (Dunod, Paris, 1964) p. 725.
39. D. Volm, K. Oettinger, T. Streibl, D. Kovalev, M. Ben-Chorin, J. Diener, B. K. Meyer, J. Majewski, L. Eckey, A. Hoffmann, H. Amano, I. Akasaki, K. Hiramatsu, and T. Detchprohm, *Phys. Rev. B*, **53**, 16,543 (1996).
40. L. Eckey, J.-Ch. Holst, P. Maxim, R. Heitz, A. Hoffmann, I. Broser, B. K. Meyer, C. Wetzel, E. N. Mokhov, and P. G. Baranov, *Appl. Phys. Lett.*, **68**, 415 (1996).
41. J. Jayapalan, B. J. Skromme, R. P. Vaudo, and V. M. Phanse, *Appl. Phys. Lett.*, **73**, 1188 (1998).
42. A. Kasi Viswanath, J. I. Lee, S. Yu, D. Kim, Y. Choi, and C. H. Hong, *J. Appl. Phys.*, **84**, 3848 (1998).
43. M. Smith, G. D. Chen, J. Y. Lin, H. X. Jiang, A. Salvador, B. N. Sverdlov, A. Botchkarev, H. Morkoc, and B. Goldenberg, *Appl. Phys. Lett.*, **68**, 1883 (1996).
44. S. Nakamura, N. Iwasa, M. Senoh, and T. Mukai, *Jpn. J. Appl. Phys.*, **31**, 1258 (1992).
45. S. Nakamura, N. Iwasa, M. Senoh, and T. Mukai, *Jpn. J. Appl. Phys.*, **32**, L8 (1993).
46. B. Goldenberg, J. D. Zook, and R. Ulmer, *Appl. Phys. Lett.*, **62**, 381 (1993).
47. I. Akasaki, H. Amano, M. Kito, and K. Hiramatsu, *J. Lumin.*, **48/49**, 666 (1991).
48. M. Smith, G. D. Chen, J. Y. Lin, H. X. Jiang, A. Salvador, B. N. Sverdlov, A. Botchkarev, and H. Morkoc, *Appl. Phys. Lett.*, **66**, 3477 (1995).
49. R. Dingle and M. Ilegems, *Solid State Commun.*, **9**, 175 (1971).
50. D. G. Thomas, J. J. Hopfield, and W. M. Augustyniak, *Phys. Rev.*, **140**, 202 (1965).
51. P. Avouris and T. N. Morgan, *J. Chem. Phys.*, **74**, 4347 (1981).
52. K. C. Zeng, J. Y. Lin, H. X. Jiang, and W. Yang, *Appl. Phys. Lett.*, **74**, 3821 (1999).
53. W. Rieger, T. Metzger, H. Angerer, R. Dimitrov, O. Ambacher, and M. Stutzmann, *Appl. Phys. Lett.*, **68**, 970 (1996).
54. N. V. Edwards, S. D. Yoo, M. D. Bremser, T. W. Weeks Jr., O. H. Nam, R. F. Davis, H. Liu, R. A. Stall, M. N. Horton, N. R. Perkins, T. F. Kuech, and D. E. Aspnes, *Appl. Phys. Lett.*, **70**, 2001 (1997).
55. R. A. Abram, G. J. Rees, and B. L. H. Wilson, *Adv. Phys.*, **27**, 799 (1978).
56. M. Smith, J. Y. Lin, H. X. Jiang, and M. A. Khan, *Appl. Phys. Lett.*, **71**, 635 (1997).
57. D. J. Wolford, G. D. Gilliland, T. F. Kuech, L. M. Smith, J. Martinsen, J. A. Bradley, C. F. Tsang, R. Venkatasubramanian, S. K. Ghandi, H. P. Hjalmarson, *J. Vac. Sci. Technol., B*, **9**, 4 (Jul/Aug 1991).
58. H. X. Jiang, L. Q. Zu, and J. Y. Lin, *Phys. Rev. B*, **42**, 7284 (1990).
59. S. Permogorov, A. Reznitsky, and V. Lysenko, *Solid State Commun.*, **47**, 5 (1983).
60. S. Lai and M. V. Klein, *Phys. Rev. Lett.*, **44**, 1087 (1980).
61. E. Cohen and M. D. Sturge, *Phys. Rev. B*, **25**, 3828 (1982).
62. J. Y. Lin, A. Dissanayake, and H. X. Jiang, *Phys. Rev. B*, **46**, 3810 (1992).
63. J. D. Baranovskii and A. L. Efros, *Sov. Phys. Semicon.*, **12**, 1328 (1978).
64. S. Nakamura, M. Senoh, S. Nahahama, N. Iwasa, T. Yamada, T. Matsushita, H. Kiyoku, and Y. Sugimoto, *Jpn. J. Appl. Phys.*, **35**, L217 (1996).
65. Y. Narukawa, Y. Kawakami, M. Funato, Sz. Fujita, Sg. Fujita, and S. Nakamura, *Appl. Phys. Lett.*, **70**, 981 (1997).
66. G. Steude, B. K. Meyer, A. Göldner, A. Hoffmann, F. Bertram, J. Christen, H. Amano, and I. Akasaki, *Appl. Phys. Lett.*, **74**, 2456 (1999).
67. M. Suzuki and T. Uenoyama, *Appl. Phys. Lett.*, **69**, 3378 (1996); *Jpn. J. Appl. Phys.*, **35**, 1420 (1996).
68. G. Mohs, T. Aoki, M. Nagai, R. Shimano, M. Kuwata-Gonokami, and S. Nakamura, *Solid State Commun.*, **104**, 643 (1997).

69. A. Satake, Y. Masumoto, T. Miyajima, T. Asatsuma, F. Nakamura, and M. Ikeda, *Phys. Rev. B*, **57**, R2041 (1998).
70. F. A. Majumder, S. Shevel, V. G. Lyssenko, H. E. Swoboda, and C. Klingshirn, *Z. Phys. B: Condens. Matter*, **66**, 409 (1987).
71. J. Ding, H. Jeon, T. Ishihara, M. Hagerott, A. V. Nurmikko, H. Luo, N. Samarth, and J. Furdyna, *Phys. Rev. Lett.*, **69**, 1707 (1992).
72. Y. Yamada, Y. Masumoto, J. T. Mullins, and T. Taguchi, *Appl. Phys. Lett.*, **61**, 2190 (1992).
73. M. Smith, G. D. Chen, J. Y. Lin, H. X. Jiang, M. Asif Khan, and Q. Chen, *Appl. Phys. Lett.*, **69**, 2837 (1996).
74. H. S. Kim, R. A. Mair, J. Li, J. Y. Lin, and H. X. Jiang, *Appl. Phys. Lett.*, **76**, 1252 (2000).
75. H. S. Kim, R. A. Mair, J. Li, J. Y. Lin, and H. X. Jiang, *Proc. Society of Photo-Optical Instrumentation Engineers (SPIE)*, **3940**, 139 (2000).
76. M. Oueslati, C. Benoit a'la Gillaume, and M. Zouaghi, *Phys. Rev. B*, **37**, 3037 (1998).
77. C. Gourdon and P. Lavallard, *Phys. Status Solidi B*, **153**, 641 (1989).
78. C. K. Sun, S. Keller, G. Wang, M. S. Minsky, J. E. Bowers, and S. P. DenBaars, *Appl. Phys. Lett.*, **69**, 1936 (1996).
79. E. S. Jeon, V. Kozlov, Y. K. Song, A. Vertikov, M. Kuball, A. V. Nurmikko, H. Liu, C. Chen, R. S. Kern, C. P. Kuo, and M. G. Craford, *Appl. Phys. Lett.*, **69**, 4194 (1996).
80. C. K. Sun, S. Keller, T. L. Chiu, G. Wang, M. S. Minsky, J. E. Bowers, S. P. DenBaars, *J. Quantum Electronics*, **3**, 731 (1997).
81. J. S. Im, A. Moritz, F. Steuber, V. Harle, F. Scholz, and A. Hangleiter, *Appl. Phys. Lett.*, **70**, 631 (1997).
82. J. S. Im, V. Harle, F. Scholz, and A. Hangleiter, *MRS Internet J. Nitride Semicond. Res.*, **1**, art. 37, (1997).
83. J. Allegre, P. Lefebvre, S. Juillabuet, W. Knap, J. Camassel, Q. Chen, and M. A. Khan, *MRS Internet J. Nitride Semicond. Res.*, **2**, art. 34, (1997).
84. Y. Narukawa, Y. Kawakami, S. Fujita, S. Fujita, and S. Nakamura, *Phys. Rev. B*, **55**, R1938 (1997).
85. M. S. Minsky, S. B. Fleischer, A. C. Abare, J. E. Bowere, E. L. Hu, S. Keller, and S. P. DenBaars, *Appl. Phys. Lett.*, **72**, 1066 (1998).
86. K. C. Zeng, M. Smith, J. Y. Lin, and H. X. Jiang, *Appl. Phys. Lett.*, **73**, 1724 (1998).
87. Y. H. Cho, J. J. Song, S. Keller, M. S. Minsky, E. Hu, U. K. Mishra, and S. P. DenBaars, *Appl. Phys. Lett.*, **73**, 1128 (1998).
88. X. Zhang, D. H. Rich, J. T. Kobayashi, N. P. Kobayashi, and P. D. Dapkus, *Appl. Phys. Lett.*, **73**, 1430 (1998).
89. Y. H. Cho, G. H. Gainer, A. J. Fischer, J. J. Song, S. Keller, U. K. Mishra, and S. P. DenBaars, *Appl. Phys. Lett.*, **73**, 1370 (1998).
90. M. Pophristic, F. H. Long, C. Tran, R. F. Karlicek, Jr., Z. C. Feng, and I. T. Ferguson, *Appl. Phys. Lett.*, **73**, 815 (1998).
91. M. Pophristic, F. H. Long, C. Tran, I. T. Ferguson, and R. F. Karlicek, Jr., *Appl. Phys. Lett.*, **73**, 3550 (1998).
92. S. F. Chichibu, H. Marchand, M. S. Minsky, S. Keller, P. T. Fini, J. P. Ibbetson, S. B. Fleischer, J. S. Speck, J. E. Bowers, E. Hu, U. K. Mishra, S. P. DenBaars, T. Deguichi, T. Sota, and S. Nakamura, *Appl. Phys. Lett.*, **74**, 1460 (1999).
93. F. D. Sala, A. D. Carlo, P. Lugli, F. Bernardini, V. Fiorentini, R. Scholz, and J. M. Jancu, *Appl. Phys. Lett.*, **74**, 2002 (1999).
94. Y. Narukawa, Y. Kawakami, S. Fujita, and S. Nakamura, *Phys. Rev. B*, **59**, 10,283 (1999).
95. M. Pophristic, F. H. Long, C. Tran, I. T. Ferguson, and R. F. Karlicek, Jr., *J. Appl. Phys.*, **86**, 1114 (1999).
96. J. Dalfors, J. P. Bergman, P. O. Holtz, B. E. Sernelius, B. Monemar, H. Amano, and I. Akasaki, *Appl. Phys. Lett.*, **74**, 3299 (1999).
97. A. Vertikov, I. Ozden, and A. V. Nurmikko, *J. Appl. Phys.*, **86**, 4697 (1999).
98. Y. H. Cho, T. J. Schmidt, S. Bidnyk, G. H. Gainer, J. J. Song, S. Keller, U. K. Mishra, and S. P. DenBaars, *Phys. Rev. B*, **61**, 7571 (2000).
99. A. Satake, Y. Masumoto, T. Miyajima, T. Asatsuma, and M. Ikeda, *Phys. Rev. B*, **60**, 16,660 (1999).
100. H. X. Jiang and J. Y. Lin, *Appl. Phys. Lett.*, **74**, 1066 (1999).

101. M. Smith, J. Y. Lin, H. X. Jiang, and A. Asif Khan, *Appl. Phys. Lett.*, **71**, 635 (1997).
102. K. Domen, A. Kuramata, and T. Tanahashi, *Appl. Phys. Lett.*, **72**, 1359 (1998).
103. W. W. Chow, A. F. Wright, and J. S. Nelson, *Appl. Phys. Lett.*, **68**, 296 (1996).
104. T. Mukai, D. Morita, and S. Nakamura, *J. Cryst. Growth*, **189/190**, 778 (1998).
105. Y. Narukawa, S. Saijou, Y. Kawakami, S. Fujita, T. Mukai, and S. Nakamura, *Appl. Phys. Lett.*, **74**, 558 (1999).
106. R. C. Miler, D. A. Kleinman, W. A. Nordland Jr., and A. C. Gossard, *Phys. Rev. B*, **22**, 863 (1980).
107. J. Feldman, G. Peter, E. O. Gobel, P. Dawson, K. Moore, C. Foxon, and R. J. Elliot, *Phys. Rev. Lett.*, **20**, 2337 (1987); G. W. 't Hooft, W. A. J. A. van der Poel, L. W. Molenkamp, and C. T. Foxon, *Phys. Rev. B*, **35**, 8281 (1987).
108. M. Sugawara, *Phys. Rev. B*, **51**, 10743 (1995).
109. H. Kudo, H. Ishibashi, R. S. Zheng, Y. Yamada, and T. Taguchi, *Appl. Phys. Lett.*, **76**, 1546 (2000).
110. H. Kudo, H. Ishibashi, R. S. Zheng, Y. Yamada, and T. Taguchi, *Phys. Status Solidi B*, **216**, 163 (1999).
111. B. Monemar, J. P. Bergman, J. Dalfors, G. Pozina, P. O. Holtz, A. Amano, and I. Akasaki, *MRS Internet J. Nitride Semicond. Res.*, **4**, 16 (1999).
112. Y. H. Cho, J. J. Song, S. Keller, M. S. Minsky, E. Hu, U. K. Mishra, and S. P. DenBaars, *Appl. Phys. Lett.*, **73**, 1128 (1998).
113. E. Berkowicz, D. Gershoni, G. Bahir, E. Lakin, D. Shilo, E. Zolotoyabko, A. C. Abare, S. P. Denbaars, and L. A. Goldren, *Phys. Rev. B*, **61**, 10,994 (2000).
114. J. S. Im, H. Kollmer, J. Off, A. Sohmer, F. Scholz, and A. Hangleiter, *Phys. Rev. B*, **57**, R9436 (1998).
115. A. Hangleiter, J. S. Im, H. Kollmer, S. Heppel, J. Off, and F. Scholz, *MRS Internet J. Nitride Semicond. Res.*, **3**, 15 (1990).
116. E. Berkowicz, D. Gershoni, G. Bahir, A. C. Abare, S. P. Denbaars, and L. C. Goldern, *Physics of Semiconductors*, edited by D. Gershoni (World Scientific, Singapore, 1999) p. 251.
117. M. Colocci, M. Gurioli, and J. Martinez-Pastor, *J. Phys. IV*, **3**, 3 (1993).
118. C. Cooper, D. I. Westwood, and P. Blood, *Appl. Phys. Lett.*, **69**, 2415 (1996).
119. X. Zhang, S. J. Chua, S. Xu, K. B. Chong, and K. Onabe, *Appl. Phys. Lett.*, **71**, 1840 (1997).
120. V. Ortiz, N. T. Pelekanos, and G. Mula, *Appl. Phys. Lett.*, **72**, 963 (1998).
121. M. Shur, *Compound Semicond.*, **Spring I**, 12 (1998).
122. T. Takeuchi, S. Sota, M. Katsuragawa, M. Komori, H. Takeuchi, H. Amano, and I. Akasaki, *Jpn. J. Appl. Phys. Part 2*, **36**, L382 (1997).
123. A. Hangleiter, J. S. Im, H. Kollmer, S. Heppel, J. Off, and F. Scholz, *MRS Internet J. Nitride Semicond. Res.*, **3**, 15 (1998).
124. Jin Seo Im, H. Kollmer, J. Off, A. Sohmer, F. Scholtz, and A. Hangleiter, *Phys. Rev. B*, **57**, R9435 (1998).
125. S. H. Park and S. L. Chuang, *Appl. Phys. Lett.*, **72**, 3103 (1998).
126. H. S. Kim, J. Y. Lin, H. X. Jiang, W. W. Chow, A. Botchkarev, and H. Morkoç, *Appl. Phys. Lett.*, **73**, 3426 (1998).
127. M. Smith, J. Y. Lin, H. X. Jiang, A. Salvador, A. Botchkarev, H. Kim, and H. Morkoc, *Appl. Phys. Lett.*, **69**, 2453 (1996); K. C. Zeng, J. Y. Lin, H. X. Jiang, A. Salvador, G. Popovici, H. Tang, W. Kim, and H. Morkoc, *Appl. Phys. Lett.*, **71**, 1368 (1997).
128. P. Lefebve, J. Allègre, B. Gil, H. Mathieu, N. Grandjean, M. Leroux, and J. Massies, *Phys. Rev. B*, **59**, 15363 (1999).
129. K. C. Zeng, J. Li, J. Y. Lin, and H. X. Jiang, *Appl. Phys. Lett.*, **76**, 864 (2000).
130. K. C. Zeng, R. Mair, J. Y. Lin, H. X. Jiang, W. W. Chow, A. Botchkarev, and H. Morkoc, *Appl. Phys. Lett.*, **73**, 2476 (1998).
131. K. C. Zeng, J. Li, J. Y. Lin, and H. X. Jiang, *Appl. Phys. Lett.*, **76**, 3040 (2000).
132. Eun-joo Shin, J. Li, J. Y. Lin, and H. X. Jiang, *Appl. Phys. Lett.*, in press.
133. M. Lindberg and S. W. Koch, *Phys. Rev. B*, **38**, 3342 (1988).
134. W. W. Chow and S. W. Koch, *Semiconductor–Laser Fundamentals: Physics of the Gain Materials* (Springer, Berlin, 1999).

135. R. Binder and S. W. Koch, *Prog. Quantum Electronics*, **19**, 307 (1995).
136. M. Sargent III, M. O. Scully, and W. E. Lamb Jr., *Laser Physics* (Addison-Wesley, Reading, MA, 1974).
137. C. H. Henry, R. A. Logan, and F. R. Merritt, *J. Appl. Phys.*, **51**, 3042 (1980).
138. W. Chow, M. Kira, and S. W. Koch, *Phys. Rev. B*, **60**, 1947 (1999).
139. A. Bykhovski, B. Gelmont, and M. Shur, *J. Appl. Phys.*, **74**, 6734 (1993).
140. O. Ambacher, *J. Phys. D: Appl. Phys.*, **31**, 2653 (1998).
141. W. W. Chow, A. F. Wright, A. Girndt, F. Jahnke, and S. W. Koch, *Appl. Phys. Lett.*, **71**, 2608 (1997).
142. T. Kuroda, A. Tackeuchia, and T. Sota, *Appl. Phys. Lett.*, **76**, 3753 (2000).
143. F. G. McIntosh, J. C. Roberts, M. E. Aumer, V. A. Joshkin, S. M. Bedair, and N. A. El-Masry, *MRS Internet J. Nitride Semicond. Res.*, **1**, 43 (1996).
144. M. Ashif Khan, J. W. Yang, G. Simin, R. Gaska, M. S. Shur, Hans-Conrad zur Loye, G. Tamulaitis, A. Zukauskas, D. J. Smith, D. Chandrasekhar, and R. Bicknell-Tassius, *Appl. Phys. Lett.*, **76**, 1161 (2000).
145. T. N. Oder, J. Li, J. Y. Lin, and H. X. Jiang, *Appl. Phys. Lett.*, in press.
146. R. K. Chang and A. J. Campillo, *Optical Processes in Microcavities*, (World Scientific, Singapore, 1996).
147. R. A. Mair, K. C. Zeng, J. Y. Lin, H. X. Jiang, B. Zhang, L. Dai, H. Tang, A. Botchkarev, W. Kim, and H. Morkoc, *Appl. Phys. Lett.*, **71**, 2898 (1997).
148. R. A. Mair, K. C. Zeng, J. Y. Lin, H. X. Jiang, B. Zhang, L. Dai, A. Botchkarev, W. Kim, H. Morkoc, and M. A. Khan, *Appl. Phys. Lett.*, **72**, 1530 (1998).
149. K. C. Zeng, L. Dai, J. Y. Lin, and H. X. Jiang, *Appl. Phys. Lett.*, **75**, 2503 (1999).
150. H. X. Jiang, J. Y. Lin, K. C. Zeng, and W. Yang, *Appl. Phys. Lett.*, **75**, (1999).
151. S. Bidnyk, B. D. Little, Y. H. Cho, J. Karasinski, J. J. Song, W. Yang, and S. A. McPherson, *Appl. Phys. Lett.*, **73**, 2242 (1998).
152. S. Chang, N. B. Rex, R. K. Chang, G. Chong, and L. J. Guido, *Appl. Phys. Lett.*, **75**, 166 (1999).
153. T. Someya, R. Werner, A. Forchel, M. Catalano, R. Cingolani, and Y. Arakawa, *Science*, **285**, 1905 (1999).
154. S. X. Jin, J. Li, J. Z. Li, J. Y. Lin, and H. X. Jiang, *Appl. Phys. Lett.*, **76**, 631 (2000).

CHAPTER 3

Time-Resolved Raman Studies of Wide Bandgap Wurtzite GaN

K.T. TSEN

Department of Physics and Astronomy, Arizona State University, Tempe, AZ 85287, USA
Tel: (480)965-5206; Fax: (480)965-7954; tsen@asu.edu

3.1 INTRODUCTION

The recent surge of activity in wide-bandgap semiconductors [1] has arisen from the need for electronic devices capable of operation at high power levels, at high temperatures and in caustic environments, and separately, from a need for optical materials, especially emitters, that are active in the blue and UV wavelengths. Electronics based on existing Si and GaAs semiconductor device technologies cannot tolerate greatly elevated temperatures or chemically hostile environments. The wide-bandgap semiconductors with their excellent thermal conductivities, large breakdown fields, and resistance to chemical corrosion will be the materials of choice for these applications. Among the wide-bandgap semiconductors, the III–V nitrides have long been viewed as a very promising semiconductor system for device applications in the blue and UV wavelengths. The wurtzite polytypes of GaN, AlN, and InN form a continuous alloy system whose direct bandgaps range from 1.9 eV for InN, to 3.4 eV for GaN, to 6.2 eV for AlN. Thus, III–V nitrides could potentially be fabricated into optical devices that are active at wavelengths ranging from the red all the way into the UV. Because of lower ohmic contact resistances and larger predicted electron saturation velocity [2] research in GaN is of particular interest.

Although much progress has been made in device-oriented applications with wide-bandgap semiconductors, very little information concerning their dynamical properties has yet been obtained. Device engineers meet to understand both carrier and phonon dynamical properties to design better and faster devices. For example, carrier energy loss rate is primarily determined by electron–phonon scattering rates [3]; electron relaxation may be greatly influenced by hot phonon effects [4], which in turn are governed by the population relaxation time of optical phonons. This chapter describes the use of Raman spectroscopy to study various properties in wurtzite GaN, with special emphasis on the dynamical properties such as electron–phonon, phonon–phonon interactions, and nonequilibrium electron distributions.

3.2 RAMAN SPECTROSCOPY IN SEMICONDUCTORS

We first present a comprehensive theory of Raman scattering from carriers in semiconductors, which is particularly useful in situations where electron distributions are nonequilibrium. We then discuss the theory of Raman scattering by lattice vibrations in semiconductors.

3.2.1 Theory of Raman Scattering from Carriers in Semiconductors

3.2.1.1 A Simple Model

In order to understand how Raman spectroscopy can be used to probe electron distribution function in semiconductors, we start with the simplest physical concept—Compton scattering. As shown in Figure 3.1, let's consider an incident photon with wavevector $\vec{k}_i$ and angular frequency ω_i interacting with an electron of mass m_e^* traveling at a velocity $\vec{V}$. After the scattering event, the scattered photon is characterized with wavevector $\vec{k}_s$ and angular frequency ω_s. The scattered electron is then moving at a velocity $\vec{V}'$. From the conservation of energy and momentum, we can write the

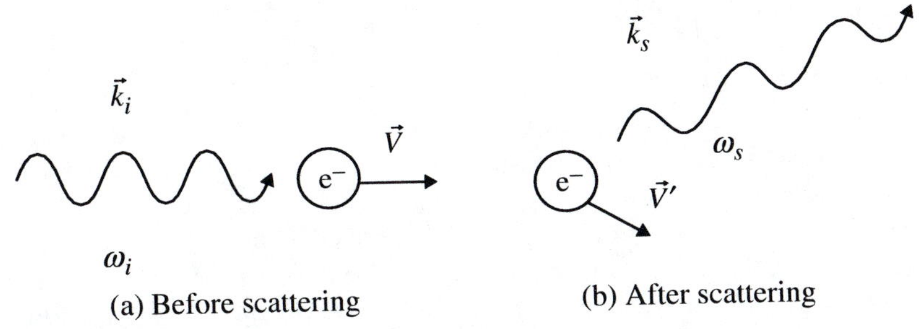

Figure 3.1 A simple model of Compton scattering, demonstrating how the electron distribution function in semiconductors can be probed by Raman spectroscopy.

following equations:

$$\hbar\omega_i + \frac{1}{2}m_e^*\vec{V}^{\prime 2} = \hbar\omega_s + \frac{1}{2}m_e^*\vec{V}^{\prime 2}, \tag{3.1}$$

$$\hbar\vec{k}_i + m_e^*\vec{V} = \hbar\vec{k}_s + m_e^*\vec{V}'. \tag{3.2}$$

If we define the energy transfer and the wavevector transfer of the photon to be $\omega \equiv \omega_i - \omega_s$, and $\vec{q} \equiv \vec{k}_i - \vec{k}_s$ respectively, then from Eqs. (3.1) and (3.2) we have

$$\omega = \vec{V} \cdot \vec{q} + \frac{\hbar\vec{q}^2}{2m_e^*}. \tag{3.3}$$

This important equation states that the energy transfer of an incident photon is (apart from a constant term, $\hbar\vec{q}^2/2m_e^*$) directly proportional to the electron velocity along the direction of wavevector transfer. In other words, it implies that Raman scattering intensity, measured at an angular frequency ω, is proportional to the number of electrons that have a velocity component along the direction of wavevector transfer given by Eq. (3.3), irrespective of their velocity components perpendicular to $\vec{q}$.

Therefore, if electron distribution function is Maxwell–Boltzmann–like, then the lineshape of the Raman scattering spectrum will be Gaussian-like centered around $\omega \cong 0$, whereas a drifted Maxwell–Boltzmann distribution with an electron drift velocity $\vec{V}_d$ will result in a Raman scattering spectrum that is a shifted Gaussian centered around $\omega \cong \vec{q} \cdot \vec{V}_d$.

We note, however, that, strictly speaking, this simple picture is correct only for a system of noninteracting electron gas in vacuum. For an electron gas in a semiconductor such as GaAs or GaN, many-body effects and the effects of band structure have to be considered. The former is usually taken into account by the random-phase approximation (RPA) [5] and the latter by sophisticated band structure calculations such as the $\vec{k} \cdot \vec{p}$ approximation [6–8].

3.2.1.2 A Full Quantum Mechanical Approach

We shall now use a quantum mechanical method to calculate a Raman scattering cross section or single-particle scattering (SPS) spectrum for a single-component plasma in a direct-bandgap semiconductor such as GaAs or GaN, probed by an ultrafast laser having pulse width t_p. For simplicity, we assume that the probe pulse is a square pulse from $-t_p/2$ to $+t_p/2$, and the electron elastic scattering is the dominant scattering process in the solid-state system. For the more general case in which inelastic scattering plays a role in the electron scattering processes, see Chia et al. [9].

We start with a typical electron–photon interaction Hamiltonian that has been shown in the equilibrium case to be [10]

$$\begin{aligned} H &= H_0 + \sum_i \left[\frac{-e}{2m_e^* c} [\vec{p}_i \cdot \vec{A}(\vec{r}_i) + \vec{A}(\vec{r}_i) \cdot \vec{p}_i] + \frac{e^2}{2m_e^* c} \vec{A}^2(\vec{r}_i) \right] \\ &\equiv H_0 + H_1 + H_2, \end{aligned} \tag{3.4}$$

where H_0 is the total Hamiltonian of the system in the absence of radiation field; e is the charge of an electron; c is the speed of light; $\vec{p}$ is the electron momentum; $\sum_i$ refers to summation over electrons; the second and third terms describe the interactions of electrons with the radiation field; and $\vec{A}$ is the vector potential of the radiation field:

$$\vec{A}(\vec{r}_i) \equiv \frac{1}{\sqrt{V}} \sum_j \left(\frac{2\pi \hbar c^2}{\omega_j} \right)^{1/2} \left(e^{i\vec{k}_j \cdot \vec{r}_i} b_{\vec{k}_j} + e^{-i\vec{k}_j \cdot \vec{r}_i} b^+_{\vec{k}_j} \right) \hat{e}_j, \tag{3.5}$$

where V is the volume of the semiconductor, and $b_{\vec{k}_j}, b^+_{\vec{k}_j}$ are photon annihilation and creation operators, respectively. Since the vector potential $\vec{A}$ is a linear combination of photon creation and annihilation operators, and the Raman scattering process involves the annihilation of an incident photon and the creation of a scattered photon, $\vec{p} \cdot \vec{A}$ and $\vec{A} \cdot \vec{p}$ terms in Eq. (3.4) will contribute to the scattering matrix element in the second order, and $\vec{A}^2$ terms will contribute in the first order in the perturbation-theory calculations of the Raman scattering cross section.

The transition amplitude from an electron–photon state $|I\rangle$ to a state $|F\rangle$ is given by

$$\begin{aligned} c_{IF} &= \frac{1}{i\hbar} \int_{-t_p/2}^{t_p/2} e^{i\omega_{FI} t'} dt' \langle F | \left[H_2 - \frac{i}{\hbar} \int_0^{t'+t_p/2} H_1 e^{-(i/\hbar) H_0 t''} \right. \\ &\quad \left. \times H_1 e^{(i/\hbar) H_0 t''} dt'' \right] |I\rangle, \end{aligned} \tag{3.6}$$

where $\omega_{FI} \equiv (\varepsilon_F - \varepsilon_I)/\hbar$. For convenience, we have chosen the zero of time so that the probe pulse runs from $-t_p/2$ to $+t_p/2$.

We let the initial state $|I\rangle$ be a product of an electron–phonon state $|i\rangle$ and the photon state $|n_i, 0_s\rangle$ consisting of n_i incident photons and zero scattered photons. The final state $|F\rangle$ is the product of an electron–phonon state $|f\rangle$

and the photon state $|n_i - 1, 1_s\rangle$. Thus, we can write the transition amplitude in a more convenient form:

$$c_{IF} = \frac{1}{i\hbar} \int_{-t_p/2}^{t_p/2} e^{i\omega_{FI} t'} C(\omega_i, \omega_s) M_{fi}(t')\, dt', \tag{3.7}$$

with

$$C(\omega_i, \omega_s) \equiv \frac{-e^2}{m_e^* c^2} \sqrt{n_i} \left[\frac{2\pi \hbar c^2}{(\omega_i \omega_s)^{1/2}} \right], \tag{3.8}$$

and

$$M_{fi}(t') \equiv (\hat{e}_i \cdot \hat{e}_s) \langle f | n_{\vec{q}}^c | i \rangle + \frac{1}{i\hbar m_e^*}$$
$$\times \left[\int_{-t'-t_p/2}^{0} dt'' e^{i\omega_s t''} \langle f | \hat{e}_i \cdot \vec{j}(\vec{k}_i) \vec{j}(-\vec{k}_s, t'') \cdot \hat{e}_s | i \rangle \right.$$
$$\left. + \int_0^{t'+t_p/2} dt'' e^{i\omega_s t''} \langle f | \hat{e}_s \cdot \vec{j}(-\vec{k}_s, t'') \vec{j}(\vec{k}_i) \cdot \hat{e}_i | i \rangle \right]. \tag{3.9}$$

The matrix element $M_{fi}(t')$ contains, in order, three terms represented by three Feynman diagrams; $\vec{k}_i, \vec{k}_s$ are the wavevectors of incident and scattered photons; $\hat{e}_i, \hat{e}_s$ are the polarization vectors of incident and scattered photons; and ω_i, ω_s are the angular frequencies of incident and scattered photons, respectively. $\vec{q} \equiv \vec{k}_i - \vec{k}_s$ is the momentum transfer of the photon to the electronic system. The operator $n_{\vec{q}}^c \equiv \sum_{j=1}^{N} e^{i\vec{q}\cdot\vec{r}_j}$ is the Fourier transform of the density operator of an electron in the conduction band. In the second quantized notation, the current operator $\vec{j}(\vec{k})$ [11,12] is

$$\vec{j}(\vec{k}) = \sum_{n,n',\vec{k}',\alpha,\alpha'} \langle \vec{k} + \vec{k}', n, \alpha | e^{i\vec{k}\cdot\vec{r}} \vec{p} | \vec{k}', n', \alpha' \rangle a^+_{\vec{k}+\vec{k}',n,\alpha} a_{\vec{k}',n',\alpha'}, \tag{3.10}$$

where $a_{\vec{p},n,\sigma}$, $a^+_{\vec{p},n,\sigma}$ are the annihilation and creation operators for a Bloch state $|\vec{p}, n, \sigma\rangle$, respectively.

We assume that energy is nearly conserved, so that $\omega_i \cong \omega_s$. We define

$$\vec{j}(\vec{k}, t) \equiv e^{(i/\hbar) H_e^0 t} \vec{j}(\vec{k}) e^{-(i/\hbar) H_e^0 t}, \tag{3.11}$$

where H_e^0 is the Hamiltonian for the unperturbed electronic system.

We are specifically interested in the case where scattering is off the electrons in the conduction band, and hole contributions are neglected.

The neglect of the hole contributions is a good approximation, because the coefficient $C(\omega_i, \omega_s)$ has the factor $e^2/m_e^* c^2$, which means that the larger the effective mass, the smaller the Raman scattering cross section. Since the effective mass of electrons in the conduction band of III–V semiconductors is typically much smaller than that of holes in the valence band, neglect of the hole contributions is justified.

We use Kane's two-band model [8] with a full set of valence bands and also use the free-electron approximation between different bands, i.e., for $n \neq n'$

$$a^+_{\vec{p},n,\sigma}(t) a_{\vec{p}',n',\sigma'}(t) = e^{(i/\hbar)(\varepsilon_{p,n,\sigma} - \varepsilon_{p',n',\sigma'})t} a^+_{\vec{p},n,\sigma} a_{\vec{p}',n',\sigma'}, \tag{3.12}$$

where $\varepsilon_{p,n,\omega}$ is the energy of the Bloch state $|\vec{p}, n, \sigma\rangle$. With this approximation and letting $\vec{k} \to 0$, and $\omega_i \cong \omega_s$, we have

$$M_{fi}(t) = \left\langle f \left| \sum_{\vec{p},\alpha,\beta} \gamma_{\alpha\beta}(\vec{p}, t, t_p, \omega_i) a^+_{\vec{p}+\vec{q},\alpha} a_{\vec{p},\beta} \right| i \right\rangle, \tag{3.13}$$

where

$$\begin{aligned} \gamma_{\alpha\beta}(\vec{p}, t, t_p, \omega_i) \equiv\ & (\hat{e}_i \cdot \hat{e}_s)\delta_{\alpha\beta} + \frac{1}{\hbar m_e^*} \sum_{n,\sigma} [\langle c, \alpha | \vec{p} \cdot \hat{e}_i | n, \sigma \rangle \\ & \times \langle n, \sigma | \vec{p} \cdot \hat{e}_s | c, \beta \rangle \Theta(t) + \langle c, \alpha | \vec{p} \cdot \hat{e}_s | n, \sigma \rangle \\ & \times \langle n, \sigma | \vec{p} \cdot \hat{e}_i | c, \beta \rangle \phi(t)]. \end{aligned} \tag{3.14}$$

Equation (3.14) is the generalization of the time-independent $\gamma_{\alpha\beta}$ derived by Hamilton and McWhorter for the equilibrium case [12]. The summation is taken over the three valence bands (heavy-hole, light-hole, and split-off-hole bands), and c represents the conduction band. The spin indices are α, β, and σ. The steplike functions $\Theta(t)$ and $\phi(t)$ appear in the expression because of the finite pulse width of the probe pulse and are given by

$$\begin{aligned} \Theta(t) &= \frac{1}{i} \int_{-t-t_p/2}^{0} dt' e^{i\omega_i t'} e^{(i/\hbar)(\varepsilon_{p,n} - \varepsilon_{p,c})t'}; \\ \phi(t) &= \frac{1}{i} \int_{0}^{t+t_p/2} dt' e^{i\omega_i t'} e^{(1/\hbar)(\varepsilon_{p,c} - \varepsilon_{p,n})t'}. \end{aligned} \tag{3.15}$$

For simplicity, we evaluate $\gamma_{\alpha\beta}(\vec{p}, t, t_p, \omega_i)$ near the center of the Brillouin zone with $\vec{k} \cdot \vec{p}$ wave functions provided by Kane and obtain

$$\gamma_{\alpha\beta}(\vec{p}, t, t_p \omega_i) = \hat{e}_i \cdot \overleftrightarrow{C}_p(t, \omega_i) \cdot \hat{e}_s \delta_{\alpha\beta} + i(\hat{e}_i \times \hat{e}_s) \cdot \overleftrightarrow{S}_p(t, \omega_i) \cdot \vec{\sigma}_{\alpha\beta}, \tag{3.16}$$

where $\overleftrightarrow{\sigma}_{\alpha\beta}$ are Pauli matrices; and $\overleftrightarrow{C}_p(t,\omega_i)$ and $\overleftrightarrow{S}_p(t,\omega_i)$ are time-dependent dyadic tensors, involving charge-density, energy-density, and spin-density fluctuations, respectively. They can be written as

$$\overleftrightarrow{C}_p(t,\omega_i) \equiv \overleftrightarrow{I}\left\{1 + \frac{2P^2}{3m_e^*}\cdot\sum_{n=1}^{3} A_c(n)\cdot\left[E_{g_n} - e^{(i/\hbar)E_{gn}(t+t_p/2)}\right.\right.$$
$$\times\left\{E_{g_n}\cos[\omega_i(t+t_p/2)] - i\hbar\omega_i\sin[\omega_i(t+t_p/2)]\right\}\big]/$$
$$\left.\left((E_{g_n} - i\Gamma_n)^2 - (\hbar\omega_i)^2\right)\right\} - \left(\hat{p}\hat{p} - \frac{\overleftrightarrow{I}}{3}\right)\frac{P^2}{m_e^*}$$
$$\times\sum_{n=1}^{2} B_c(n)\cdot\left[E_{g_n} - e^{(i/\hbar)E_{gn}(t+t_p/2)}\{E_{g_n}\cos[\omega_i(t+t_p/2)]\right.$$
$$\left. - i\hbar\omega_i\sin[\omega_i(t+t_p/2)]\}\right]/\left((E_{g_n} - i\Gamma_n)^2 - (\hbar\omega_i)^2\right); \tag{3.17}$$

$$\overleftrightarrow{S}_p(t,\omega_i) \equiv -\overleftrightarrow{I}\left\{\frac{P^2}{3m_e^*}\cdot\sum_{n=1}^{3} A_s(n)\cdot\left[\hbar\omega_i - e^{(i/\hbar)E_{gn}(t+t_p/2)}\right.\right.$$
$$\{\hbar\omega_i\cos[\omega_i(t+t_p/2)] - iE_{g_n}\sin[\omega_i(t+t_p/2)]\}\big]/$$
$$\left.\left((E_{g_n} - i\Gamma_n)^2 - (\hbar\omega_i)^2\right)\right\} - \left(\hat{p}\hat{p} - \frac{\overleftrightarrow{I}}{3}\right)\frac{P^2}{m_e^*}$$
$$\times\sum_{n=1}^{2} B_s(n)\cdot\left[\hbar\omega_i - e^{(i/\hbar)E_{gn}(t+t_p/2)}\{\hbar\omega_i\cos[\omega_i(t+t_p/2)]\right.$$
$$\left. - iE_{g_n}\sin[\omega_i(t+t_p/2)]\}\right]/[(E_{g_n} - i\Gamma_n)^2 - (\hbar\omega_i)^2], \tag{3.18}$$

where $A_c(1) = A_c(2) = A_c(3) = B_c(1) = -B_c(2) = 1$; $A_s(1) = A_s(2) = B_s(1) = -B_s(2) = 1$, and $A_s(3) = -2$; Γ_1, Γ_2, and Γ_3 are the damping constants involved in the Raman scattering processes; E_{g_1}, E_{g_2}, and E_{g_3} are the energy difference between the conduction band and the heavy-hole, light-hole and split-off-hole bands evaluated at wavevector $\vec{k}$, respectively; and $P \equiv -i\langle S|p_z|Z\rangle$ is the momentum matrix element between the conduction and valence bands at the Γ-point in Kane's notations.

The Raman scattering cross section is proportional to the transition probability $|c_{IF}|^2$ averaged over all the possible initial states (which may be far from equilibrium). Since the coefficient $C(\omega_i, \omega_s)$ is a constant if $\omega_i \cong \omega_s$, we finally obtain

$$\frac{d^2\sigma}{d\omega\, d\Omega} \propto \left\langle \sum_f \left| \int_{-t_p/2}^{t_p/2} dt\, e^{i\omega_{FI} t} \right.\right.$$

$$\left.\left. \times \left\langle f \left| \sum_{\vec{p},\alpha,\beta} \gamma_{\alpha\beta}(\vec{p}, t, t_p, \omega_i) a^{+}_{\vec{p}+\vec{q},\alpha} a_{\vec{p},\beta} \right| i \right\rangle \right|^2 \right\rangle$$

$$= \int_{-t_p/2}^{t_p/2} dt \int_{-t_p/2-t}^{t_p/2-t} dt' e^{i\omega t'} \sum_{\vec{p},\alpha,\beta} \gamma_{\alpha\beta}(\vec{p}, t, t_p, \omega_i)$$

$$\times \sum_{\vec{p},\alpha',\beta'} \gamma^{+}{\alpha'\beta'}(\vec{p}', t' + t, t_p, \omega_i) \cdot$$

$$\times \left\langle a^{+}_{\vec{p}',\beta'}(t') a_{\vec{p}'+\vec{q},\alpha'}(t') a^{+}_{\vec{p}+\vec{q},\alpha} a_{\vec{p},\beta} \right\rangle. \tag{3.19}$$

Our result for the Raman scattering cross section shows that the cross section is a double time integral of a product of two $\gamma_{\alpha\beta}(\vec{p}, t, t_p, \omega_i)$ terms (which contain information about the polarization of incident and scattered photons, the probe pulse width, and the characteristics of the band structure near the Γ-point) and a dynamical electron correlation function. We note that $\gamma_{\alpha\beta}(\vec{p}, t, t_p, \omega_i)$ is completely known from Eqs. (3.17) and (3.18), once the polarizations of incident and scattered photons, the probe pulse width, and the near-bandgap energies are given. The only unknown quantity in Eq. (3.19) is the time-dependent electron correlation function $\langle a^{+}_{\vec{p}',\beta'}(t') a_{\vec{p}'+\vec{q},\alpha'}(t') a^{+}_{\vec{p}+\vec{q},\alpha} a_{\vec{p},\beta} \rangle$, which we now determine for a nonequilibrium electron distribution.

To evaluate the electron correlation function $\langle a^{+}_{\vec{p}',\beta'}(t') a_{\vec{p}+\vec{q},\alpha'}(t') a^{+}_{\vec{p}+\vec{q},\alpha}$ $a_{\vec{p},\beta} \rangle$, we need to use the equation of motion [13] and the RPA. The electrons of interest are in the conduction band, and the time dependence of the operators is for times during the probe pulse of duration t_p. It is convenient to find the correlation function by defining its frequency Fourier transform function, that is,

$$\int_{-\infty}^{+\infty} dt' e^{i\omega' t'} \left\langle a^{+}_{\vec{p}',\beta'}(t') a_{\vec{p}'+\vec{q},\alpha'}(t') a^{+}_{\vec{p}+\vec{q},\alpha} a_{\vec{p},\beta} \right\rangle$$

$$\equiv -i\hbar \left[g^{+}_{\alpha',\beta',\alpha,\beta} (\vec{p}', \vec{p}, \omega', \vec{q}) + g^{-}_{\alpha',\beta',\alpha,\beta} (\vec{p}', \vec{p}, \omega', \vec{q}) \right]. \tag{3.20}$$

In Eq. (3.20) we have defined the retarded and advanced correlation function:

$$g^{+}_{\alpha',\beta',\alpha,\beta}\left(\vec{p}', \vec{p}, t', \vec{q}\right) \equiv \frac{i}{\hbar}\varphi(t')\left\langle a^{+}_{\vec{p}',\beta'}(t')a_{\vec{p}'+\vec{q},\alpha'}(t')a^{+}_{\vec{p}+\vec{q},\alpha}a_{\vec{p},\beta}\right\rangle \quad (3.21)$$

and

$$g^{-}_{\alpha',\beta',\alpha,\beta}(\vec{p}', \vec{p}, t', \vec{q}) \equiv \frac{i}{\hbar}\varphi\left(-t'\right)\left\langle a^{+}_{\vec{p}',\beta'}(t')a_{\vec{p}'+\vec{q},\alpha'}(t')a^{+}_{\vec{p}+\vec{q},\alpha}a_{\vec{p},\beta}\right\rangle \quad (3.22)$$

where $\varphi(t)$ is a steplike function with $\varphi(0) = 1/2$. The relationship between the Fourier transform of the retarded and advanced functions can be found to be

$$g^{-}_{\alpha',\beta',\alpha,\beta}(\vec{p}', \vec{p}, \omega', \vec{q}) = -\left[g^{+}_{\alpha',\beta',\alpha',\beta}(\vec{p}', \vec{p}, \omega', \vec{q})\right]^{*}. \quad (3.23)$$

We use the equation-of-motion technique within the RPA to derive the time dependence of $g^{+}_{\alpha',\beta',\alpha,\beta}(\vec{p}', \vec{p}, \omega', \vec{q})$. From the equation of motion in Heisenberg representation, we have

$$\begin{aligned}
&\left(\hbar\omega' + \varepsilon_{\vec{p}'} - \varepsilon_{\vec{p}'+\vec{q}} + i\hbar\delta\right)g^{+}_{\alpha',\beta',\alpha,\beta}(\vec{p}', \vec{p}, \omega', \vec{q}) \\
&\quad = -\left\langle a^{+}_{\vec{p}',\beta'}a_{\vec{p}'+\vec{q},\alpha'}a^{+}_{\vec{p}+\vec{q},\alpha}a_{\vec{p},\beta}\right\rangle \\
&\qquad +\delta_{\alpha'\beta'}\left[\frac{4\pi e^{2}}{q^{2}\varepsilon(\omega)}\right]\left[n(\vec{p}') - n(\vec{p}' + \vec{q})\right]\sum_{\vec{k},\sigma} g^{+}_{\sigma,\sigma,\alpha,\beta}(\vec{k}, \vec{p}, \omega', \vec{q}) \\
&\qquad -\frac{i\hbar}{\tau}\left[g^{+}_{\alpha',\beta',\alpha,\beta}(\vec{p}', \vec{p}, \omega', \vec{q}) - \oint \frac{d\Omega_{\vec{p}'}}{4\pi} g^{+}_{\alpha',\beta',\alpha,\beta}(\vec{p}', \vec{p}, \omega', \vec{q})\right],
\end{aligned} \quad (3.24)$$

where δ is an infinitesimal positive quantity, and $\varepsilon(\omega) = \varepsilon_{\infty} + [(\varepsilon_{0} - \varepsilon_{\infty})\omega_{T}^{2}]/(\omega_{T}^{2} - \omega^{2})$ is the frequency-dependent dielectric function. The terms ε_{0}, ε_{∞} are the static and high-frequency dielectric constants, respectively; and ω_{T} is the angular frequency of a transverse optical phonon. Here, we have phenomenologically added the last term of Eq. (3.24), which conserves the particle number, spin, and energy. This way of introducing a phenomenological collision time is exactly the same as in the work of Hamilton and McWhorter [12]. It essentially introduces an elastic collision time τ in the electronic system, because the correlation function relaxes to its average value at a constant energy $\varepsilon_{\vec{p}}$. Specifically, τ includes electron–electron, electron–impurity, and electron–acoustic phonon scattering processes.

Equation (3.24) allows us to find the closed form of the retarded electron correlation function in the frequency domain. The Raman scattering cross section is proportional to the imaginary part of $g^{+}_{\alpha',\beta',\alpha,\beta}(\vec{p}',\vec{p},\omega',\vec{q})$ through Eqs. (3.19)–(3.21), so the cross section becomes

$$\begin{aligned}
\frac{d^2\sigma}{d\omega\, d\Omega} \propto{}& \int_{-\infty}^{\infty} d\omega' \int_{-t_p/2}^{t_p/2} dt \int_{-t_p/2-t}^{t_p/2-t} dt' e^{i(\omega-\omega')t'} \\
&\times \sum_{\vec{p},\alpha,\beta} \gamma_{\alpha\beta}(\vec{p},t,t_p,\omega_i) \cdot \sum_{\vec{p}',\alpha',\beta'} \gamma^{+}_{\alpha',\beta'}(\vec{p}',t'+t,t_p,\omega_i) \\
&\times \operatorname{Im}\Bigg\{\Bigg[-\langle a^{+}_{\vec{p}',\beta'} a_{\vec{p}'+\vec{q},\alpha'} a^{+}_{\vec{p}+\vec{q},\alpha} a_{\vec{p},\beta}\rangle + \delta_{\alpha'\beta'}\left[\frac{4\pi e^2}{q^2\varepsilon(\omega)}\right] \\
&\times [n(\vec{p}') - n(\vec{p}'+\vec{q})] \sum_{\vec{k},\sigma} g^{+}_{\sigma,\sigma,\alpha,\beta}(\vec{k},\vec{p},\omega',\vec{q})\Bigg] \\
&\times \frac{1}{\hbar\omega' + \varepsilon_{\vec{p}'} - \varepsilon_{\vec{p}'+\vec{q}} + i\hbar/\tau'} \\
&\times \left[1 - \frac{i\hbar}{\tau}\left\langle \frac{1}{\hbar\omega' + \varepsilon_{\vec{p}'} - \varepsilon_{\vec{p}'+\vec{q}} + i\hbar/\tau'}\right\rangle_{\Omega_{\vec{p}'}}\right]^{-1}\Bigg\}, \qquad (3.25)
\end{aligned}$$

where $\langle 1/(\hbar\omega' + \varepsilon_{\vec{p}'} - \varepsilon_{\vec{p}'+\vec{q}} + i\hbar/\tau')\rangle_{\Omega_{\vec{p}'}}$ represents the average over the solid angle $\Omega_{\vec{p}'}$, and $1/\tau' = 1/\tau + \delta$. $n(\vec{p})$ is the electron distribution function. "Im" means taking the imaginary part of.

We point out we did not use the fluctuation–dissipation theorem in our derivation, so Eq. (3.25) is valid irrespective of the nature of electron distribution, i.e., equilibrium or nonequilibrium.

The Raman scattering cross section given by equation (3.25) has three parts, namely, the charge-density fluctuations (CDF), energy-density fluctuations (EDF), and spin-density fluctuations (SDF). The spin of the electron is coupled with the incident light through the second-order $\vec{p}_i \cdot \vec{A}$ perturbation terms. SDF is the dominant contribution for higher electron concentrations. In addition, the EDF contribution dominates when the CDF contribution is screened at high electron densities. These three contributions are related to different polarization selection rules as pointed out later.

Applying the RPA to the zero-time electron correlation function $\langle a^{+}_{\vec{p}',\beta'} a_{\vec{p}'+\vec{q},\alpha'} a^{+}_{\vec{p}+\vec{q},\alpha} a_{\vec{p},\beta}\rangle$, we obtain a simple and useful relation:

$$\langle a^{+}_{\vec{p}',\beta'} a_{\vec{p}'+\vec{q},\alpha'} a^{+}_{\vec{p}+\vec{q},\alpha} a_{\vec{p},\beta}\rangle = n(\vec{p})[1-n(\vec{p}+\vec{q})]\delta_{\vec{p}\vec{p}'}\delta_{\alpha\alpha'}\delta_{\beta\beta'}. \quad (3.26)$$

From Eqs. (3.24) and (3.26), we have

$$\sum_{\vec{k},\sigma} g^{+}_{\sigma,\sigma,\alpha,\beta}(\vec{k},\vec{p},\omega',\vec{q}) = \frac{-n(\vec{p})[1-n(\vec{p}+\vec{q})]/\left(\hbar\omega'+\varepsilon_{\vec{p}}-\varepsilon_{\vec{p}+\vec{q}}+i\hbar\delta\right)\cdot\delta_{\alpha\beta}}{1-2\sum_{\vec{k}}\left([4\pi e^2/q^2\varepsilon(\omega)][n(\vec{k})-n(\vec{k}+\vec{q})]/\left(\hbar\omega'+\varepsilon_{\vec{k}}-\varepsilon_{\vec{k}+\vec{q}}+i\hbar\delta\right)\right)}. \quad (3.27)$$

The Kronecker delta in Eqs. (3.26) and (3.27) are very important in separating the SDF contributions from those of CDF and EDF.

The product of $\gamma_{\alpha\beta}(\vec{p},t,t_p,\omega_i)$ and $\gamma^{+}_{\alpha'\beta'}(\vec{p}',t'+t,t_p,\omega_i)$ that appears in equation (3.25) is in general very complicated. For simplicity, we drop the anisotropic part; that is, we ignore the part containing $(\hat{p}\hat{p}-\overleftrightarrow{I}/3)$ in Eqs. (3.17) and (3.18). Then,

$$\begin{aligned}\gamma_{\alpha\beta}(\vec{p},t,t_p,\omega_i)\gamma^{+}_{\alpha'\beta'}(\vec{p}',t'+t,t_p,\omega_i) &\cong \left[(\hat{e}_i\cdot\hat{e}_s)C_p(t,\omega_i)\delta_{\alpha\beta}+i(\hat{e}_i\times\hat{e}_s)\cdot\overleftrightarrow{\sigma}_{\alpha\beta}S_p(t,\omega_i)\right]\\ &\quad\times\left[(\hat{e}_i\cdot\hat{e}_s)C^{*}_{p'}(t'+t,\omega_i)\delta_{\alpha\beta}+i(\hat{e}_i\times\hat{e}_s)\cdot\overleftrightarrow{\sigma}^{+}_{\alpha\beta}S^{*}_{p'}(t'+t,\omega_i)\right],\end{aligned} \quad (3.28a)$$

where $C_p(t,\omega_i)$ and $S_p(t,\omega_i)$ are the remaining scalars resulting from the isotropic ($\overleftrightarrow{I}$) part of Eqs. (3.17) and (3.18), respectively. We notice that equation (3.28a) gives rise not only to terms containing $(\hat{e}_i\cdot\hat{e}_s)$ and $(\hat{e}_i\times\hat{e}_s)$ but also to cross terms involving $(\hat{e}_i\cdot\hat{e}_s)(\hat{e}_i\times\hat{e}_s)$. Fortunately, the cross terms exactly vanish when the Kronecker deltas and a property of Pauli matrices ($\sum_\alpha \overleftrightarrow{\sigma}_{\alpha,\alpha}=0$) are used. Equation (3.28a) becomes

$$\begin{aligned}\gamma_{\alpha\beta}(\vec{p},t,t_p,\omega_i)\gamma^{+}_{\alpha'\beta'}(\vec{p}',t'+t,t_p,\omega_i) &\cong (\hat{e}_i\cdot\hat{e}_s)^2\delta_{\alpha\beta}\delta_{\alpha'\beta'}C_p(t,\omega_i)C^{*}_{p'}(t'+t,\omega_i)\\ &\quad+(\hat{e}_i\times\hat{e}_s)\cdot\overleftrightarrow{\sigma}_{\alpha\beta}(\hat{e}_i\times\hat{e}_s)\cdot\overleftrightarrow{\sigma}^{+}_{\alpha'\beta'}S_p(t,\omega_i)S^{*}_{p'}(t'+t,\omega_i).\end{aligned} \quad (3.28b)$$

We now combine the results of Eqs. (3.26), (3.27), and (3.28) to obtain the desired form of the Raman scattering cross section. The contribution proportional to $(\hat{e}_i \cdot \hat{e}_s)^2$ corresponds to that of CDF and EDF, and is given by

$$\left(\frac{d^2\sigma}{d\omega d\Omega}\right)_{CDF,EDF} \propto \sum_{\vec{p}} -n(\vec{p})[1-n(\vec{p}+\vec{q}](\hat{e}_i \cdot \hat{e}_s)^2$$

$$\times \int_{-\infty}^{\infty} d\omega' \int_{-t_p/2}^{t_p/2} dt \int_{-t_p/2-t}^{t_p/2-t} dt' e^{i(\omega-\omega')t'}$$

$$\times \sum_{\vec{p}'} \Bigg[C_p(t,\omega_i) C^*_{p'}(t'+t,\omega_i)\delta_{\vec{p}\vec{p}'} + 2C_p(t,\omega_i)C^*_{p'}(t'+t,\omega_i)$$

$$\times \frac{[4\pi e^2/q^2\varepsilon(\omega)][(n(\vec{p}')-n(\vec{p}'+\vec{q}))/(\hbar\omega'+\varepsilon_{\vec{p}'}-\varepsilon_{\vec{p}'+\vec{q}})]}{1-2\sum_{\vec{k}'}[4\pi e^2/q^2\varepsilon(\omega)][n(\vec{k}')-n(\vec{k}'+\vec{q})]/(\hbar\omega'+\varepsilon_{\vec{k}'}-\varepsilon_{\vec{k}'+\vec{q}})} \Bigg]$$

$$\times \mathrm{Im}\Bigg\{ \frac{1}{\hbar\omega'+\varepsilon_{\vec{p}'}-\varepsilon_{\vec{p}'+\vec{q}}+i\hbar/\tau}$$

$$\times \left[1-\frac{i\hbar}{\tau}\left\langle\frac{1}{\hbar\omega'+\varepsilon_{\vec{p}'}-\varepsilon_{\vec{p}'+\vec{q}}+i\hbar/\tau}\right\rangle_{\Omega_{\vec{p}'}}\right]^{-1}\Bigg\}. \tag{3.29}$$

The CDF and EDF contributions are very complicated. They contain both the screening effects of Coulomb interaction and the electron–phonon interaction through the dielectric constant. The double time integrals in Eq. (3.29) show the effects of probing with an ultrashort laser pulse—broadening of the Raman spectra; and the imaginary part, which contains the collision time, gives rise to the narrowing of Raman spectra.

The other contributions to the Raman scattering cross section are from the SDF. The SDF contributions are proportional to $(\hat{e}_i \times \hat{e}_s)^2$ and are independent of both the Coulomb interaction and electron–phonon interactions. They are therefore much simpler than either CDF or EDF and are most conveniently used to probe electron distribution functions in semiconductors. The expression for SDF is given by

$$\left(\frac{d^2\sigma}{d\omega d\Omega}\right)_{SDF} \propto \sum_{\vec{p}} -n(\vec{p})[1-n(\vec{p}+\vec{q})](\hat{e}_i \times \hat{e}_s)^2$$

$$\times \int_{-\infty}^{\infty} d\omega' \int_{-t_p/2}^{t_p/2} dt \int_{-t_p/2-t}^{t_p/2-t} dt' e^{i(\omega-\omega')t'} S_p(t,\omega_i) S^*_p(t'+t,\omega_i)$$

$$\times \operatorname{Im}\left\{\frac{1}{\hbar\omega' + \varepsilon_{\vec{p}} - \varepsilon_{\vec{p}+\vec{q}} + i\hbar/\tau}\right.$$
$$\left.\times\left[1 - \frac{i\hbar}{\tau}\left\langle\frac{1}{\hbar\omega' + \varepsilon_{\vec{p}} - \varepsilon_{\vec{p}+\vec{q}} + i\hbar/\tau}\right\rangle_{\Omega_{\vec{p}}}\right]^{-1}\right\}. \tag{3.30}$$

Equations (3.29) and (3.30) represent results for the CDF, EDF, and SDF contributions to the Raman scattering cross section, respectively. These equations are useful for a direct bandgap semiconductor for which elastic scattering processes such as electron–electron, electron–hole, electron–impurity, and electron–acoustic phonon scattering are dominant. These results are derived from a full time-dependent quantum mechanical treatment for a square probing laser pulse of duration t_p and frequency ω_i. In general, the effects of probing with a short laser pulse broadening of the SPS spectrum; and the effects of electron scattering processes are narrowing of the SPS spectrum. In particular, the fluctuation–dissipation theorem was not used in the derivation of Eqs. (3.29), (3.30), so the results are applicable to both equilibrium and nonequilibrium systems. A phenomenological electron scattering time τ was introduced in a way similar to that used in the equilibrium case. We note that in the limit of a very long probe pulse ($t_t \to \infty$) and equilibrium electron distributions, the results can be shown to reduce to expressions previously given for the Raman scattering cross section in the equilibrium case [12].

It is very instructive to note that if we assume that the pulse width of the probe pulse is sufficiently wide, collision effects are negligible, the electron distribution function is nondegenerate, and the term involving matrix elements "S_p" does not depend on the electron momentum, then Eq. (3.30) can be shown to become

$$\left(\frac{d^2\sigma}{d\omega\, d\Omega}\right)_{SDF} \propto \int d^3p \cdot n(\vec{p}) \cdot \delta\left[\omega - \vec{V}\cdot\vec{q} - \frac{\hbar q^2}{2m_e^*}\right]. \tag{3.31}$$

The δ-function in Eq. (3.31) ensures that both the energy and momentum are conserved.

We note that Eq. (3.31) has reached the same conclusion as was obtained previously, simply by considering the conservation of energy and momentum in the Compton scattering process. It shows that the measured Raman scattering cross section at a given solid angle $d\Omega$ (which determines $\vec{q}$) provides *direct* information about the electron distribution function in the direction of wavevector transfer $\vec{q}$.

An intriguing point for probing carrier distributions with Raman spectroscopy is that since the Raman scattering cross section is inversely proportional to the effective mass of the carrier, it preferentially probes electron transport even if holes are simultaneously present. This unique feature makes the interpretation of electron distribution in Raman scattering experiments much simpler than with other techniques.

3.2.2 Theory of Raman Scattering by Lattice Vibrations in Semiconductors

Consider an incident laser beam of angular frequency ω_i that is scattered by a semiconductor, and the scattered radiation is analyzed spectroscopically, as shown in Figure 3.2. In general, the scattered radiation consists of a laser beam of angular frequency ω_i accompanied by weaker lines of angular frequencies $\omega_i \pm \omega$. The line at an angular frequency $\omega_i - \omega$ is called a *Stokes line*, and the line at an angular frequency $\omega_i + \omega$ is usually referred to as an *anti-Stokes line*. The important aspect is that the angular frequency shifts ω are independent of ω_i. In this way, this phenomenon differs from that of luminescence, in which it is the angular frequency of the emitted light that is independent of ω_i. The effect just described is called the *Raman effect*. It was predicted by Smekal [14] and is implicit in the radiation theory of Kramers and Heisenberg [15]. It was discovered experimentally by Raman [16] and by Landsberg and Mandel' shtam [17] in 1928. It can be understood as an inelastic scattering of light in which an internal form of motion of the scattering system is either excited or absorbed during the process.

3.2.2.1 *Simple Classical Theory*

Imagine a crystalline lattice having an internal mode of vibration characterized by a normal coordinate,

$$Q = Q_0 \cos \omega t. \tag{3.32}$$

The electronic polarizability α is generally a function of Q, and since in general, $\omega \ll \omega_i$, at each instant we can regard Q as fixed compared with

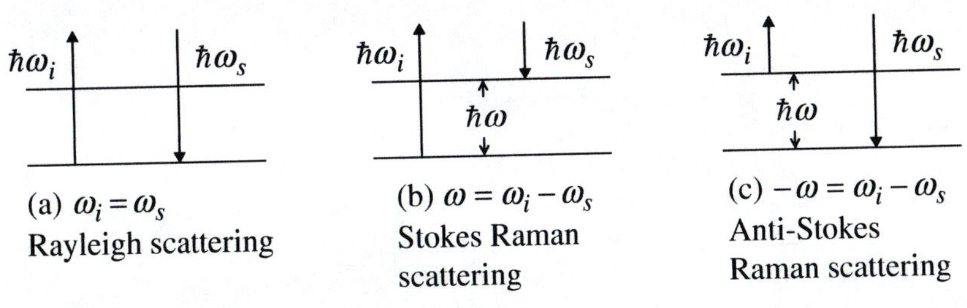

Figure 3.2 A diagram showing the (a) Rayleigh scattering process; (b) Stokes Raman scattering process; and (c) anti-Stokes Raman scattering process.

the variation of the external field $\vec{E}$, i.e., at angular frequency ω_i the induced dipole moment $\vec{P}$ is

$$\vec{P} = \alpha\vec{E} = \alpha(Q)\vec{E}. \tag{3.33}$$

Let $\alpha_0 = \alpha(0)$ be the polarizability in the absence of any excitation. We can write

$$\begin{aligned} \alpha(Q) &= \alpha_0 + \left(\frac{\partial\alpha}{\partial Q}\right)_0 Q + \frac{1}{2}\left(\frac{\partial^2\alpha}{\partial Q^2}\right)_0 Q^2 + \cdots \\ &= \alpha_0 + \alpha_1 Q + \frac{1}{2}\alpha_2 Q^2 + \cdots, \end{aligned} \tag{3.34}$$

where $[\partial\alpha/\partial Q]_0 \equiv \alpha_1$, $[\partial^2\alpha/\alpha Q^2]_0 \equiv \alpha_2$, and the derivative is to be evaluated at zero excitation field.

If we assume that $\vec{E} = \vec{E}_0 \cos\omega_i t$, we find that

$$\begin{aligned} \vec{P}(t) &= \left(\alpha_0\vec{E}_0 + \frac{1}{4}\alpha_2 Q_0^2\vec{E}_0\right)\cos\omega_i t \\ &+ \frac{\vec{E}_0}{2}\alpha_1 Q_0[\cos(\omega_i + \omega)t + \cos(\omega_i - \omega)t] \\ &+ \frac{1}{8}\alpha_2 Q_0^2\vec{E}_0[\cos(\omega_i + 2\omega) + \cos(\omega_i - 2\omega)] + \cdots. \end{aligned} \tag{3.35}$$

For an oscillating dipole moment, the magnetic and electric fields of the emitted electromagnetic wave are given by [18]

$$\vec{B} = \frac{1}{c^2 r}\left[\frac{\partial^2\vec{P}(t - r/c)}{\partial^2 t}\right] \times \hat{n} \tag{3.36a}$$

and

$$\vec{E} = \vec{B} \times \hat{n}, \tag{3.36b}$$

where $\vec{r}$ is the position vector connecting the center of dipole moment to the point of observation, and $\hat{n} = \vec{r}/|\vec{r}|$.

Therefore, the first term in Eq. (3.35) gives rise to Rayleigh scattering; the second term gives the anti-Stokes and Stokes first-order Raman lines, respectively; the third term takes into account the anti-Stokes and Stokes second-order Raman lines, and so on. We notice that in Eq. (3.35), the intensities of the Stokes and anti-Stokes lines are equal. This is because all classical theories neglect the possibility of spontaneous emission.

3.2.2.2 *Quantum Mechanical Theory*

In the quantum mechanical treatment of scattering of light by lattice vibrations, we consider the total Hamiltonian of the system, including the radiation field:

$$H = H_0' + H_{el-ph} + H', \tag{3.37}$$

where H_0' includes contributions from the electronic system, lattice vibrations (or phonons), and the radiation field; $H_{el-ph} = -e\varphi(\vec{r}_i)$ describes the interaction of electrons with phonons, where $\varphi(\vec{r}_i)$ is the potential due to, say, deformation potential and/or Fröhlich interactions; and

$$\begin{aligned} H' &= \sum_i \frac{-e}{2m_e c}\left[\vec{A}(\vec{r}_i)\cdot\vec{p}_i + \vec{p}_i\cdot\vec{A}(\vec{r}_i)\right] + \sum_i \frac{e^2}{2m_e c^2}\vec{A}^2(\vec{r}_i) \\ &= \sum_i \frac{-e}{m_e c}\left[\vec{p}_i\cdot\vec{A}(\vec{r}_i)\right] + \sum_i \frac{e^2}{2m_e c^2}\vec{A}^2(\vec{r}_i) \equiv H_1' + H_2' \end{aligned} \tag{3.38}$$

takes into account the electron–photon interactions, where $\vec{A}(\vec{r}_i)$ is the vector potential of the radiation field given by Eq. (3.5).

We notice that for a typical Raman scattering process in which $\omega_i \gg \omega$, photons do not interact directly with phonons but through electron–phonon interactions, i.e., the H_{el-ph} term in the total Hamiltonian. Since the Raman scattering process involves the annihilation of an incident photon and the creation of a scattered photon, the $\vec{p}\cdot\vec{A}$ and $\vec{A}\cdot\vec{p}$ terms in Eq. (3.38) will contribute to the scattering matrix element in the third order, and $\vec{A}^2$ terms will contribute in the second order in perturbation-theory calculations of the Raman scattering cross section. If we neglect nonlinear processes, then only the $\vec{p}\cdot\vec{A}$ and $\vec{A}\cdot\vec{p}$ terms in Eq. (3.38) are important and need to be considered.

From time-dependent perturbation theory and the Fermi golden rule, we obtain for the scattering probability (which is proportional to the Raman scattering cross section) for a one-phonon Stokes Raman process [19]

$$\begin{aligned} P(\omega_s) = \frac{2\pi}{\hbar}\Bigg| &\sum_{n,n'} \frac{\langle i|H_1'|n\rangle\langle n|H_{el-ph}|n'\rangle\langle n'|H_1'|i\rangle}{[\hbar\omega_i - (E_n - E_i)][\hbar\omega_i - \hbar\omega - (E_{n'} - E_i)]} \\ &+ \sum_{n,n'} \frac{\langle i|H_1'|n\rangle\langle n|H_1'|n'\rangle\langle n'|H_{el-ph}|i\rangle}{[\hbar\omega_i - (E_n - E_i)][\hbar\omega_i - \hbar\omega_s - (E_{n'} - E_i)]} \\ &+ \sum_{n,n'} \frac{\langle i|H_1'|n\rangle\langle n|H_{el-ph}|n'\rangle\langle n'|H_1'|i\rangle}{[-\hbar\omega_s - (E_n - E_i)][-\hbar\omega_s - \hbar\omega - (E_{n'} - E_i)]} \end{aligned}$$

$$+\sum_{n,n'} \frac{\langle i|H_1'|n\rangle\langle n|H_1'|n'\rangle\langle n'|H_{el-ph}|i\rangle}{[-\hbar\omega_s - (E_n - E_i)][-\hbar\omega_i + \hbar\omega - (E_{n'} - E_i)]}$$

$$+\sum_{n,n'} \frac{\langle i|H_{el-ph}|n\rangle\langle n|H_1'|n'\rangle\langle n'|H_1'|i\rangle}{[-\hbar\omega - (E_n - E_i)][-\hbar\omega + \hbar\omega_i - (E_{n'} - E_i)]}$$

$$\left.+\sum_{n,n'} \frac{\langle i|H_{el-ph}|n\rangle\langle n|H_1'|n'\rangle\langle n'|H_1'|i\rangle}{[-\hbar\omega - (E_n - E_i)][-\hbar\omega - \hbar\omega_s - (E_{n'} - E_i)]}\right|^2$$

$$\times\delta(\hbar\omega_i - \hbar\omega_s - \hbar\omega), \tag{3.39}$$

where $|i\rangle$ is the initial state of the system, and E_i is its energy; $|n\rangle$, $|n'\rangle$ are intermediate states with energies E_n, E_n', respectively.

We note that there are three processes involved in one-phonon Raman scattering: the incident photon is annihilated, the scattered photon is emitted, and a phonon is annihilated (or created). Since these three processes can occur in any time order in time-dependent perturbation-theory calculations of scattering probability, we expect that there will be six terms or contributions to $P(\omega_s)$, which is consistent with Eq. (3.39). The δ-function here ensures that energy is conserved in the Raman scattering process.

3.3 SAMPLES, EXPERIMENTAL SETUP, AND APPROACH

3.3.1 General Considerations

The measurement of a Raman spectrum requires at least the following equipment:

(1) a collimated and monochromatic light source
(2) a spectrometer to analyze the spectral content of the scattered radiation
(3) a sensitive optical system to collect and detect the generally weak scattered radiation

Since a Raman signal is typically very small (as is clear from the fact that third-or higher-order time-dependent perturbation theory is involved in the calculation of the Raman scattering cross section), every component has to be optimized. We shall consider these components individually:

3.3.1.1 Light Source

Before the invention of lasers, the light source was typically a high-power discharge lamp. Discrete emission lines of a gas or vapor were used, and only transparent samples could be studied because of their much larger scattering volumes. Since many common semiconductors are opaque, Raman scattering studies of semiconductors became feasible only after the advent of lasers. The continuous-wave (cw) He–Ne laser was the first to be used in Raman spectroscopy, and it was soon replaced by Nd:YAlG, Ar^+, and Kr^+ ion lasers. The latter two produce several high-power (>1 W in a single line) discrete emission lines covering the red, yellow, green, blue, and violet regions of the visible spectrum. With these high average power cw lasers it became feasible to obtain not only one-phonon but also two-phonon Raman spectra in semiconductors. With continuously tunable cw lasers based on dyes, on color centers in ionic crystals, and more recently on Ti-doped sapphire it became possible to perform Raman excitation spectroscopies, i.e., resonant Raman scattering in which one monitors the Raman signal as a function of the excitation laser wavelength.

3.3.1.2 Spectrometer

In most Raman experiments on semiconductors the Raman signal is typically four to six orders of magnitude weaker than the elastically scattered laser light. At the same time the difference in energy between the Raman scattered photons and the excitation laser photons is only about 1% of the laser energy. This percentage is even smaller when Raman spectroscopy is used to measure an SPS spectrum. In order to observe this very weak sideband in the vicinity of a strong laser line, the spectrometer must satisfy several stringent conditions. First, it must have a good spectral resolving power. Modern Raman spectrometers typically have resolving power $(\Delta\lambda/\lambda) \geq 10^4$, which can be obtained easily with diffraction gratings. It is, however, important that these gratings not produce ghosts and/or satellites, which can be confused with the Raman signal. A Raman spectrometer must have an excellent stray light rejection ratio, which is defined as the ratio of the background stray light (light at all wavelengths other than the nominal one specified by the spectrometer) to the signal. Stray light can be produced either by imperfections in the optical system or by the scattering of light off walls and dust particles inside the spectrometer. Most simple spectrometers have a rejection ratio of 10^{-4}–10^{-6}. As a result, the background stray light can be orders of magnitude larger than the Raman signal. This issue can be resolved by (a) making the sample surface as smooth as possible and therefore minimizing the elastically scattered laser light, (b) employing a notch filter to block out the elastically scattered laser light, (c) putting two or even three simple spectrometers in tandem. A properly designed double monochromator, in which two simple spectrometers are placed in series, can have a rejection ratio as small as 10^{-14}, equal to the product of the ratio for the two simple monochromators. This rejection ratio is adequate for Raman studies in most semiconductors. Currently, triple spectrometers are popularly used with multichannel detectors, which are described next. In these spectrometers two simple monochromators are placed in tandem for use as a notch filter. The third spectrometer provides all the dispersion required for separating the Raman signal from the elastically scattered laser light.

3.3.1.3 Detector and Photon-Counting Electronics

Raman recorded the weak inelastically scattered light in his pioneering experiment in 1928 by using photographic plates. These detectors actually have many of the desirable characteristics of modern systems: they have the sensitivity to detect individual photons; they are multichannel detectors in that they can measure many different wavelengths at the same time; finally, they can integrate the signal over long periods of time, from hours to even days. Their one big advantage over modern detection systems is that they are relatively inexpensive; however, they also have some serious drawbacks: their outputs are not linear in light intensity, and it is also difficult to convert the recorded signal into digital form for analysis.

The first major advance in photoelectric recording of Raman signals was the introduction of photon counting methods. Instead of integrating all the photocurrent pulses arriving at the anode of a photomultiplier tube as the signal, a discriminator selects and counts only those pulses with large enough amplitude to have originated from the photocathode. The background pulses (noise) remaining in such systems are those generated by thermionic emission of electrons at the photocathode. This noise can be minimized by cooling the entire photomultiplier tube to $-20\,^{\circ}$C (through thermoelectric coolers) or to liquid nitrogen temperature. One of the most popular photomultipliers for Raman scattering experiments has a GaAs photocathode cooled to $-20\,^{\circ}$C. When coupled to properly designed counting electronics, such a detection system has a background noise or dark counts of a few counts per second and a dynamic range of 10^6.

The major disadvantage of the aforementioned detector system compared with the photographic plate is that it counts the total number of photons emerging from the spectrometer without spatially resolving the positions (and hence the wavelengths) of the photons. Consequently, the Raman spectrum is obtained only after scanning the spectrometer output over a wavelength range containing the spectral range of interest. Recently, several multichannel detection systems have become available commercially. These systems are based on either charge-coupled devices (CCDs) or position-sensitive imaging photomultiplier tubes.

3.3.2 Current Studies

The wurtzite structure GaN sample used in our Raman experiments was a 2μm-thick undoped GaN (with a residual electron density of $n \cong 10^{16}/\text{cm}^3$)

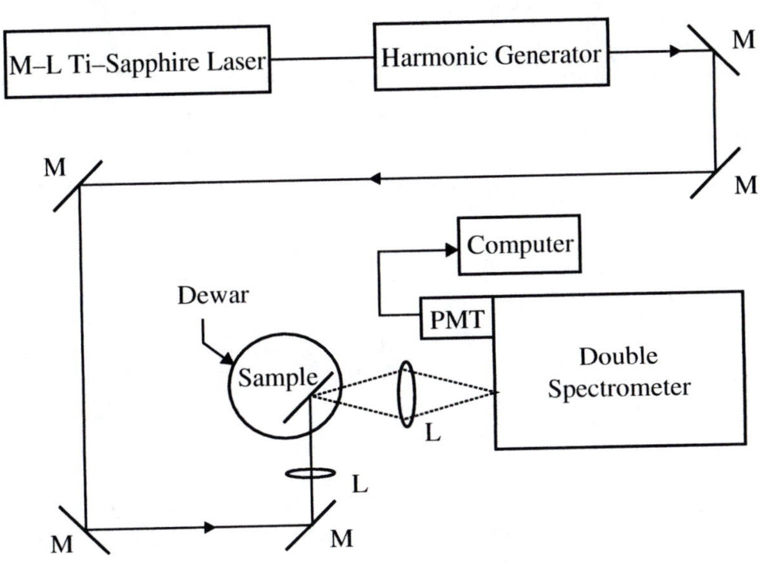

Figure 3.3 Experimental setup for transient Raman scattering experiments in which a laser single pulse is used to both excite and probe nonequilibrium excitations. M, mirror; L, lens; PMT, photomultiplier tube.

grown by molecular beam epitaxy on a (0001)-oriented sapphire substrate. The z-axis of this wurtzite structure GaN is perpendicular to the sapphire substrate plane.

The experimental setup for transient/time-resolved picosecond/ subpicosecond Raman spectroscopy is shown in Figures 3.3 and 3.4, respectively [20–31]. The output of the higher harmonic of a cw mode-locked Ti:sapphire laser is used to either carry out transient Raman measurements or perform pump/probe experiments. In the case of transient experiments, the output from the higher harmonics is directed into the sample, whereas for time-resolved Raman experiments, the output is split into two equally intense but perpendicularly polarized beams. One of them is used to excite the electron–hole pair density in the sample and the other, after being suitably delayed, is used to probe the evolution of nonequilibrium electron/phonon distributions. The 90° and/or backward-scattered Raman signal is collected and analyzed by a standard Raman system, which consists of a double spectrometer, a photomultiplier tube, and/or a CCD detector.

One important advantage of probing nonequilibrium excitations with Raman spectroscopy in semiconductors is that since a Raman signal is detected only when excitation photons are present, the time resolution is essentially limited by the pulse width of the excitation laser and not

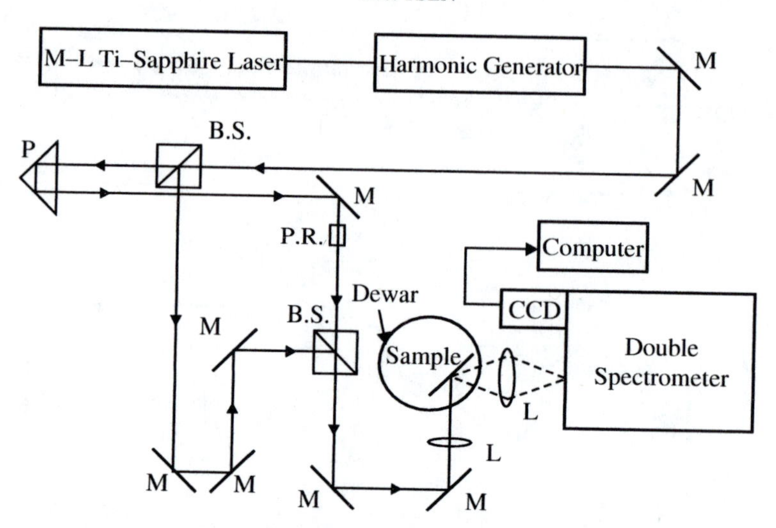

Figure 3.4 Experimental setup for time-resolved Raman scattering experiments in which a laser pulse is used to excite nonequilibrium excitations, and the time-delayed pulse is employed to probe the time evolution of nonequilibrium excitations. M, mirror; L, lens; B.S., beam splitter; P.R., polarization rotator; P, variable delay line; CCD, charge-coupled device.

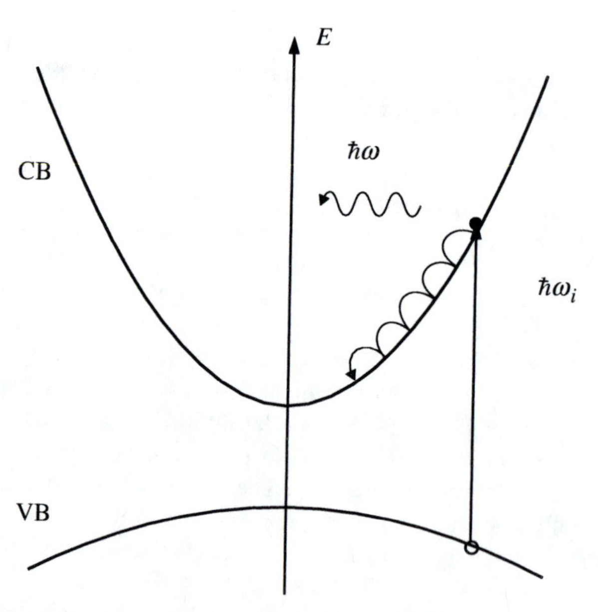

Figure 3.5 A two-band diagram showing how nonequilibrium excitations such as phonons can be created and studied by transient/time-resolved Raman spectroscopy. CB, conduction band; VB, valence band; $\hbar\omega$ is the energy of generated nonequilibrium excitations, in this case LO phonons; and $\hbar\omega_i$ is the energy of the excitation photons.

by the response of the detection system. This explains why our detection system has a time resolution of the order of nanoseconds, whereas the time resolution in our Raman experiments is typically on the scale of subpicoseconds.

We use picosecond/subpicosecond time-resolved Rman spectroscopy for studying nonequilibrium electron distributions, electron–phonon, and phonon–phonon interactions in wurtzite GaN. As shown in Figure 3.5, electron–hole pairs are photoexcited by the excitation photons either across the bandgap of GaN either directly (in the case of above-bandgap excitation) or through nonlinear processes (in the case of below-bandgap excitations). These energetic electron–hole pairs relax to the bottom of the conduction band (for electrons) and to the top of the valence band (for holes) by emitting nonequilibrium phonons through electron–phonon interactions. By monitoring the occupation number of these emitted nonequilibrium phonons, information such as the strength of electron–phonon interactions and phonon–phonon interactions can be readily obtained.

3.4 PHONON MODES IN WURTZITE STRUCTURE GaN

Phonon modes in wurtzite structure GaN are complicated due to the fact that GaN contains a large number of atoms in the unit cell. We limit our discussions here to the phonon modes at the zone center that are directly accessible by Raman spectroscopy. Furthermore, we illustrate only the optical phonon modes. Since the wavevector is zero, the acoustic modes are just translations. The space group at the Γ-point is represented by C_{6v}. This full, reducible representation can be decomposed into irreducible representations according to Hayes and Loudon [32]

$$\Gamma_{opt} = A_1(Z) + 2B_1 + E_1(X, Y) + 2E_2, \tag{3.40}$$

where the X, Y, and Z in parentheses represent the polarization directions. $X = (001)$, $Y = (010)$, $Z = (001)$. The atomic vibrations of various modes are illustrated in Figure 3.6. The E_2 modes are Raman active, the A_1 and E_1 modes are both Raman and infrared active, and the B_1 modes are silent. It is worthwhile pointing out that there is a close relation between the cubic and hexagonal wurtzite structure—the difference in the neighboring atoms begins only in the third shell; however, the unit cell in the wurtzite structure contains twice as many atoms as that in the zinc-blende structure. This results in a folding of the Brillouin zone of the zinc-blende structure in the $\Gamma \rightarrow L$ direction, and then in a doubling of the number of modes. Therefore, instead of the TO mode in the zinc-blende structure, there are

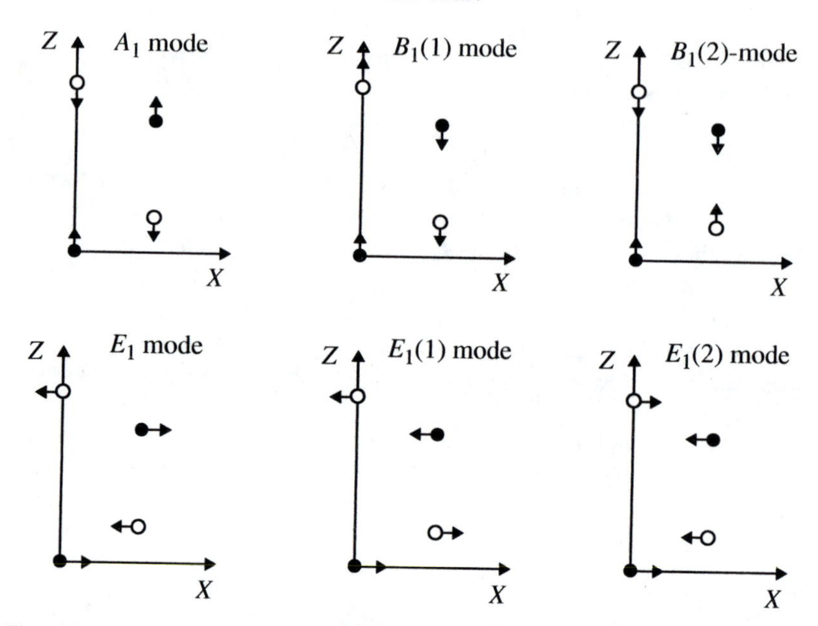

Figure 3.6 The atomic vibrations of various phonon modes in wurtzite GaN. The solid circles represent Ga atoms, and open ones correspond to N atoms.

two modes in the wurtzite structure: E_2^1 and TO modes. At the Γ-point of the Brillouin zone, the latter is split into A_1 and E_1 modes.

3.5 EXPERIMENTAL RESULTS

3.5.1 Studies of Electron–Phonon Interactions in Wurtzite GaN by Nonlinear Laser Excitation Processes [33]

A below-bandgap picosecond laser system was used to excite electron–hole pairs through a two-photon absorption process. The highly energetic electrons and holes then relaxed by emitting various phonon modes. The same laser was employed to detect the emitted nonequilibrium phonon populations by Raman spectroscopy. From the analysis of experimental results we concluded that high-energy electrons relax primarily through the emission of longitudinal optical phonons and that Fröhlich interaction is much stronger than deformation potential interaction in wurtzite structure GaN.

3.5.1.1 Samples and Experimental Technique

The sample used in this work was described in detail in Section 3.3. Very intense ultrashort laser pulses were generated by the second harmonic of a cw mode-locked Ti:sapphire laser operating at a repetition rate of about

80 MHz. The photon energy was about 2.8 eV. The average output power of the laser was about 80 mW. The laser had a pulse width of about 1 ps. A laser power density as high as 1 GW/cm^2 can be achieved by focusing the laser beam with a microscope objective. We note that due to the wide bandgap of GaN, the photons in our laser beam did not have sufficient energy to directly excite electron–hole pairs across the bandgap. A two-photon absorption process was used to create electron–hole pairs with large excess energies (≥ 2 eV). The same high-power laser beam was also used to study the strength of electron–phonon interaction by monitoring the emitted nonequilibrium phonon populations with Raman scattering, as depicted in Figure 3.3. Polarized Raman scattering experiments were carried out in a variety of scattering geometries as specified in the following section. The sample was kept in contact with a cold fingertip of a closed-cycle refrigerator. The temperature of the sample was estimated to be about 25 K. The scattered signal was collected and analyzed with a standard Raman setup equipped with a CCD detection system.

3.5.1.2 Experimental Results and Analysis

Figures 3.7a and 3.8a show two polarized Stokes Raman spectra of a GaN sample taken at $T \cong 25$ K and at a laser power density of 1 GW/cm^2, and in $Z(X, X)\overline{Z}$, $Y(X, X)\overline{Y}$ scattering configurations, respectively, where $X = (100)$, $Y = (010)$, $Z = (001)$. Similar cw Raman spectra have been reported by Azuhata et al. at $T = 300$ K [34]. In Figure 3.7a, the sharp peak around 757 cm^{-1} comes from scattering of light by the E_g phonon mode of sapphire, and the shoulder close to 741 cm^{-1} belongs to the A_1(LO) phonon mode in GaN. In Figure 3.8a, the sharp peak around 574 cm^{-1} corresponds to the E_2 phonon mode in GaN, the peak centered about 538 cm^{-1} represents the A_1(TO) phonon mode in GaN, and the small structure around 422 cm^{-1} is from the A_{1g} phonon mode of sapphire.

The anti-Stokes Raman spectra corresponding to Figures 3.7a and 3.8a are shown in Figures 3.7b and 3.8b, respectively. Since at a sample temperature as low as $T \cong 25$ K, thermal occupations of phonons are varnishingly small, any Raman signal observed in the anti-Stokes Raman spectra must arise from nonequilibrium phonon modes. With the help of the Stokes Raman spectra in Figures 3.7a and 3.8a, we made the following identifications: the broad structure centered around -741 cm^{-1} comes from Raman scattering from nonequilibrium A_1(LO) phonon mode. Strikingly, we noticed that within experimental accuracy there were no detectable nonequilibrium phonon populations for either A_1(TO) or E_2 phonon modes of GaN, or for E_g or A_{1g} phonon modes of sapphire.

We also carried out similar Raman experiments at scattering geometries of $A(Z, X)Y$ and $Y(Z, X)\overline{Y}$, where E_1(LO) and E_1(TO) phonon modes could

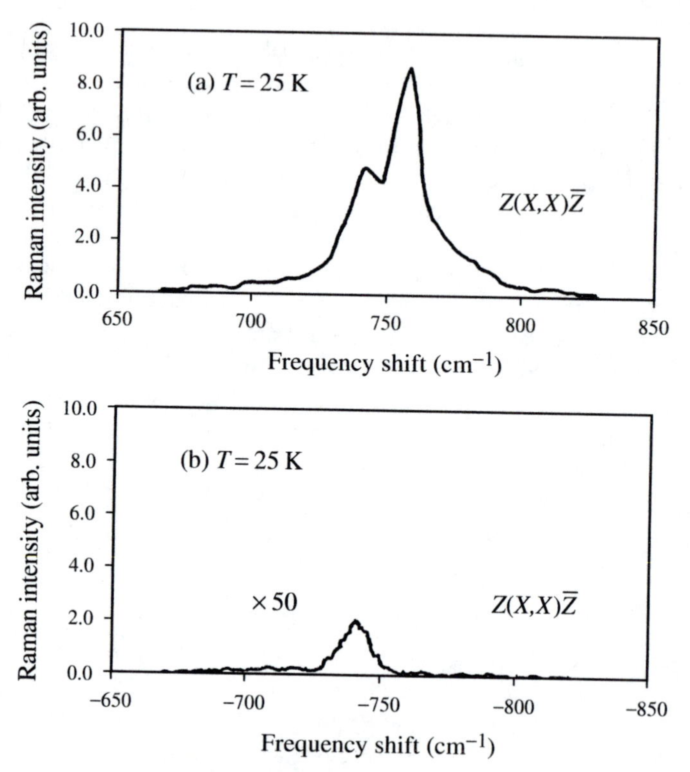

Figure 3.7 Transient (a) Stokes and (b) anti-Stokes Raman spectra for a wurtzite GaN sample taken by an ultrashort pulse laser having a pulse width of 1 ps and photon energy of 2.8 eV. The scattering configuration was $Z(X, X)\bar{Z}$. The anti-Stokes signal has been magnified by a factor of 50.

be observed; here, $A = (\sin\theta, -\cos\theta, 0)$ with $\theta \cong 108°$. These experimental results indicated that a substantial occupation of nonequilibrium phonons was observed for the E_1(LO) mode but not for the E_1(TO).

Therefore, our experimental results demonstrated directly that hot electrons thermalize primarily through the emission of polar longitudinal optical phonons in wurtzite structure GaN. Electrons interact with LO phonons through Fröhlich interaction as well as deformation potential interaction, and they interact with TO phonons via deformation potential interaction only. Since only the LO phonon modes are driven out of equilibrium, our experimental results showed that Fröhlich interaction is much stronger than deformation potential interaction in this wide bandgap semiconductor.

From the measured Raman spectra we estimated that the phonon occupation number of the observed nonequilibrium A_1(LO) and E_1(LO) phonon modes at the laser power density of 1 GW/cm^{-1} was about $\Delta n \cong 0.16 \pm 0.01$, 0.14 ± 0.01, respectively. This suggests that electrons interact

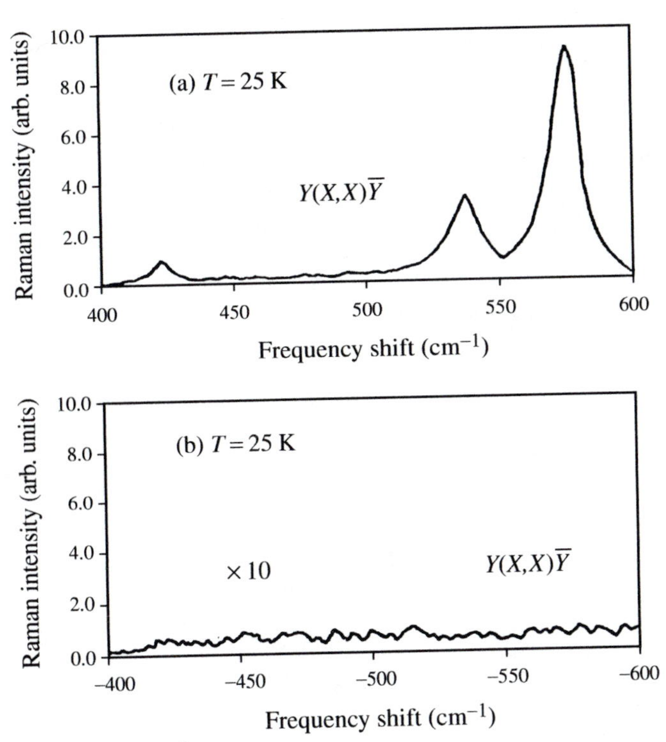

Figure 3.8 Transient (a) Stokes and (b) anti-Stokes Raman spectra for a wurtzite GaN sample taken by an ultrashort pulse laser having a pulse width of 1 ps and photon energy of 2.8 eV. The scattering configuration was $Y(X, X)\overline{Y}$. The anti-Stokes signal has been magnified by a factor of 10.

almost as strongly with A_1(LO) as with E_1(LO) phonon modes in wurtzite structure GaN.

Figure 3.9 shows the measured nonequilibrium population of the A_1(LO) phonon mode as a function of the square of excitation laser power density. The experimental data fit a straight line very well. The fact that the observed nonequilibrium phonon population increases linearly with the square of the laser power density confirms that the hot electron–hole pairs are photoexcited by the two-photon absorption process.

3.5.2 Nonequilibrium Electron Distributions and Electron–Longitudinal Optical Phonon Scattering Rates in Wurtzite GaN Studied by an Ultrashort Ultraviolet Laser [35,36]

An ultrashort ultraviolet laser system was employed to investigate the nature of nonequilibrium electron distributions as well as to directly measure

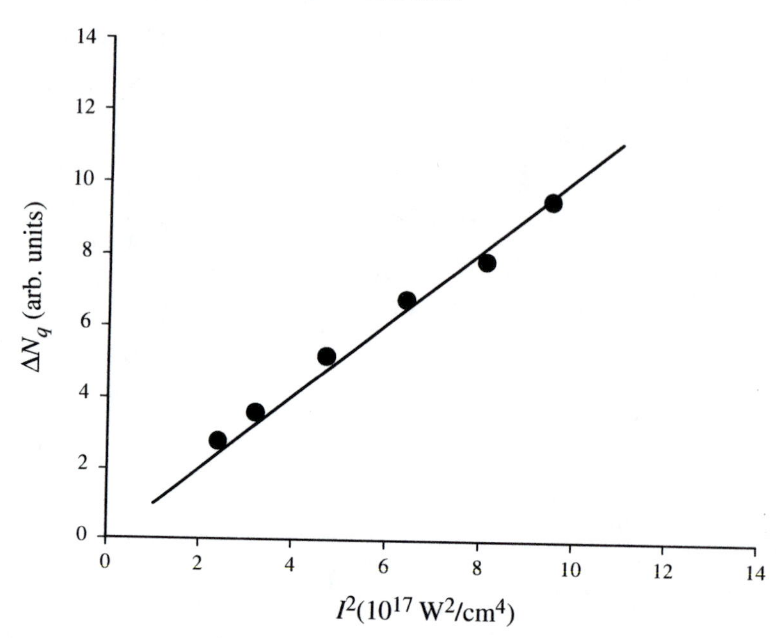

Figure 3.9 The nonequilibrium A_1(LO) phonon population as a function of the square of excitation laser intensity. The fact that the data are fit very well by a straight line suggests that the nonequilibrium phonons are generated by a nonlinear process, e.g., a two-photon absorption process.

electron–longitudinal phonon scattering rates in wurtzite GaN. The total electron–LO phonon scattering rate was determined to be $(4 \pm 0.8) \times 10^{13}$/s, which is about an order of magnitude larger than that found in GaAs. The very much larger electron–LO phonon scattering rate is mostly due to the large ionicity in wurtzite GaN. In addition, we show that for electron densities $n \geq 5 \times 10^{17}$/cm, as a result of efficient electron–electron scattering, nonequilibrium electron distributions photoexcited in wurtzite GaN can be very well described by Fermi–Dirac distribution functions with effective electron temperatures much higher than the lattice temperature.

3.5.2.1 *Samples and Experimental Technique*

The wurtzite structure GaN used in this study was described in detail in Section 3.3. The third harmonic of a cw mode-locked Ti:sapphire laser was used to photoexcite electron–hole pairs across the bandgap of GaN. For the study of nonequilibrium electron distributions, a photon energy of 4.5 eV was chosen, so that electron–hole pairs were photoexcited with an excess energy of $\cong 1$ eV. On the other hand, in the investigation of electron–LO phonon interactions, a photon energy of 5 eV was selected, so that electron–hole pairs possessed an excess energy of $\cong 1.5$ eV. The pulse width of the laser could be varied from about 50 fs to 2 ps. For investigating nonequilibrium electron distributions, the same laser pulse was used to both excite and probe nonequilibrium electron distributions, as shown in Figure 3.3; therefore, the experimental results presented here represent an average over the laser pulse duration. In the pump/probe configuration, which was used to study electron–phonon interactions, the laser beam was split into two equal but perpendicular polarized beams, as depicted in Figure 3.4. One was used to excite electron–hole pairs and the other to probe nonequilibrium LO phonon populations through Raman scattering. The density of photoexcited electron–hole pairs was determined by fitting the time-integrated luminescence spectrum and was about $n \cong 10^{16}\,\text{cm}^{-3}$ for the experiments involving the measurements of electron–LO phonon scattering rates. The sample was kept in contact with the cold fingertip of a closed-cycle refrigerator. The temperature of the sample was estimated to be 25 K. The backward-scattered Raman signal was collected and analyzed with a standard Raman setup equipped with a photomultiplier tube (for the detection of electron distributions) and a CCD (for the detection of phonons) detection system.

3.5.2.2 *Experimental Results and Discussion*

Figures 3.10a and 3.10b show two transient single-particle scattering (SPS) spectra of a wurtzite structure GaN sample taken at $T \cong 25$ K by an ultrafast laser system having a pulse width of about 600 fs and for electron densities $n \cong 3 \times 10^{18}\,\text{cm}^{-3}$ and $5 \times 10^{17}\,\text{cm}^{-3}$, respectively. Here the scattering geometry $Z(X, Y)\overline{Z}$ was used, which ensured that the SPS signal came from scattering of light associated with spin-density fluctuations. One intriguing aspect of using a Raman scattering technique to probe carrier distribution functions needs to be emphasized here: the Raman scattering cross section is inversely proportional to the square of the effective mass; since the effective mass of electrons is usually much smaller than that of holes in semiconductors, Raman spectroscopy effectively probes electron distributions even if holes are simultaneously present.

In order to obtain better insight, we used Eq. (3.30), which we developed in Section 3.2.1, to fit the lineshapes of these SPS spectra. We note that in fitting the experimental results, we used three parameters: Γ, τ, and T_{eff} (the effective electron temperature). In addition, because the photon energy used in the experiments was sufficiently far from any relevant energy gaps in GaN, the fitting processes were not sensitive to the detailed band structure of GaN. The solid curves in Figures 3.10a and 3.10b are theoretical calculations based on Eq. (3.30) with Fermi–Dirac distribution functions and parameter sets that best fit the data: $T_{eff} = 800$ K, $\tau = 15$ fs, $\Gamma = 20$ meV for $n \cong 3 \times 10^{18}\,\text{cm}^{-3}$; and $T = 500$ K, $\tau = 20$ fs, $\Gamma = 20$ meV for $n \cong 5 \times 10^{17}\,\text{cm}^{-3}$. The effective temperature of the electrons was found to be much higher than that of the lattice, as expected. In addition, as the electron density increased,

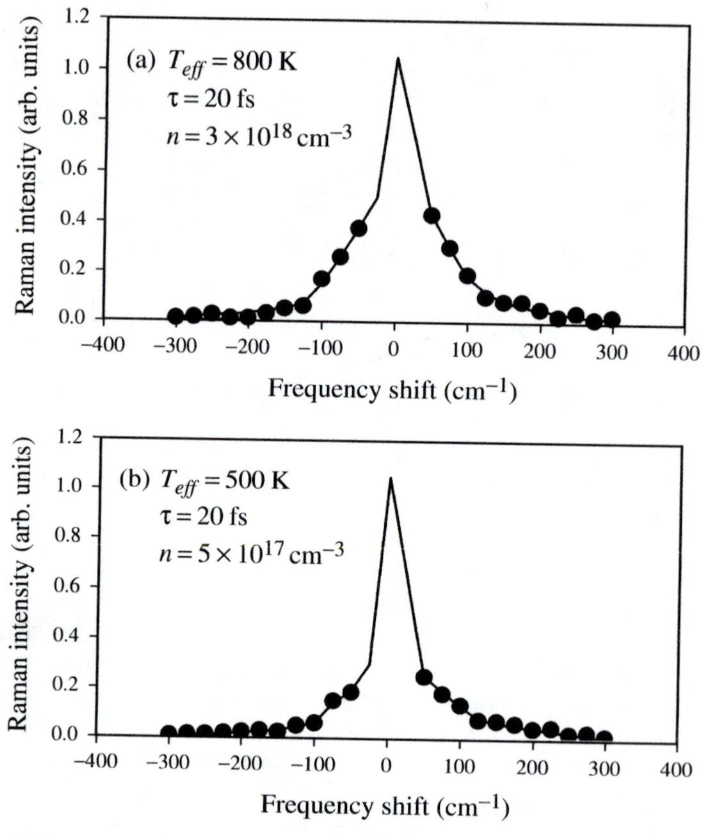

Figure 3.10 Transient single-particle scattering spectra for a wurtzite GaN sample, taken by an ultrafast laser having a pulse width of 600 fs and for electron–hole pair densities of (a) $n \cong 3 \times 10^{18}\,\text{cm}^3$ and (b) $n \cong 5 \times 10^{17}\,\text{cm}^3$, respectively. The solid circles are data. The curves are a theoretical fit based on Eq. (3.30).

the effective electron temperature increased and the electron collision time became shorter. We also noticed that the value of the damping constant $\Gamma(= 20\,\text{meV})$ involved in the Raman scattering process was very close to that $(\cong 13\,\text{meV})$ obtained from an analysis of a resonance Raman profile under equilibrium conditions [37]. From the quality of the fit, we concluded that for electron densities $n \geq 5 \times 10^{17}\,\text{cm}^{-3}$, as a result of efficient momentum randomization, electron distribution functions in wurtzite GaN can be very well described by Fermi–Dirac distributions with the effective temperature of the electrons much higher than that of the lattice.

Figure 3.11 shows time-resolved nonequilibrium populations of the A_1(LO) phonon mode in wurtzite GaN taken at $T \cong 25$ K with an ultrafast laser having a pulse width of 100 fs. The rise and fall of the signal reflect

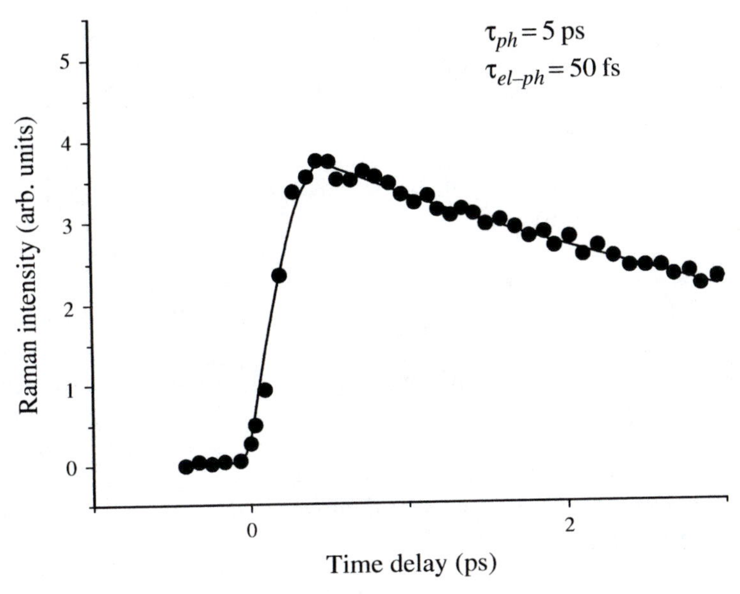

Figure 3.11 Nonequilibrium populations of the A_1(LO) phonon mode as a function of time delay. The solid circles are data. The curve is based on a simple cascade model with electron–phonon scattering rate as the only adjustable parameter.

the generation and decay, respectively, of the nonequilibrium A_1(LO) phonon mode.

We used an electron cascade model to fit the experimental data in Figure 3.11 [38]. The use of this model was appropriate under our experimental conditions because (1) the electron–hole pair density excited was low ($n \cong 10^{16}\,\text{cm}^{-3}$), and (2) the electrons did not have sufficient excess energy to scatter to other satellite valleys in GaN [39].

In the electron cascade model, nonequilibrium the LO phonon occupation number $n_{ph}(t)$ is given by the following expression:

$$\frac{dn_{ph}(t)}{dt} = G(t) + \frac{n_{ph}(t)}{\tau_{ph}}, \tag{3.41}$$

where $G(t)$ is the LO phonon generation rate by the excitation pulse laser, and τ_{ph} is the LO phonon lifetime.

For our experimental conditions, the LO phonon generation rate was given by

$$G(t) = \frac{f(t)}{\tau_{el-ph}}, \tag{3.42}$$

where $f(t) = 1$ for $0 \leq t \leq m\tau_{el-ph}$, and $f(t) = 0$ for $t \leq 0$ or $t \geq m\tau_{el-ph}$; τ_{el-ph} is the electron–LO phonon scattering rate; and m is an integer determined by the photon energy, LO phonon energy, bandgap, and bandstructure of wurtzite GaN.

For wurtzite GaN [40,41], if we take $m_e^* = 0.2m_e$, $m_h^* = 0.8m_e$, $\hbar\omega_i = 5\,\text{eV}$, $E_g = 3.49\,\text{eV}$, the index of refraction $n = 2.8$, and $\hbar\omega_{LO} = 0.09263\,\text{eV}$ (corresponding to A_1(LO) phonon mode energy), then the phonon wavevector probed in our backscattered Raman experiments is $q = 2nk_i = 1.42 \times 10^6\,\text{cm}^{-1}$, and electron excess energy is given by $\Delta E_e = (\hbar\omega_i - E_g) \times [m_h^*/(m_e^* + m_h^*)] = 1.21\,\text{eV}$. This means that the energetic electrons are capable of emitting 13 LO phonons during their thermalization to the bottom of the conduction band; however, because of the conservation of both energy and momentum for the electron–LO phonon interaction process, there exists a range of LO phonon wavevectors that electrons can emit. As depicted in Figure 3.12, for an electron with wavevector $\vec{k}_e$ and excess energy ΔE_e, the minimum and maximum LO phonon wavevectors it can interact with are given by

$$k_{min} = \frac{\sqrt{2m_e^*}}{\hbar}\left(\sqrt{\Delta E_e} - \sqrt{\Delta E_e - \hbar\omega_{LO}}\right) \tag{3.43}$$

and

$$k_{max} = \frac{\sqrt{2m_e^*}}{\hbar}\left(\sqrt{\Delta E_e} + \sqrt{\Delta E_e - \hbar\omega_{LO}}\right). \tag{3.44}$$

Because of the nature of the energy–wavevector relationship of the electron, the lower the electron energy, the larger the k_{min} and the smaller the k_{max}. Therefore, at some electron energy, the k_{min} of an LO phonon will be larger than the wavevector probed by our Raman scattering experiments, $q = 1.42 \times 10^6\,\text{cm}^{-1}$. When that happens during the relaxation process, the energetic electrons can no longer emit LO phonons with wavevectors detectable in our Raman scattering experiments. By taking this factor into consideration, we found that although in principle the energetic electrons were capable of emitting 13 LO phonons during their thermalization to the bottom of the conduction band, only 7 of them could be detected in our Raman experiments because of the conservation of both energy and momentum. In other words, $m = 7$ under our experimental conditions.

Therefore, there are two adjustable fitting parameters in this electron cascade model—phonon lifetime (τ_{ph}) and electron–phonon scattering time (τ_{el-ph}). Since the LO phonon lifetime τ_{ph} can be independently measured from the decaying part of the Raman signal, the electron–LO phonon scattering time τ_{el-ph} was used as the only adjustable parameter in the fitting process.

We found that $\tau_{el-ph} = 50 \pm 10$ fs, which gave rise to an electron–LO phonon scattering rate of $\Gamma_{el-ph} = 1/\tau_{el-ph} = (2.2 \pm 0.4) \times 10^{13}\text{s}^{-1}$, provided the best fit to the experimental data in Figure 3.11.

Similar experiments for the E_1(LO) phonon mode were also carried out. Our experimental results in this case showed that the electron–LO phonon scattering rate for the E_1(LO) phonon mode was given by $\tau_{el-ph} = (2.4 \pm 0.4) \times 10^{13}\text{s}^{-1}$, which was very close to the value found for the A_1(LO) phonon mode.

Consequently, the total electron–LO phonon scattering rate was given by $\Gamma_{total} = (4.6 \pm 0.8) \times 10^{13}\ \text{s}^{-1}$. Since the electron–LO phonon scattering rate in GaAs is about $5 \times 10^{12}\ \text{s}^{-1}$, the observed total electron–LO phonon scattering rate in wurtzite GaN was almost one order of magnitude larger than that in GaAs. The much larger LO–TO phonon energy splitting in wurtzite GaN provides a clue to this mystery. We attribute this enormous increase in electron–LO phonon scattering rate in GaN to its much larger ionicity. In general, the strength of electron–LO phonon coupling is set by the lattice–dipole interactions and is expressed by [42]

$$\frac{1}{\gamma} = \omega_{LO}^2 \left[\frac{1}{\varepsilon_\infty} - \frac{1}{\varepsilon_0} \right] = \left[\frac{\omega_{LO}^2 - \omega_{TO}^2}{\varepsilon_\infty} \right]. \tag{3.45}$$

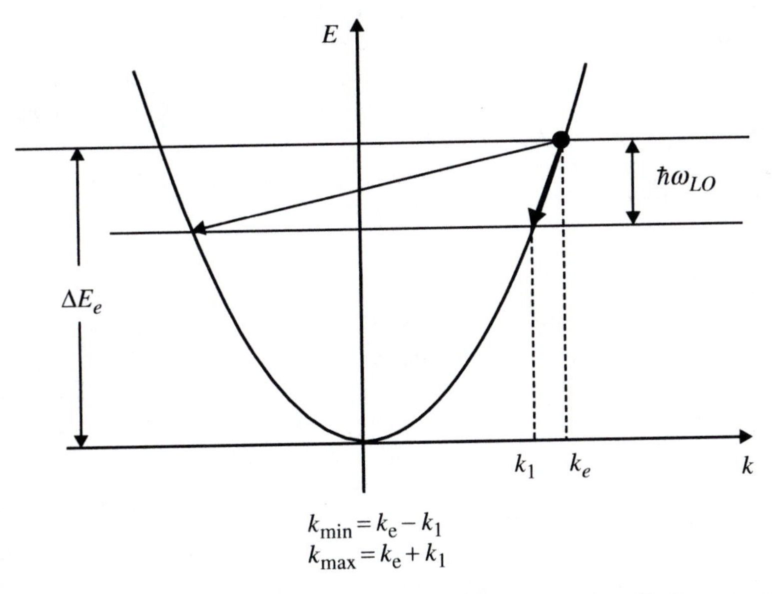

Figure 3.12 A diagram demonstrating that because of the conservation of both energy and momentum the minimum and maximum phonon wavevectors with which an electron can interact during its thermalization toward the bottom of the conduction band are given by Eqs. (3.43) and (3.44), respectively.

This splitting is directly proportional to the lattice polarization and is a measure of the effective charge. In GaAs, the LO–TO energy splitting is about 3 meV, whereas it is about 25 meV in wurtzite GaN; however, in the two-mode system of hexagonal GaN, this energy must be split between the two modes to compute the contribution of each to the dielectric function. This splitting together with the much smaller dielectric constant of GaN, leads to an expected increase of a factor of 9 in the electron–LO phonon scattering strength, which is quite close to that observed experimentally. Phillips estimates that the effective charge of GaN is about twice that of GaAs, and the dielectric constant is almost half [43]. These two factors together would suggest a factor of 8 increase in the electron–LO phonon scattering strength. This result provides a consistent interpretation for the increase in electron–LO phonon scattering rate as being due to the ionicity of wurtzite GaN relative to GaAs.

3.5.3 Anharmonic Decay of the Longitudinal Optical Phonons in Wurtzite GaN Studied by Subpicosecond Time-Resolved Raman Spectroscopy [44]

Decay of the longitudinal optical phonons in wurtzite GaN has been studied by subpicosecond time-resolved Raman spectroscopy. It is commonly believed that LO phonons in other semiconductors use the 2LA decay channel [45]. In contrast, our experimental results showed that among the various possible decay channels, the LO phonons in wurtzite GaN decay primarily into a large-wavevector TO and a large-wavevector LA or TA phonon. These experimental results are consistent with recent theoretical calculations of the phonon dispersion curves in wurtzite structure GaN.

3.5.3.1 Samples and Experimental Technique

The GaN sample used in this study was described in details in Section 3.3. In our time-resolved Raman experiments, ultrashort ultraviolet laser pulses were generated by a frequency-quadrupled cw mode-locked Ti: sapphire laser system, as shown in Figure 3.4. The laser pulse width could be varied from 50 fs to about 2 ps. The photon energy was chosen to be about 5 eV, so that electron–hole pairs could be excited with an excess energy of more than 1 eV. In the investigation of phonon–phonon interactions in GaN, a pump/probe configuration was employed. The laser beam was split into two equal intensity but perpendicularly polarized beams. One was used to excite electron–hole pairs and the other to probe nonequilibrium LO phonon populations through Raman scattering. The density of the photoexcited electron–hole pairs was determined by fitting the time-integrated luminescence spectrum, and was about $n \cong 5 \times 10^{16}\,\mathrm{cm}^{-3}$. The sample was kept in contact with the cold fingertip of a closed-cycle refrigerator. The temperature of the laser-irradiated sample was estimated to be 25 K. The scattering geometry was taken to be $Z(X, X)\overline{Z}$. The backward-scattered Raman signal was collected and analyzed with a standard Raman setup equipped with a CCD detection system.

3.5.3.2 Experimental Results and Discussion

Figures 3.13a and 3.13b show a typical set of transient polarized Stokes and anti-Stokes Raman spectra for a GaN sample taken at $T \cong 25$ K with an ultrafast laser having a 2-ps pulse width and a photon energy of 5 eV, in $Z(X, X)\overline{Z}$ scattering geometry. These spectra are very similar to Figures 3.7a and 3.7b of Section 3.5.1.2., where a photon energy below the bandgap of GaN was used for Raman scattering experiments. We demonstrated here again that a nonequilibrium LO phonon mode can be generated with an above-bandgap laser excitation of hot electron–hole pair density. By arguments similar to those made in Section 3.5.1.2., we concluded that the broad structure around -741 cm^{-1} in the anti-Stokes Raman spectrum of Figure 3.13b came from scattering of light by a nonequilibrium A_1(LO) phonon mode in wurtzite GaN.

In order to obtain better insight into the generation of nonequilibrium optical phonons in semiconductors, we used a simple cascade, two-parabolic-band model [46] to calculate the nonequilibrium phonon occupation number as a function of phonon wavevector under our experimental conditions. Although both electrons and holes are excited, we assume that the electron mass is much smaller than the hole mass, so that most of the excess energy of the photon goes to the electron. For simplicity, we neglect the nonequilibrium phonons emitted by the relaxation of the holes. With these assumptions we have only two populations to consider: $f_{\vec{k}}(t)$, the occupation number of electrons with wavevector $\vec{k}$, and $n_{\vec{q}}(t)$, the occupation number of LO phonons with wavevector $\vec{q}$. The two distributions can be calculated as a function of time by solving two Boltzmann equations:

$$\frac{df_{\vec{k}}}{dt} = G_{\vec{k}} + \left(\frac{\partial f_{\vec{k}}}{\partial t}\right)_{LO} - \frac{f_{\vec{k}}}{\tau_{\vec{k}}}; \tag{3.46}$$

$$\frac{dn_{\vec{q}}}{dt} = \left(\frac{\partial n_{\vec{q}}}{\partial t}\right)_e - \frac{n_{\vec{q}}}{\tau_{LO}}. \tag{3.47}$$

In Eq. (3.46), $G_{\vec{k}}$ is the rate of generation of electrons due to optical excitation; $(\partial f_{\vec{k}}/\partial t)_{LO}$ is the rate of change in electron occupation number due to scattering with LO phonons; and $\tau_{\vec{k}}$ is the electron lifetime due to any other decay processes, such as radiative recombination. In Eq. (3.47), $(\partial n_{\vec{q}}/\partial t)_e$ is the rate of change of the LO phonon occupation number due to

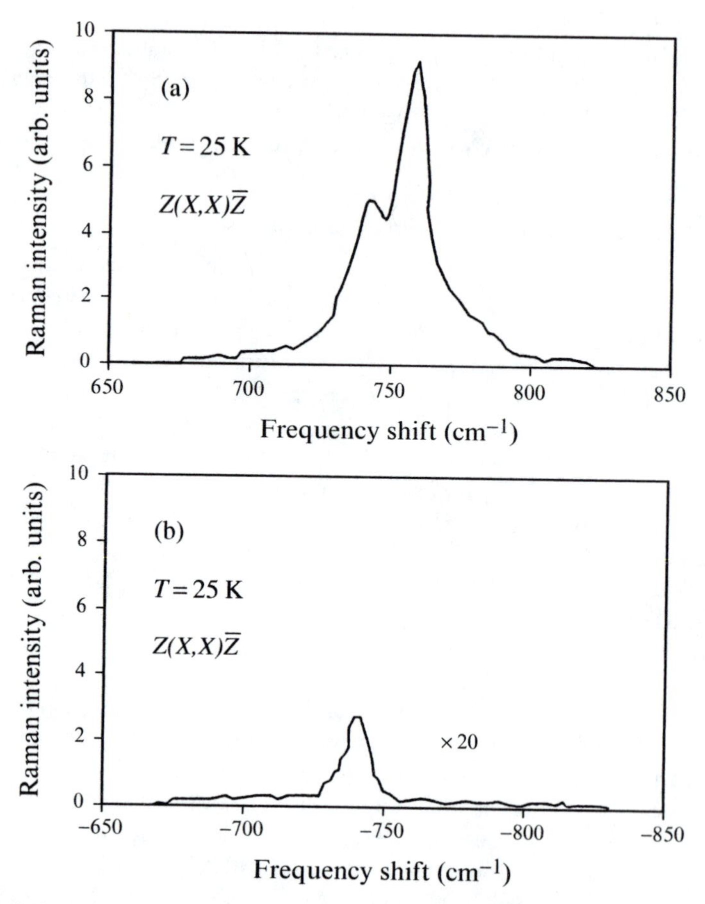

Figure 3.13 Transient (a) Stokes and (b) anti-Stokes Raman spectra for a wurtzite GaN sample taken by an ultrashort pulse laser having a pulse width of 2 ps and photon energy of 5.0 eV. The scattering configuration was $Z(X, X)\overline{Z}$. The anti-Stokes signal has been magnified by a factor of 20.

its interaction with the electron; τ_{LO} is the lifetime of LO phonons due to their decay into lower-energy phonons. For simplicity, τ_{LO} was assumed to be independent of $\vec{q}$. The rates $(\partial f_{\vec{k}}/\partial t)_{LO}$ and $(\partial n_{\vec{q}}/\partial t)_e$ can be evaluated from time-dependent perturbation theory as follows.

It has been shown that the electron–LO phonon interaction in polar semiconductors such as GaAs, GaN is dominated by the Fröhlich interaction $H_{ele-ph} = -e\varphi(\vec{r}_i)$. If we represent the matrix element of H_{ele-ph} for scattering an electron in state $\vec{k}$ to $\vec{k}+\vec{q}$ (or $\vec{k}-\vec{q}$) with absorption (or emission) of a LO phonon with wavevector $\vec{q}$ as

$$M_{\vec{k},\vec{k}\pm\vec{q}} = \left|\left\langle \vec{k}\,\middle|\,H_{ele-ph}\,\middle|\,\vec{k}\pm\vec{q}\right\rangle\right|^2, \tag{3.48}$$

then using Fermi's golden rule, we can write $(\partial f_{\vec{k}}/\partial t)_{LO}$ as

$$\begin{aligned}\left(\frac{\partial f_{\vec{k}}}{\partial t}\right)_{LO} = \frac{2\pi}{\hbar}\sum_{\vec{q}}\Big[&-M_{\vec{k},\vec{k}-\vec{q}}(n_{\vec{q}}+1)f_{\vec{k}}(1-f_{\vec{k}-\vec{q}})\\ &\times\delta(E_{\vec{k}}-E_{\vec{k}-\vec{q}}-\hbar\omega_{LO})-M_{\vec{k},\vec{k}+\vec{q}}n_{\vec{q}}f_{\vec{k}}(1-f_{\vec{k}+\vec{q}})\\ &\times\delta(E_{\vec{k}}-E_{\vec{k}+\vec{q}}+\hbar\omega_{LO})+M_{\vec{K}+\vec{q},\vec{k}}(n_{\vec{q}}+1)(1-f_{\vec{k}})\\ &\times f_{\vec{k}+\vec{q}}\delta(E_{\vec{k}+\vec{q}}-E_{\vec{k}}-\hbar\omega_{LO})+M_{\vec{k}-\vec{q},\vec{k}}n_{\vec{q}}(1-f_{\vec{k}})\\ &\times f_{\vec{k}+\vec{q}}\delta(E_{\vec{k}-\vec{q}}-E_{\vec{k}}+\hbar\omega_{LO})\Big]. \end{aligned} \tag{3.49}$$

The first two terms in Eq. (3.47) arise from the decay of the electron in state $\vec{k}$ with emission or absorption of a LO phonon. The last two terms are due to scattering of electrons in states $\vec{k}\pm\vec{q}$ into the state $\vec{k}$ through emission or absorption of an LO phonon.

Similarly, we find that

$$\begin{aligned}\left(\frac{\partial n_{\vec{q}}}{\partial t}\right)_{e} = \frac{2\pi}{\hbar}\sum_{\vec{k}}\Big[&M_{\vec{k},\vec{k}-\vec{q}}(n_{\vec{q}}+1)f_{\vec{k}}(1-f_{\vec{k}-\vec{q}})\\ &\times\delta(E_{\vec{k}}-E_{\vec{k}-\vec{q}}-\hbar\omega_{LO})-M_{\vec{k},\vec{k}+\vec{q}}n_{\vec{q}}f_{\vec{k}}(1-f_{\vec{k}+\vec{q}})\\ &\times\delta(E_{\vec{k}}-E_{\vec{k}+\vec{q}}+\hbar\omega_{LO})\Big], \end{aligned} \tag{3.50}$$

where $E_{\vec{k}}=\hbar^2k^2/2m_e^*$; $M_{\vec{k},\vec{k}\pm\vec{q}}=((2\pi\hbar^2e)/(Vm_e^*q^2))[(m_e^*e\omega_{LO}/\hbar)\times(1/\varepsilon_\infty-1/\varepsilon_0)]$, with ε_∞, ε_0 being high-frequency and static dielectric constants.

By assuming that the electron distribution is nondegenerate and that $(\partial f/\partial E)\hbar\omega_{LO}\ll f(E)$ [47], we have

$$\begin{aligned}\frac{df(E,t)}{dt} =& \frac{e}{(2m_e^*E)^{1/2}}\left(\frac{m_e^*e\omega_{LO}}{\hbar}\right)\left(\frac{1}{\varepsilon_\infty}-\frac{1}{\varepsilon_0}\right)\\ &\times\left[f(E^+,t)\ln\left[\frac{(E^+)^{1/2}+E^{1/2}}{(E^+)^{1/2}-E^{1/2}}\right]-f(E,t)\right.\\ &\left.\times\ln\left[\frac{E^{1/2}+(E^-)^{1/2}}{E^{1/2}-(E^-)^{1/2}}\right]\right]+G(E,t)-\frac{f(E,t)}{\tau(E)} \end{aligned} \tag{3.51}$$

and

$$\frac{dn_{\vec{q}}}{dt} = -\frac{n_{\vec{q}}}{\tau_{LO}}+\frac{2m_e^*e}{\hbar^3q^3}\left(\frac{m_e^*e\omega_{LO}}{\hbar}\right)\left(\frac{1}{\varepsilon_\infty}-\frac{1}{\varepsilon_0}\right)\int_{E_{\max}}^{\infty}f(E,t)\,dE, \tag{3.52}$$

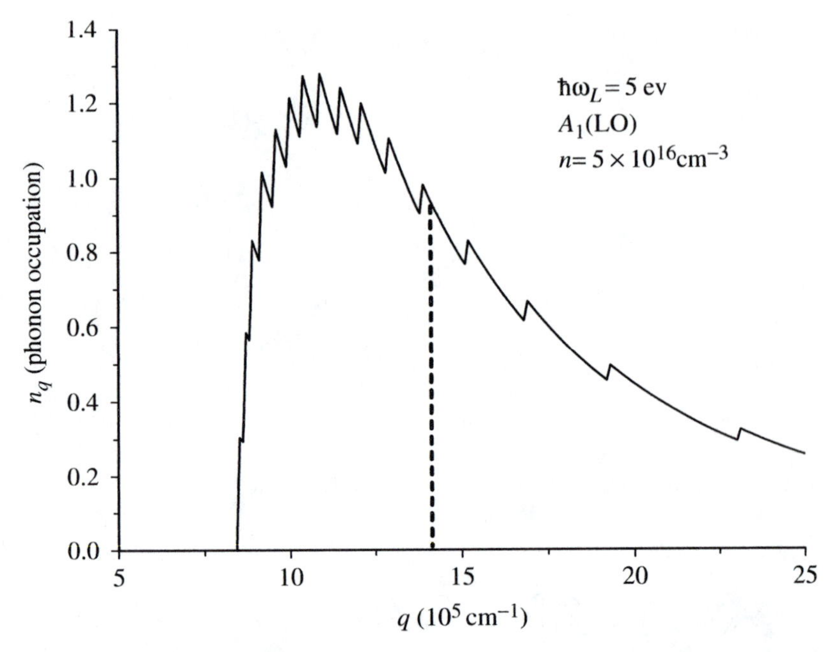

Figure 3.14 The calculated A_1(LO) phonon occupation number as a function of phonon wavevector for a GaN sample. The stimulated ultrafast laser had a pulse width of 300 fs, and the phonon occupation was monitored at a time delay of 300 fs. The excitation photon energy was 5 eV. The photoexcited electron–hole pair had a density of $n \cong 5 \times 10^{16}$ cm^{-3}. The dashed vertical line indicates the value of the wavevector probed by the Raman scattering experiments.

where $E_{max} = (q'/2 - (\hbar\omega_{LO}/2q'))^2 + \hbar\omega_{LO}$; $E_{min} = (q'/2 - (\hbar\omega_{LO}/2q'))^2$; $q' = (\hbar q/\sqrt{2m_e^*})$; $E^+ = E + \hbar\omega_{LO}$; $E^- = E - \hbar\omega_{LO}$.

Equations (3.51) and (3.52), can be solved very easily to obtain both electron distributions and nonequilibrium phonon distributions.

Figure 3.14 shows the calculated A_1(LO) phonon occupation number as a function of phonon wavevector for a GaN sample taken with an ultrafast laser having a pulse width of 300 fs and at a time delay of 300 fs after the peak of the excitation laser pulse. The excitation photon energy was 5 eV. The electron–hole pair density was 5×10^{16} cm^3. The dashed vertical line indicates the wavevector probed by Raman spectroscopy. The sharp cutoff in phonon occupation number at around 8×10^5 cm^{-1} was due to the conservation of both energy and momentum during the electron–phonon interaction processes.

Figure 3.15 shows several time-resolved nonequilibrium populations of the A_1(LO) phonon mode in wurtzite GaN taken at a lattice temperature of $T = 10$, 150, and 300 K, respectively with an ultrafast laser having a pulse width of 300 fs and a photon energy of 5 eV. The population relaxation time of A_1(LO) phonons as a function of the lattice temperature ranging from 10

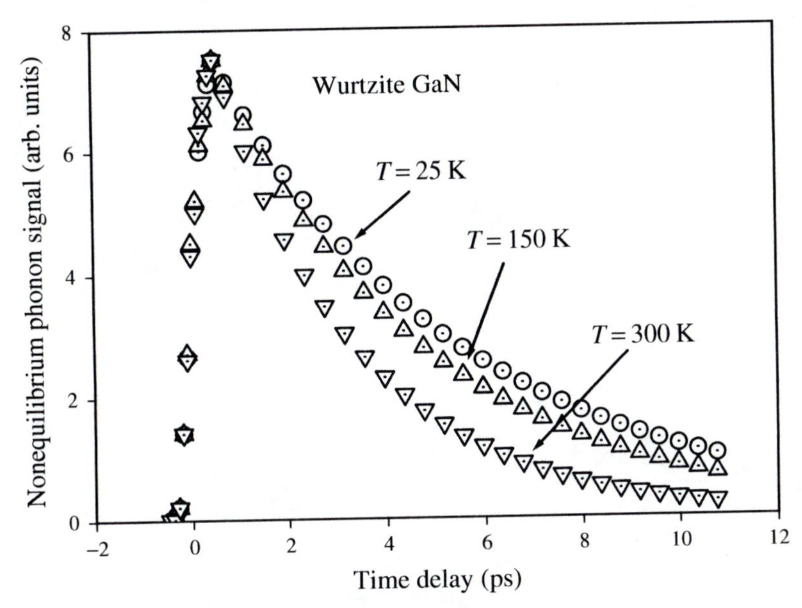

Figure 3.15 A typical set of nonequilibrium populations of the A_1(LO) phonon mode as a function of time delay, taken with an ultrafast laser having a pulse width of 300 fs and photon energy of 5 eV, at lattice temperatures of $T \cong 25, 150, 300$ K, respectively.

to 300 K can be obtained from the decaying parts of the Raman signals, as shown in Figure 3.16.

Klemens [48] and Ridley [49] used perturbation theory to show that the temperature-dependence part of the decay of the LO phonon population $n_{ph}(\omega, T)$ in semiconductors can be expressed as

$$\frac{dn_{ph}(\omega, T)}{dt} = -n_{ph}(\omega, T)\left[\frac{1 + n_{ph1}(\omega_1, T) + n_{ph2}(\omega_2, T)}{\tau_0}\right], \quad (3.53)$$

where τ_0 is the decay time of an LO phonon at $T = 0$ K (in this case, τ_0 was measured to be about 5 ps); $\omega = \omega_1 + \omega_2$; and $n_{ph1}(\omega_1, T) = 1/(e^{\hbar\omega_1/k_BT} - 1)$, $n_{ph2}(\omega_2, T) = 1/(e^{\hbar\omega_2/k_BT} - 1)$ are the occupation numbers of decayed lower-energy phonons at the lattice temperature T, and k_B is the Boltzmann' constant. Equation (3.53) says that the temperature dependence of LO phonon lifetime is given by

$$\frac{1}{\tau(T)} = \frac{1}{\tau_0}[1 + n_{ph1}(\omega_1, T) + n_{ph2}(\omega_2, T)]. \quad (3.54)$$

We noted that the population relaxation time for LO phonons in GaAs is about 8 ps at $T = 10$ K; on the other hand, that of LO phonons in wurtzite

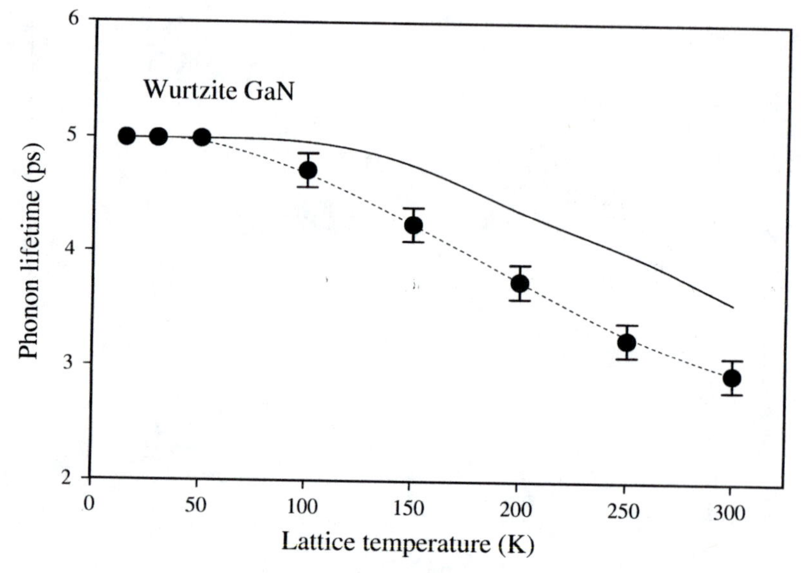

Figure 3.16 The temperature dependence of the population relaxation time of the A_1(LO) phonon mode for a GaN sample. The solid circles are experimental data. The solid curve is for the decay channel $\omega_{LO} \rightarrow 2\omega_{LA(TA)}$ with $\hbar\omega_{LA(TA)} = 370\,\text{cm}^{-1}$; the dashed curve is for the decay channel $\omega_{LO} \rightarrow \omega_{TO} + \omega_{LA(TA)}$, as predicted by Eq. (3.54). The energies of the TO and LA or TA phonons involved in the fitting process are $\omega_{TO} = 540\,\text{cm}^{-1}$ and $\omega_{LA(TA)} = 201\,\text{cm}^{-1}$, respectively.

GaN is about 5 ps at $T = 10$ K. This decrease in LO phonon lifetime may be due either to defects and dislocations present in GaN sample or to a larger density of state for the decay of LO phonons in GaN. Further investigation is now under way to clarify this point.

There are various possible channels for the decay of zone-center LO phonons in wurtzite GaN. First, from the experimental data in Figure 3.16, the decay of zone-center LO phonons into a small-wavevector LO and a small-wavevector LA or TA phonon does not seem likely, because this channel predicts a much larger temperature dependence of the lifetime of LO phonons than the experimental data indicate. The solid curve in Figure 3.16 represents the decay channel $\omega_{LO} \rightarrow 2\omega_{LA(TA)}$, with $\hbar\omega_{LA(TA)} = 370\,\text{cm}^{-1}$, which is half the energy of zone-center LO phonons, as predicted by Eq. (3.54). We found that, within our experimental uncertainty, the zone-center LO phonons cannot decay into large-wavevector, equal-energy but opposite-momentum LA or TA phonons, which is usually assumed in the decay of LO phonons in other III–IV compound semiconductors. On the other hand, we found that the decay channel of zone-center LO phonons in wurtzite GaN into a large-wavevector TO phonon (assuming that the TO phonon dispersion curve is relatively flat across the Brillouin zone) and a large-wavevector LA or TA phonon fits our experimental data very well. The dotted curve in Figure 3.16 corresponds to such a decay channel, $\omega_{LO} \rightarrow \omega_{TO} + \omega_{LA(TA)}$, as predicted by Eq. (3.54). The energies of the TO and LA or TA phonons involved in the fitting process are $\omega_{TO} = 540\,\text{cm}^{-1}$ and $\omega_{LA(TA)} = 201\,\text{cm}^{-1}$, respectively.

Recently, Azuhata et al. calculated a phonon dispersion curve for wurtzite GaN by using a rigid-ion model [50]. Their calculations indicate that the zone-center LO phonons in wurtzite GaN cannot decay into two large-wavevector lower-branch LA or TA phonons, because $2\omega_{LA(TA)} < \omega_{LO}$, whereas it is possible for a zone-center LO phonon to decay into a large-wavevector TO and a large-wavevector LA or TA phonon. These theoretical predictions agree very well with our experimental analysis.

3.6 ELECTRONIC RAMAN SCATTERING FROM Mg-DOPED WURTZITE GaN

p-Type doping in nitride-based wide bandgap semiconductors has presented a big challenge to researchers. The recently successful *p*-type doping in metalorganic vapor-phase epitaxy GaN using Mg has allowed the realization of blue LEDs [51,52]. Although much work has been devoted to the study of the properties of Mg impurities in GaN samples, the electronic and optical properties of Mg impurities remain controversial and inconclusive. Electronic Raman scattering experiments were carried out on both MBE- and MOCVD-grown Mg-doped wurtzite GaN samples. Aside from the expected Raman lines, a broad structure (FWHM $\cong 15\,\text{cm}^{-1}$) observed for the first time at around $841\,\text{cm}^{-1}$ was attributed to the electronic Raman scattering from neutral Mg impurities in Mg-doped GaN. From the analysis of the temperature dependence of this electronic Raman scattering signal, the binding energy of the Mg impurities in wurtzite GaN was found to be $E_b \cong 172 \pm 20\,\text{meV}$. These experimental results demonstrated that the energy between the ground and first excited states of Mg impurities in wurtzite GaN is about three-fifths of its binding energy.

3.6.1 Samples and Experimental Technique

The Mg-doped wurtzite GaN samples studied in this work were grown by MBE and MOCVD techniques on (0001)-oriented sapphire substrates with about 1 μm-thick AlN buffer layers. The GaN layers were about 2 μm thick. The *z*-axis of this wurtzite structure GaN is perpendicular to the sapphire substrate plane. The Mg concentrations were $N_a = 10^{19}\,\text{cm}^3$, $5 \times 10^{18}\,\text{cm}^3$ for MBE- and MOCVD-grown samples, respectively, as determined by SIMS experiments.

Raman scattering experiments were carried out by using the second harmonic of a mode-locked cw YAG laser operating at a repetition rate of about

76 MHz. The laser pulse width was about 80 ps, and photon energy was about 2.34 eV. This excitation laser, which had an average laser power of about 200 mW, was focused into the samples with a spot size of about 100 μm. Raman scattering measurements were made in $Z(X, X)\overline{Z}$ and $Y(X, X)\overline{Y}$ scattering geometry, where $X = (100)$, $Y = (010)$, and $Z = (001)$. The Raman scattering signal was detected and analyzed by a standard Raman system equipped with a CCD as a multichannel detector.

3.6.2 Experimental Results and Analysis

Figures 3.17 and 3.18 show two typical Raman scattering spectra for an MBE-grown, Mg-doped GaN sample and an MOCVD-grown, Mg-doped GaN sample, respectively, taken at $T = 25$ K in $Z(X, X)\overline{Z}$ configuration. These two spectra are very similar. They are normalized to their $A_1(LO)$ phonon mode intensities and are shown on the same scale. As a result, we found that the intensity of 841 cm^{-1} corresponds reasonably well to the Mg concentration. The structure centered around 757 cm^{-1} comes from scattering of light by the E_g phonon mode of sapphire [53,54]; the shoulder close to 741 cm^{-1} belongs to the $A_1(LO)$ phonon mode of GaN. In addition

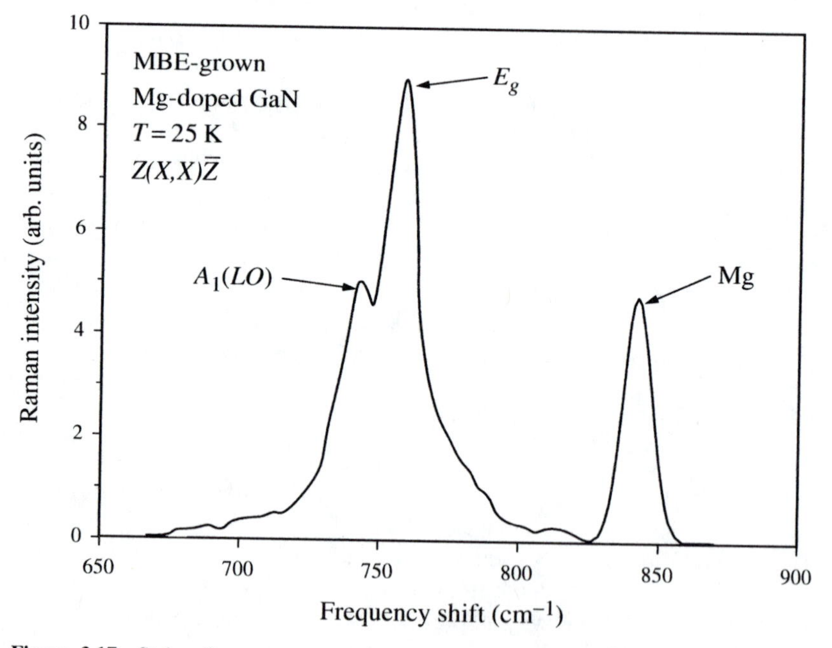

Figure 3.17 Stokes Raman spectrum for an MBE-grown Mg-doped GaN sample. The structure around 841 cm^{-1} is attributed to the electronic Raman scattering from neutral Mg impurities in Mg-doped GaN.

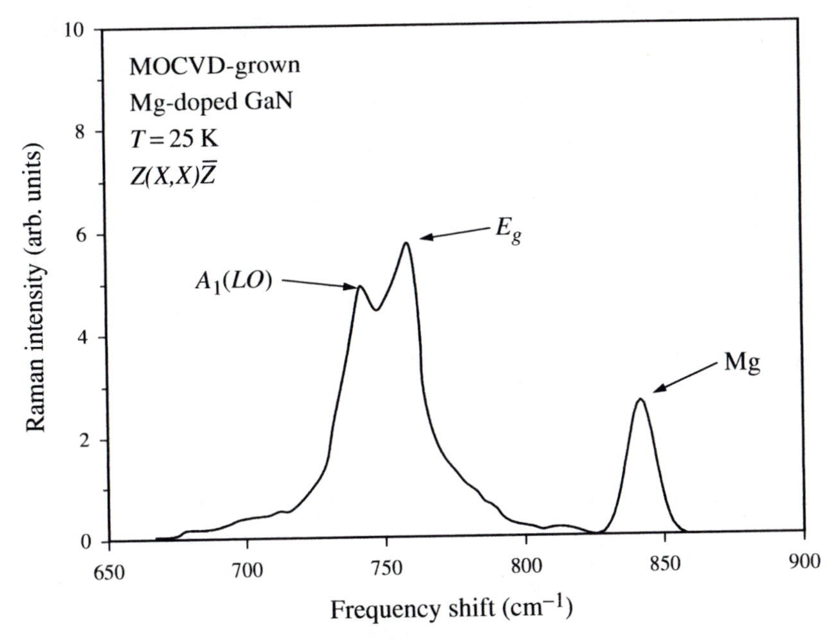

Figure 3.18 Stokes Raman spectrum for an MOCVD-grown Mg-doped GaN sample. The structure around 841 cm^{-1} is attributed to the electronic Raman scattering from neutral Mg impurities in Mg-doped GaN.

to these expected phonon modes, a well-defined structure showed up at about 841 cm^{-1} ($\cong$105 meV). This additional structure was attributed to electronic Raman scattering (ERS) from neutral Mg impurities in GaN, for the following reasons:

(1) It is well known that photons can excite holes from the ground state to excited states in neutral acceptors in semiconductors—the so-called electronic Raman scattering [55].

(2) This structure was observed for both MBE- and MOCVD-grown Mg-doped GaN samples but not for either MBE- or MOCVD-grown undoped GaN samples.

(3) The intensity of the structure was found to decrease with increasing lattice temperature, which is consistent with our assignment attribution of the structure to electronic Raman scattering from neutral Mg impurities and rules out any possible assignments of either overtone or combination modes, whose signals would be expected to increase with temperature.

(4) The observed signal correlated quite well with the Mg concentration, as indicated by Figures 3.17 and 3.18.

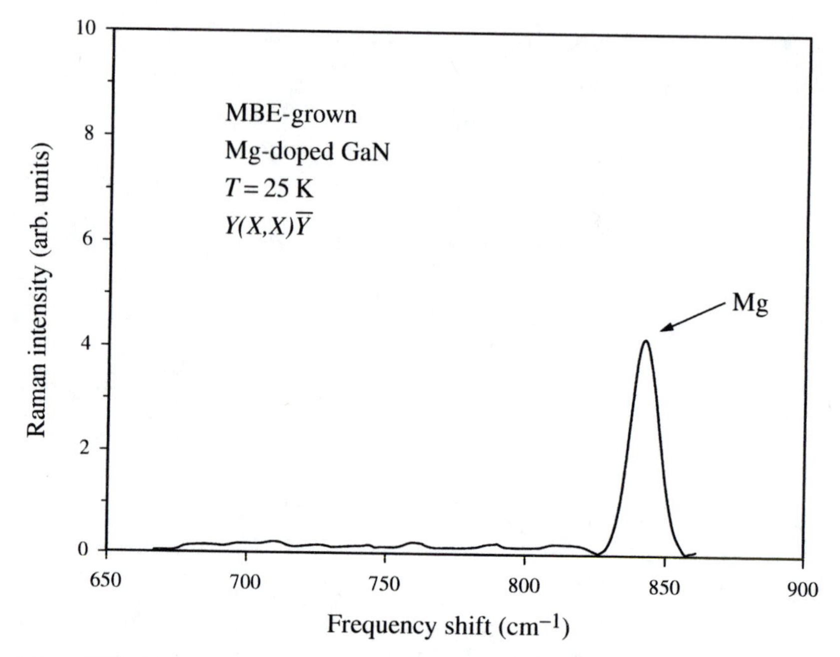

Figure 3.19 A Raman spectrum for an MBE-grown Mg-doped GaN sample, taken under the same experimental conditions as Figure 3.17 but at scattering configuration $Y(X, X)\overline{Y}$.

Figure 3.19 shows a Raman spectrum for an MBE-grown Mg-doped GaN sample, taken under the same experimental conditions as in Figure 3.17 but in $Y(X, X)\overline{Y}$ scattering configuration. No phonon modes associated with either GaN or sapphire were observed in $Y(X, X)\overline{Y}$ configuration, as expected. The ERS signal associated with Mg impurities was also observed in $Y(X, X)\overline{Y}$, with its intensity comparable to that observed in $Z(X, X)\overline{Z}$ configuration.

Since ERS occurs only when the acceptor is neutral (and therefore holes are available in impurities for ERS by photons), our observed ERS signal associated with Mg impurities can serve as a measure of the number of neutral Mg impurities in the samples. Figure 3.20 shows the measured fraction of the hole population on Mg acceptors as a function of lattice temperatures ranging from 300 K to 800 K for an MBE-grown Mg-doped GaN sample. The hole population was found to decrease by about a factor of 2 as the lattice temperature increased from 300 K to 800 K. These experimental results can be used to obtain the binding energy of Mg impurities in wurtzite GaN as follows.

The fraction of the hole population in Mg impurities in GaN is given by [56]

$$\frac{p_a}{N_a} = \frac{1}{1 + \frac{1}{2}e^{[\mu(T)-\varepsilon_a]/k_B T}}, \tag{3.55}$$

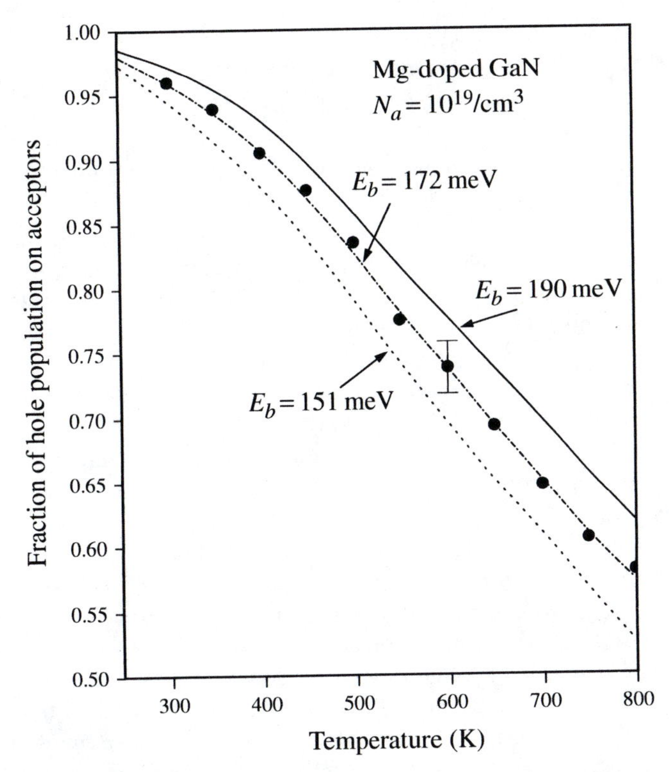

Figure 3.20 Fraction of the hole population on Mg impurities in MBE-grown Mg-doped GaN as a function of the lattice temperature. The solid circles represent experimental data. The three curves are theoretical predictions based on Eq. (3.55) with fitting parameter $E_b = 151$, 172, and 190 meV, respectively.

where p_a is the number of neutral Mg acceptors per unit volume, and N_a is the total number of Mg impurities per unit volume; $\mu(T)$ is the chemical potential; ε_a is the energy of the acceptor; and k_B is the Boltzmann constant. We note that N_a becomes the concentration of neutral acceptors p_a when the lattice temperature approaches 0 K. Because the Mg concentration is quite high, for simplicity we also assume that the effects of compensation (if there are any) are minimal.

Under our experimental conditions, the chemical potential can be shown to be [57]

$$\mu(T) = \frac{1}{2}(\varepsilon_a + \varepsilon_V) - \frac{k_B T}{2} \ln\left[\frac{N_a}{2U_V(T)}\right] + k_B T \sinh^{-1}\left[\sqrt{\frac{U_V(T)}{8N_a}} \cdot e^{-(\varepsilon_a - \varepsilon_V)/2k_B T}\right], \quad (3.56)$$

where ε_v is the energy at the top of the valence band and $U_V(T) = 2 \cdot ((2\pi m_p^* k_B T)/\hbar^2)^{3/2}$, where m_p^* is the effective mass of a hole.

Because under our experimental conditions $N_a \cong 10^{19}\,\mathrm{cm}^{-3}$, and $m_p^* = 0.8m_e$, where m_e is the mass of an electron, when use Eq. (3.55) to fit our experimental results in Figure 3.20, $E_b \equiv \varepsilon_a - \varepsilon_v$—the binding energy of the impurities, is the only adjustable parameter. We found that $E_b = 172 \pm 20\,\mathrm{meV}$ best fits our experimental data. The fact that this value lies in the range (150–250 meV) [58,59] of the binding energy of Mg impurities reported in the literature by other experimental methods confirms indirectly that our assignment of the structure around $841\,\mathrm{cm}^{-1}$ to ERS of holes from neutral Mg impurities in Mg-doped GaN was correct.

Since the contribution of the ERS signal in semiconductors comes primarily from the transition of holes from the ground state to the first excited state [55], our experimental results provide for the first time important information for the electronic structure of Mg impurities in GaN, i.e., the energy difference between the ground and first excited states of Mg impurities in Mg-doped GaN is about three-fifths of its binding energy.

We note that in addition to the expected Raman signal from GaN, sapphire, and ERS, we also observed two Raman signals very close to $260\,\mathrm{cm}^{-1}$ and $657\,\mathrm{cm}^{-1}$ which have been reported by Harima et al. as local modes in Mg-doped GaN [60]. The intensity of these two modes was found in our experiments to be at least one order of magnitude smaller than that of the optical phonon modes in GaN. We believe that the reason why an ERS signal from Mg-doped GaN was not observed by previous researchers is most likely due to the quality of the Mg-doped GaN samples and/or the pulse laser employed in our experimental study. More work is apparently needed in order to obtain a conclusive explanation.

3.7 CONCLUSIONS AND FUTURE EXPERIMENTS

We have presented experimental results on the dynamical properties of wurtzite structure GaN such as electron–phonon scattering rates, phonon–phonon interaction times, and nonequilibrium electron distributions as well as on the electronic properties of Mg impurities in wurtzite GaN by using Raman spectroscopy. A comprehensive theory of Raman scattering in semiconductors was also given, which is particularly useful under the circumstances of nonequilibrium conditions. We demonstrated that the total electron–LO phonon scattering rate in wurtzite GaN is $(4 \pm 0.8) \times 10^{13}\mathrm{s}^{-1}$, which is about an order of magnitude larger than that found in GaAs. The very larger electron–LO phonon scattering rate is mostly due to the large ionicity in wurtzite GaN. In addition, we showed that for electron densities $n \geq 5 \times 10^{17}\,\mathrm{cm}^{-1}$, as a result of efficient electron–electron scattering, nonequilibrium electron distributions photoexcited in wurtzite GaN can be very well described by Fermi–Dirac distribution functions with an effective electron temperature much higher than the lattice temperature. We found that the population relaxation time of LO phonons in wurtzite GaN is about 5 ps, a value comparable to that in GaAs. In contrast to the 2LA decay channel usually assumed for LO phonons in other semiconductors, our experimental results showed that among the various possible decay channels, the LO phonons in wurtzite GaN decay primarily into a large-wavevector TO and a large-wavevector LA or TA phonon. These experimental results are consistent with recent theoretical calculations of the phonon dispersion curves in wurtzite structure GaN. From analysis of the temperature dependence of electronic Raman scattering signal, the binding energy of the Mg impurities in wurtzite GaN was found to be $E_b \cong 172 \pm 20\,\text{meV}$. These experimental results demonstrate that the energy between the ground and first excited states of Mg impurities in wurtzite GaN is about three-fifths of its binding energy.

Although these experimental results have helped establish a solid foundation for the understanding of various properties in nitride-based III–V semiconductors, there are still many important aspects that remain to be explored. One of them is the investigation of transient nonequilibrium electron transport in nitride based III–V semiconductors. Because of the much larger electron effective mass in nitride-based III–V semiconductors, the drift velocity (and therefore device operation speed) of electronic devices based on these nitride materials is inherently much smaller than that of devices based on other semiconductors such as GaAs. This will tremendously limit the application range of the former materials. In order to enhance the speed of devices based on these nitride materials, novel transient transport properties such as electron velocity overshoot and ballistic electron transport, which have been studied in the case of GaAs and InP, will have to be utilized.

Acknowledgments

I would like to thank D. K. Ferry for his helpful discussions and H. Morkoç for GaN samples. This work was supported by the National Science Foundation under Grant No. DMR-9301100.

References

1. J. I. Pankove and T. D. Moustakas, "Gallium Nitride," in *Semiconductors and Semimetals*, edited by R. K. Willardson and E. R. Weber (Academic Press, New York, 1997) Vol. 50.
2. D. K. Ferry, *Phys. Rev. B*, **12**, 2361 (1975).
3. J. Shah, "Ultrafast Spectroscopy of Semiconductors and Semiconductor Nanostructures" in *Springer Series in Solid-State Sciences* (Springer, New York, 1996) Vol. 115.
4. C. Hamaguchi and M. Inoue, *Proc. of the 7th Int. Conf. on Hot Carriers in Semiconductors* (Adam Hilger, New York, 1992).

5. P. M. Platzman and P. A. Wolff, "Waves and Interactions in Solid State Plasmas" (Academic Press, New York, 1973), Solid State Supplement Vol. 13.
6. D. K. Ferry, *Semiconductors* (MacMillan, New York, 1991) Chap. 13.
7. B. K. Ridley, *Quantum Processes in Semiconductors*, 3d ed. (Oxford Press, London, 1993).
8. E. O. Kane, *J. Phys. Chem. Solids*, **1**, 249 (1957).
9. C. Chia, O. F. Sankey, K. T. Tsen, *Mod. Phys. Lett. B*, **7**, 331 (1993).
10. S. S. Jha, *Nuovo Cimento*, **63B**, 331 (1969).
11. A. Mooradian and A. L. McWhorter, in *Light Scattering Spectra of Solids*, edited by G. B. Wright (Springer, New York, 1969) p. 297.
12. D. C. Hamilton and A. L. McWhorter, in *Light Scattering Spectra of Solids*, edited by G. B. Wright (Springer, New York, 1969) p. 309.
13. D. Pines, *The Many-Body Problem* (Benjamin, New York, 1962) p. 44.
14. A. Smekal, *Naturwissensch.*, **11**, 873 (1923).
15. H. A. Kramers and W. Heisenburg, *Zeit. F. Physik*, **31**, 681 (1925).
16. C. V. Raman, *Ind. J. Phys.*, **2**, 387 (1928).
17. G. Lansberg and L. Mandel'shtam, *Naturwissensch.*, **16**, 57; ibid., **16**, 772 (1928).
18. J. D. Jackson, *Classical Electrodynamics*, 2d ed. (Wiley, New York, 1975) pp. 391–397.
19. P. Y. Yu and M. Cardona, *Fundamentals of Semiconductors* (Springer, New York, 1996).
20. K. T. Tsen, G. Halama, O. F. Sankey, S.-C.Y. Tsen, and H. Morkoç, *Phys. Rev. B*, **40**, 8103 (1989).
21. K. T. Tsen, K. R. Wald, T. Ruf, P. Y. Yu, and H. Morkoç, *Phys. Rev. Lett.*, **67**, 2557 (1991).
22. T. Ruf, K. R. Wald, P. Y. Yu, K. T. Tsen, H. Morkoç, K. T. Chan, *Superlattices and Microstruct.*, **13**, 203 (1993).
23. K. T. Tsen, *Int. J. Modern Phys. B*, **7**, no. 25, 4165 (1993).
24. E. D. Grann, S. J. Sheih, C. Chia, K. T. Tsen, O. F. Sankey, S. Gunser, D. K. Ferry, G. Maracus, R. Droopad, A. Salvador, A. Botchkarev, and H. Morkoç, *Appl. Phys. Lett.*, **64**, 1230–1232 (1994).
25. E. D. Grann, S. J. Shieh, K. T. Tsen, O. F. Sankey, S. Gunser, D. K. Ferry, A. Salvador, A. Botchkarev, and H. Morkoç, *Phys. Rev. B*, **51**, 1631 (1995).
26. E. D. Grann, K. T. Tsen, O. F. Sankey, D. K. Ferry, A. Salvador, A. Botchkarev, and H. Morkoç, *Appl. Phys. Letts.*, **67**, 1760 (1995).
27. K. T. Tsen, E. D. Grann, S. Guha, and J. Menendez, *Appl. Phys. Lett.*, **68**, 1051 (1996).
28. E. D. Grann, K. T. Tsen, D. K. Ferry, A. Salvador, A. Botchkarev, and H. Morkoç, *Phys. Rev. B*, **53**, 9838 (1996).
29. K. T. Tsen, D. K. Ferry, J. S. Wang, C. H. Huang, and H. H. Lin, *Appl. Phys. Lett.*, **69**, 3575 (1996).
30. E. D. Grann, K. T. Tsen, D. K. Ferry, A. Salvador, A. Botchkarev, and H. Morkoç, *Phys. Rev. B*, **56**, 9539 (1997).
31. K. T. Tsen, D. K. Ferry, A. Salvador, and H. Morkoç, *Phys. Rev. Lett.*, **80**, 4807 (1998).
32. W. Hayes and R. Loudon, *Scattering of Light by Crystals* (Wiley, New York, 1978).
33. S. J. Shieh, K. T. Tsen, D. K. Ferry, A. Botchkarev, B. Sverdlov, A. Salvador, and H. Morkoç, *Appl. Phys. Lett.*, **67**, 1757 (1995).
34. T. Azuhata, T. Sota, K. Suzuki, S. Nakamura, *J. Phys.: Condens. Matter*, **7**, L129 (1995).
35. K. T. Tsen, R. P. Joshi, D. K. Ferry, A. Botchkarev, B. Sverdlov, A. Salvador, and H. Morkoç, *Appl. Phys. Lett.*, **68**, 2990 (1996).
36. K. T. Tsen, D. K. Ferry, A. Botchkarev, B. Sverdlov, A. Salvador, and H. Morkoç, *Appl. Phys. Lett.*, **71**, 1852 (1997) .
37. A. Pinczuk, G. A. Abstreiter, R. Trommer, and M. Cardona, *Solid State Commun.*, **30**, 429 (1979).
38. K. T. Tsen and H. Morkoç, *Phys. Rev. B*, **38**, 5615 (1988).
39. S. Bloom, G. Harbeke, E. Meier, and I. B. Ortenburger, *Phys. Status. Solidi B*, **66**, 161 (1974).
40. S. Strite and H. Morkoç, *J. Vac. Sci. Technol., B*, **10**, 1237 (1992).
41. S. J. Pearton, J. C. Zolper, R. J. Shul, and F. Ren, *Appl. Phys. Rev.*, **86**, 1 (1999).
42. E. M. Conwell, *High Field Transport in Semiconductors* (Academic Press, New York, 1967).
43. J. C. Phillips, *Bonds and Bands in Semiconductors* (Academic Press, New York, 1973).

44. K. T. Tsen, R. P. Joshi, D. K. Ferry, A. Botchkarev, B. Sverdlov, A. Salvador, and H. Morkoç, *Appl. Phys. Lett.*, **72**, 2132 (1998).
45. J. A. Kash, J. C. Tsang, *Light Scattering in Solids VI*, edited by M. Cardona and G. Guntherodt (Springer, New York, 1991) p. 423.
46. C. L. Colins and P. Y. Yu, *Phys. Rev. B*, **30**, 4501 (1984).
47. Y. B. Levinson and B. N. Levinsky, *Solid State Commun.*, **16**, 713 (1975).
48. P. G. Klemens, *Phys. Rev.*, **148**, 845 (1966).
49. B. K. Ridley, *J. Phys.: Condens. Matter*, **8**, L511 (1996).
50. T. Azuhata, T. Matsunaga, K. Shimada, K. Yoshida, T. Sota, K. Suzuki, and S. Nakamura, *Physica B: Condens. Matter*, **219**, 493 (1996).
51. S. Nakamura, T. Mukai, M. Senoh, and N. Iwasa, *Jpn. J. Appl. Phys.*, **31**, L139 (1992).
52. H. Amano, M. Kito, K. Hiramatsu, and I. Akasaki, *Jpn. J. Appl. Phys.*, **28**, L2112 (1989).
53. T. Azuhata, T. Sota, K. Suzuki, and S. Nakamura, *J. Phys.: Condens. Matter*, **7**, L129 (1995).
54. S. J. Shieh, K. T. Tsen, D. K. Ferry, A. Botchkarev, B. Sverdlov, A. Savador, and H. Morkoç. *Appl. Phys. Lett.*, **67**, 1757 (1995).
55. M. V. Klein, in Light Scattering in Solids I, edited by M. Cardona and G. Guntherodt. Vol. 8 of *Topics in Applied Physics*. (Springer, New York, 1983)
56. R. Kubo, *Statistical Mechanics*, 5th ed. (North Holland, Amsterdam, 1965) p. 246.
57. J. P. Mckelvey, *Solid State and Semiconductor Physics* (Harper and Row, N.Y. 1966) Chap. 9.
58. S. Fisher, C. Wetzel, E. E. Haller, and B. K. Meyer, *Appl. Phys. Lett.*, **67**, 1298 (1995).
59. T. Tanaka, A. Watanabe, H. Amano, Y. Kobayashi, I. Akasaki, S. Yamazaki, and M. Koike, *Appl. Phys. Lett.*, **65**, 593 (1994).
60. H. Harima, T. Inoue, S. Nakashima, M. Ishida, and M. Taneya, *Appl. Phys. Lett.*, **75**, 1383 (1999).

CHAPTER 4

Optical Properties of InGaN-Based III-Nitride Heterostructures

Y.-H. CHO and J.-J. SONG[1]

[1] *Center for Laser and Photonics Research and Department of Physics Oklahoma State University, Stillwater, OK 74078, USA*

4.1 INTRODUCTION

With the recent progress in nitride growth technology, much research has been devoted to group III nitride wide-bandgap semiconductors for their applications, such as in UV visible light emitters, solar-blind UV detectors, and high-temperature and high-power electronic devices. In particular, InGaN-based structures are of great scientific interest for the commercialization of high-brightness visible UV LEDs [1] and cw violet current-injection laser diodes with lifetimes in excess of 10,000 hours [2,3]. The (In, Al)GaN alloys, consisting of InN, GaN, and AlN, cover the entire visible to deep-UV region of the spectrum (1.9–6.2 eV). In spite of the poor structural quality (e.g., the large dislocation density) of nitride epitaxial layers compared with that of other III–V semiconductors, high-brightness LEDs [4] and cw violet laser diodes [5] based on InGaN structures have been constructed with high performance and high quantum efficiency [6]. Understanding the physical mechanisms giving rise to spontaneous emission and stimulated emission (SE) in InGaN–based structures is crucial not only from the viewpoint of scientific interest but also for designing practical devices. To explain the spontaneous and stimulated emission characteristics of InGaN-based structures, several groups have proposed different mechanisms (alloy potential fluctuations, quantum dot–like In phase separations, spontaneous polarization, strain-induced piezoelectric polarization, etc.); however, there has been much disagreement in the literature as to which of these mechanisms is dominant.

This chapter presents recent studies of the optical properties of InGaN-based structures with emphasis on those aspects relevant to photonic applications. Experimental aspects of linear and nonlinear optical processes are discussed with special attention to InGaN thin films and various InGaN/GaN quantum structures. The experimental data presented were taken from various sample structures by a wide range of techniques to better understand both the spontaneous and stimulated emission properties. This chapter is organized as follows. Section 4.2 surveys the general optical properties of InGaN-based materials. The results of luminescence, absorption, temporal evolution, and stimulated emission, as well as the influence of In composition in InGaN wells and of Si doping concentration in GaN barriers on both the optical and structural properties are discussed. Section 4.3 focuses on the dependence of the emission spectra on temperature, excitation energy, excitation density, and excitation length to clarify the critical role of the energy band tail states in InGaN-based structures. Section 4.4 is devoted to nonlinear optical studies of InGaN-based structures by means of variable stripe optical gain and nondegenerate pump-probe spectroscopy.

4.2 OPTICAL PROPERTIES OF InGaN-BASED STRUCTURES

To explain the optical processes in InGaN-based structures, many mechanisms have been proposed, such as alloy potential fluctuations, quantum dot–like In phase separations, spontaneous polarization, and strain-induced piezoelectric polarization. For this reason, the responsible emission mechanisms given in the literature are still varied and controversial. In the first subsection we discuss general aspects of spontaneous and stimulated recombination in InGaN-based structures. In the next subsection we present experimental optical properties generally observed by a number of authors, and we discuss mechanisms proposed in the literature to explain the experimental data. In the final subsection we discuss the influence of In composition and Si doping on the characteristics of InGaN quantum structures, with emphasis on their optical characteristics.

4.2.1 General Aspects of Optical Transitions in InGaN-Based Structures

In this subsection we survey the general aspects associated with spontaneous and stimulated emission in InGaN-based structures. Among a number of mechanisms proposed for the spontaneous emission characteristics of InGaN-based structures, a brief description is given of (i) carrier localization in potential variations caused by alloy fluctuations and/or phase separation and (ii) built-in internal electric fields (caused by spontaneous polarization and strain-induced piezoelectric polarization). Finally, possible mechanisms for stimulated emission in the InGaN-based system are discussed.

4.2.1.1 Carrier Localization and Spatial Variations of Optical Emission

Several groups have stated that the recombination of carriers localized at band-tail states of potential fluctuations is an important spontaneous emission mechanism in InGaN/GaN QWs [7–10]. The potential fluctuations can be induced by alloy disorder, impurities, interface irregularities, and/or self-formed quantum dot–like regions in the QW active regions. It has been argued that the incorporation of In atoms could play a crucial role in suppressing nonradiative recombination rates by the capture of carriers in localization centers originating from quantum dot–like and phase-separated In-rich regions [11–13]. Narukawa et al. have observed self-formed quantum dot–like features associated with In composition inhomogeneity in the well region of $In_{0.2}Ga0.8N/In_{0.05}Ga_{0.95}N$ purple laser diode structures using cross-sectional transmission electron microscopy and energy-dispersive X-ray microanalysis [12]. The main radiative recombination in these QWs was attributed to excitons highly localized at deep traps probably originating from phase-separated In-rich regions acting as quantum dots [12,13]. It was suggested that the self-formation of quantum dots may be a result of the intrinsic nature of InGaN ternary alloys to have In compositional modulation due to phase separation. It is likely that this carrier localization formed in the plane of the layers enhances the quantum efficiency by suppressing lateral carrier diffusion, thereby reducing the probability that carriers will enter nonradiative recombination centers. Moreover, the effective potential fluctuation due to the In composition modulation can be enhanced by the large bowing of the InGaN bandgap. The bandgap (E_g) of the $In_xGa_{1-x}N$ ternary alloy is given by $E_{g,\text{InGan}}(x) = (1-x)E_{g,\text{GaN}} + xE_{g,\text{InN}} - bx(1-x)$, where $E_{g,\text{GaN}}(E_{g,\text{InN}})$ is E_g of GaN (InN), and b is a bowing parameter. The bowing parameter for the $In_xGa_{1-x}N$ ternary alloy was found to be much larger than generally assumed and is strongly composition dependent, as seen in experimental and theoretical studies of the bandgaps of strained $In_xGa_{1-x}N$ epilayers (e.g., $b \approx 3.2\,\text{eV}$ at $0 < x < 0.2$ [14,15], and $b \approx 3.8\,\text{eV}$ at $x = 0.1$ [16]).

Understanding the origin and the detailed nature of the carrier localization in $In_xGa_{1-x}N$-based structures is of considerable scientific interest and technological importance. Recently, spatial variations in the optical properties of GaN- and $In_xGa_{1-x}N$-related materials have been more directly investigated by a number of groups using submicrometer spatial resolution luminescence techniques, such as cathodoluminescence (CL), near-field scanning optical microscopy (NSOM), and scanning tunneling microscope–induced luminescence (STL). The optical information investigated by CL, NSOM, or STL complements high-resolution topographs simultaneously obtained by scanning electron microscopy, shear-force microscopy, or scanning tunneling microscopy, respectively. These local characterizations of optical properties from the spatially nonuniform samples provide spectral and spatial mapping of several chemical and physical quantities of interest.

The spatial variations of luminescence from InGaN/GaN QW structures [17,18], as well as GaN films [19–21], have been studied by CL measurements. Chichibu et al. obtained spatially resolved CL map images of MOCVD-grown InGaN/GaN single QWs at 10 K [17]. These 3-nm-thick InGaN layers had In contents of 5, 20, and 50%. Both uncapped samples and samples capped by a 6-nm-thick undoped GaN layer were studied. Chichibu et al. estimated an In-cluster size of less than 60 nm (the spatial resolution of the CL mapping technique, which is essentially limited by the diffusion length in the matrix). Zhang et al. performed spatially resolved and time-resolved CL experiments for an MOCVD-grown InGaN/GaN single QW at 93 K to study the carrier relaxation and recombination dynamics [18]. The InGaN layer had a 15% average In composition, was 4 nm thick, and was capped by a 70-nm-thick GaN layer. Local In alloy fluctuations on a scale of less than ~ 100 nm were observed for low excitation conditions, indicating strong carrier localization at In-rich regions. Time-resolved CL experiments revealed that carriers generated in the boundary regions diffuse toward and recombine at In-rich centers, with strong lateral excitonic localization prior to radiative recombination.

NSOM is an alternative approach to optical spectroscopy and mapping that is used to study the optical properties of semiconductors with spatial resolution well beyond the diffraction limit [22–26]. NSOM has various modes: (i) *illumination mode*, in which the near-field probe acts as a tiny light source, (ii) *collection mode*, in which the near-field probe acts as a tiny light detector, and (iii) *illumination-collection mode*, in which the near-field probe acts as both a tiny light source and a detector. In the case of illumination mode, the spectrum can be influenced by carrier diffusion and relaxation, resulting in poorer spatial resolution than with other modes. Crowell et al. carried out an NSOM study of larger defects in an InGaN single QW, with illumination through a tapered silver-coated optical fiber of ~ 100-nm-diameter aperture [27]. They found no spectroscopic signature of localized carrier recombination at temperatures above 50 K, in striking contrast to the aforementioned CL studies [17,18]. A plausible explanation is the larger carrier diffusion lengths in the higher-temperature and higher-injection NSOM experiments. Vertikov et al. reported reflection NSOM results for InGaN epilayers and

QWs in the illumination mode using an aluminum-coated fiber probe with a 100-nm-size aperture [28] and in the collection mode using an aluminum-coated fiber probe with a $\sim$50-nm-size aperture [29]. Using the collection mode, they reported that the range of In alloy fluctuations reaches the 100-nm lateral scale [29].

STL is another alternative method that uses the localized filament current of an STM tip to generate photon emission locally [30]. STL studies of HVPE-grown GaN [31] and an MOCVD-grown InGaN/GaN MQW [32] have recently been reported. Evoy et al. observed that their STL images exhibited 30–100-nm-scale fluctuations for an MOCVD-grown InGaN/GaN MQW [32]; however, the STL images of the InGaN/GaN MQW exhibited a smooth variation of the luminescence intensity instead of distinctively dark spots, which may be expected in the case of local In composition modulation.

Some of the spatially resolved luminescence results are still controversial, depending on the samples, the spatial resolution provided by the spectroscopic techniques, and other measurement conditions. Further work is needed to better understand the phase-separated and/or quantum dot–like features, the correlation between local and macroscopic emission properties, and their roles in the recombination dynamics and lasing processes of InGaN-based structures.

4.2.1.2 Built-in Spontaneous and Strain-Induced Piezoelectric Polarization Field

The built-in macroscopic polarization, which consists of (i) the spontaneous polarization due to interface charge accumulations between two constituent materials and (ii) the piezoelectric polarization due to lattice-mismatch-induced strain, plays a significant role in the wurtzite III-nitrides [33–43]. Spontaneous polarization has long been observed in ferroelectric materials. It has been shown that wurtzite III-nitrides can have a nonzero macroscopic polarization even in equilibrium (with zero strain). From ab initio studies of the wurtzite III-nitrides, it was found that due to their low-symmetry crystal structure, they have very large spontaneous polarization fields (e.g., -0.081, -0.029, and $-0.032\,\mathrm{C/m^2}$ for AlN, GaN, and InN, respectively) [33]. The spontaneous polarization increases with increasing nonideality of the structure, from GaN to InN to AlN, because of the sensitive dependence of the polarization on the structural parameters. In particular, the spontaneous polarization of AlN was found to be only about three to five times smaller than that of typical ferroelectric perovskites. On the other hand, electric polarization fields can be generated by lattice-mismatch-induced strain in strained-layer superlattices of III–V semiconductors [44]. Whether polarization fields are generated by the strain depends on the symmetry of the strain components, and the orientation of the polarization fields is determined by the superlattice (SL) growth axis. For the most commonly studied case of strained-layer superlattices made from zinc-blende structures grown along the [100] orientation, the piezoelectric effect does not occur, since only diagonal strain components are generated, and diagonal strains do not induce an electric polarization vector in these zinc-blende structures. If any other growth axis is chosen, however, off-diagonal strain components are generated, resulting in electric polarization fields in zinc-blende superlattices. For instance, with a [111] growth axis in the zinc-blende structures, the polarization vectors point along the growth axis (perpendicular to the layers). Since one of the constituent materials is in biaxial tension, whereas the other is in biaxial compression, the polarization polarity changes at the SL interfaces. These large alternating electric fields significantly change the SL electronic structure and optical properties. The wurtzite III-nitride structures grown along the [0001] orientation provide a similar interesting situation. It has been reported that if wurtzite nitride layers are under biaxial strain due to lattice mismatch, large piezoelectric fields can be generated, since nitrides have very large piezoelectric constants (e.g., $e_{31} = -0.60$, -0.49, and $-0.57\ C/m^2$ for AlN, GaN, and InN, respectively) [33].

An electric field applied to a QW structure changes the subband energy levels and bound-state wavefunctions in the QW and, hence, the optical transition energies and oscillator strengths. For bulk semiconductors, this is known as the Franz–Keldysh effect, which is associated with photon-assisted tunneling. The effective energy gap is reduced, since the energy bands are tilted by the applied electric field, and therefore the conduction and valence band wavefunctions have tails in the energy gap with some overlapping. Consequently, a small shift of the absorption edge to lower energies occurs in the presence of an electric field, resulting in a low-energy absorption tail below the fundamental energy gap. Since excitons are ionized by the electric field in bulk semiconductors, the excitons in bulk semiconductors play little part in this effect. The situation is quite different in QWs, since the quantum confinement is strong enough to prevent significant ionization for the applied field parallel to the growth direction [45]. The electric field produces a considerable redshift of the confined-state energy levels, a phenomenon known as the *quantum-confined Stark effect*. In this case, the excitons are not dissociated but only polarized under a large electric field. The quantum-confined Stark effect can be employed in a number of optical devices, including low-energy switches and high-speed electroabsorptive modulators.

Takeuchi et al. proposed that piezoelectric fields due to lattice-mismatch-induced strain generate the quantum-confined Stark effect in InGaN/GaN QWs, resulting in a blueshift behavior with increasing excitation intensity and a well-width dependence of the luminescence peak energy [38]. They calculated a strain-induced electric field of 1.08 MV/cm for strained $In_{0.13}Ga_{0.87}N$ grown on GaN, assuming $e_{31} = -0.22 C/m^2$. Since alloy composition fluctuations in the ternary InGaN could also cause the luminescence redshift, Im et al. excluded this possibility in their time-resolved photoluminescence (TRPL) experiments by investigating the binary GaN in GaN/$Al_{0.15}Ga_{0.85}N$ QWs [40]. Based on their observations of (i) a decrease in the photoluminescence (PL) peak position and an increase in decay times with increasing well width from 1.3 to 10 nm and (ii) a PL peak shift to energies well below the GaN bandgap in the thicker layers (5 nm and 10 nm), they proposed that the piezoelectric field effect is significant. Similar experimental results were also observed for $In_{0.05}Ga_{0.95}N$/GaN QWs by the same group, so they concluded that the piezoelectric field effect is prominent in both GaN/AlGaN and InGaN/GaN QWs [41]. These polarization charges generate a built-in internal electric field directed along the growth direction and modify both electronic energy levels and wavefunctions. These internal polarization fields give rise to changes in optical matrix elements and can be screened by photogenerated electron–hole pairs. It has been argued that the spontaneous polarization and/or the strain-induced piezoelectric polarization play an important role in carrier recombination in both GaN/AlGaN and InGaN/GaN QWs [40,41,43]; however, such experimental results are not fully understood for InGaN/GaN QWs with the larger In composition used in state-of-the-art commercialized blue violet laser diodes. In order to understand the influence of a rather large In composition on the characteristics of InGaN-based structures, a comparison study of structural and optical properties is often required, since any growth parameter changes (especially in InGaN-based structures) may unexpectedly affect other structural (e.g., quality and interface) properties and hence the optical and electrical properties.

4.2.1.3 Stimulated Emission and Lasing Mechanisms

Since the first observation of optically pumped lasing in needlelike GaN single crystals at 2 K by Dingle et al. [46], a number of studies on optically pumped SE and lasing have been performed for GaN and related heterostructures in recent years. The optical pumping method, which generates high photoexcited carrier densities required for the onset of SE and lasing, has important advantages, since SE and lasing phenomena can be investigated without complicated doping, device processing, and electrical contact procedures. Details of the experimental optical pumping scheme and various stimulated recombination mechanisms were given in Chapter 3.

There have been several attempts to account for the SE gain and identify the lasing mechanism in III–V and II–VI semiconductors. The responsible gain mechanisms for GaN were shown to be exciton–exciton scattering and electron–hole plasma recombination (or band-to-band transition) in Chapter 3. Several SE and lasing mechanisms for the InGaN-based laser diode structures have recently been discussed by several groups [12,47–51]. Optical gain experiments were also performed on InGaN-based structures [52–58]. Whereas electron–hole plasma recombination is accepted as the gain mechanism in most III–V semiconductors, such as GaAs and InP, there is a debate over whether the SE in InGaN-based structures is caused mostly by electron–hole plasma recombination.

A considerable amount of attention has been given to the influence of strongly localized band-tail states on the SE and lasing processes in InGaN MQWs. Although many data support carrier localization recombination as the mechanism leading to spontaneous emission in these materials, the results for SE behavior in the literature are varied and often controversial. This has led some research groups to assign the spontaneous emission peak to recombination of localized carriers, and the SE peak to a more traditional electron–hole plasma recombination originating from free carriers, whereas others claim that recombination originating from strong carrier localization is the origin of both spontaneous and stimulated emission. We will resolve this issue in Sections 4.3 and 4.4.

4.2.2 Fundamental Optical Properties of InGaN-Based Structures

This subsection is devoted to the optical properties generally observed for InGaN epilayers and InGaN/GaN QWs, in conjunction with proposed interpretations in the literature. Recombination in InGaN-based materials is characterized by (i) a large Stokes shift between the emission peak and the absorption band edge, (ii) a blueshift of the emission peak with increasing excitation density, and (iii) a redshift of the emission peak with time (a rise in lifetime with decreasing emission energy). Interestingly, these phenomena have been attributed to carrier localization at potential fluctuations and/or built-in internal electric fields, as described in the previous subsection. The data presented in this subsection were taken by a wide range of techniques to better understand both the spontaneous and stimulated emission. The optical emission, absorption, excitation power emission dependence, and emission temporal dynamics were investigated by PL, PL excitation (PLE), optically pumped emission, and TRPL, respectively. Finally, we discuss the possible explanations for the optical properties generally observed in InGaN-based structures.

4.2.2.1 Optical Emission and Absorption Properties

Figure 4.1 shows typical 10 K PL and PLE spectra of the InGaN-related emission of Figure 4.1a an $In_{0.18}Ga_{0.82}N$/GaN MQW and Figure 4.1b an $In_{0.18}Ga_{0.82}N$ epilayer, respectively. Both the MQW and epilayer were grown on *c*-plane sapphire films by MOCVD, following the deposition of a 1.8-μm-thick GaN buffer layer. The MQW sample consisted of a 12-period SL with 3-nm-thick $In_{0.18}Ga_{0.82}N$ wells and 4.5-nm-thick GaN barriers, and a 100-nm-thick $Al_{0.07}Ga_{0.93}N$ capping layer. The InGaN epilayer was

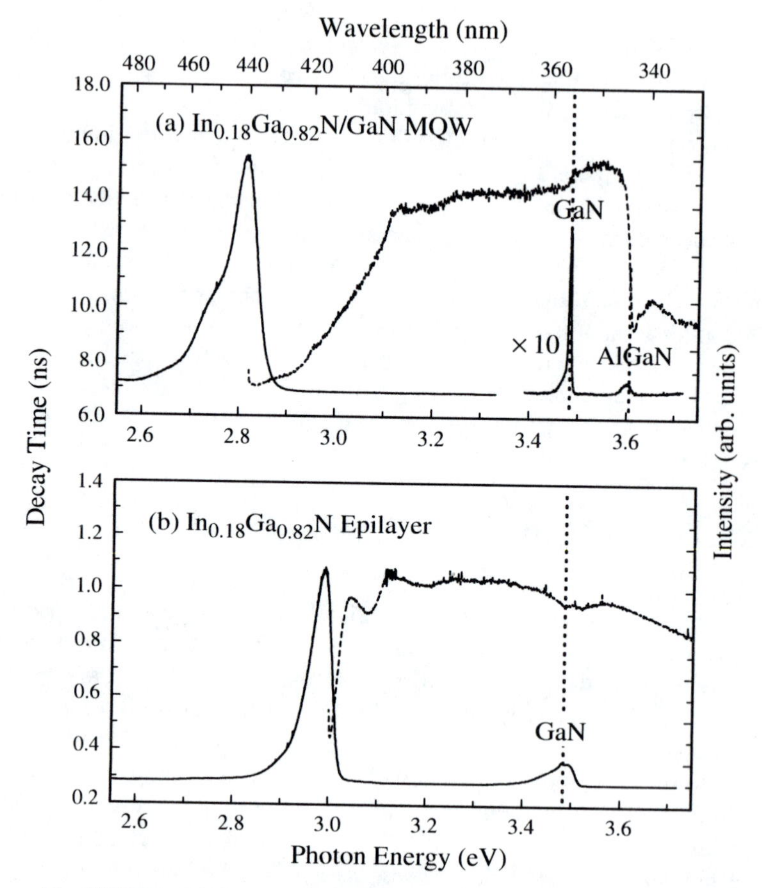

Figure 4.1 10 K PL (solid lines) and PLE (dashed lines) spectra for (a) an $In_{0.18}Ga_{0.82}N$/GaN MQW and (b) an $In_{0.18}Ga_{0.82}N$ epilayer. Both were grown by MOCVD on *c*-plane sapphire. A large Stokes shift of the PL emission from the InGaN layers with respect to the band edge measured by PLE spectra is observed. The near-band-edge emission from the GaN and $Al_{0.07}Ga_{0.93}N$ layers was observed at 3.48 and 3.6 eV, respectively. The PLE contributions from the GaN layers [in (a) and (b)] and the $Al_{0.07}Ga_{0.93}N$ layer [in (a)] are clearly seen. (Y. H. Cho)

100 nm thick and capped by a 50-nm-thick GaN layer. The average In composition was about 18% from X-ray diffraction analysis, assuming Vegard's law. It was observed that the MQW sample was fully pseudomorphic from symmetric and asymmetric reciprocal space mapping [59]. The PL spectra (solid lines) were measured using the 325-nm line of a cw He–Cd laser. The PLE spectra (dashed lines) were measured using quasi-monochromatic light dispersed by a (1/2)-m monochromator from a xenon lamp.

In Figure 4.1, the InGaN-related emission has a peak energy of ~2.8 and ~2.99 eV at 10 K for Figure 4.1a the $In_{0.18}Ga_{0.82}N$/GaN MQW and

Figure 4.1b the $In_{0.18}Ga_{0.82}N$ epilayer, respectively. The near-band-edge emission from the $Al_{0.07}Ga_{0.93}N$ cladding layer [in Figure 4.1a] and the GaN layers [in Figure 4.1a and b] are also clearly seen at 3.60 and 3.48 eV, respectively. With the PLE detection energy set at the InGaN-related emission peak, the contributions from the GaN layers [in Figure 4.1a and b] and from the $Al_{0.07}Ga_{0.93}N$ capping layer [in Figure 4.1a] are clearly distinguishable, and the energy positions of the absorption edges are well matched to the PL peak positions. The absorption of the InGaN wells in the MQW increases monotonically, reaching a maximum at $\sim$3.1 eV and remains almost constant until absorption by the GaN barriers occurs at 3.48 eV. A large Stokes shift of the InGaN-related spontaneous emission peak with respect to the absorption edge measured by low-power PLE spectroscopy is clearly observed for both samples. The Stokes shift is much larger for the MQW than for the epilayer, indicating that the band-tail states responsible for the "soft" absorption edge are significantly larger in the MQW than in the epilayer, probably due to the influence of the MQW interfaces on the overall potential fluctuations.

The large Stokes shift between the emission and absorption edge is a common feature in InGaN epilayers and QWs. Recently, the Stokes shift and the broadening of the absorption edge from InGaN epilayers and diodes measured by different groups were plotted as a function of emission energy [60,61]. Both the Stokes shift and the absorption broadening increase with increasing In composition, which causes the emission energy to decrease. These phenomena have been widely observed by several groups and attributed to either carrier localization, the piezoelectric field effect, or both.

4.2.2.2 Excitation Density Dependence and Optically Pumped Stimulated Emission

Figure 4.2 shows the evolution of the 10 K emission spectra with increasing excitation pump density (I_{exc}) for Figure 4.2a the $In_{0.18}Ga_{0.82}N$/GaN MQW and Figure 4.2b the $In_{0.18}Ga_{0.82}N$ epilayer. The spectra shown in Figure 4.2 were taken with an excitation energy of 3.49 eV and collected in a surface emission geometry to minimize the effects of reabsorption on the emission spectra. (The SE peak is due to a leak of the in-plane SE at the sample edge. No SE was observed from the middle of the sample in this geometry, indicating the high quality of the sample structure [62].) The pump spot size was $\sim$1 mm^2, and the excitation wavelength was the third harmonic (355 nm) of a Nd : YAG (yttrium aluminum garnet) laser. As I_{exc} is increased, the spontaneous emission peak of the MQW blueshifts until it reaches $\sim$2.9 eV, and after this point it increases only in intensity until the SE threshold is reached. The SE, indicated by the arrows in Figure 4.2, develops on the low-energy side of the spontaneous emission peak. The blueshift of the spontaneous emission with increasing I_{exc} is attributed to band filling of localized states due to the intense optical pump. With

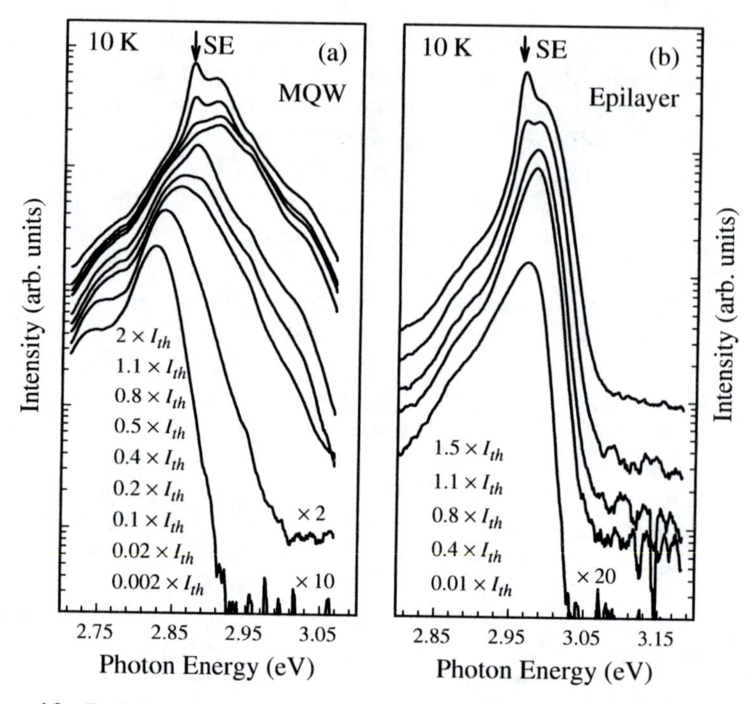

Figure 4.2 Evolution of InGaN emission spectra from below to above the SE threshold, I_{th}, at 10 K for (a) an $In_{0.18}Ga_{0.82}N$/GaN MQW and (b) an $In_{0.18}Ga_{0.82}N$ epilayer. The emission was collected in a surface emission geometry. I_{th} for the MQW and epilayer was ~170 and 130 kW/cm^2, respectively, for the experimental conditions. A large blueshift of the emission is cleary seen with increasing excitation density for the MQW sample, showing band filling of localized states. (Y. H. Cho)

increasing I_{exc}, the filling level increases and the PL maximum shifts to higher energies until sufficient population inversion is achieved and net optical gain results in the observed SE peak. The I_{exc}-induced blueshift is significantly larger for the MQW than for the epilayer (~80 meV for the MQW compared with ~14 meV for the epilayer), further indicating that the potential fluctuations are significantly larger in the MQW than in the epilayer. Details of the SE features and optical nonlinearities of InGaN-based structures will be given in Sections 4.3 and 4.4, respectively.

4.2.2.3 Temporal Evolution and Dynamics of Optical Emission

There have been a number of TRPL studies of InGaN epilayers and quantum wells. The measured recombination decay times for InGaN-related emission were generally in the range of several hundreds of picoseconds to a few tens of nanoseconds, depending on the In composition of InGaN layer, the number of QWs, and/or other growth conditions [8,63,64,66].

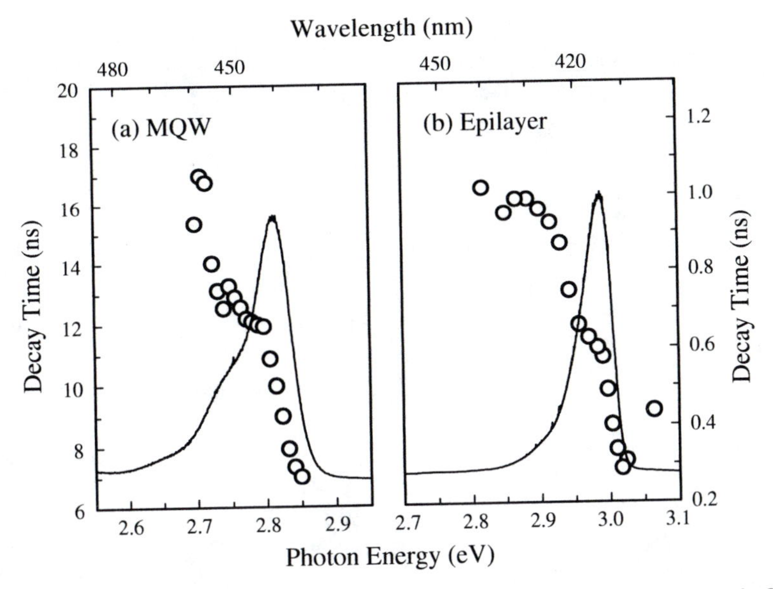

Figure 4.3 10 K decay time (open circles) as a function of detection energy across the PL spectrum for (a) an $In_{0.18}Ga_{0.82}N$/GaN MQW and (b) an $In_{0.18}Ga_{0.82}N$ epilayer. A rise in lifetime with decreasing emission energy, resulting in a redshift behavior of the emission with time, reflects that the InGaN-related emission is due to radiative recombination of carriers localized at potential fluctuations. (Y. H. Cho)

Shown in Figure 4.3 is the effective recombination lifetime as a function of detection energy across the 10 K PL spectrum of Figure 4.3a an $In_{0.18}Ga_{0.82}N$/GaN MQW and Figure 4.3b an $In_{0.18}Ga_{0.82}N$ epilayer. The TRPL measurements were carried out using a tunable picosecond pulsed laser system consisting of a cavity-dumped dye laser synchronously pumped by a frequency-doubled mode-locked Nd : YAG laser as an excitation source and a streak camera for detection. The output laser pulses from the dye laser had a duration of less than 5 ps and were frequency doubled into the UV spectral region by a nonlinear crystal. The overall time resolution of the system was better than 15 ps.

Figure 4.3 shows a rise in the effective lifetime τ_{eff} with decreasing emission energy across the PL spectrum, resulting in a redshift of the emission peak energy with time. This is evidence that the InGaN-related emission is due to radiative recombination of carriers localized at potential fluctuations. It is well known that the recombination of localized carriers is governed not only by radiative recombination but also by the transfer to and trapping in the energy tail states [13]. The differences in τ_{eff} between the two samples indicate that the potential fluctuations localizing the carriers are significantly smaller in the epilayer than in the MQW. The longer lifetime for the MQWs in this work compared with those reported by other groups is probably due to a relatively larger degree of carrier localization caused by a larger number of QWs and/or different growth conditions used in this work [8,63–66].

An important parameter associated with the PL efficiency of photoexcited carriers is τ_{eff}, given by:

$$\frac{1}{\tau_{eff}} = \frac{1}{\tau_r} + \frac{1}{\tau_{nr}}. \tag{4.1}$$

where τ_r is the radiative lifetime, and τ_{nr} is the nonradiative lifetime. The nonradiative processes include multiphonon emission, capture and recombination at impurities and defects, the Auger recombination effect, and surface recombination, as well as diffusion of carriers away from the region of observation. Assuming a radiative efficiency η of 100% at $T = 10\,\mathrm{K}$ and using the equation $\eta(T) = \tau_{eff}(T)/\tau_r(T) \approx I_{PL}(T)/I_{PL}(10\,\mathrm{K})$, one can determine the radiative and nonradiative lifetimes as a function of temperature.

Figure 4.4 shows τ_{eff} as a function of temperature for Figure 4.4a an $In_{0.18}Ga_{0.82}N$/GaN MQW and Figure 4.4b an $In_{0.18}Ga_{0.82}N$ epilayer. The

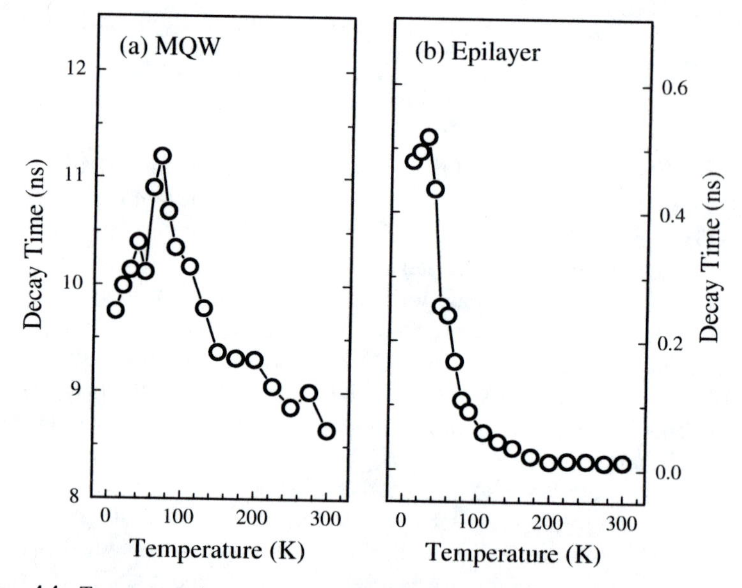

Figure 4.4 Temperature dependence of the lifetimes for (a) an $In_{0.18}Ga_{0.82}N$/GaN MQW and (b) an $In_{0.18}Ga_{0.82}N$ epilayer. The rise in effective recombination lifetime with increasing temperature observed from 10 to ~70 (30) K for the MQW (epilayer) is indicative of recombination dominated by radiative recombination channels, whereas the decrease in lifetime with increasing temperature for $T > 70$ (30) K indicates the increasing dominance of nonradiative recombination. It was observed that the MQW has a significantly larger lifetime than the epilayer for all temperatures studied, and the lifetimes of both are significantly larger than that of GaN epilayers and heterostructures. (Y. H. Cho)

rise in τ_{eff} with increasing temperature from 10 to $\sim$ 70 (30) K for the MQW (epilayer) is indicative of recombination dominated by radiative recombination channels, whereas the decrease in τ_{eff} with increasing temperature for $T > 70$ (30) K indicates the increasing dominance of nonradiative recombination channels [67,68]. The MQW has a significantly larger τ_{eff} than the epilayer for all temperatures studied, and the lifetimes of both are significantly larger than that of GaN epilayers and heterostructures [63,69]. This is attributed to suppression of nonradiative recombination by the localization of carriers at statistical potential fluctuations arising from the nonrandom nature of this alloy.

For GaN/$Al_{0.07}Ga_{0.93}N$ QWs, Lefebvre et al. observed a significant spectral distribution of decay times from TRPL experiments, which they interpreted in terms of localization of carriers by potential fluctuations due to alloy disorder and to well thickness variations [70]. The inhomogeneous potential fluctuations can be caused by alloy composition fluctuation, well size irregularity, and/or other crystal imperfections such as point defects and dislocations, resulting in a spatial bandgap variation in the plane of the layers.

4.2.2.4 Discussion

Recently, the influence of internal electric fields and carrier localization on optical properties in both GaN/AlGaN and InGaN/GaN QWs has been reported by several authors. In the case of InGaN/GaN QWs, unfortunately, both internal electric field and carrier localization effects may be generally enhanced with increasing In alloy composition (at least up to 50%), and it is difficult to tell which effect is predominant. Although one may try to exclude (or reduce) the influence of alloy compositional fluctuation by introducing sample configurations with nonalloy (or small-alloy) active regions, such as GaN/AlGaN QWs, the results obtained from these samples may not necessarily be extended to the case of samples with alloy active regions, such as the blue-light-emitting InGaN/GaN QWs.

Not only for InGaN/GaN but also for GaN/AlGaN [70] QW structures, most of the earlier work by several groups using optical properties to assign the responsible recombination mechanism relied mainly on the following experimental observations: (a) a large Stokes shift between the emission peak energy and absorption edge (Figure 4.1), (b) a blueshift emission behavior with increasing optical excitation power density (Figure 4.2), and (c) a redshift emission behavior with decay time (equivalently, a rise in decay time with decreasing emission energy) (Figure 4.3). Interestingly, although a number of research groups have observed similar experimental findings for the QWs, their interpretations on the observations have been quite different. These phenomena have been interpreted in terms of either the built-in internal electric field effect or carrier localization, so careful consideration is required to interpret these observations. This is because these observations may be–at least qualitatively–explained in terms of both the internal electric field and carrier localization models, as follows.

In the presence of an internal electric field, the electrons and holes are separated to opposite sides within the InGaN wells. An overlap between the wavefunctions of these spatially separated electrons and holes allows their radiative recombination within the wells at a lower energy than if there was no internal electric field [observation (a)]. When carriers are either optically or electrically generated, the generated carriers partially screen the internal electric field, and thus separated electrons and holes come spatially closer together. Accordingly, the effective bandgap, which was reduced by the internal electric field, approaches the bandgap without the internal electric field [observation (b)]. As time passes after an excitation pulse, the generated carriers recombine, and hence the internal electric field (which was screened by the generated carriers) recovers [observation (c)].

In the presence of carrier localization, on the other hand, large potential fluctuations result in the emission from localized carriers at potential minima, whereas the absorption mostly occurs at the average potential energy [observation (a)]. As the excitation power increases, the carriers can fill the band-tail states of the fluctuations and thus the average emission energy increases [observation (b)]. As time passes after an excitation pulse, the carriers go down to the lower band-tail states, resulting in the redshift behavior with time [observation (c)]. Accordingly, the trend of experimental observations (a), (b), and (c) can be plausibly explained by the internal electric field and/or carrier localization models, especially for the case of InGaN/GaN QWs. Because of this ambiguity, one cannot strongly argue which effect is predominant without more direct experimental evidence.

We note that a larger Stokes shift, a larger spontaneous emission blueshift with excitation density, and a longer lifetime for the MQW compared with those for the epilayer strongly reflect that the potential fluctuations are significantly larger in the MQW than in the epilayer. In general, the following effects can cause an emission shift in MQWs but not in epilayers: (i) the quantum confinement (blueshift), (ii) the built-in internal electric field (redshift), and (iii) the potential fluctuations (redshift) related to the presence of MQW interfaces. As seen in Figure 4.1, the redshifting influence is larger than the blueshifting influence in the MQW. This indicates that the MQW emission properties are predominantly affected by the interfaces through (ii) and/or (iii). Although internal electric fields may be present in the MQWs, potential fluctuations are more likely to dominate the emission properties of the MQWs, since the 3-nm InGaN well widths are less than the exciton Bohr radius and too thin to cause substantial electron–hole separation. The large PL peak energy difference of $\sim$190 meV between the MQW ($\sim$2.8 eV) and the epilayer ($\sim$2.99 eV) in Figure 4.1 shows that the potential fluctuations are larger in the MQW. Since the quality of upper layers is affected by that of lower layers or interface properties, the presence of interfaces may change the properties of the InGaN active region. In addition, the different growth conditions between the MQW and the epilayer may also affect the degree of potential fluctuations. In Section 4.3, we will see direct experimental evidence that carrier localization caused by potential fluctuations is the predominant effect for both spontaneous and stimulated emission in state-of-the-art InGaN QW devices.

4.2.3 In Composition and Si Doping Effects on Optical Transitions

The emission characteristics of the III-nitride system are often influenced by the material combination (e.g., GaN/$Al_xGa_{1-x}N$ or $In_xGa_{1-x}N$/GaN) and the alloy composition in the active region (e.g., $x < 0.1$ or $x > 0.15$ in $In_xGa_{1-x}N$/GaN QWs), as well as other structural properties (e.g., well thickness, doping concentration, number of QWs). There have been several studies on the dependence of structural, electrical, and optical properties on the growth conditions for the AlInGaN material system. Studies on the influence of In composition on the optical and structural characteristics of $In_xGa_{1-x}N$/GaN-based structures are very important for the design of optical device applications and for extending their wavelength range from UV to greenyellow. The effects of Si doping on both structural and optical properties are also crucial for both physical and practical aspects, since Si doping changes not only the carrier concentration but also the growth mode and thus influences optical performance. Therefore, in order to fully understand the influence of any growth parameter changes (e.g., In composition and Si doping) on the characteristics of InGaN-based structures, a comparison study of structural and optical properties is required, since the structural quality may be very sensitively affected by the change in the growth parameters. In this subsection we focus on the effects of In composition and Si doping on optical properties in conjunction with structural properties.

4.2.3.1 Influence of In Composition in InGaN Layers

An increase in the In composition of InGaN results in an enhancement of both the strain and alloy potential fluctuations due to the increase in the degree of lattice mismatch and In phase segregation, respectively. Recently, the emission and absorption spectra of a range of commercial InGaN LEDs and high-quality epilayers were summarized by Martin et al. [60]. They demonstrated a linear dependence of the Stokes shift on the emission peak energy, using experimental spectra of both diode and epilayer samples, supplemented by data from the literature. The broadening of the absorption edge was shown to increase with decreasing emission peak energy. The authors discussed these results in terms of the localization of excitons at highly-In-rich quantum dots within a phase-segregated alloy. More recently, Narukawa et al. investigated the temperature dependence of radiative and nonradiative recombination lifetimes in undoped $In_{0.02}Ga_{0.98}N$ UV LEDs and observed that the incorporation of a small amount of In into the GaN layer improved the external quantum efficiency by the suppression of nonradiative recombination processes [71]. A recent study on the structural and optical properties of $In_xGa_{1-x}N$/GaN MQWs with different In compositions is presented here [72]. Room-temperature (RT) SE wavelengths of the samples used for this work were between 395 nm and 405 nm, which is close to the operational wavelength of state-of-the-art violet current-injection lasers [3].

Figures 4.5a and 4.5b show 10 K PL and PLE, and RT SE spectra of MOCVD-grown $In_xGa_{1-x}N$/GaN MQWs with In compositions of 8.8, 12.0, and 13.3%. These samples consisted of (i) a 2.5-μm-thick GaN buffer layer doped with Si at $3 \times 10^{18}\,cm^{-3}$, (ii) a five-period SL consisting of 3-nm-thick undoped $In_xGa_{1-x}N$ wells and 7-nm-thick GaN barriers doped with Si at $\sim 5 \times 10^{18}\,cm^{-3}$, and (iii) a 100-nm-thick GaN capping layer. In order to obtain samples with different In compositions in the $In_xGa_{1-x}N$ wells, trimethylindium fluxes of 13, 26, and 39 μmol/min were used for the different samples while the $In_xGa_{1-x}N$ well growth time was kept constant. The PLE detection energy was set at the main $In_xGa_{1-x}N$-related PL peak. With increasing In composition, the $In_xGa_{1-x}N$-related PLE band edge redshifted and showed broadened features. By applying the sigmoidal formula $\alpha = \alpha_o/[1 + \exp(E_{eff} - E)/\Delta E)]$, to the PLE spectra [60], the authors obtained "effective bandgap" values of 3.256, 3.207, and 3.165 eV and broadening parameter values of 23, 36, and 40 meV for the samples with In compositions of 8.8, 12.0, and 13.3%, respectively. This broadening of the PLE spectra with increasing In indicates that the absorption states are distributed over a wider energy range due to an increase in the degree of fluctuations in dot size and/or shape [60], or due to interface imperfection as observed in X-ray diffraction (XRD) patterns [72]. The Stokes shift increased from about 135 to 180 meV as the In composition increased from 8.8 to 13.3%. The large Stokes shifts and their increase with In composition can be explained by carrier localization [10,13] or the piezoelectric effect [42], or a combined effect of both mechanisms [11,60].

In order to check device applicability, SE experiments were performed on the $In_xGa_{1-x}N$/GaN MQWs at RT (the normal device operating temperature). The SE spectra shown in Figure 4.5b were obtained for a pump density of $1.5 \times I_{th}$, where I_{th} represents the SE threshold for each sample. Below I_{th}, as the excitation power density was raised the spontaneous emission peak blueshifted due to the band filling of localized states by the intense optical pump, as shown earlier. As the excitation power density was raised above I_{th} a considerable spectral narrowing occurred [49]. The emission

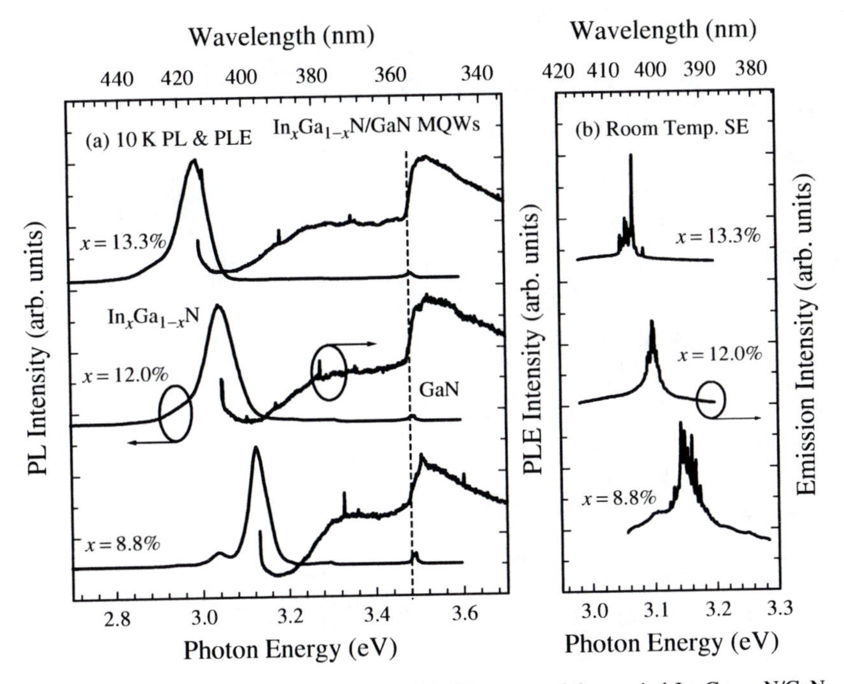

Figure 4.5 (a) 10 K PL and PLE, and (b) RT SE spectra of five-period $In_xGa_{1-x}N/GaN$ MQWs with In compositions of 8.8, 12.0, and 13.3%. (Y. H. Cho)

spectra are composed of many narrow peaks of less than 0.1 nm FWHM, which is on the order of the instrument resolution. The SE threshold was 150, 89, and 78 kW/cm^2, for the samples with In contents of 8.8, 12.0, and 13.3%, respectively. These thresholds are approximately an order of magnitude lower than that of a high-quality nominally undoped single-crystal GaN film measured under the same experimental conditions [73].

With increasing In composition from 8.8 to 13.3%, the SE threshold decreased while the FWHM of the high-resolution XRD SL-1 satellite peaks and the PLE band edge broadening increased. The increase in the SL-1 peaks FWHM with increasing In composition indicates the deterioration of interface quality due to the difficulty of uniform In incorporation into GaN layers. This interface imperfection or composition inhomogeneity is also reflected in the broadened band edge of PLE spectra. This interface fluctuation may be a source of scattering loss, and the absorption states distributed over a wider energy range can broaden the gain spectrum. Interestingly, although both these factors are disadvantageous to SE, a lower SE threshold density is observed for higher In composition. This behavior is contrary to that of traditional III–V semiconductors such as GaAs and InP, for which the FWHM

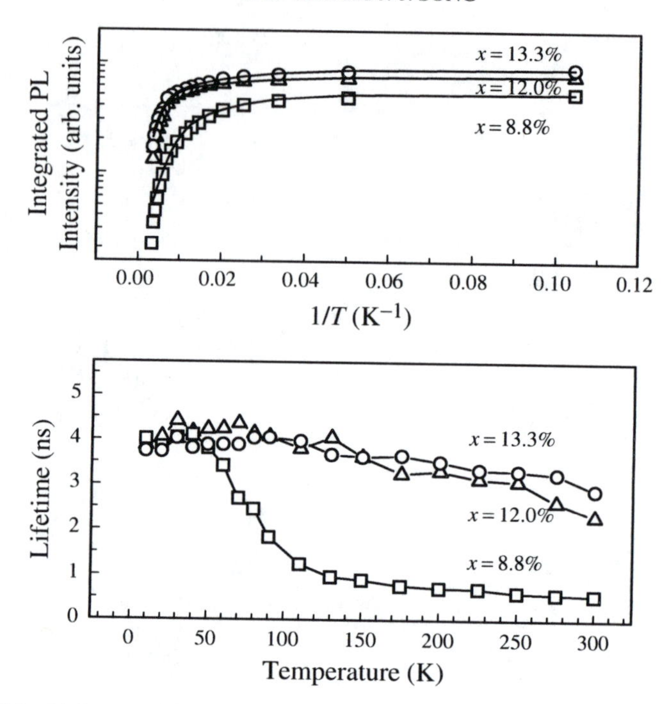

Figure 4.6 (a) Integrated PL intensity as a function of $1/T$ and (b) carrier lifetime as a function of T for InGaN-related emission in $In_xGa_{1-x}N$/GaN MQWs with In compositions of 8.8, 12.0, and 13.3%. (Y. H. Cho)

of the SL diffraction peaks is closely related to the optical quality of MQWs and the performance of devices using MQWs as an active layer [74,75].

Figure 4.6a shows that as the temperature was increased from 10 K to 300 K, the integrated PL intensities decreased by a factor of 25, 6, and 5 for the $In_xGa_{1-x}N$/GaN MQW samples with In compositions of 8.8, 12.0, and 13.3%, respectively. These results indicate that samples with a higher In composition are less sensitive to temperature change, possibly because of less thermally activated nonradiative recombination. In order to clarify this, temperature-dependent carrier lifetimes were measured by TRPL, as illustrated in Figure 4.6b. For $T < 50$ K, the lifetimes increase, indicating that radiative recombination dominates in these samples at low temperatures. With further increases in temperature, the lifetime decreases, reflecting that nonradiative processes predominantly influence the emission at higher temperatures. The lifetime starts to drop steeply at $\sim$ 50 K for the 8.8% In sample. The lifetime for the 12.0% In sample crosses over that for the 13.3% In sample at $\sim$ 150 K and reaches a lower value at 300 K. From an analysis of the temperature dependence of the integrated PL intensities and carrier lifetimes, one can extract 300 K nonradiative recombination lifetimes of 0.6,

2.7, and 3.6 ns for samples with 8.8, 12.0, and 13.3% In, respectively. These results are consistent with the temperature-dependent PL data and indicate the suppression of nonradiative recombination for higher In composition samples.

The possible mechanism for these phenomena can be argued as follows. First, the effect of localization, which keeps carriers away from nonradiative pathways, can be enhanced with the increase of In, as shown by the increase in the Stokes shift in Figure 4.5a [11,12]. Second, the incorporation of more In into the $In_xGa_{1-x}N$ well layer can reduce the density of nonradiative recombination centers [71]. The RT SE threshold can be lowered by suppressing nonradiative recombination, since only the radiative recombination contributes to gain. In addition, a lower RT SE threshold for samples with higher In compositions indicates that this effect of suppressing nonradiative recombination overcomes the drawbacks associated with increasing interface imperfections. It should be noted that this favorable effect of lowering the SE threshold seems to saturate with higher In composition, since there is not much difference in carrier lifetimes and therefore SE threshold densities for the 12.0 and 13.3% In samples. This presents challenges in developing laser diodes with longer emission wavelengths.

As the In composition increases, the FWHM of SL diffraction peaks broadens due to the spatial fluctuation of interfaces; however, the RT SE threshold densities decrease with increasing In composition, and this is attributed to the suppression of nonradiative recombination. The explanation for these phenomena may be the role of In atoms in keeping carriers away from nonradiative pathways and/or in reducing the density of nonradiative recombination centers.

4.2.3.2 Influence of Si Doping in GaN Barriers

The effects of Si doping on both structural and optical properties are crucial, especially for designing practical devices based on III-nitride MQWs. The influence of Si doping on the optical properties of GaN epilayers [76,77], $In_xGa_{1-x}N$/GaN QWs [10,49,78–81], and GaN/$Al_xGa_{1-x}N$ QWs [82,83] has been widely studied. Assuming an ideal case without structural variations caused by Si doping, the Si doping dependence of the emission properties of QW structures provides useful information on the screening of the built-in internal electric field in QW structures; however, it has been reported that Si doping may not only dramatically screen the internal electric field but also seriously change the structural and hence optical properties.

Recently, the influence of Si doping on both structural and optical properties (spontaneous and stimulated emission) in InGaN/GaN MQW structures has been intensively studied [10,49,59,79]. A series of InGaN/GaN MQW samples were grown on *c*-plane sapphire substrates by MOCVD, specifically to study the influence of Si doping in the GaN barriers. The samples consisted of a 1.8-μm-thick GaN buffer layer and a 12-period MQW consisting of 3-nm-thick $In_{0.18}Ga_{0.82}N$ wells and 4.5-nm-thick GaN barriers, with a 100-nm-thick $Al_{0.07}Ga_{0.93}N$ capping layer. To study the influence of Si doping in the GaN barriers, the disilane doping precursor flux was systematically varied from 0 to 4 nmol/min during GaN barrier growth. Accordingly, a doping concentration in the range of $n < 1 \times 10^{17}$ to $3 \times 10^{19}\,\mathrm{cm}^{-3}$ was achieved for the different samples, as determined by secondary ion mass spectroscopy and Hall measurements.

Figure 4.7 shows 10 K PL (solid lines), PLE (dashed lines), and decay times plotted as a function of emission energy (open circles) for the main InGaN-related PL peak. The near-band-edge emission from the AlGaN cladding layer and the GaN barriers are also clearly seen at 3.60 and 3.48 eV, respectively. A large Stokes shift of the PL emission from the InGaN wells with respect to the band edge measured by PLE is clearly observed. As the Si doping concentration increases, the Stokes shift decreases. The Stokes shift for the sample with $n = 3 \times 10^{19}\,\mathrm{cm}^{-3}$ is $\sim$ 120 meV smaller than that of the nominally undoped sample. The effect of Si doping on the decay time of the MQWs was also explored using TRPL measurements. Figure 4.7 (open circles) shows the 10 K decay time (τ_d) monitored at different emission energies. The measured lifetime becomes longer with decreasing emission energy, and hence the peak energy of the emission shifts to the low-energy side with time. This behavior is characteristic of localized states, which in this case are most likely due to alloy fluctuations and/or interface irregularities in the MQWs. A decrease in τ_d with increasing Si doping, from $\sim$ 30 ns (for $n < 1 \times 10^{17}\,\mathrm{cm}^{-3}$ to $\sim$ 4 ns (for $n = 3 \times 10^{19}\,\mathrm{cm}^{-3}$), is clearly seen and can be attributed to a decrease in the potential fluctuations leading to recombination.

One should be very careful in interpreting the low-temperature lifetime results because nonradiative recombination processes (which can be caused by poor sample quality) may affect the measured lifetime. To clarify this point, the temperature dependence of the lifetime was investigated, as shown in Figure 4.8. For the nominally undoped sample with $n < 1 \times 10^{17}\,\mathrm{cm}^{-3}$ (open symbols), an increase of τ_d with temperature (up to around 40 ns for the higher-energy-side emission) was observed at temperatures below 70 K, in qualitative agreement with the temperature dependence of radiative recombination. As the temperature is further increased beyond a certain crossover temperature, T_c, which is determined by the radiative and nonradiative recombination rates, the lifetime starts to decrease, because nonradiative processes predominantly influence the emission. Further evidence of this is given by the fact that the lifetimes become independent of emission energy at higher temperatures. Note that T_c gradually increases as n increases: $T_c \sim 70$ K for $n < 1 \times 10^{17}\,\mathrm{cm}^{-3}$, $T_c \sim 100$ K for $n = 2 \times 10^{18}\,\mathrm{cm}^{-3}$ (not shown here), and $T_c \sim 140$ K for $n = 3 \times 10^{19}\,\mathrm{cm}^{-3}$. These data indicate

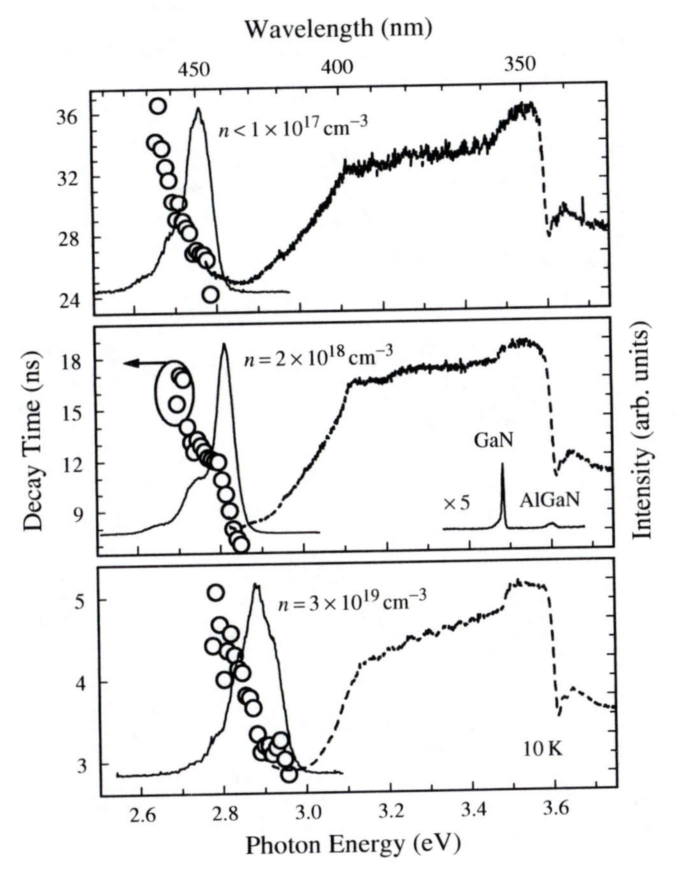

Figure 4.7 10 K PL (solid lines), PLE (dashed lines), and decay time (open circles) of 12-period $In_{0.18}Ga_{0.82}N$/GaN MQWs with Si doping concentrations ranging from $< 1 \times 10^{17}$ to $3 \times 10^{19}\ cm^{-3}$ in the GaN barriers. The Stokes shift between the PL emission peak and the absorption band edge observed from PLE spectra decreases significantly as the Si doping concentration increases. Luminescence decay times measured by TRPL are shown as a function of emission energy. With increasing Si doping concentration, the 10 K lifetimes decrease from ~30 ns (for $n < 1 \times 10^{17}\ cm^{-3}$) to ~4 ns (for $n = 3 \times 10^{19}\ cm^{-3}$). (Y. H. Cho)

that the decrease in lifetime with increasing n is due to a decrease of the radiative recombination lifetime itself rather than an increased influence of nonradiative recombination processes. Therefore, the decrease in lifetime with increasing n is mainly due to a decrease in potential localization and hence a decrease in the carrier migration time into the lower tail states in the MQW active regions.

To investigate the effect of Si doping on GaN surface morphology, three reference 7-nm-thick GaN epilayers were also grown at 800 °C with disilane flux rates of 0, 0.2, and 2 nmol/min during growth [78,79]. The surface

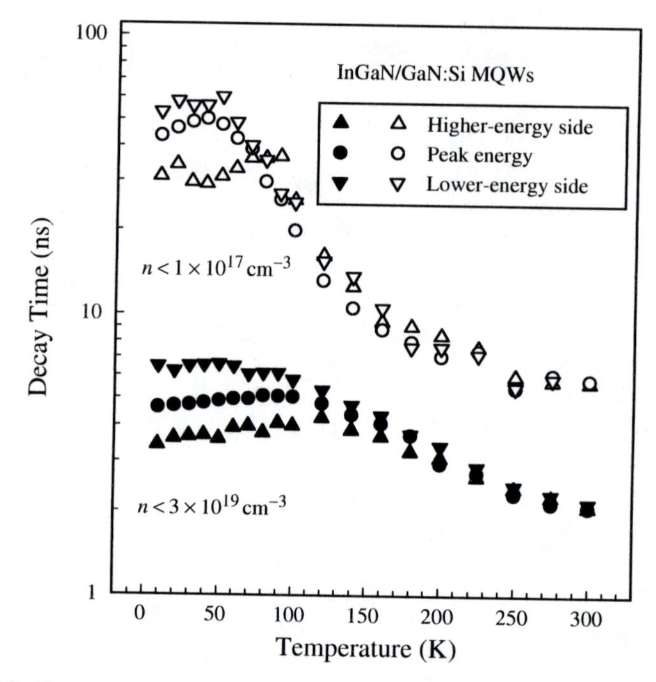

Figure 4.8 Temperature dependence of decay times monitored above (up triangles), below (down triangles), and at (circles) the emission peak for the $In_{0.18}Ga_{0.82}N$/GaN MQWs with $n < 1 \times 10^{17}$ cm^{-3} (open symbols) and $n = 3 \times 10^{19}$ cm^{-3} (closed symbols) in the GaN barriers. The characteristic crossover temperature, T_c, (which is determined by the radiative and nonradiative recombination rates) gradually increases as n increases: $T_c \sim 70$ K for $n < 1 \times 10^{17}$ cm^{-3}, $T_c \sim 100$ K for $n = 2 \times 10^{18}$ cm^{-3} (not shown), and $T_c \sim 140$ K for $n = 3 \times 10^{19}$ cm^{-3}. (Y. H. Cho)

morphology of these reference samples was investigated using atomic force microscopy (AFM) in the tapping mode. Figure 4.9 shows 2-μm × 2-μm AFM images of the three reference GaN epilayers. The root-mean-square surface roughness estimated from the AFM images was 0.49, 0.42, and 0.22 nm for GaN epilayers with disilane flow rates of (a) 0, (b) 0.2, and (c) 2 nmol/min during growth, respectively. Thus, smoother GaN surfaces with a more homogeneous terrace length were achieved at higher Si doping levels.

High-resolution XRD experiments revealed that Si doping of GaN barriers improves the interface properties of the InGaN/GaN MQW samples [59], in good correlation with the optical properties. Figure 4.10 shows (0002) reflection high-resolution XRD ω-2θ scans measured from the 12-period InGaN/GaN MQW structures with different Si doping concentrations ranging from $<1 \times 10^{17}$ to 3×10^{19} cm^{-3} in the GaN barriers. The strongest peak is due to the GaN buffer layer, and the high-angle shoulder of the GaN peak is due to the AlGaN capping layer. All the spectra clearly show higher-order SL diffraction peaks, indicating good layer periodicity. The SL

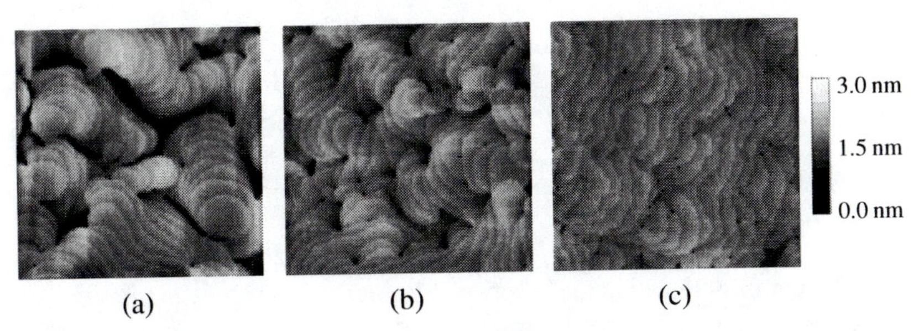

Figure 4.9 2-μm × 2-μm AFM images of three 7-nm-thick GaN films with different disilane flow rates during growth. The root-mean-square surface roughness estimated from AFM images was 0.49, 0.42, and 0.22 nm for GaN epilayers with a disilane flow rate of (a) 0, (b) 0.2, and (c) 2 nmol/min, respectively. A higher-quality GaN surface morphology and a larger average terrace length were achieved by increasing Si incorporation. (Y. H. Cho)

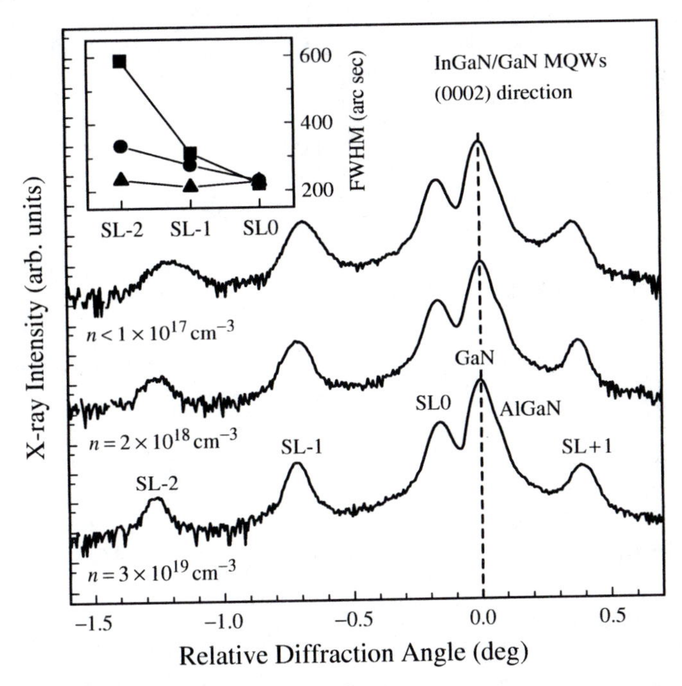

Figure 4.10 (0002) Reflection high-resolution XRD curves of 12-period $In_{0.18}Ga_{0.82}N$/GaN MQWs having different Si doping levels in the GaN barrier layers. The variation in the FWHM of the higher-order SL satellite peaks is shown in the inset as a function of the Si doping level (squares: $< 1 \times 10^{17}$ cm^{-3}; circles: 2×10^{18} cm^{-3}; triangles: 3×10^{19} cm^{-3}).

period was determined from the positions of the SL satellite peaks. Note that the SL peaks for the nominally undoped sample ($n < 1 \times 10^{17}$ cm^{-3}) exhibit broadening with an asymmetric lineshape, whereas the SL peaks for the Si-doped samples (e.g., $n = 3 \times 10^{19}$ cm^{-3}) show a more symmetric lineshape. The broadening mechanism is partially due to spatial variation of the SL period, possibly caused by intermixing and/or interface roughness. The inset in Figure 4.10 depicts the variation in the FWHM of the higher-order SL satellite peaks as a function of Si doping level. As n increases, the FWHM of the higher-order SL satellite peaks narrows: the FWHM of the second-order SL peaks was observed to be 589, 335, and 234 arcsec for $n < 1 \times 10^{17}$, $n = 2 \times 10^{18}$, and $n = 3 \times 10^{19}\,cm^{-3}$, respectively. The ratio of integrated intensity between SL peaks is almost the same for all three samples, as expected for nominally identical SL structures that differ principally only in their degree of structural perfection. In addition, symmetric and asymmetric reciprocal space mapping scans were similar for the different samples, regardless of the Si doping concentration. Accordingly, it was concluded that Si doping in the GaN barriers significantly improves the structural and interface quality of the InGaN/GaN MQWs. Since the incorporation of Si atoms can change the quality and/or surface free energy of the GaN barriers, it may affect the growth condition (or mode) of the subsequent InGaN wells and interfaces [76,84,85]. The observed (i) decrease in the Stokes shift and (ii) decrease in radiative recombination lifetime with increasing Si doping can be explained by a decrease in potential fluctuations and an enhancement of the interfacial structural qualities with the incorporation of Si atoms into the GaN layers. Therefore, although Si doping of the GaN barriers does not change the overall strain state (i.e., it does not cause relaxation of the lattice-mismatch strain), it does significantly affect the structural and interface quality of InGaN/GaN MQWs.

Moreover, Bidnyk et al. reported the influence of Si doping on the emission efficiency and SE threshold pump density of InGaN/GaN MQWs with different Si doping concentrations in the GaN barriers [49]. All the samples showed a strong blueshift of the spontaneous emission with increasing excitation density, mainly due to band filling of the energy-tail states. The blueshift ceased shortly before the onset of SE. The SE spectrum for the MQW sample with $n = 2 \times 10^{18}\,cm^{-3}$ was composed of many narrow peaks of ~ 1 Å FWHM, and there was no noticeable broadening of the SE peaks when the temperature was tuned from 175 to 575 K. Interestingly, only moderately Si doped MQW samples exhibited enhanced luminescence efficiency and a reduction of the SE threshold. That is, the maximum emission intensity and the lowest SE threshold ($55\,kW/cm^2$) were achieved for the sample with $n = 2 \times 10^{18}\,cm^{-3}$, and a SE threshold of $165\,kW/cm^2$ was obtained for the sample with $n = 3 \times 10^{19}\,cm^{-3}$. This fact, combined with the preceding results, indicates that although Si doping improves the interface properties, it does not necessarily enhance the SE properties. This strongly reflects that localized carriers at potential fluctuations caused by interface irregularities play an important role in emission. We shall later see more evidence that recombination of localized carriers is the predominant emission mechanism in InGaN/GaN MQW structures used for current state-of-the-art laser diodes.

Similar effects have been found for Si doping in InGaN barriers of $In_xGa_{1-x}N/In_yGa_{1-y}N$ QW structures [81,86]. Spatially resolved CL studies have showed that as the Si doping in InGaN barriers increases, the density of dot–like CL bright spots increases, in good agreement with the increased density of nanoscale islands observed by AFM [86].

4.3 OPTICAL TRANSITIONS AT VARIOUS TEMPERATURES AND EXCITATION CONDITIONS

The fundamental properties of optical transitions in InGaN-based heterostructures were discussed in the previous section. This section focuses mainly on the optical phenomena associated with strong carrier localization in InGaN-based heterostructures. The localized band-tail states are identified by analyzing their emission spectra dependence on temperature, excitation energy, and excitation density. The temperature and excitation condition dependence of the emission features gives important insight into the recombination mechanisms responsible for the spontaneous and stimulated emission, and provides relevant information for both physical and practical aspects. This section is divided into two subsections. In the first subsection, the temperature dependence of PL spectra, integrated PL intensity, temporal evolution, and optically pumped SE spectra from InGaN structures is examined. In the second subsection, energy-selective spontaneous and stimulated emission studies of InGaN-based structures are presented. Both spontaneous and stimulated emission spectra have been observed for excitation photon energies over a wide spectral range above the emission peak position, illustrating the presence of localized band-tail states in both spontaneous and stimulated emission. The excitation density and length dependence of stimulated emission is also discussed.

4.3.1 Temperature Dependence of Optical Transition

4.3.1.1 Temperature-Induced Anomalous Optical Transition

It is known that the temperature dependence of the fundamental energy gap is mainly caused by the changes of band structure induced by lattice thermal expansion and electron–phonon interactions. The temperature-induced change of the fundamental energy gap E_g can be generally given by the

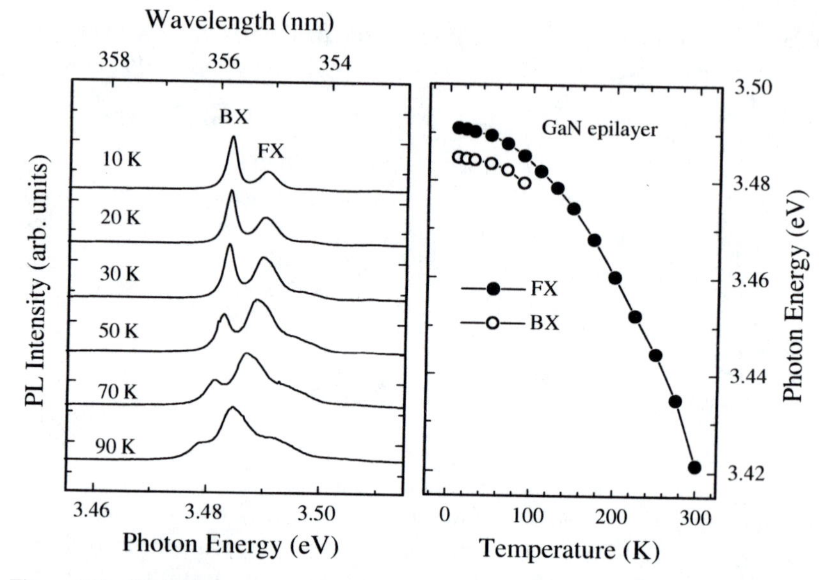

Figure 4.11 Typical temperature dependence of near-band-edge free-exciton (FX) and bound-exciton (BX) luminescence spectra and peak positions taken from an MOCVD-grown GaN epilayer. The energy difference between the BX and the FX energies is about 6 meV, which corresponds to the BX-to-FX transition temperature of ~70 K.

Varshni empirical equation [87],

$$E_g(T) = E_g(0) - \frac{aT^2}{(b+T)}, \tag{4.2}$$

where $E_g(T)$ is the transition energy at a temperature T, and a and b are known as Varshni thermal coefficients. The parameters $a = 8.32 \times 10^{-4}$ eV/K (1×10^{-5} eV/K) and $b = 835.6$ K (1196 K) for the GaN $\Gamma_9^v - \Gamma_7^c$ ($In_{0.14}Ga_{0.86}N$) transition were obtained from photoreflectance studies [88,89]. Figure 4.11 shows a typical temperature dependence of the free-exciton (FX) and the bound-exciton (BX) emissions from an MOCVD-grown GaN epilayer. The temperature-dependent PL peak shift for the GaN is consistent with the estimated energy decrease of about 70 meV between 10 and 300 K. The energy difference between the BX and the FX energies is about 6 meV, which corresponds to the BX-to-FX transition temperature of ~ 70 K.

On the other hand, it has recently been reported that the PL emission from InGaN-based structures does not follow the typical temperature dependence of the energy gap shrinkage. An anomalous temperature-induced luminescence blueshift has been observed in InGaN QW structures by several groups and has been attributed to an involvement of band-tail states

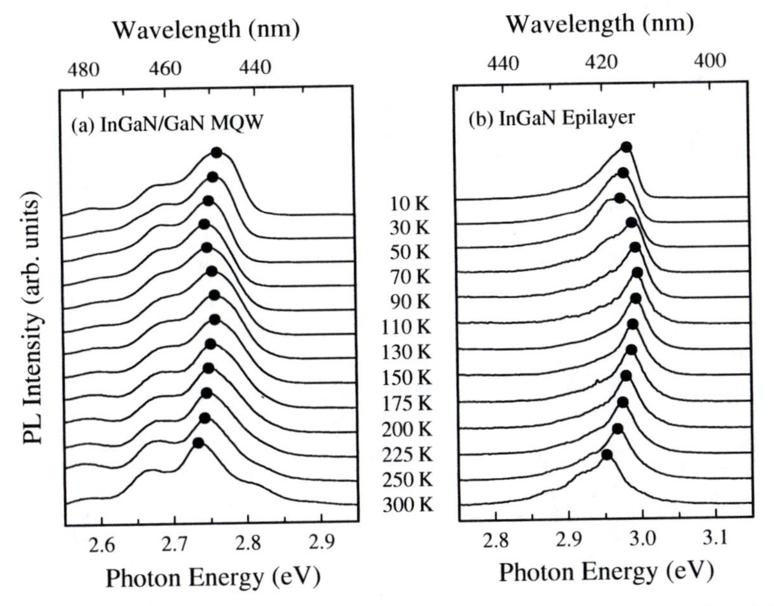

Figure 4.12 Typical InGaN-related PL spectra for (a) an $In_{0.18}Ga_{0.82}N$/GaN MQW and (b) an $In_{0.18}Ga_{0.82}N$ epilayer in the temperature range 10 to 300 K. The main emission peak of both samples shows an S-shaped shift with increasing temperature (solid circles). All spectra are normalized and shifted in the vertical direction for clarity. Note that the turning temperature from redshift to blueshift is about 70 and 50 K for the $In_{0.18}Ga_{0.82}N$/GaN MQW and the $In_{0.18}Ga_{0.82}N$ epilayer, respectively.

caused by potential fluctuations [90–92]. We shall present studies of the correlation between the temperature-induced anomalous emission behavior with its carrier dynamics for both InGaN epilayers and InGaN/GaN MQWs [92,93]. A similar deviation from the typical temperature-induced energy gap shrinkage has been reported in ordered (Al)GaInP [94,95] and disordered (Ga)AlAs/GaAs superlattices [96,97].

Figure 4.12 shows the evolution of the InGaN-related PL spectra for Figure 4.12a an $In_{0.18}Ga_{0.82}N$/GaN MQW and Figure 4.12b an $In_{0.18}Ga_{0.82}N$ epilayer over the temperature range 10 to 300 K. As the temperature increases from 10 K to T_I, where T_I is 70 (50) K for the MQW (epilayer), the peak energy position E_{PL} redshifts 19 (10) meV. This value is about five times as large as the expected bandgap shrinkage of ~4 (2) meV for the MQW (epilayer) over this temperature range [89]. With a further increase in temperature, E_{PL} blueshifts 14 (22.5) meV from T_I to T_{II}, where T_{II} is 150 (110) K for the MQW (epilayer). Considering the estimated temperature-induced bandgap shrinkage of ~13 (7) meV for the MQW (epilayer), the actual blueshift of the PL peak with respect to the band edge is about 27 (29.5) meV over this temperature range. When the temperature is further increased

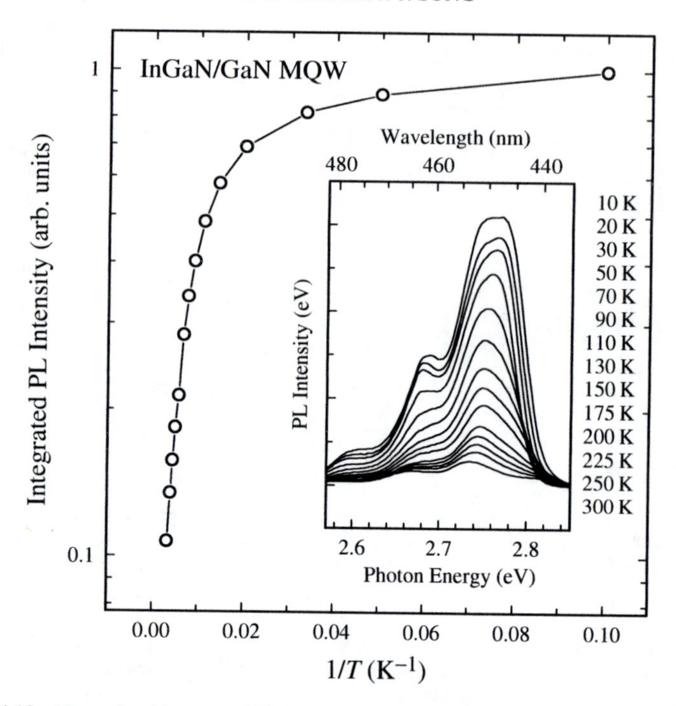

Figure 4.13 Normalized integrated PL intensity as a function of $1/T$ for the InGaN-related emission in $In_{0.18}Ga_{0.82}N$/GaN MQWs (open circles). The inset shows InGaN-related PL spectra for the temperature range 10 to 300 K. An activation energy of ~35 meV is obtained from the Arrhenius plot.

above T_{II}, the peak positions redshift again. From the observed redshift of 16 (45) meV and the expected bandgap shrinkage of ~ 43 (51) meV from T_{II} to 300 K for the MQW (epilayer), an actual blueshift of the PL peak relative to the band edge is estimated to be about 27 (6) meV in this temperature range.

Figure 4.13 shows an Arrhenius plot of the normalized integrated PL intensity of the InGaN-related PL emission of the $In_{0.18}Ga_{0.82}N$/GaN MQW. The total luminescence intensity from this sample is reduced by only one order of magnitude from 10 to 300 K, indicating a high PL efficiency even at high temperatures. (For the MOCVD-grown GaN epilayer seen in Figure 4.11, the change in integrated PL intensity between 10 and 300 K is two to three orders of magnitude.) For $T > 70$ K, the integrated PL intensity of the InGaN-related luminescence is thermally activated with an activation energy of about 35 meV. In general, the quenching of the luminescence with temperature can be explained by thermal emission of carriers out of confining potentials with an activation energy correlated with the depth of the confining potentials. Since the observed activation energy is much less than the band offsets as well as the bandgap difference between the wells and the barriers, the thermal quenching of the InGaN-related emission is *not* due to the thermal activation of electrons and/or holes from the InGaN wells into the GaN barriers. Instead, the dominant mechanism for the quenching of the InGaN-related PL is thermionic emission of photogenerated carriers out of the potential minima caused by potential variations, such as alloy and interface fluctuations, as will be discussed later.

The carrier dynamics of the InGaN-related luminescence was studied by TRPL over the same temperature range. Figure 4.14 shows E_{PL}, the relative energy difference (ΔE) between E_{PL} and E_g at each temperature, and the decay times (τ_d) monitored at the peak energy, lower-energy side, and higher-energy side of the peak as a function of temperature. A comparison clearly shows that the temperature dependence of ΔE and E_{PL} is strongly correlated with the change in τ_d. For both the $In_{0.18}Ga_{0.82}N$/GaN MQW and the $In_{0.18}Ga_{0.82}N$ epilayer, an overall increase of τ_d is observed with increasing temperature for $T < T_I$, in qualitative agreement with the temperature dependence of radiative recombination [67,68]. As seen in the previous section, in this temperature range, τ_d becomes longer with decreasing emission energy, and hence the peak energy of the emission shifts to the low energy side with time. This behavior is characteristic of carrier localization, most likely due to alloy fluctuations (and/or interface roughness in MQWs). As the temperature is further increased beyond T_I, the lifetime of the MQW (epilayer) quickly decreases to less than 10 (0.1) ns and remains almost constant between T_{II} and 300 K, indicating that nonradiative processes predominantly affect the emission. This is further evidenced by the fact that the difference between the lifetimes monitored above, below, and at the peak energy disappears for $T > T_I$, in contrast with the observations for $T < T_I$. This characteristic temperature T_I is also where the turnover occurs from redshift to blueshift for ΔE and E_{PL} with increasing temperature. Furthermore, in the temperature range between T_I and T_{II}, where a blueshift of E_{PL} is detected, τ_d dramatically decreases from 35 to 8 (0.4 to 0.05) ns for the MQW (epilayer). Above T_{II} , where a redshift of E_{PL} is observed, no sudden change in τ_d occurs for both the MQW and the epilayer.

From these results, the InGaN-related recombination mechanism for different temperature ranges can be explained as follows: (i) For $T < T_I$, since the radiative recombination process is dominant, the carrier lifetime increases, giving the carriers more opportunity to relax down into lower energy-tail states caused by the inhomogeneous potential fluctuations before recombining. This reduces the higher-energy side emission intensity and thus produces a redshift in the peak energy position with increasing temperature. (ii) For $T_I < T < T_{II}$, since the dissociation rate is increased and other nonradiative processes become dominant, the carrier lifetimes decrease greatly with increasing temperature and also become independent of emission energy. Due to the decreasing lifetime, the carriers recombine before reaching the lower energy-tail states. This gives rise to an apparent

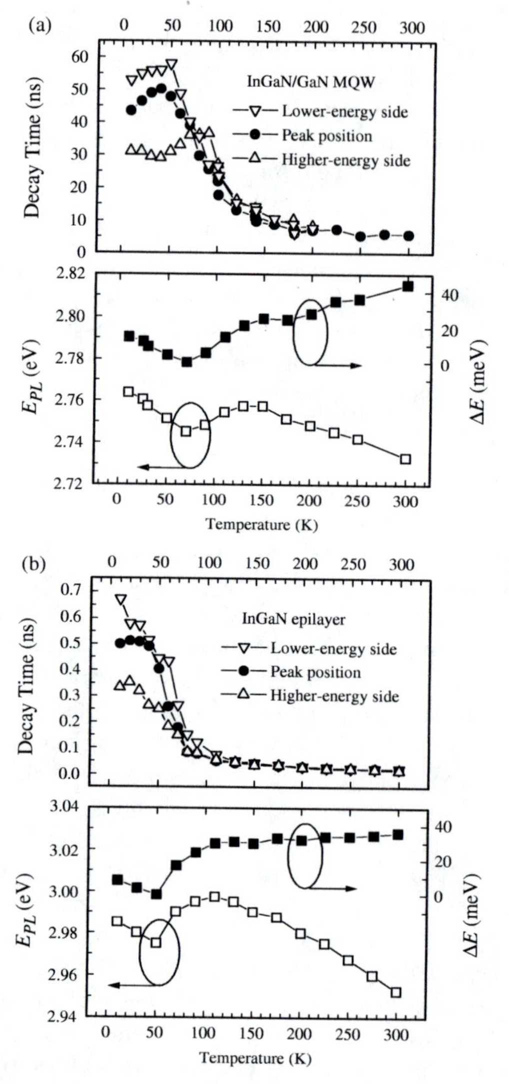

Figure 4.14 InGaN-related PL spectral peak position E_{PL} (open squares) and decay time as a function of temperature for (a) an $In_{0.18}Ga_{0.82}N$/GaN MQW and (b) an $In_{0.18}Ga_{0.82}N$ epilayer. ΔE (closed squares) is the relative energy difference between E_{PL} and E_g at each temperature. The minimum value of ΔE is designated as zero for simplicity. Note that the lower-energy side of the PL peak has a much longer lifetime than the higher-energy side below a certain temperature T_I, whereas there is little difference between lifetimes monitored above, below, and at the peak energy above T_I, where T_I is about 70 (50) K for the MQW (epilayer). This characteristic temperature T_I is also where the turnover of the InGaN PL peak energy from redshift to blueshift occurs with increasing temperature. A blueshift behavior of emission peak energy with increasing temperature is still seen at room temperature for the $In_{0.18}Ga_{0.82}N$/GaN MQW, whereas this behavior is much less for the $In_{0.18}Ga_{0.82}N$ epilayer.

broadening of the higher-energy side emission and leads to a blueshift in the peak energy. (iii) For $T > T_{II}$, since nonradiative recombination processes are dominant and the lifetimes are almost constant [in contrast with case (ii)], the photogenerated carriers are less affected by the change in carrier lifetime, so the blueshift behavior becomes smaller. Note that the slope of ΔE is very sensitive to the change in τ_d with temperature for both the InGaN/GaN MQW and the InGaN epilayer. Since this blueshift behavior is smaller than the temperature-induced bandgap shrinkage in this temperature range, the peak position exhibits an overall redshift behavior. Consequently, the change in carrier recombination mechanism with increasing temperature causes the S-shaped redshift–blueshift–redshift behavior of the peak energy for the InGaN-related luminescence, and the anomalous temperature dependence of the emission is attributed to optical transitions from "localized" to "extended" energy-tail states in the InGaN-based structures.

An interesting difference in the emission shift behavior of the InGaN/GaN MQW compared with that of the InGaN epilayer is the greater effective blueshifting behavior even near RT, probably due to a different degree of carrier localization for the two structures. Except for this, a similar temperature-induced S-shaped emission behavior was observed for both the InGaN/GaN MQW and the InGaN epilayer, even though τ_d of the former is about two orders of magnitudes longer than that of the latter. This fact indicates that the anomalous temperature-induced emission shift mainly depends on the change in carrier recombination dynamics rather than on the absolute value of τ_d.

4.3.1.2 Temperature Dependence of Optically Pumped Stimulated Emission

The optical pumping method has been widely used for studying stimulated emission and lasing phenomena, since it can generate sufficiently high carrier densities necessary for the onset of stimulated emission and lasing without electrical contacts on samples so that somewhat complicated doping and device processing procedures are not required. The stimulated recombination process in InGaN/GaN quantum wells is better understood by investigating the temperature dependence of the SE spectral shape, integrated SE intensity, and SE threshold density I_{th} [49]. Optically pumped SE was performed in the side-pumping geometry where edge emission from the samples was collected into a 1-m spectrometer and recorded by an optical multichannel analyzer or a UV-enhanced gated CCD. The third harmonic (355 nm) of an injection-seeded Nd : YAG laser with a pulse width of 6 ns and a repetition rate of 30 Hz was used as the pumping source. The laser beam was focused on the sample surface using a cylindrical lens to form an excitation spot in the form of a line [73]. The laser light intensity could be attenuated continuously using a variable neutral density filter.

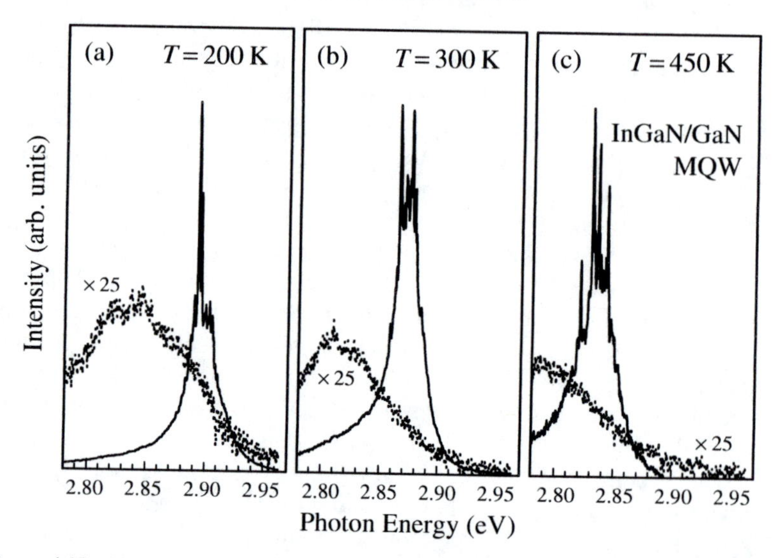

Figure 4.15 SE spectra of an $In_{0.18}Ga_{0.82}N$/GaN MQW with barrier Si doping of $n = 2 \times 10^{18}\,cm^{-3}$ at (a) 200 K, (b) 300 K, and (c) 450 K, illustrating that the SE peak is composed of a multitude of narrow (<0.1 nm) peaks that do not noticeably broaden with increasing temperature. These SE spectra were collected for excitation densities twice the SE threshold for the respective temperatures. The spontaneous emission spectra (dotted lines) are also shown for excitation densities half the SE threshold at each temperature. The SE spectra are normalized for clarity.

Figure 4.15 shows SE spectra for an $In_{0.18}Ga_{0.82}N$/GaN MQW sample with barrier Si doping of $n = 2 \times 10^{18}\,cm^{-3}$. The SE peak has a statistical distribution of a multitude of narrow (< 0.1 nm FWHM) emission lines. No noticeable broadening of these narrow emission lines was observed as the temperature was tuned from 10 K to 575 K. The dotted lines are the broad spontaneous emission spectra taken at pump densities approximately half that of the SE threshold for each temperature. As the excitation power density was raised above the SE threshold, a considerable spectral narrowing occurred (solid lines in Figure 4.15). The major effect of the temperature change from 200 K [Figure 4.15a] to 450 K [Figure 4.15c] was a shift of the spontaneous and stimulated emission peaks toward lower energy.

An increase in the temperature leads to a decrease in PL intensity, indicating the onset of losses and a decrease in quantum efficiency of MQWs. At high temperatures, only a small fraction of carriers reach conduction-band minima, and most of them recombine nonradiatively. Therefore, the temperature increase reduces modal gain and leads to an increase in the SE threshold. To evaluate the relative number of carriers that recombine radiatively, the integrated emission intensity was measured as a function of excitation power for different temperatures, as shown in Figure 4.16.

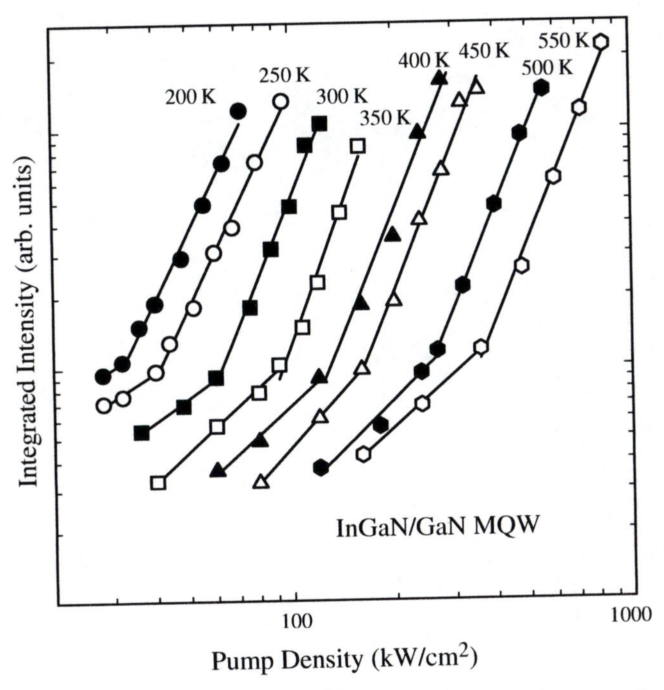

Figure 4.16 Integrated intensity of $In_{0.18}Ga_{0.82}N$/GaN MQW emission as a function of optical excitation density for different temperatures. The slope change from 0.8–1.3 to 2.2–3.0 indicates the transition from spontaneous to stimulated emission.

For the temperature range studied, it was found that under low excitation densities, the integrated intensity from the sample increases almost linearly with pump density I_p (i.e., $\propto I_p^{\alpha}$, where $\alpha = 0.8 - 1.3$), whereas at high excitation densities, this dependence becomes superlinear (i.e., $\propto I_p^{\beta}$, where $\beta = 2.2 - 3.0$). The excitation pump power at which the slope of the integrated intensity changes corresponds to the SE threshold density I_{th} at a given temperature. Interestingly, the slopes of the integrated intensity below and above the SE threshold did not significantly change over the temperature range involved in this study. This indicates that the SE mechanism in InGaN/GaN MQWs at RT is the same at hundreds of degrees above RT.

The temperature dependence of the SE threshold is shown in Figure 4.17 (solid dots). SE was observed throughout the entire temperature range studied, from 175 K to 575 K. The SE threshold was $\sim 25\,\text{kW/cm}^2$ at 175 K, $\sim 55\,\text{kW/cm}^2$ at 300 K, and $\sim 300\,\text{kW/cm}^2$ at 575 K, and roughly followed an exponential dependence. It is likely that such a low SE threshold is due to the large localization of carriers in MQWs. The solid line in Figure 4.17 is a least-squares fit of the experimental data to the empirical form $I_{th}(T) = I_o \exp(T/T_c)$, for the temperature dependence of I_{th}. The

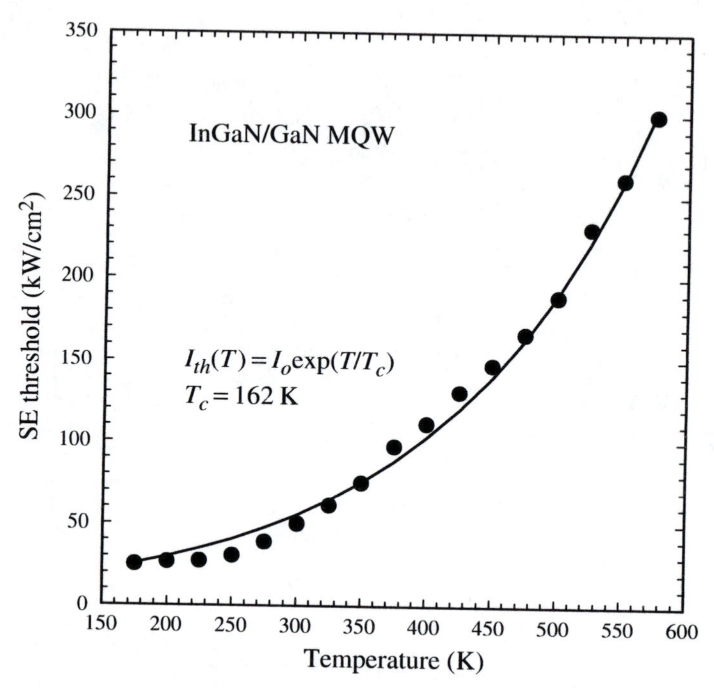

Figure 4.17 Temperature dependence of the SE threshold in the temperature range 175–575 K for the $In_{0.18}Ga_{0.82}N$/GaN MQW sample. The solid line represents the best least-squares fit to the experimental data (solid dots). A characteristic temperature of 162 K is derived from the fit over the temperature range of 175–575 K.

characteristic temperature T_c, is about 162 K in the temperature range 175–575 K for this sample. This characteristic temperature is considerably larger than the near-RT values reported for laser structures based on other III–V (Refs. [98,99]) and II–VI (Refs. [100,101]) materials, where relatively small values of T_c were a strong limiting factor for high-temperature laser operation. Such a low sensitivity of I_{th} to temperature change in InGaN/GaN MQWs opens up enormous opportunities for their high-temperature applications. Laser diodes with InGaN/GaN lasing mediums can potentially operate at temperatures exceeding RT by a few hundred degrees Kelvin.

4.3.2 Excitation Condition Dependence of Optical Transition

4.3.2.1 Energy-Selective Luminescence Properties

Energy-selective spontaneous and stimulated emission studies of InGaN-based structures provide important information about the energy boundary between localized and extended (or delocalized) band-tail states as well as the different carrier generation/transfer dynamics. In this section we review

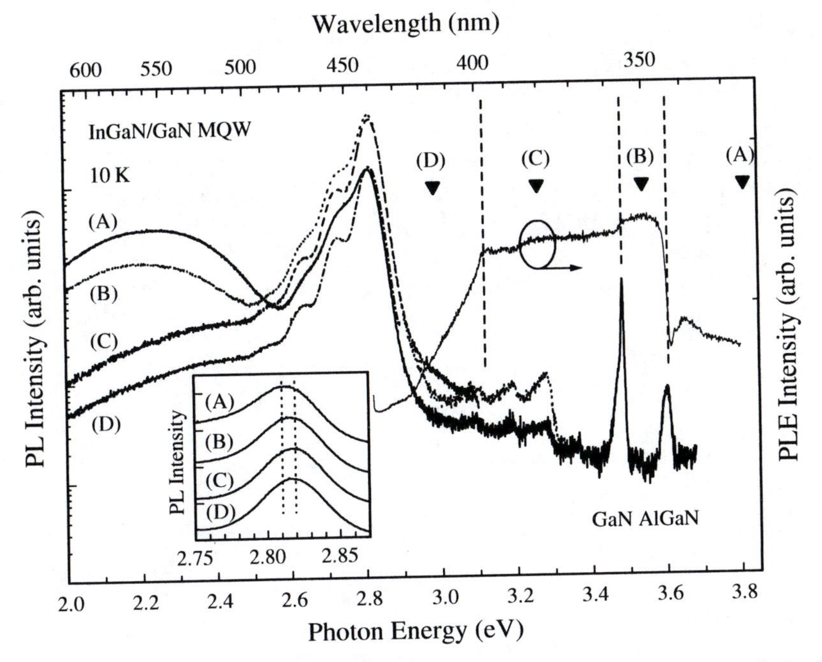

Figure 4.18 Evolution of the PL emission of the $In_{0.18}Ga_{0.82}N$/GaN MQW for excitation energies of (A) 3.81, (B) 3.54, (C) 3.26, and (D) 2.99 eV. Note the change in the intensity ratio of the main peak to the secondary peak between spectra. The inset shows a linear-scale plot of the normalized main emission spectra, which are vertically shifted for clarity. The excitation photon energies are indicated over the PLE spectrum for reference.

recent studies on the excitation energy dependence of the InGaN-related emission in InGaN/GaN MQWs investigated by energy-selective PL and PLE spectroscopy [102,103].

Figure 4.18 shows 10 K $In_{0.18}Ga_{0.82}N$-related luminescence spectra measured with four different excitation photon energies E_{exc} of (A) 3.81, (B) 3.54, (C) 3.26, and (D) 2.99 eV. Each E_{exc} is indicated over the PLE spectrum for reference. The InGaN-related *main* and *secondary peaks* are shown at energies of 2.80 and $\sim$ 2.25 eV, respectively. The oscillations on the main PL peak are due to Fabry–Perot interference fringes. When E_{exc} varies from above (curve A) to below (curve B) the near-band-edge emission energy E_g of the AlGaN capping layer ($E_{g,\text{AlGaN}}$), the relative intensity ratio of the main peak to the secondary peak noticeably changes. For $E_{exc} < E_{g,\text{GaN}}$, the secondary peak has nearly disappeared while the main peak remains (curve C). When the excitation energy is further decreased to just above the InGaN main emission peak (curve D), no noticeable change is observed for the PL shape except for a decrease in the overall intensity of the emission. The inset in Figure 4.18 shows the normalized spectra for the main peak on a linear scale. As the excitation energy is decreased (from curve A to curve D) the intensity of the lower-energy side of the main peak is reduced, whereas that of the higher-energy side is enhanced. This results in a ~7-meV blueshift of the peak energy and a narrower spectral width for the main peak with decreasing excitation energy. These facts strongly reflect that the recombination mechanism of the InGaN-related emission is significantly affected by the excitation (or carrier generation) conditions, as we shall see later.

The upper part of Figure 4.19 shows the 10 K PLE spectra for the $In_{0.18}Ga_{0.82}N$-related main PL emission monitored at (a) 2.87, (b) 2.81, (c) 2.75, and (d) 2.68 eV. The PL spectrum for $E_{exc} = 3.81$ eV is also shown in this figure for reference. When the PLE detection energy is set

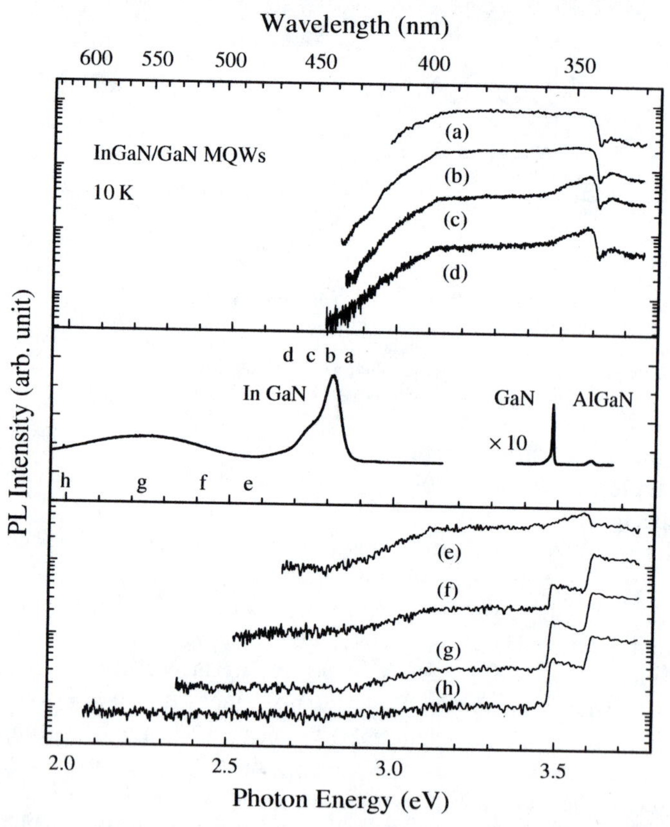

Figure 4.19 10 K PLE spectra taken at detection energies of (a) 2.87, (b) 2.81, (c) 2.75, (d) 2.68, (e) 2.57, (f) 2.41, (g) 2.24, and (h) 2.01 eV. The PL spectrum for an excitation energy of 3.81 eV is also shown for reference. The spectra are shifted in the vertical direction for clarity, and the respective detection energies are indicated on the PL plot. As the detection energy decreases, the contribution of the $Al_{0.07}Ga_{0.93}N$ capping layer noticeably increases. When the detection energy is lower than 2.24 eV, the contribution of the $In_{0.18}Ga_{0.82}N$ wells is almost negligible.

below the peak energy of the main emission (curves c and d), the contributions from the InGaN wells, the GaN barriers, and the AlGaN capping layer are clearly distinguishable, whereas for the detection energy above and at the peak position of the main InGaN emission $E_{p,\mathrm{InGaN}}$ (curves a and b), the PLE signal below $E_{g,\mathrm{AlGaN}}$ shows almost a constant intensity across the $E_{g,\mathrm{GaN}}$ region, indicating that carrier generation in the InGaN rather than in the GaN plays an important role. In both cases, the PLE signal above $E_{g,\mathrm{AlGaN}}$ is suddenly diminished, obviously due to the absorption of the $Al_{0.07}Ga_{0.93}N$ capping layer. When the detection energy is below $E_{p,\mathrm{InGaN}}$ (curves c and d), the contributions of both the GaN and AlGaN regions are enhanced compared to the curves a and b: as the detection energy decreases, the PLE signal above $E_{g,\mathrm{GaN}}$ is monotonically raised with respect to the almost flat region of the PLE signal between 3.15 and 3.4 eV. These facts imply that for $E_{exc} > E_{g,\mathrm{GaN}}$, the lower-energy side of the InGaN main emission peak is governed mainly by carrier generation in the GaN barriers and subsequent carrier transfer to the InGaN wells. From the different PLE contributions for the higher- and lower-energy sides of $E_{p,\mathrm{InGaN}}$, one can expect different recombination mechanisms for various excitation energies, as will be described later.

The lower part of Figure 4.19 also shows the 10 K PLE spectra for the secondary peak taken for detection at (e) 2.57, (f) 2.41, (g) 2.24, and (h) 2.01 eV. When the detection energy is higher than 2.24 eV, the InGaN wells still partly contribute to the secondary peak emission (curves e and f), whereas for the detection energy below 2.24 eV, the contribution of the InGaN wells almost disappears (curves g and h). Note that as the detection energy decreases, the contribution of the AlGaN capping layer is noticeably increased. These facts indicate that the main source of the secondary peak does not originate from the InGaN wells but originates predominantly from the AlGaN capping layer and partly from the GaN layers (consistent with the so-called yellow luminescence band). This observation was also confirmed by PL measurements using the 325-nm line of a He–Cd laser with varying excitation intensities. As the He–Cd laser excitation intensity increased, the relative emission intensity ratio of the InGaN main peak to the secondary peak increases. That is, the intensity of the main peak increased linearly, whereas that of the secondary peak saturated with increasing excitation intensity. This is another indication that the secondary peak is defect-related emission. Therefore, the main peak is due to the InGaN wells, and the secondary peak is mainly from the AlGaN capping layer and the GaN barriers rather than the InGaN wells. For $E_{exc} > E_{g,\mathrm{AlGaN}}$, most carriers are generated in the AlGaN capping layer, and these photogenerated carriers partly migrate into the MQW region (corresponding to the main peak) and partly recombine via defect-related luminescence in the AlGaN layer itself and in the GaN barriers (corresponding to the secondary peak).

The PLE observations in the frequency domain are closely related to the carrier dynamics in the time domain. To clarify the temporal dynamics of the luminescence, TRPL measurements were performed at 10 K for different E_{exc}. Time-integrated PL spectra showed the same behavior as observed in the described cw PL measurements for the InGaN-related main emission (see Figure 4.18): a blueshift of the peak energy and a spectral narrowing of the lower-energy side as E_{exc} decreased from above $E_{g,\mathrm{AlGaN}}$ to below $E_{g,\mathrm{GaN}}$. The carrier recombination lifetime becomes longer with decreasing emission energy, and therefore the peak energy of the emission shifts to the low-energy side with time, as shown in Figure 4.3. No significant change in lifetime at and above the peak energy position ($\tau_d \sim 12$ ns at the peak position) was observed when E_{exc} was varied; however, the peak position reached the lower-energy side faster for the $E_{exc} > E_{g,\mathrm{GaN}}$ case than for the $E_{exc} < E_{g,\mathrm{GaN}}$ case, and after ~20 ns, the peak position was almost the same for both cases. The starting peak position is lower for the $E_{exc} > E_{g,\mathrm{GaN}}$ case than for the $E_{exc} < E_{g,\mathrm{GaN}}$ case, so for the $E_{exc} > E_{g,\mathrm{GaN}}$ case, the redshifting behavior with time is smaller and most carriers recombine at relatively lower energies. The carriers generated from the GaN barriers (or AlGaN capping layer) migrate toward the InGaN wells, and the carrier transfer allows the photogenerated carriers to have a larger probability of reaching the lower energy states at the MQW interfaces. This may be due to more binding and scattering of carriers by interface defects and roughness near the MQWs interface region, since the carriers go through the MQW interfaces during the carrier transfer, enhancing trapping and recombination rates at the interface-related states. Therefore, the lower emission peak energy position and the spectral broadening to lower energies for $E_{exc} > E_{g,\mathrm{GaN}}$ indicate that the lower (or deeper) energy-tail states are more related to the interface-related defects and roughness at the MQW interfaces than to the alloy fluctuations and impurities within the InGaN wells, whereas for $E_{exc} < E_{g,\mathrm{GaN}}$, the carriers responsible for the emission are directly generated within the InGaN wells and thus recombine at relatively higher emission energies.

Another interesting feature in the PLE spectra of Figure 4.19 is the different slopes for $E_{exc} < 3.0$ eV in curves b, c, and d. In order to investigate details of this phenomenon, energy-selective PL measurements were performed for $E_{exc} < E_{g,\mathrm{GaN}}$. For the energy-selective PL study, the second harmonic of a mode-locked Ti : sapphire laser was used as a tunable excitation source to excite the sample normal to the sample surface. The emission was collected normal to the sample surface, coupled into a 1-m spectrometer, and spectrally analyzed using a CCD. The excitation photon energy from the frequency-doubled Ti : sapphire laser was tuned across the states responsible for the "soft" absorption edge of the InGaN layers. Spontaneous emission was observed for excitation photon energies over a wide spectral range above the spontaneous emission peak position (2.8 eV at 10 K). As the excitation photon energy was tuned from just below the bandgap of the GaN barrier layers to the high-energy side of the InGaN absorption edge no significant changes in the spontaneous emission spectra were observed. The only change was a decrease in the emission intensity as the excitation photon energy was tuned below the higher-energy side of the 'soft' InGaN absorption edge, consistent with the reduction in the absorption coefficient with decreasing photon energy in this spectral region; however, as the excitation photon energy was tuned below approximately 2.98 eV significant changes in the spontaneous emission spectra were observed. For excitation photon energies below ~ 2.98 eV, emission from the low-energy wing of the main InGaN PL peak became more and more pronounced with decreasing excitation photon energy, as illustrated in Figures 4.20 and 4.21. The spectra in Figure 4.20 have been normalized at 2.79 eV for clarity. Figure 4.21 clearly shows the onset of this behavior for excitation photon energies below approximately 2.98 eV.

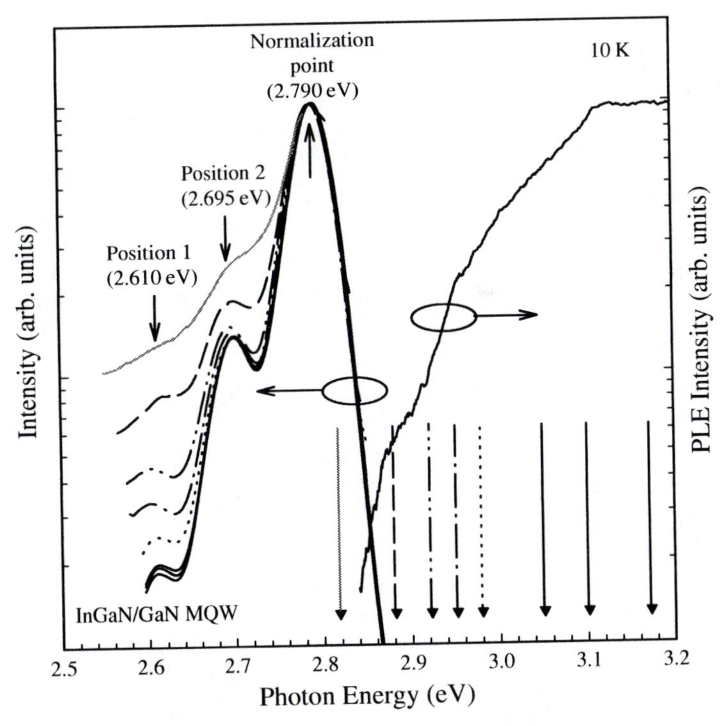

Figure 4.20 Spontaneous emission spectra from an $In_{0.18}Ga_{0.82}N$/GaN MQW with 2×10^{18} cm^{-3} Si doping in the GaN barriers as a function of excitation photon energy. The excitation photon energy for a given spectrum is indicated by the corresponding arrow on the x-axis. The spectra have been normalized at 2.79 eV for clarity. The absorption edge measured by PLE is also shown.

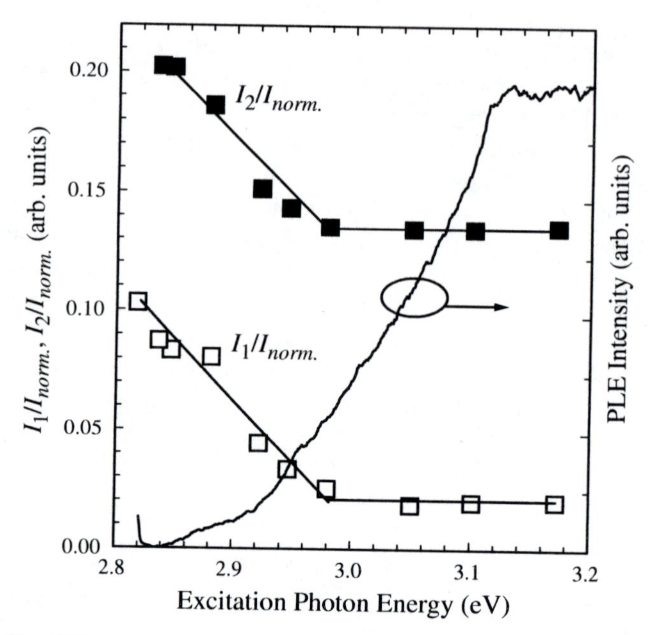

Figure 4.21 10 K spontaneous emission intensity from an $In_{0.18}Ga_{0.82}N$/GaN MQW at 2.61 eV (I_1) and 2.695 eV (I_2) relative to the peak emission intensity (at 2.79 eV) as a function of excitation photon energy. The solid lines are shown only as a visual guide. A clear shift in the emission lineshape is seen for excitation photon energies below ~2.98 eV. The absorption edge measured by PLE is also given (solid line) for comparison.

This behavior indicates that the transition from localized to extended band-tail states is located at ~2.98 eV. As such, this energy defines the "mobility edge" of the band-tail states in this structure, where carriers with energy above this value are free to migrate and those of lesser energy are spatially localized by (large) potential fluctuations in the InGaN layers.

This redshift in the spontaneous emission is explained as follows: when E_{exc} is higher than the mobility edge, the photogenerated carriers can easily populate the tail states by their migration, but their lifetimes are relatively short due to the presence nonradiative recombination channels. As E_{exc} is tuned below the mobility edge the nonradiative recombination rate is significantly reduced due to the capture of the carriers in small volumes. This increase in lifetime with decreasing E_{exc} results in increased radiative recombination from lower energy states. The position of the mobility edge is seen to be ~80 meV above the spontaneous emission peak and ~130 meV below the start of the InGaN absorption edge. Its spectral position indicates that extremely large potential fluctuations are present in the InGaN active layers of the MQW, leading to carrier confinement and resulting in efficient radiative recombination.

4.3.2.2 *Energy-Selective Optically Pumped Stimulated Emission*

Energy-selective optically pumped SE experiments similar to the preceding energy-selective PL experiments were performed to elucidate whether the localized carrier recombination responsible for the spontaneous emission is also responsible for SE in these materials [50]. For energy-selective SE experiments, the second harmonic of an injection-seeded, Q-switched Nd:YAG laser (532 nm) pumped an amplified dye laser. The deep red to near-infrared radiation from the dye laser was frequency doubled in a nonlinear crystal to achieve the near-UV to violet laser radiation needed to optically excite the InGaN/GaN MQWs in the spectral region of interest. The frequency-doubled radiation ($\sim$ 4-ns pulse width, 10-Hz repetition rate) was focused to a line on the sample surface. The excitation spot size was approximately $100 \times 5000\,\mu$m. The emission from the sample was collected from one edge of the sample, coupled into a 1-m spectrometer, and spectrally analyzed using a UV enhanced CCD.

The experiments were performed on $In_{0.18}Ga_{0.82}N$/GaN MQW samples with undoped ($n < 1 \times 10^{17}$ cm^{-3}) and Si-doped ($n \sim 2 \times 10^{18}$ cm^{-3}) GaN barriers. Figure 4.22 illustrates the change in the SE spectra with decreasing E_{exc} for the InGaN/GaN MQW with undoped GaN barrier layers, and Figure 4.23 shows the behavior of the SE peak for the Si-doped MQW, which is presented in Figure 4.24. Figures 4.22 and 4.23 show the behavior of the SE peak as E_{exc} is tuned from above $E_{g,\mathrm{AlGaN}}$ to below the absorption edge of the InGaN active layers. As E_{exc} is tuned to lower energies no noticeable change is observed in the SE spectrum until E_{exc} crosses a certain value ($\sim$ 3.0 eV and $\sim$ 2.95 eV for undoped and Si-doped MQWs, respectively), at which point the SE peak redshifts quickly with decreasing E_{exc}. The redshift of the SE peak as E_{exc} is tuned below $\sim$2.95 eV (for the Si-doped MQW) is consistent with the mobility edge behavior observed for the spontaneous emission, as described previously. This behavior of the SE peak is due to enhanced population inversion at lower energies as the carriers are confined more efficiently with decreasing E_{exc}. The mobility edge measured in these experiments lies $\sim$ 110 meV above the spontaneous emission peak, $\sim$62 meV above the SE peak, and $\sim$ 185 meV below the absorption edge of the InGaN well regions. The location of the mobility edge with respect to the spontaneous and stimulated emission peaks further indicates that large potential fluctuations are present in the InGaN active regions, resulting in strong carrier localization. This explains the efficient radiative recombination (stimulated and spontaneous) observed from these structures as well as the small temperature sensitivity of the SE.

As a measure of the coupling efficiency of the exciting photons to the gain mechanism responsible for the SE peak, I_{th} was measured as a function of E_{exc} for the InGaN/GaN MQW with Si-doped GaN barrier layers.

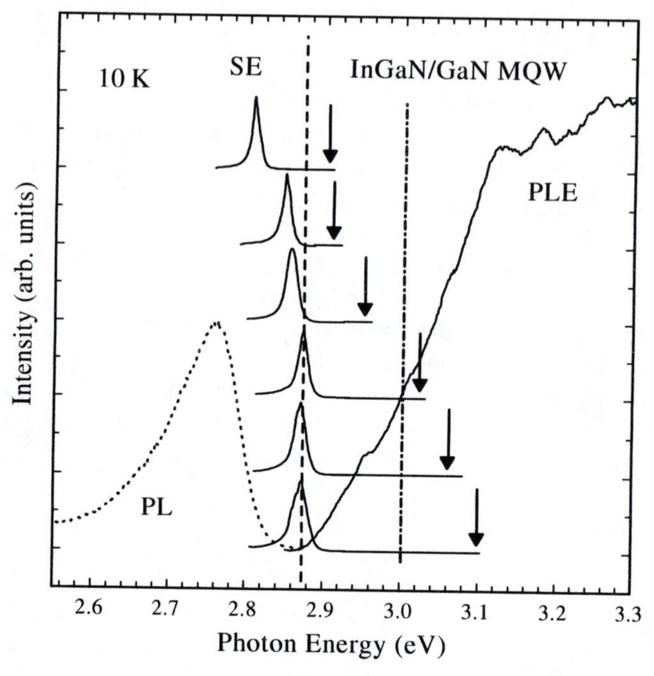

Figure 4.22 SE spectra as a function of excitation photon energy, E_{exc}, for the $In_{0.18}Ga_{0.82}N$/GaN MQW with undoped ($n < 1 \times 10^{17}$ cm^{-3}) GaN barriers. "Mobility edge"–type behavior is clearly seen in the SE spectra with decreasing E_{exc}. The redshift of the SE peak is shown with decreasing E_{exc} as E_{exc} is tuned below the mobility edge (the dashed-dotted line). The excitation photon energies for the given SE spectra are represented by the arrows. The dashed line is given as a reference for the unshifted SE peak position. The SE spectra are normalized and displaced vertically for clarity. PL and PLE spectra are also given for comparison.

A comparison between this and the coupling efficiency obtained for the spontaneous emission peak measured by PLE is given in Figure 4.24, where $1/I_{th}$ is plotted as a function of E_{exc} to give a better measure of the coupling efficiency and to afford an easier comparison with the PLE measurements. Four distinct slope changes are seen in the PLE spectrum. The first, at ~3.12 eV, marks the beginning of the "soft" absorption edge of the InGaN active region, whereas the other three, located at ~2.96 eV, 2.92 eV, and 2.87 eV suggest varying degrees of localization. The change in $1/I_{th}$ at ~3.1 eV is due to a decrease in the absorption coefficient below the absorption edge, and is an expected result. The change in $1/I_{th}$ is coincident with a slope change in the PLE spectrum and is attributed to a significant decrease in the effective absorption cross section for excitation photon energies below the mobility edge. The inset in Figure 4.24 shows the same comparison over a wider energy range for both the Si-doped and undoped MQWs. A strong correlation between

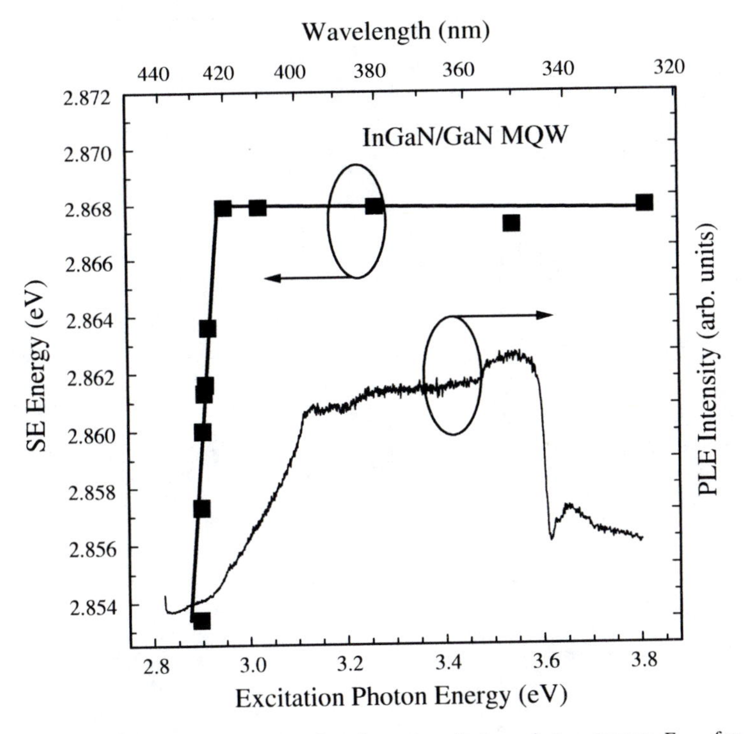

Figure 4.23 SE peak position as a function of excitation photon energy E_{exc} for the $In_{0.18}Ga_{0.82}N$/GaN MQW with $2 \times 10^{18}\ cm^{-3}$ Si doping in the GaN barriers. "Mobility edge"–type behavior is clearly seen in the SE spectra with decreasing E_{exc}. The solid lines are given only as a visual guide. The PLE spectrum is also given for comparison.

the SE threshold and PLE measurements is clearly seen over the entire range for both samples. The correlation between the high carrier density behavior and the cw PLE results indicates that carrier localization plays a significant role in both the spontaneous and stimulated emission processes.

4.3.2.3 Excitation-Length Dependence of Stimulated Emission

Recently, separate research groups have observed two different SE peaks from $In_xGa_{1-x}N$/GaN MQWs grown under different growth conditions, illustrating dramatically different SE behavior in $In_xGa_{1-x}N$ MQWs for relatively small changes in the experimental conditions [104–106], which also indicates that this two-peak SE behavior is a general property of present state-of-the-art $In_xGa_{1-x}N$-based blue LDs. These results suggest that some of the varied results reported in the literature may be due to slightly different experimental conditions, resulting in significant changes in the SE behavior. We shall present the detailed results of the excitation length and the excitation

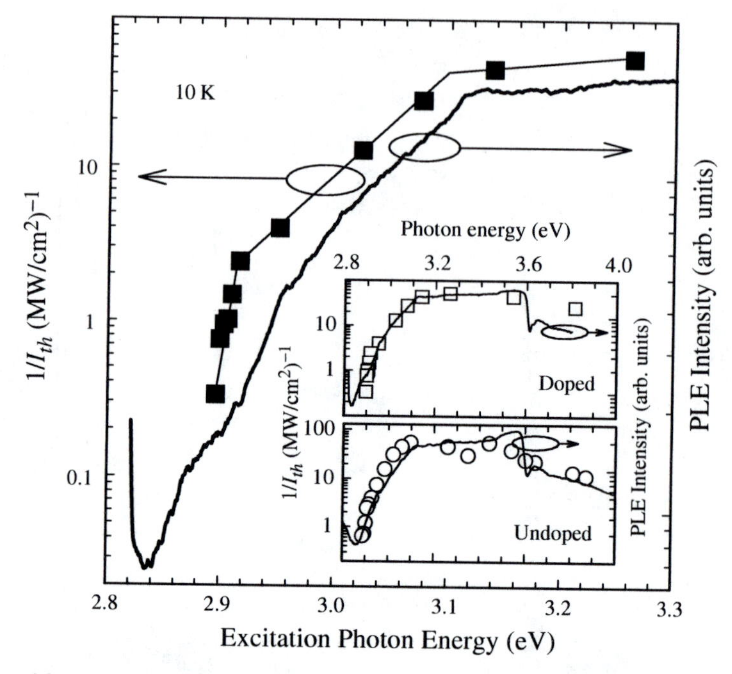

Figure 4.24 Inverse SE threshold as a function of excitation photon energy E_{exc} (solid squares) shown in comparison with the results of low-power PLE experiments (solid line) for the Si-doped $In_{0.18}Ga_{0.82}N$/GaN MQW. The inset shows the same comparison over a wider energy range for both the undoped and Si-doped InGaN/GaN MQWs, illustrating the similarities over the entire energy range.

density dependence of SE behaviors in $In_xGa_{1-x}N$/GaN MQWs. These experiments provide a better understanding of the SE and lasing behavior of these structures, which is important for the development and optimization of future LD structures.

Typical power-dependent emission spectra at 10 K for an $In_{0.18}Ga_{0.82}N$/GaN MQW sample with barrier Si doping of $n = 2 \times 10^{18}\,cm^{-3}$ are shown in Figure 4.25 for an excitation length L_{exc} of 1300 μm. At low I_{exc}, a broad spontaneous emission peak is observed at ~2.81 eV, consistent with low-power cw PL spectra. As I_{exc} is increased, a new peak emerges at ~2.90 eV [designated here as SE peak (1)] and grows superlinearly with increasing I_{exc}. If I_{exc} is further increased, another new peak is observed at ~2.86 eV [designated here as SE peak (2)], which also grows superlinearly with increasing I_{exc}. SE peak (1) is the statistical distribution of a multitude of narrow (~0.1 nm) emission lines (see Figure 4.15), and no noticeable broadening of these narrow emission lines in SE peak (1) was observed as the temperature was increased from 10 K to 575 K, as illustrated in Figure 4.15. Both SE peaks (1) and (2) originate on the high-energy side of the low-power spontaneous emission peak (given by

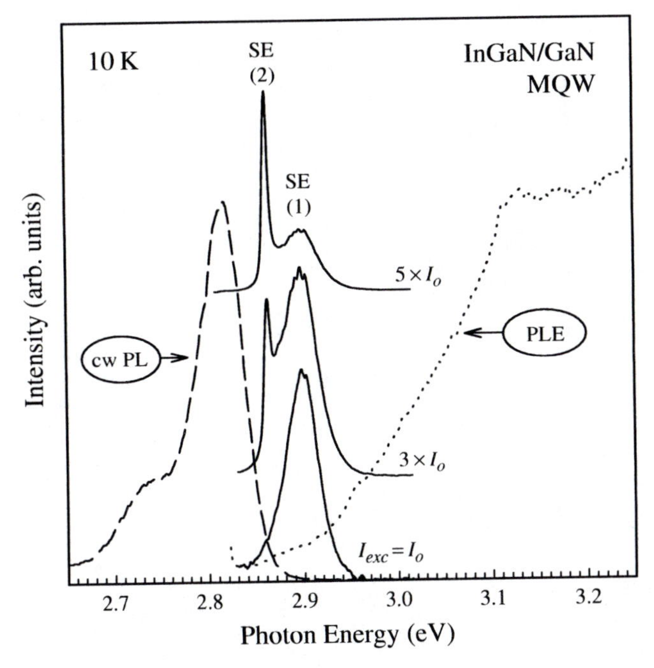

Figure 4.25 10 K SE spectra (solid lines) from an $In_{0.18}Ga_{0.82}N$/GaN MQW sample subjected to several excitation densities, where $I_o = 100\,kW/cm^2$. The low-power PL (dashed line) and PLE (dotted line) spectra are also shown for comparison. The SE spectra are normalized and displaced vertically for clarity.

the dashed line in Figure 4.25) and are redshifted by more than 0.2 eV below the "soft" absorption edge. Both SE peaks are highly TE polarized, with a TE to TM ratio of ~200. SE peak (2) was studied in previous sections and attributed to stimulated recombination of localized states because of energy-selective optically pumped SE studies showing mobility edge–type behavior in the SE spectra with varying excitation photon energy.

Figure 4.26 shows I_{th} of SE peaks (1) and (2) as a function of L_{exc} at 10 K and RT. Note that I_{th} for peak (2) is larger than that of peak (1) for all excitation lengths employed but approaches that of peak (1) with increasing L_{exc} in an asymptotic fashion. The high I_{th} of peak (2) with respect to peak (1) and its increased presence for longer L_{exc} suggest that it results from a lower-gain process than that of peak (1). Figure 4.27 shows the peak positions of SE peaks (1) and (2) as a function of L_{exc} at 10 K and RT. For $L_{exc} < 500\,\mu m$, only SE peak (1) is observed and has a peak emission photon energy at 10 (300) K of ~2.92 (2.88) eV and an I_{th} of ~100 (475) KW/cm^2. As I_{exc} is increased and/or Lexc is increased SE peak (2) emerges at 2.86 (2.83) eV at 10 (300) K. The peak positions were measured for I_{exc} fixed relative to the SE thresholds of the respective peaks; i.e., $I_{exc} = 2 \times I_{th}$. As L_{exc} is increased

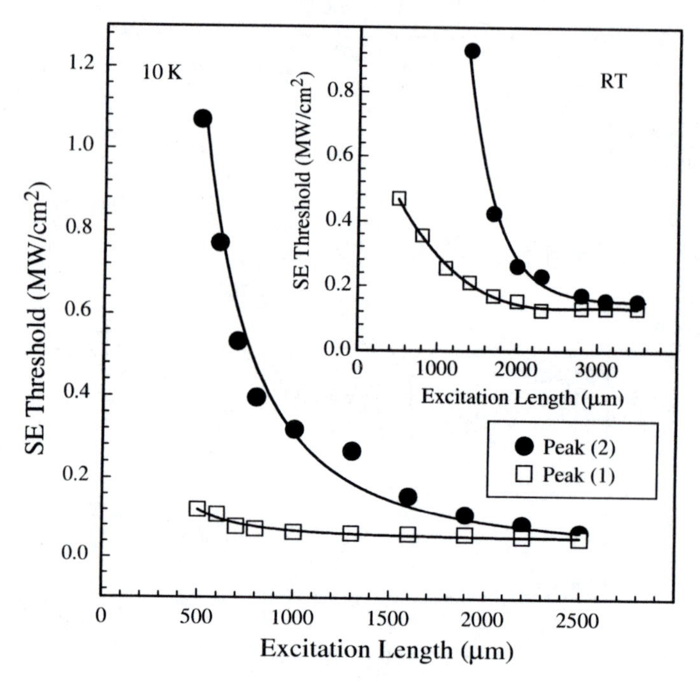

Figure 4.26 SE threshold as a function of excitation length for SE peaks (1) and (2) at 10 K for the $In_{0.18}Ga_{0.82}N$/GaN MQW. The solid lines are given only as visual guides.

SE peak (1) shifts to lower energies due to a reabsorption process, and the SE peak (2) position is observed to be weakly dependent on L_{exc}. The apparent blueshift of SE peak (2) with increasing L_{exc} is a result of the experimental conditions. Since the I_{th} of SE peak (2) is a strong function of L_{exc}, the peak positions shown for small L_{exc} are for I_{exc} considerably higher than for large L_{exc}. The slight redshift of SE peak (2) with increasing I_{exc} due to many-body effects and lattice heating then manifests itself as the apparent blueshift with increasing L_{exc}. This phenomenon is also observed at 300 K, as shown in the inset in Figure 4.27. The redshift of SE peak (1) with increasing L_{exc} can be explained by gain and absorption competition in the "soft" absorption edge of the InGaN active regions, where gain saturation with longer L_{exc} combined with the background absorption tail leads to the observed redshift. The fact that SE peak (2) does not experience a reabsorption-induced redshift with increasing L_{exc} is explained by the significant reduction of the absorption tail in this spectral region (see Figure 4.25).

The gain saturation behavior of SE peak (1) is consistent with the observation of Kuball et al. [54] of a high-gain mechanism in the band-tail region of MQWs with similar active regions. The large spectral range exhibiting gain is explained by compositional fluctuations inside the active region. The redshift of SE peak (1) with increasing L_{exc} is consistent with observations

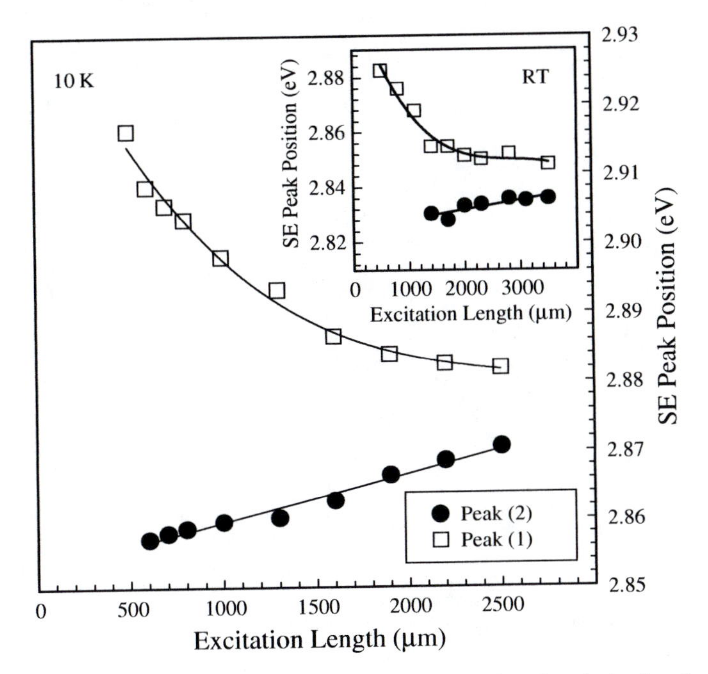

Figure 4.27 Peak position of SE peaks (1) and (2) as a function of excitation length at 10 K for the $In_{0.18}Ga_{0.82}N$/GaN MQW. The inset shows the behavior observed at room temperature. The solid lines are given only as visual guides.

of a redshift in the optical gain spectrum with increasing L_{exc} reported by Mohs et al. [107]. It is also consistent with the observation by Nakamura [55] that the external quantum efficiency of his cw blue LDs *decreases* with increasing cavity length. These similarities, combined with the relatively low I_{th} of SE peak (1) with respect to SE peak (2) and its similar spectral position with laser emission from diodes of similar structure [108], suggest that lasing in current state-of-the-art cw blue LDs originates from the gain mechanism responsible for SE peak (1). Its origin may lie in an entirely different degree of carrier localization than is responsible for SE peak (2). Further experiments are needed to clarify this issue.

The dependence of the emission intensity of peaks (1) and (2) on I_{exc} is shown in Figure 4.28 for $L_{exc} = 1300\,\mu m$ at 10 K. The emission of peak (1) increases in a strongly superlinear fashion ($\sim I_{exc}^{3.8}$) until the I_{th} of peak (2) is reached, at which point it turns linear, indicating that peak (2) competes for gain with peak (1). This is most likely a result of competition for carriers or reabsorption of the emitted photons. The presence of SE peak (2) is therefore seen to be deleterious to SE peak (1). The same process is observed at RT and for various excitation lengths. This gain competition may limit this material's performance in high-power LD applications, where

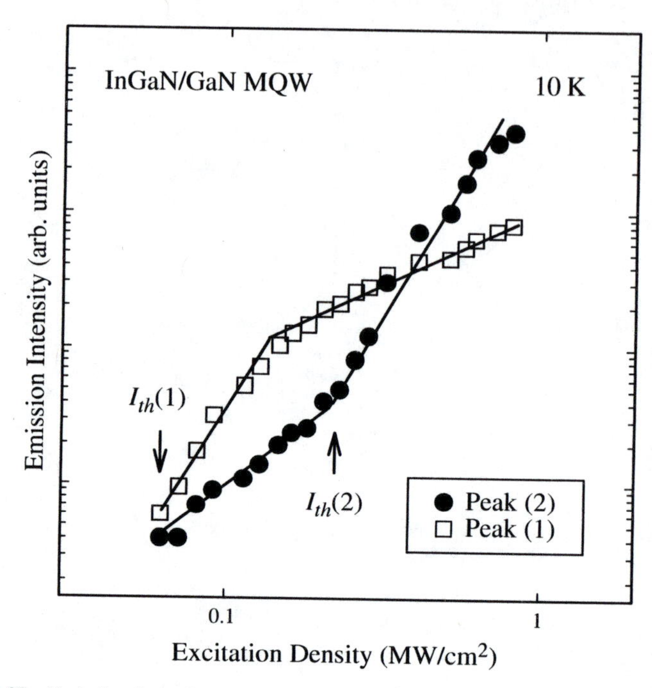

Figure 4.28 Emission intensity of SE peaks (1) and (2) as a function of optical excitation density at 10 K, illustrating gain competition between SE peaks (1) and (2). The excitation length is 1300 μm. The respective SE thresholds I_{th} of SE peaks (1) and (2) are indicated for completeness. The solid lines are given only as visual guides.

increased driving current and/or longer cavity lengths may result in a shift of the dominant gain mechanism and a drastic change in the emission behavior.

4.3.3 Summary

In this section, the optical emission properties associated with localized band-tail states were investigated by their temperature, excitation energy, excitation density, and excitation length dependence. A relationship between an abnormal temperature-induced S-shaped emission shift in InGaN MQWs and epilayers was studied with its carrier dynamics as a function of temperature. It was found that the InGaN-related spontaneous emission features are significantly affected by different carrier recombination dynamics that vary with temperature, because of band-tail states arising from large In alloy inhomogeneity, layer thickness variations, and/or defects in the MQWs. Increasing the temperature strongly affects the recombination features by means of thermal filling of band-tail states in the presence of carrier localization. In other words, thermally populated carriers at band-tail states recombine with different average emission peak energies, depending on the change in carrier lifetime and their thermal energy. Even at RT, the temperature-induced blueshift behavior of the spontaneous emission peak of InGaN/GaN MQWs is still observable, and the emission peak energy is still lower than the mobility edge obtained by energy-selective PL and SE experiments. This suggests that carrier localization plays an important role in InGaN/GaN MQW structures, even at RT operation. Optically pumped SE properties of InGaN/GaN MQWs were studied over a wide temperature range, from 175 K to 575 K. The characteristic temperature derived from the temperature dependence of the SE threshold was 162 K. The following SE features indicate that SE in InGaN/GaN quantum wells is strongly related to carrier localization, rather than to electron–hole plasma: (i) a narrow FWHM of SE spectra and its temperature invariance in the temperature range 175–575 K, (ii) an extremely low SE threshold density, and (iii) a rather weak temperature dependence of the SE threshold. Also, the slopes of the integrated emission intensity below and above the SE threshold are not sensitive to temperature changes. A low SE threshold and a weak temperature sensitivity of the SE threshold make InGaN MQWs an attractive material for the development of LDs operable well above RT.

Excitation energy dependence studies of spontaneous and stimulated emission have provided important information about the energy boundary between localized and extended band tail states. An interesting change in PL spectra was observed with varying excitation photon energy above and below the GaN bandgap. The PL spectra were interpreted by noting the evolution of PLE spectra as a function of detection energy for the InGaN emission. From these results, the lower-energy tail states are more related to the interface-related detects and roughness at the MQW interfaces rather than to alloy fluctuations and impurities within the InGaN wells. A mobility edge–type behavior was found in both PL and SE spectra as the excitation photon energy was tuned across the states responsible for the broadened absorption edge of the InGaN active regions. The mobility edge is well above the PL and SE peak positions but well below the absorption edge, indicating that both the spontaneous and stimulated emissions originate from carriers localized at low-energy band-tail states due to extremely large potential fluctuations in the MQW InGaN active layers. Moreover, the dependence of the SE threshold density on excitation energy is very consistent with the PLE spectra, indicating that the coupling efficiency of the exciting photons to the gain mechanism responsible for the SE peak is correlated with that of the spontaneous emission mechanism. The correlation between the high-density behavior and the low-density cw PLE indicates that carrier localization plays a significant role in both the spontaneous and stimulated emission processes. The SE features in InGaN/GaN QW structures were further investigated through the dependence of the SE on excitation length and density. Two distinctly different SE peaks were observed with different dependencies on excitation length. The high-energy SE peak exhibits a strong redshift with increasing excitation length due to competition between an easily saturable gain mechanism and a background absorption tail, whereas the lower-energy SE peak does not exhibit this reabsorption-induced redshift with increasing excitation length. The presence of the lower-energy SE peak has been shown to be detrimental to the higher-energy SE peak due to gain competition in the InGaN active region. This competition may prove to be an obstacle in the design of InGaN-based high-power LDs, where high current densities and/or long cavity lengths lead to a shift in the dominant gain mechanism and a change in emission characteristics.

Many observations presented in the previous section can be interpreted in terms of a built-in internal electric field and/or carrier localization; however, it should be noted that the experimental observations presented in this section for low (high) carrier densities are not explainable in the framework of internal electric field (electron–hole plasma) alone without consideration of carrier localization associated with potential fluctuation.

4.4 OPTICAL GAIN AND NONLINEARITIES OF InGaN-BASED STRUCTURES

Optical nonlinear properties are of technological importance for optoelectronics applications, since they enable the absorption edge to be easily modified with optical excitation. Optical gain and pump-probe experiments provide significant insight into the mechanisms responsible for stimulated emission and lasing phenomena. The results of nanosecond nondegenerate optical pump-probe absorption experiments on GaN and InGaN epilayers were compared in Chapter 3. This section presents optical gain mechanism studies and nanosecond nondegenerate pump-probe experiments of InGaN QW structures [58]. The most distinctive nonlinear optical feature of InGaN-based structures is the pronounced band filling in their band-tail states. In general, crystal imperfections, impurities, phonons, and carrier-carrier scattering broaden the absorption spectrum and generate absorption-tail states. Nonlinear absorption in the absorption tail is a subtle combination of the effects from broadening, bandgap renormalization, state filling, Coulomb screening, and impurity absorption. Both optical gain and density-dependent band-tail-state filling nonlinearities in InGaN-based structures are discussed in this section. An overview of the recent optically pumped SE and lasing properties in InGaN-based laser structures is given in Refs. [109] and [110].

4.4.1 Optical Gain Measurement by the Variable-Stripe Method

In this subsection, optical gain measurements of InGaN/GaN MQWs and epilayers are presented. Knowledge of the dependence of optical gain on generated carrier density is important for understanding the SE processes and for designing and optimizing practical laser diode structures.

For the optical gain measurements, the variable stripe excitation length method was used [111,112]. The samples were optically excited by the third harmonic (355 nm) of an injection-seeded, Q-switched Nd : YAG laser (~ 5 ns FWHM, 10-Hz repetition rate). The excitation beam was focused to a line on the sample surface using a cylindrical lens, and the excitation length was precisely varied using a mask connected to a computer-controlled stepper motor. The emission was collected from one edge of the samples, coupled into a (1/4)-m spectrometer, and spectrally analyzed using a UV enhanced CCD. The modal gain $g_{\mathrm{mod}}(E)$ at energy E is extracted from

$$\frac{I_1(E, L_1)}{I_2(E, L_2)} = \frac{\exp[g_{\mathrm{mod}}(E)L_1] - 1}{\exp[g_{\mathrm{mod}}(E)L_2] - 1}. \tag{4.3}$$

with $L_{1,2}$ two different stripe excitation lengths and $I_{1,2}$ the respectively measured emission intensities. Special care was taken to avoid gain saturation effects in the optical gain spectra by using stripe lengths shorter than those for which saturation effects occur.

The modal gain spectra at 10 K as a function of above-gap optical excitation density are shown in Figures 4.29a and 4.29b for an $In_{0.18}Ga_{0.82}N$/GaN MQW and an $In_{0.18}Ga_{0.82}N$ epilayer, respectively. The excitation densities in Figure 4.29 are given with respect to the SE threshold measured for long (> 2000 μm) excitation lengths. A clear blueshift in the gain peak with increasing optical excitation is seen for the MQW. This blueshift was observed to stop for $I_{exc} > 12 \times I_{th}$. Further increases in I_{exc} result only in an increase in the modal gain maximum. The maxima of the gain spectra in Figure 4.29a are redshifted by more than 160 meV with respect to the soft absorption edge of the InGaN well layers. The large shift in the gain maximum to higher energy with increasing I_{exc} is consistent with band filling of localized states in the InGaN active layers. Similar behavior was observed at RT. The blueshift in the gain spectra of the InGaN epilayer is seen to be considerably smaller than that of the MQW and to stop at considerably lower excitation densities. The modal gain spectra of both samples correspond spectrally with the low-energy tail of the band-tail-state absorption bleaching spectra, with the crossover from absorption to gain corresponding approximately with the maximum in the observed bleaching.

For completeness, the relevant 10 K PL, PLE, SE, and modal gain spectra are shown together in Figure 4.30 for (a) the $In_{0.18}Ga_{0.82}N$/GaN MQW and (b) the $In_{0.18}Ga_{0.82}N$ epilayer. Note that the x-axis of Figure 4.30b covers half that of Figure 4.30a. Representative SE spectra are shown (solid

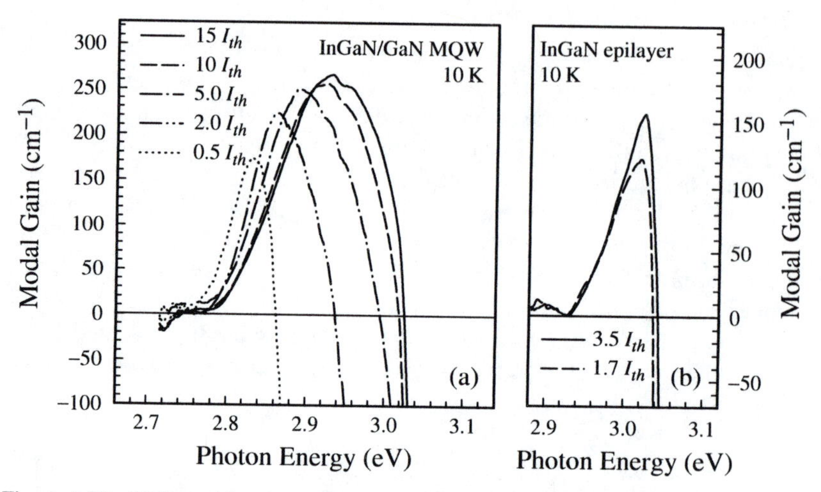

Figure 4.29 10 K modal gain spectra of (a) an $In_{0.18}Ga_{0.82}N$/GaN MQW and (b) an $In_{0.18}Ga_{0.82}N$ epilayer as a function of above-gap optical excitation density. The excitation densities are given with respect to the SE threshold I_{th} measured for long (>2000 μm) excitation lengths. A clear blueshift in the gain maximum and gain/absorption crossover point is seen with increasing excitation density for the $In_{0.18}Ga_{0.82}N$/GaN MQW sample. This trend is much less obvious for the $In_{0.18}Ga_{0.82}N$ epilayer.

line) for a pump density of $I_{exc} = 1.5 \times I_{th}$ (I_{th} denotes the SE threshold) and an excitation spot size of $\sim 100 \times 5000\,\mu m$. Note that the SE peak is situated at the end of the absorption tail for both samples. The SE is seen to occur on the high-energy side of the low-power spontaneous emission peak for the MQW and slightly on the low-energy side for the epilayer. The modal gain spectra were taken with I_{exc} much greater than I_{th} and the excitation length less than 200 μm to minimize reabsorption-induced distortions in the spectra. The modal gain maxima are 250 and 150 cm^{-1} for the MQW and the epilayer, respectively. The SE peak for long excitation lengths (> 5000 μm) is situated on the low-energy tail of the gain curve measured for small excitation lengths (< 200 μm). This is explained by gain and absorption competition in the band-tail region of this alloy, where gain saturation with longer excitation lengths combined with the background absorption tail leads to a redshift of the SE peak with increasing excitation length. The modal gain spectrum for the MQW is seen to be significantly broader (~ 24 nm FWHM) than that of the epilayer (~ 7 nm FWHM), although both peak significantly below the onset of the soft absorption edge. It should also be noted that the optical gain maximum occurs only at photon energies below the mobility edge measured in energy-selective studies described in the previous section, giving further evidence that localized states are the origin of optical gain in the InGaN/GaN MQW structures.

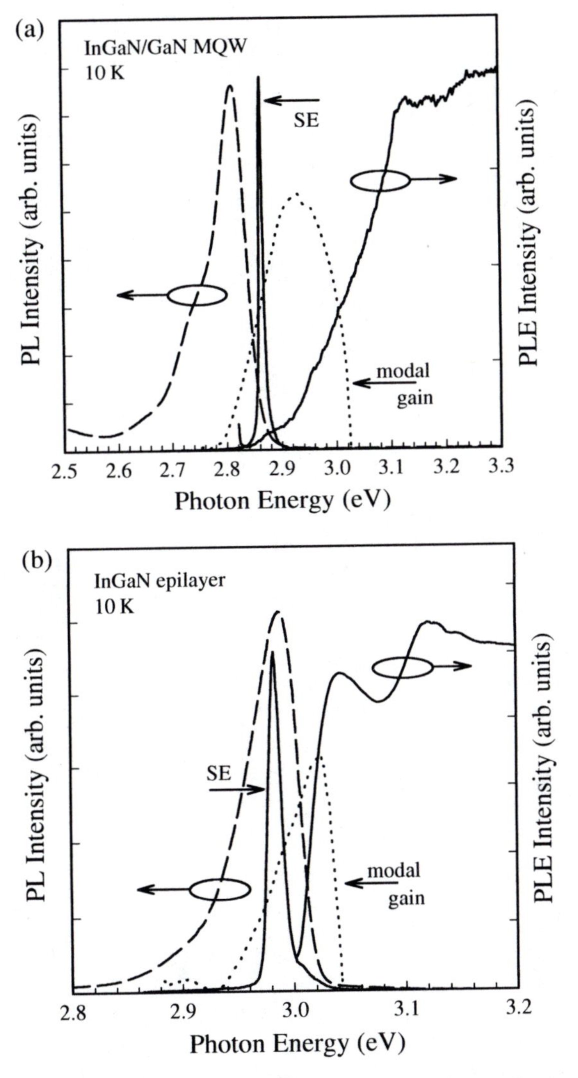

Figure 4.30 PL (dashed lines), SE (solid lines), and modal gain (dotted lines) spectra taken at 10 K from (a) an $In_{0.18}Ga_{0.82}N$/GaN MQW and (b) an $In_{0.18}Ga_{0.82}N$ epilayer. The SE and gain spectra were measured for excitation lengths of >5000 and >200 μm, respectively. The maximum modal gain is 250 and 150 cm^{-1} for the $In_{0.18}Ga_{0.82}N$/GaN MQW and $In_{0.18}Ga_{0.82}N$ epilayer, respectively. The low-density PLE spectra are also shown for reference.

4.4.2 Nondegenerate Pump–Probe Spectroscopy

In this subsection, we present nanosecond nondegenerate optical pump–probe absorption experiments on InGaN/GaN MQWs and InGaN epilayers.

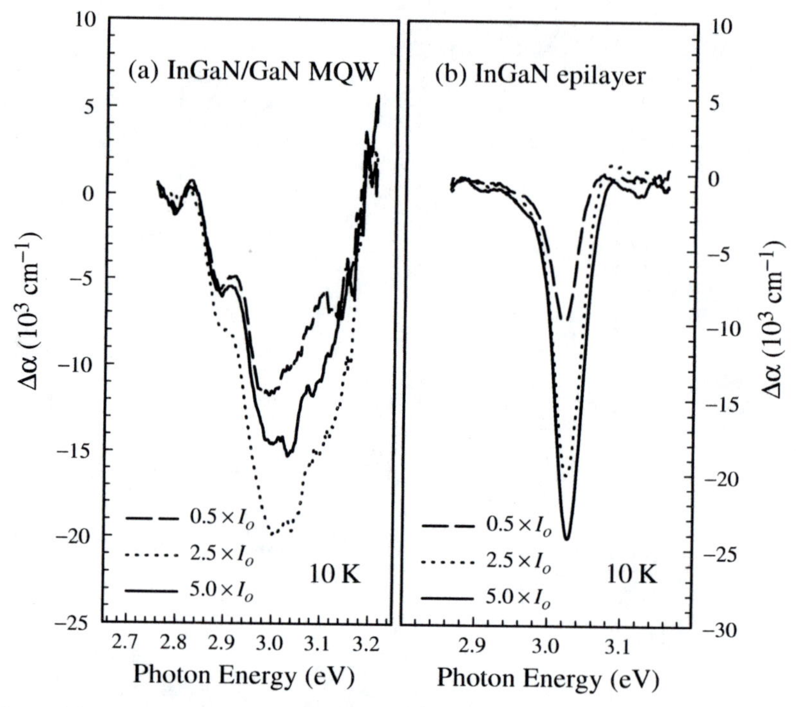

Figure 4.31 10 K nanosecond nondegenerate pump–probe experimental results for (a) an $In_{0.18}Ga_{0.82}N$/GaN MQW and (b) an $In_{0.18}Ga_{0.82}N$ epilayer showing absorption bleaching ($\Delta\alpha$ negative) of band-tail states with increasing excitation density, I_{exc}. $\Delta\alpha(I_{exc}) = \alpha(I_{exc}) - (0)$, and $I_o = 100\,\text{kW/cm}^2$.

These experiments provide further insight into the SE mechanisms associated with InGaN active regions and how the soft absorption edge changes with increasing optical excitation density. The third harmonic of the Nd : YAG laser previously described (355 nm) was used to synchronously pump the individual samples and a dye solution. The probe was the superradiant emission from the dye solution (covering the entire spectral range of the localized states) collected and focused onto the samples, coincidental with the pump beam. The intensity of the probe was kept several orders of magnitude lower than the pump beam to avoid any nonlinear effects due to the probe. The spot size of the probe was kept at approximately one-third that of the pump to minimize the role of variations in the pump intensity across the excitation spot. The transmitted (broadband) probe (with and without the pump beam) was then collected and coupled into a $\frac{1}{4}$-m spectrometer and spectrally analyzed using a UV-enhanced, gated CCD.

Figure 4.31 shows differential absorption spectra, $\Delta\alpha(I_{exc}) = \alpha(I_{exc}) - \alpha(0)$, that were obtained from the absorption spectra $\alpha(I_{exc})$ of Figure 4.31a the $In_{0.18}Ga_{0.82}N/GaN$ MQW and Figure 4.31b the $In_{0.18}Ga_{0.82}N$ epilayer near the fundamental absorption edge. (The oscillatory structure is a result of thin-film interference.) With increasing I_{exc} of the above-gap pump pulse, $\alpha(I_{exc})$ in the band-tail region decreases significantly. This absorption bleaching of the band-tail states is clearly seen for both structures with increasing (I_{exc}), where the bleaching is seen to cover the entire spectral range of the absorption tails of the samples. This bleaching saturated for (I_{exc}) exceeding ~ 2 MW/cm^2 at 10 and 300 K for the $In_{0.18}Ga_{0.82}N$ epilayer. Similar behavior was observed for the $In_{0.18}Ga_{0.82}N/GaN$ MQW sample, the only difference being a larger spectral region exhibiting absorption bleaching because of the larger band tailing of the MQW sample. Note that the induced transparency associated with the absorption bleaching is quite large, exceeding 2×10^4 cm^{-1} at both 10 and 300 K. Both InGaNH based structures exhibited markedly different behavior in the $\alpha(I_{exc})$ and $\Delta\alpha(I_{exc})$ spectra than has been observed in GaN thin films [113,114].

For both samples, the bleaching maximum is at ~ 3.02 eV and is significantly blueshifted with respect to the luminescence maximum. This is explained by the intraband relaxation of photogenerated carriers, where the carriers are created with energies above the mobility edge shown in the previous section, but are quickly caught in the potential wells of the band-tail states. Further intraband relaxation at low temperatures can then occur only by phonon-assisted tunneling to (deeper) neighboring potential wells or by further relaxation to lower-energy states within the same wells until the potential minima are reached. Radiative recombination of these states is expected to be mainly from these potential minima, whereas higher-energy states are temporarily occupied by the relaxing carriers. The latter results in the observed absorption bleaching.

An interesting difference between the two InGaN-based structures is the behavior of the absorption bleaching as I_{th} is exceeded. As the pump density increases, the bleaching also increases for both samples, but as I_{th} is exceeded, the bleaching of the MQW tail states *decreases* significantly with increasing excitation, whereas the bleaching of the epilayer tail states continues to increase with increasing excitation density. This is shown in Figures 4.31a and 4.31b, where the dotted and dashed lines show the bleaching spectra for excitation densities below I_{th}, and the solid lines show the bleaching spectra for excitation densities above I_{th}. The differences in behavior are explained by considering the recombination lifetimes shown in Figures 4.3a and 4.3b. Although radiative recombination from these samples occurs from potential minima, states of higher energy are temporarily occupied as the carriers excited by the pump beam relax, resulting in the observed bleaching "peaking" at higher energy than the luminescence maxima. The bleaching of the InGaN/GaN MQW decreases for excitation densities above I_{th} because of the fast depopulation of the states from which SE originates. Carriers with energies above the SE peak now have a much greater number of available lower-energy states, which, in turn, results in a decrease in states occupied at higher energies and therefore a decrease in the observed bleaching for excitation densities above I_{th}. This behavior was previously observed in other materials in which intrinsic disorder through compositional fluctuations led to carrier localization similar to that observed here [115]. It is important to note that the modal gain shown in Figures 4.30a and 4.30b (dotted lines) corresponds spectrally with the low-energy tail of the localized state absorption bleaching, with the crossover from absorption to gain corresponding approximately to the maximum in the observed bleaching for both samples, indicating that the gain originates from localized states. The induced transparency maximum of the MQW sample corresponds spectrally to the mobility edge, as seen in the previous section.

4.4.3 Summary

The nonlinear optical properties of band-tail states in highly excited InGaN/GaN MQWs and InGaN epilayers were presented in this section, by means of variable stripe gain spectroscopy and nanosecond nondegenerate optical pump–probe spectroscopy. A large blueshift was observed in the gain maximum as the pump power density increased in variable stripe gain spectroscopy. The blueshift of the gain maximum with increasing above-gap optical excitation was attributed to the filling of localized band-tail states due to the intense optical pump. The large spectral region covered by the blueshift evidences the large magnitude of the potential fluctuations present in the InGaN active layers. Nanosecond nondegenerate optical pump–probe spectroscopy of the near-band-edge transitions showed strong absorption bleaching (induced transparency) of band-tail states with increasing above-gap optical excitation. The magnitude of the bleaching was found to be significantly affected by the onset of SE, indicating that the carriers responsible for bleaching and SE share the same recombination channels. These results provide strong evidence for the dominance of localized state recombination in the gain and SE spectra of the InGaN/GaN heterostructures.

It should be noted that the optical gain maximum of the InGaN/GaN MQW (see Figure 4.29) occurs only at photon energies below the mobility edge ($\sim$ 2.95 eV) measured in the energy-selective PL and SE experiments, and the induced transparency maximum of the MQW (see Figure 4.31) corresponds spectrally to the mobility edge. The experimental results presented in this section strongly indicate that localized carriers responsible for band-tail state bleaching share the same recombination channels as the carriers responsible for optical gain and SE in InGaN/GaN MQWs, providing strong evidence that optical gain originates from localized carriers in this material system.

Acknowledgments

Most of work presented in this chapter was accomplished at the Center for Laser and Photonics Research, Oklahoma State University and supported by AFOSR, BMDO, ONR, and DARPA. We are particularly grateful to all collaborators at Oklahoma State University, S. Bidnyk, B. D. Little, G. H. Gainer, Dr. T. J. Schmidt, Dr. A. J. Fischer, Dr. S. J. Hwang, Dr. G. H. Park, and Dr. Y. H. Kwon, for their substantial contributions and valuable discussions in this work. Special thanks to G. H. Gainer for his patient and expert editorial assistance.

References

1. S. Nakamura and G. Fasol, *The Blue Laser Diode* (Springer, Berlin, 1997).
2. S. Nakamura, M. Senoh, S. Nagahama, N. Iwasa, T. Yamada, T. Matsushita, H. Kiyoku, Y. Sugimoto, T. Kozaki, H. Umemoto, M. Sano, and K. Chocho, *Jpn. J. Appl. Phys., Part 2*, **36**, L1568 (1997).
3. S. Nakamura, M. Senoh, S. Nagahama, N. Iwasa, T. Yamada, T. Matsushita, H. Kiyoku, Y. Sugimoto, T. Kozaki, H. Umemoto, M. Sano, and K. Chocho, *Appl. Phys. Lett.*, **72**, 2014 (1998).
4. S. Nakamura, M. Senoh, N. Iwasa, S. Nagahama, T. Yamada, and T. Mukai, *Jpn. J. Appl. Phys., Part 2*, **34**, L1332 (1995).
5. S. Nakamura, M. Senoh, S. Nagahama, N. Iwasa, T. Yamada, T. Matsushita, Y. Sugimoto, and H. Kiyoku, *Appl. Phys. Lett.*, **69**, 4056 (1996).
6. S. Nakamura, M. Senoh, S. Nagahama, N. Iwasa, T. Yamada, and T. Nukai, *Appl. Phys. Lett.*, **68**, 3286 (1996); **69**, 1477 (1996); **69**, 3034 (1996); **69**, 4056 (1996).
7. S. Chichibu, T. Azuhata, T. Sota, and S. Nakamura, *Appl. Phys. Lett.*, **69**, 4188 (1996); **70**, 2822 (1997).
8. E. S. Jeon, V. Kozlov, Y.-K. Song, A. Vertikov, M. Kuball, A. V. Nurmikko, H. Liu, C. Chen, R. S. Kern, C. P. Kuo, and M. G. Craford, *Appl. Phys. Lett.*, **69**, 4194 (1996).
9. P. Perlin, V. Iota, B. A. Weinstein, P. Wiśniewski, T. Suski, P. G. Eliseev, and M. Osiǹski, *Appl. Phys. Lett.*, **70**, 2993 (1997).
10. Y. H. Cho, J. J. Song, S. Keller, M. S. Minsky, E. Hu, U. K. Mishra, and S. P. Denbaars, *Appl. Phys. Lett.*, **73**, 1128 (1998).
11. S. F. Chichibu, A. C. Abare, M. S. Minsky, S. Keller, S. B. Fleisher, J. E. Bowers, E. Hu, U. K. Mishra, L. A. Coldren, S. P. Denbaars, and T. Sota, *Appl. Phys. Lett.*, **73**, 2006 (1998).
12. Y. Narukawa, Y. Kawakami, M. Funato, Sz. Fujita, Sg. Fujita, and S. Nakamura, *Appl. Phys. Lett.*, **70**, 981 (1997).
13. Y. Narukawa, Y. Kawakami, Sz. Fujita, Sg. Fujita, and S. Nakamura, *Phys. Rev. B*, **55**, R1938 (1997).
14. T. Takeuchi, H. Takeuchi, S. Sota, H. Sakai, H. Amano, and I. Akasaki, *Jpn. J. Appl. Phys., Part 2*, **36**, L177 (1997).
15. C. Wetzel, T. Takeuchi, S. Yamaguchi, H. Katoh, H. Amano, and I. Akasaki, *Appl. Phys. Lett.*, **73**, 1994 (1998).
16. M. D. McCluskey, C. G. V. de Walle, C. P. Master, L. T. Romano, and N. M. Johnson, *Appl. Phys. Lett.*, **72**, 2725 (1998).
17. S. Chichibu, K. Wada, and S. Nakamura, *Appl. Phys. Lett.*, **71**, 2346 (1997).
18. X. Zhang, D. H. Rich, J. T. Kobayashi, N. P. Kobayashi, and P. D. Dapkus, *Appl. Phys. Lett.*, **73**, 1430 (1998).
19. F. A. Ponce, D. P. Bour, W. Goltz, and P. J. Wright, *Appl. Phys. Lett.*, **68**, 57 (1996).
20. S. J. Rosner, E. C. Carr, M. J. Ludowise, G. Girolami, and H. I. Erikson, *Appl. Phys. Lett.*, **70**, 420 (1997).
21. S. J. Rosner, G. Girolami, H. Marchand, P. T. Fini, J. P. Ibbetson, L. Zhao, S. Keller, U. K. Mishra, S. P. DeaBaars, and J. S. Speck, *Appl. Phys. Lett.*, **74**, 2035 (1999).

22. D. W. Pohl, W. Denk, and M. Lanz, *Appl. Phys. Lett.*, **44**, 651 (1984).
23. U. Dürig, D. W. Pohl, and F. Rohner, *J. Appl. Phys.*, **59**, 3318 (1986).
24. E. Betzig, J. K. Trautman, T. D. Harris, J. S. Weiner, and R. L. Kostelak, *Science*, **251**, 1468 (1991).
25. E. Betzig and J. K. Trautman, *Science*, **257**, 189 (1992).
26. M. A. Paesler and P. Moyer, *Near-Field Optics: Theory, Instrumentation, and Applications* (Wiley, New York, 1996).
27. P. A. Crowell, D. K. Young, S. Keller, E. L. Hu, and D. D. Awschalom, *Appl. Phys. Lett.*, **72**, 927 (1998).
28. A. Vertikov, M. Kuball, A. V. Nurmikko, Y. Chen, and S.-Y. Wang, *Appl. Phys. Lett.*, **72**, 2645 (1998).
29. A. Vertikov, A. V. Nurmikko, K. Doverspike, G. Bulman, J. Edmond, *Appl. Phys. Lett.*, **73**, 493 (1998).
30. D. L. Abraham, A. Veider, Ch. Schonenberger, H. P. Meier, D. J. Arent, and S. F. Alvarado, *Appl. Phys. Lett.*, **56**, 1564 (1990).
31. B. Garni, J. Ma, N. Perkins, J. Liu, T. F. Kuech, and M. G. Lagally, *Appl. Phys. Lett.*, **68**, 1380 (1996).
32. S. Evoy, C. K. Harnett, H. G. Craighead, S. Keller, U. K. Mishra, and S. P. DenBaars, *Appl. Phys. Lett.*, **74**, 1457 (1999).
33. F. Bernardini, V. Fiorentini, and D. Vanderbilt, *Phys. Rev. B*, **56**, R10024 (1997).
34. F. Bernardini and V. Fiorentini, *Phys. Rev. B*, **57**, R9427 (1998).
35. V. Fiorentini, F. Bernardini, F. D. Sala, A. D. Carlo, and P. Lugli, *Phys. Rev. B*, **60**, 8849 (1999).
36. M. Leroux, N. Grandjean, M. Laügt, J. Massies, B. Gil, P. Lefebvre, and P. Bigenwald, *Phys. Rev. B*, **58**, R13371 (1998).
37. M. Leroux, N. Grandjean, J. Massies, B. Gil, P. Lefebvre, and P. Bigenwald, *Phys. Rev. B*, **60**, 1496 (1999).
38. T. Takeuchi, S. Sota, M. Katsuragawa, M. Komori, H. Takeuchi, H. Amano, and I. Akasaki, it Jpn. J. Appl. Phys., Part 2, **36**, L382 (1997).
39. T. Takeuchi, C. Wetzel, S. Yamaguchi, H. Sakai, H. Amano, I. Akasaki, Y. Kaneko, S. Nakagawa, Y. Yamaoka, and N. Yamada, *Appl. Phys. Lett.*, **73**, 1691 (1998).
40. J. S. Im, H. Kollmer, J. Off, A. Sohmer, F. Scholz, and A. Hangleiter, *Phys. Rev. B*, **57**, R9435 (1998).
41. A. Hangleiter, J. S. Im, H. Kollmer, S. Heppel, J. Off, and F. Scholz, *MRS Internet J. Nitride Semicond. Res.*, **3**, 15 (1998).
42. H. Kollmer, J. S. Im, S. Heppel, J. Off, F. Scholz, and A. Hangleiter, *Appl. Phys. Lett.*, **74**, 82 (1999).
43. R. Langer, J. Simon, O. Konovalov, N. Pelekanos, A. Barski, and M. Leszczynski, *MRS Internet J. Nitride Semicond. Res.*, **3**, 46 (1998).
44. D. L. Smith and C. Mailhiot, *Phys. Rev. Lett.*, **58**, 1264 (1987).
45. D. A. B. Miller, D. S. Chemla, T. C. Damen, A. C. Gossard, W. Wiegmann, T. H. Wood, and C. A. Burrus, *Phys. Rev. B*, **32**, 1043 (1985).
46. R. Dingle, K. L. Shaklee, R. F. Leheny, and R. B. Zetterstrom, *Appl. Phys. Lett.*, **19**, 5 (1971).
47. S. Nakamura, M. Senoh, S. Nahahama, N. Iwasa, T. Yamada, T. Matsushita, H. Kiyoku, and Y. Sugimoto, *Jpn. J. Appl. Phys., Part 2*, **35**, L217 (1996).
48. K. Domen, A. Kuramata, and T. Tanahashi, *Appl. Phys. Lett.*, **72**, 1359 (1998).
49. S. Bidnyk, T. J. Schmidt, Y. H. Cho, G. H. Gainer, J. J. Song, S. Keller, U. K. Mishra, and S. P. Denbaars, *Appl. Phys. Lett.*, **72**, 1623 (1998).
50. T. J. Schmidt, Y. H. Cho, G. H. Gainer, J. J. Song, S. Keller, U. K. Mishra, and S. P. Denbaars, *Appl. Phys. Lett.*, **73**, 560 (1998).
51. H. X. Jiang and J. Y. Lin, it Appl. Phys. Lett., **74**, 1066 (1999).
52. G. Frankowsky, F. Steuber, V. Härle, F. Scholz, and A. Hangleiter, *Appl. Phys. Lett.*, **68**, 3746 (1996).
53. D. Wiesmann, I. Brener, L. Pfeiffer, M. A. Khan, and C. J. Sun, *Appl. Phys. Lett.*, **69**, 3384 (1996).

54. M. Kuball, E. S. Jeon, Y. K. Song, A. V. Nurmikko, P. Kozodoy, A. Abare, S. Keller, L. A. Coldren, U. K. Mishra, S. P. DenBaars, and D. A. Steigerwald, *Appl. Phys. Lett.*, **70**, 2580 (1997).
55. S. Nakamura, *MRS Internet J. Nitride Semicond. Res.*, **2**, 5 (1997).
56. G. Mohs, T. Aoki, M. Nagai, R. Shimano, M. Kuwata-Gonokami, and S. Nakamura, *Solid State Commun.*, **104**, 643 (1997).
57. Y.-K. Song, M. Kuball, A. V. Nurmikko, G. E. Bulman, K. Doverspike, S. T. Sheppard, T. W. Weeks, M. Leonard, H. S. Kong, H. Dieringer, and J. Edmond, *Appl. Phys. Lett.*, **72**, 1418 (1998).
58. T. J. Schmidt, Y. H. Cho, G. H. Gainer, J. J. Song, S. Keller, U. K. Mishra, and S. P. DenBaars, *Appl. Phys. Lett.*, **73**, 1892 (1998).
59. Y. H. Cho, F. Fedler, R. J. Hauenstein, G. H. Park, J. J. Song, S. Keller, U. K. Mishra, and S. P. DenBaars, *J. Appl. Phys.*, **85**, 3006 (1999).
60. R. W. Martin, P. G. Middleton, K. P. O'Donnel, and W. Van der Stricht, *Appl. Phys. Lett.*, **74**, 263 (1999).
61. K. P. O'Donnell, R. W. Martin, and P. G. Middleton, *Phys. Rev. Lett.*, **82**, 237 (1999).
62. S. Bidnyk, T. J. Schmidt, G. H. Park, and J. J. Song, *Appl. Phys. Lett.*, **71**, 729 (1997).
63. C. I. Harris, B. Monemar, H. Amano, and I. Akasaki, *Appl. Phys. Lett.*, **67**, 840 (1995).
64. C. K. Sun, S. Keller, G. Wang, M. S. Minsky, J. E. Bowers, and S. P. DenBaars, *Appl. Phys. Lett.*, **69**, 1936 (1996).
65. C. K. Sun, T. L. Chiu, S. Keller, G. Wang, M. S. Minsky, S. P. Denbaars, and J. E. Bowers, *Appl. Phys. Lett.*, **71**, 425 (1997).
66. J. S. Im, V. Härle, F. Scholz, and A. Hangleiter, *MRS Internet J. Nitride Semicond. Res.*, **1**, 37 (1996).
67. B. K. Ridley, *Phys. Rev. B*, **41**, 12 190 (1990).
68. J. Feldmann, G. Peter, E. O. Göbel, P. Dawson, K. Moore, C. Foxon, and R. J. Elliott, *Phys. Rev. Lett.*, **59**, 2337 (1987).
69. W. Shan, X. C. Xie, J. J. Song, and B. Goldenberg, *Appl. Phys. Lett.*, **67**, 2512 (1995).
70. P. Lefebvre, J. Allègre, B. Gil, A. Kavokine, H. Mathieu, W. Kim, A. Salvador, A. Botchkarev, and H. Morkoç, *Phys. Rev. B*, **57**, R9447 (1998).
71. Y. Narukawa, S. Saijou, Y. Kawakami, S. Fujita, T. Mukai, and S. Nakamura, *Appl. Phys. Lett.*, **74**, 558 (1999).
72. Y. H. Kwon, G. H. Gainer, S. Bidnyk, Y. H. Cho, J. J. Song, M. Hansen, and S. P. Denbaars, *Appl. Phys. Lett.*, **75**, 2545 (1999).
73. X. H. Yang, T. J. Schmidt, W. Shan, J. J. Song, and B. Goldenberg, *Appl. Phys. Lett.*, **66**, 1 (1995).
74. A. Krost, J. Böhrer, A. Dadgar, R. F. Schnabei, D. Bimberg, S. Hansmann, and H. Burkhard, *Appl. Phys. Lett.*, **67**, 3325 (1995).
75. H. Sugiura, M. Mitsuhara, H. Oohashi, T. Hirono, and K. Nakashima, *J. Cryst. Growth*, **147**, 1 (1995).
76. S. Ruvimov, Z. Liliental-Weber, T. Suski, J. W. Ager III, J. Washburn, J. Krueger, C. Kisielowski, E. R. Weber, H. Amano, and I. Akasaki, *Appl. Phys. Lett.*, **69**, 990 (1996).
77. E. F. Schubert, I. D. Goepfert, W. Grieshaber, and J. M. Redwing, *Appl. Phys. Lett.*, **71**, 921 (1997).
78. S. Keller, A. C. Abare, M. S. Minsky, X. H. Wu, M. P. Mack, J. S. Speck, E. Hu, L. A. Coldren, U. K. Mishra, and S. P. Denbaars, *Material Science Forum*, 264–268, 1157 (1998).
79. Y. H. Cho, T. J. Schmidt, S. Bidnyk, J. J. Song, S. Keller, U. K. Mishra, and S. P. DenBaars, *MRS Internet J. Nitride Semicond. Res.*, **4S1**, G6.44 (1999).
80. P. A. Grudowski, C. J. Eiting, J. Park, B. S. Shelton, D. J. H. Lambert, and R. D. Dupuis, *Appl. Phys. Lett.*, **71**, 1537 (1997).
81. S. Chichibu, D. A. Cohen, M. P. Mack, A. C. Abare, P. Kozodoy, M. Minsky, S. Fleischer, S. Keller, J. E. Bowers, U. K. Mishra, L. A. Coldren, D. R. Clarke, and S. P. DenBaars, *Appl. Phys. Lett.*, **73**, 496 (1998).
82. A. Salvador, G. Liu, W. Kim, Ö. Aktas, A. Botchkarev, and H. Morkoç, *Appl. Phys. Lett.*, **67**, 3322 (1995).
83. K. C. Zeng, J. Y. Lin, H. X. Jiang, A. Salvador, G. Popovici, H. Tang, W. Kim, and H. Morkoç, *Appl. Phys. Lett.*, **71**, 1368 (1997).

84. H. J. Osten, J. Klatt, G. Lippert, B. Dietrich, and E. Bugiel, *Phys. Rev. Lett.*, **69**, 450 (1992).
85. D. J. Eaglesham, F. C. Unterwald, and D. C. Jacobson, *Phys. Rev. Lett.*, **70**, 966 (1993).
86. K. Uchida, T. Tang, S. Goto, T. Mishima, A. Niwa, and J. Gotoh, *Appl. Phys. Lett.*, **74**, 1153 (1999).
87. Y. P. Varshni, *Physica*, **34**, 149 (1967).
88. W. Shan, T. J. Schmidt, X. H. Yang, S. J. Hwang, J. J. Song, and B. Goldenberg, *Appl. Phys. Lett.*, **66**, 985 (1995).
89. W. Shan, B. D. Little, J. J. Song, Z. C. Feng, M. Schurman, and R. A. Stall, *Appl. Phys. Lett.*, **69**, 3315 (1996).
90. K. G. Zolina, V. E. Kudryashov, A. N. Turkin, and A. E. Yunovich, *MRS Internet J. Nitride Semicond. Res.*, **1**, 11 (1996).
91. P. G. Eliseev, P. Perlin, J. Lee, and M. Osiński, *Appl. Phys. Lett.*, **71**, 569 (1997).
92. Y. H. Cho, G. H. Gainer, A. J. Fischer, J. J. Song, S. Keller, U. K. Mishra, and S. P. Denbaars, *Appl. Phys. Lett.*, **73** , 1370 (1998).
93. Y. H. Cho, B. D. Little, G. H. Gainer, J. J. Song, S. Keller, U. K. Mishra, and S. P. DenBaars, *MRS Internet J. Nitride Semicond. Res.*, **4S1**, G2.4 (1999).
94. F. A. J. M. Driessen, G. J. Bauhuis, S. M. Olsthoorn, and L. J. Giling, *Phys. Rev. B*, **48**, 7889 (1993).
95. K. Yamashita, T. Kita, H. Nakayama, and T. Nishino, *Phys. Rev. B*, **55**, 4411 (1997).
96. A. Chomette, B. Deveaud, A. Regreny, and G. Bastard, *Phys. Rev. Lett.*, **57**, 1464 (1986).
97. T. Yamamoto, M. Kasu, S. Noda, and A. Sasaki, *J. Appl. Phys.*, **68**, 5318 (1990).
98. H. Shoji, Y. Nakata, K. Mukai, Y. Sugiyama, M. Sugawara, N. Yokoyama, and H. Ishikawa, *Appl. Phys. Lett.*, **71**, 193 (1997).
99. P. D. Floyd and D. W. Treat, *Appl. Phys. Lett.*, **70**, 2493 (1997).
100. H. Jeon, J. Ding, A. V. Nurmikko, W. Xie, D. C. Grillo, M. Kobayashi, R. L. Gunshor, G. C. Hua, and N. Otsuka, *Appl. Phys. Lett.*, **60**, 2045 (1992).
101. J. M. Gaines, R. R. Drenten, K. W. Haberern, T. Marshall, P. Mensz, and J. Petruzzello, *Appl. Phys. Lett.*, **62**, 2462 (1993).
102. Y. H. Cho, J. J. Song, S. Keller, U. K. Mishra, and S. P. Denbaars, *Appl. Phys. Lett.*, **73**, 3181 (1998).
103. T. J. Schmidt, Y. H. Cho, S. Bidnyk, J. J. Song, S. Keller, U. K. Mishra, and S. P. DenBaars, *SPIE Proc. Ultrafast Phenomena in Semiconductors III*, **3625**, 57 (1999).
104. T. Deguchi, T. Azuhata, T. Sota, S. Chichibu, M. Arita, H. Nakanishi, and S. Nakamura, *Semicond. Sci. Technol.*, **13**, 97 (1998).
105. T. J. Schmidt, S. Bidnyk, Y. H. Cho, A. J. Fischer, J. J. Song, S. Keller, U. K. Mishra, and S. P. Denbaars, *Appl. Phys. Lett.*, **73**, 3689 (1998).
106. T. J. Schmidt, S. Bidnyk, Y. H. Cho, A. J. Fischer, J. J. Song, S. Keller, U. K. Mishra, and S. P. DenBaars, *MRS Internet J. Nitride Semicond. Res.*, **4S1**, G6.54 (1999).
107. G. Mohs, T. Aoki, M. Nagai, R. Shimano, M. Kuwata-Gonokami, and S. Nakamura, *Proc. 2nd International Conference on Nitride Semiconductors*, Tokushima, Japan, 234 (1997).
108. M. P. Mack, A. Abare, M. Aizcorbe, P. Kozodoy, S. Keller, U. K. Mishra, L. Coldren, and S. P. DenBaars, *MRS Internet J. Nitride Semicond. Res.*, **2**, 41 (1997).
109. J. J. Song and W. Shan, Group III Nitride Semiconductor Compounds: Physics and Applications, edited by B. Gil (Oxford Univ. Press, New York, 1998) pp. 182–241.
110. J. J. Song and W. Shan, Properties, Processing, and Applications of Gallium Nitride and Related Semiconductors, edited by J. H. Edgar, S. Strite, I. Akasaki, and H. Amano (Michael Faraday House, London, 1999) pp. 596–602.
111. K. L. Shaklee and R. F. Leheny, *Appl. Phys. Lett.*, **18**, 475 (1971).
112. K. L. Shaklee, R. E. Nahory, and R. F. Lehny, *J. Lumin.*, **7**, 284 (1973).
113. T. J. Schmidt, J. J. Song, Y. C. Chang, B. Goldenberg, and R. Horning, *Appl. Phys. Lett.*, **72**, 1504 (1998).
114. T. J. Schmidt, Y. C. Chang, and J. J. Song, *SPIE Conf. Proc.*, **3419**, 61 (1998).
115. T. Breitkopf, H. Kalt, C. Klingshirn, and A. Reznitsky, *J. Opt. Soc. Am. B*, **13**, 1251 (1996).

CHAPTER 5

Optical Properties of Homoepitaxial GaN

R. STĘPNIEWSKI[1], A. WYSMOŁEK[1,2], K.P. KORONA[1], and J.M. BARANOWSKI[1]

[1]*Institute of Experimental Physics, Warsaw University, Hoża 69, 00-681 Warsaw, Poland*
[2]*Grenoble High Magnetic Field Laboratory, MPI/FKF-CNRS, BP 166, F-38042, Grenoble Cedex 9, France*

5.1 INTRODUCTION

For a long time fundamental studies of GaN were limited by the unsatisfactory quality of the samples. The major problem that hangered the development of technology was the lack of good-quality, lattice-matched substrates for MOCVD epitaxy. Since 1995, with the work of K. Pakuła et al. [1], the growth of homoepitaxial GaN MOCVD layers has allowed substantial progress in the optical studies of this interesting material. GaN monocrystal plates, which were used as substrates, were grown from a dilute solution of atomic nitrogen in liquid gallium at a temperature of 1600 °C and a nitrogen pressure of about 15–20 kbar [2]. Crystals grown by this method form as platelets with their hexagonal **c**-axis [0001] perpendicular to the surface. The homoepitaxial layers grown on such substrates have the same orientation. The stress in such MOCVD layers is small ($\Delta c/c < 2\times10^{-4}$), as verified by X-ray measurements [3]. Luminescence and reflectivity results [4] reveal their good optical properties. The increasing quality of grown homoepitaxial GaN layers during the last 5 years, visualized in Figure 5.1, allows the observation in luminescence spectra of discrete lines with a halfwidth of 0.1 meV. Optical studies of such samples provide new insight into the band energy structure and the nature of the impurities, which dominate the optical properties of this material. Some of the results obtained for homoepitaxial layers have been reviewed in Refs. [5–9].

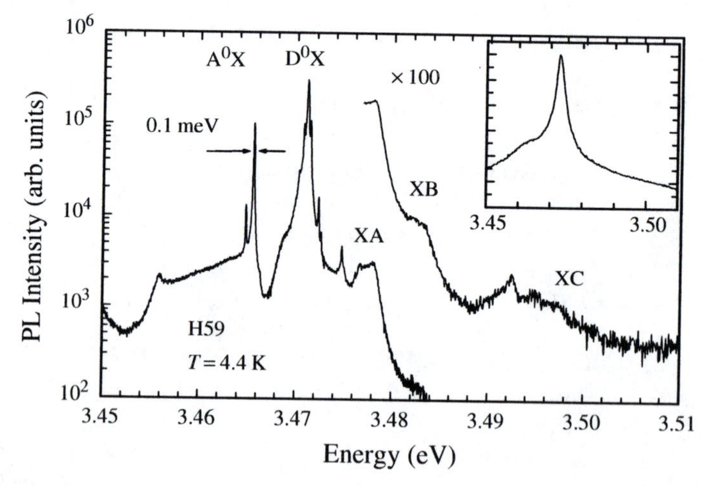

Figure 5.1 Highly resolved PL spectra of homoepitaxial GaN in the excitonic region. A^0X and D^0X denote the neutral-acceptor- and neutral-donor-bound exciton recombination, respectively. XA, XB, and XC represent free-exciton recombination related to three different valence subbands. For comparison, the luminescence result from Ref. 1 is presented in the inset.

In this chapter we concentrate on the optical properties of such layers. First, we discuss in detail the free-exciton structure observed in reflectance. The applied theoretical description, which includes the complete structure of the excited exciton states, allows us to reproduce the reflectance spectrum over the entire measured range. The band structure parameters of GaN are derived from the fitting procedure. Studies of temperature and pressure effects on reflectivity in the free-exciton range follow.

In the next section we present the properties of bound excitons that dominate near-bandgap luminescence, followed by a discussion of the luminescence phonon replica lines observed both for free and bound excitons.

We conclude our discussion of the optical properties of GaN with a presentation of donor–acceptor pair spectra emission, which is the manifestation of the influence of impurities on the recombination processes in GaN.

We also discuss the results of magnetooptical studies of GaN homoepitaxial layers. The magnetic field through the coupling to the orbital angular momentum and to the spin of the electron and holes involved into optical transitions allows for their detailed study.

5.2 FREE EXCITONS

Gallium nitride is a direct-gap semiconductor that crystallizes in the wurtzite structure. In direct-gap semiconductors, there is a strong coupling of free excitons (FE) with photons of similar energy. This phenomenon is essential for understanding the near-band-edge optical response and has been studied in different materials such as, for example, GaAs [10,11] and CdS [12]. It is generally accepted that the top of the GaN valence band is of Γ_9 symmetry. Three excitons, A, B, and C, which correspond to the split valence bands of Γ_9, Γ_7, and Γ_7 symmetry (Figure 5.2), respectively, can be observed in reflectance [13–18]. Reflectance spectra of homoepitaxial GaN have generally sharper lines (broadening parameter Γ below 1 meV [17]) than heteroepitaxial ones.

Due to the momentum conservation rule, the photon emission by the free exciton is less probable than by the bound exciton (BE) unless there are some defects or inhomogenities that can facilitate radiative recombination. For this reason the FE lines are much weaker than the BE ones [4,8,17]. A comparison of the reflectivity and luminescence spectra of homoepitaxial GaN is shown in Figure 5.3, where one can see that the dominant features of the reflectance spectrum are the free-exciton A and B lines at energies of 3.478 eV and 3.483 eV, respectively. The structure of the C exciton ($h\nu = 3.499$ eV) is less pronounced but also visible. In contrast, in the photoluminescence spectra, the free-exciton emission is much weaker than the luminescence lines observed at $h\nu = 3.472$ eV and $h\nu = 3.467$ eV. Magnetooptical experiments performed on homoepitaxial GaN [9,19],

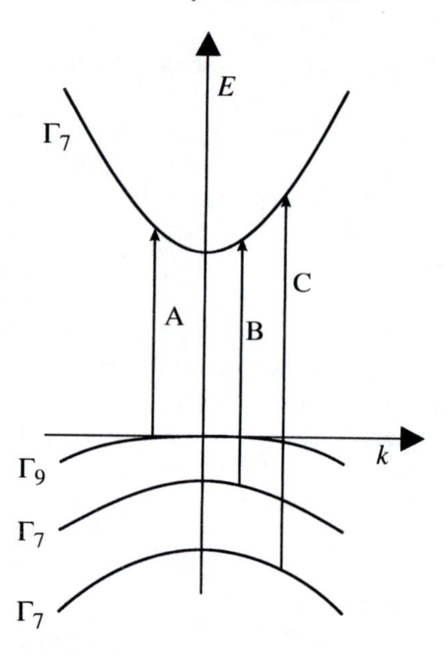

Figure 5.2 Band structure of GaN in the vicinity of the Γ point. The optically active excitonic transitions XA, XB, and XC from valence subbands of Γ_9, Γ_7, and Γ_7 symmetry, respectively, to the conduction band (Γ_7 symmetry) are marked by arrows.

which will be discussed later, show that these strong photoluminescence lines come from excitons bound to neutral donors (D^0X) and acceptors (A^0X), respectively. The photoluminescence line A seems to be a doublet, which could be due to longitudinal-transverse splitting of the A exciton.

5.2.1 Polariton Structure

In order to precisely describe the near-band-edge optical properties of a semiconductor a coupling of free excitons (FE) with photons has to be taken into account. Since in GaN the differences between excitons A and B, and between B and C are only about 5 meV and 16 meV, respectively, the different excitons cannot be treated separately, like in CdS [20,21], but coupling of an electromagnetic wave with A, B, and C excitons has to be taken into account simultaneously. Thus, polariton dispersion curves have been calculated by solving the coupled equations [22]:

$$\omega_X = \omega_{TX} + \frac{\hbar k^2}{2M}, \tag{5.1}$$

$$\frac{\partial^2 P_X}{\partial t^2} + \omega_X^2(k) P_X = \alpha_X \omega_X^2(k) E - \gamma X \frac{\partial P_X}{\partial t}, \qquad X = A,\ B,\ C, \tag{5.2}$$

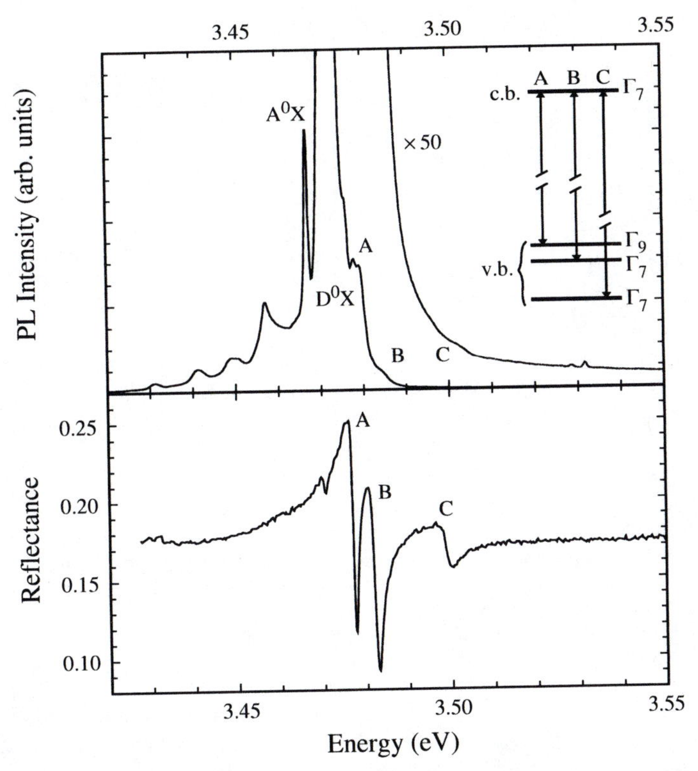

Figure 5.3 Comparison of PL and reflectance spectra of GaN. The luminescence scale is logarithmic, whereas the reflectance scale is linear. The donor-bound exciton (D^0X) peak is truncated. The signal from the free excitons is best seen in the reflectance spectrum.

$$\varepsilon^* \frac{\partial^2 E}{\partial t^2} - c^2 \, \Delta E = -4\pi \frac{\partial^2 (P_A + P_B + P_C)}{\partial t^2}, \tag{5.3}$$

where E is a photon electric field, and P_X is a polarization contribution due to exciton X. Equation (5.1) describes the center-of-mass motion of the free exciton with wavevector k and mass $M = m_e + m_h$. Equation (5.2) represents the excitonic oscillator coupled with a photon electric field through polarizability α_X. The frequency of the oscillator is ω_X, and the damping constant is γ_X. Equation (5.3) describes the electromagnetic wave in a material with dielectric constant $\varepsilon^*(\omega)$. The influence of the excitonic excited states (discrete and continuous) has been included in the dielectric function [17] by adding to the residual dielectric constant ε_R components calculated from Elliott's formulas [23].

The solution of the described polariton model is presented in Figure 5.4. Dispersion curves are shown in ω_X versus real and imaginary parts of wavevector k coordinates. Four branches arising from photons coupled to A,

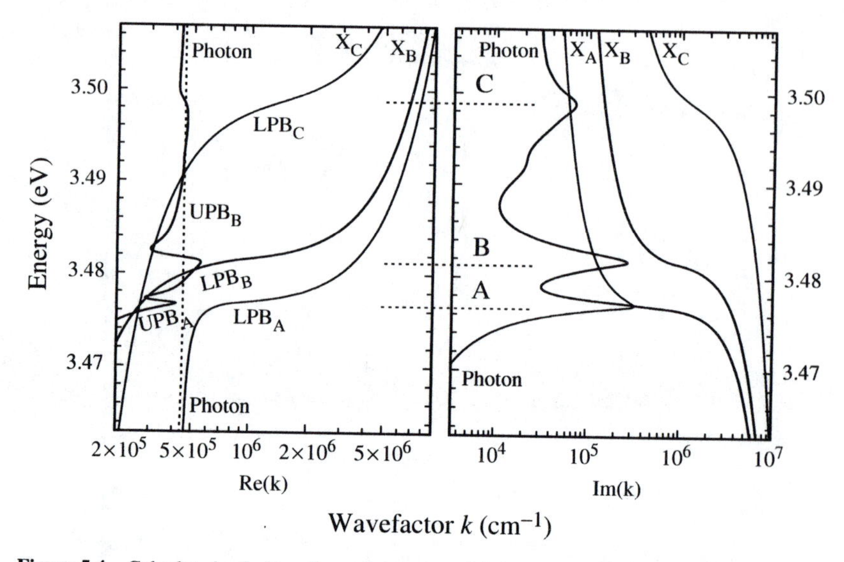

Figure 5.4 Calculated polariton dispersion curves of GaN. Four branches arising from photon coupling with three excitons A, B, and C (which originate from the crystal-field and spin-orbit splitting of the valence band) are labeled: LPB_X, lower polariton branch of exciton X; UPB_X, upper polariton branch of exciton X (for parameters see Table 5.1).

B, and C excitons are labeled LPB_X (lower polariton branch of exciton X) and UPB_X (upper polariton branch of exciton X). The calculated dispersion relations presented in Figure 5.4 were obtained assuming parameter values ω_X, γ_X, and α_X that resulted from the fit to the measured reflectance spectra (see Section 5.2.2). It is worth noting that values of the damping constant γ_X are relatively high. This is particularly noticeable for the C-exciton, for which damping is so strong that the exciton–photon coupling is hardly resolved.

In the energy range where there is more than one polariton branch, the incident electromagnetic wave will excite more than one polariton wave with different wavevectors (see Figure 5.4). In order to relate phases and amplitudes of these polariton waves, additional boundary conditions are required. Here, we apply the approach first proposed by Pekar [24], i.e., we assume that for each exciton X, its polarization contribution vanishes at the surface: $P_X = 0$. In this case, the effective index of refraction, n_{eff}, is introduced [25]:

$$n_{eff}(\omega) = \frac{\Sigma_j (n_j(\omega) E_j(\omega))}{\Sigma_j E_j(\omega)}, \tag{5.4}$$

where $E_j(\omega)$ is the electric-field amplitude of polariton j. According to the imposed boundary conditions, excitons are excluded from a surface layer of thickness l_d (the so-called dead layer). Consequently, the reflectivity R is the function of two interfering waves with phase difference $\theta = 2kn_{eff}l_d$

and is given by $(n_s - 1)^2/(n_s + 1)^2$, calculated with the complex refractive index n_s defined as [12]:

$$n_s = \frac{n_0((n_{eff} + n_0) + (n_{eff} - n_0)\exp(i\theta))}{((n_{eff} + n_0) - (n_{eff} - n_0)\exp(i\theta))}, \tag{5.5}$$

where $n_0 = \sqrt{\varepsilon_R}$ is the residual (neglecting excitonic resonances) refractive index.

5.2.2 Analysis of Reflectance Measurements

For reflectance measurements, the sample was immersed in a bath of liquid helium. The epitaxial layer with optimum surface quality was chosen for the reflectance measurements. Experiments were performed at 1.8 K using optical fibers whose ends were mounted in front of the sample with a 30° angle of incidence. This off-normal direction of the incident light plays a negligible role in the data analysis and will be neglected. Reflectance spectra were scaled with respect to the reflectivity of Al and derived from the ratio of the reflection signals measured from the sample and from the Al mirror. The reflectivity of Al was assumed to be 0.92 in the spectral range considered [26].

The low-temperature near-band-edge reflectance spectrum of GaN is shown in Figure 5.5a. According to the polariton model described in Section 5.2.1 we calculated the corresponding reflectance spectrum and fitted it to the experimental data (see Figure 5.5a). Exciton masses were fixed to be the same for all three excitons and set to $M = m_0$ (electron mass) [16], whereas exciton energies E_{TX}, linewidths Γ_X, polarizabilities $4\pi\alpha_X$, the dead-layer thickness l_d, the residual dielectric constant ε_R, and bandgap energy E_g were assumed to be fitting parameters. The best-fit values of E_{TX}, Γ_X, and α_X are given in Table 5.1. We deduced $\varepsilon_R = 5.2$, $E_g = 3.497$ eV, and $l_d = 4.1$ nm, in agreement with the expected exciton radius. As shown in Figure 5.5a, the measured reflectance was well reproduced in calculations.

A small discrepancy between the calculated curve and experimental data exists in the energy range from 3.455 eV to 3.475 eV (the region of bound excitons) and above the excitonic structures. This difference can be related to the additional dispersion of the real part of the dielectric function, induced by a strong absorption (for energies above 5 eV) present in GaN [39] but not included in our calculations.

Our analysis is quite accurate, particularly for estimating excitonic energies. It should be noted that in the case of homoepitaxial GaN layers the changes in exciton energies from sample to sample (related to a possible residual strain [3]) are rather small (less than 1 meV) [27]. Hence, the results, though obtained for a particular sample, give a fairly reliable estimate of exciton energies in hexagonal GaN.

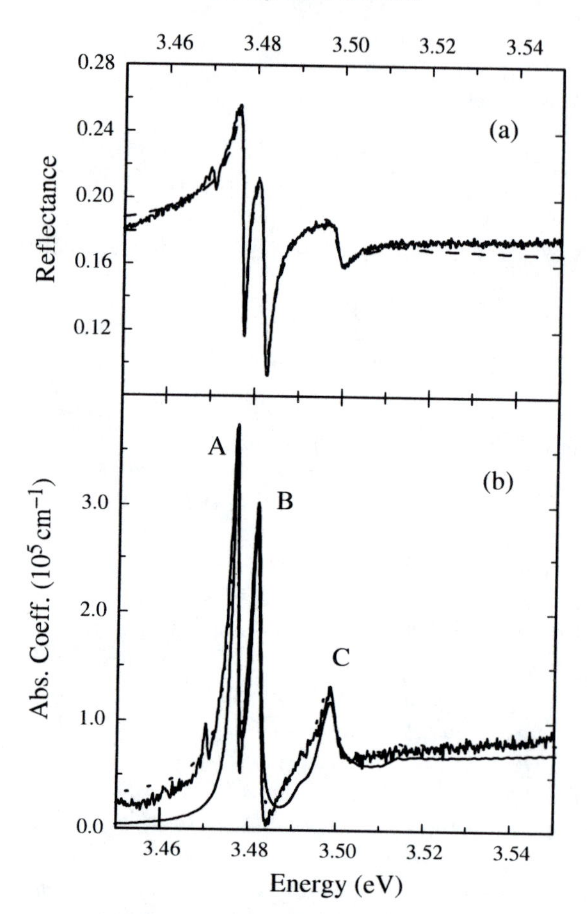

Figure 5.5 Experimental reflectance spectrum of a GaN layer at $T = 1.8$ K, solid line; fitted theoretical curve, dashed line. (b) The absorption coefficient: derived by Kramers–Kronig analysis, solid line; two curves obtained from the exciton–polariton model: "external" (with $n_s(\omega)$ from Eq. (5.5)), dotted curve; bulk (with $n_{eff}(\omega)$ from Eq. (5.4)), thin, solid line.

The polarizabilities of excitons in GaN (see Table 5.1) can be compared with the corresponding values ($4\pi\alpha_0$) reported for other semiconductors. The exciton–photon coupling is stronger in GaN than in GaAs ($4\pi\alpha_0 = 0.0012$) [10] but definitely weaker than in II–VI compounds ($4\pi\alpha_0 = 0.0094$ for CdS, and $4\pi\alpha_0 = 0.0063$ [12] in the case of ZnTe [12]), as would be expected, since GaAs is a more covalent compound, whereas II–VI materials show more ionic character with respect to GaN.

The linewidth Γ_A (or its effective dumping constant $\gamma_A = 10^{-12}\,\text{ps}^{-1}$) of exciton A is the smallest one (see Table 5.1). This dumping constant is in good agreement with the dephasing rate $1/T_2 = 0.33 \times 10^{-12}\,\text{ps}^{-1}$ [28] measured by four-wave mixing on GaN. The strongest dumping is observed

Table 5.1 Parameters of exciton polaritons in hexagonal GaN obtained from analysis of the near-band-edge reflectance spectrum.

Exciton X	E_{TX} *(eV)*	Γ_X *(meV)*	*Polarizability* $(4\pi\alpha_X)$	$\Delta_{LTX} = E_{LX} - E_{TX}$ *(meV)*
A	3.4767(3)	0.7(2)	0.0027(3)	0.90(10)
B	3.4815(3)	1.5(2)	0.0031(3)	1.04(10)
C	3.4986(8)	3.1(8)	0.0011(3)	0.37(10)

in the case of exciton C. Since the exciton C is nearly degenerate with the excited states of excitons A and B, exciton C decays of very fast. Its lifetime can be estimated at $t_C = 0.2$ ps.

In the cubic approximation for the wurtzite crystal the energy differences between A, B, and C caused by the crystal-field Δ_{cr} and the spin-orbit Δ_{so} splittings are described by [29]:

$$E_{C,B} - E_A = \frac{\Delta_{so} + \Delta_{cr}}{2} \pm \left[\left(\frac{\Delta_{so} + \Delta_{cr}}{2} \right)^2 - \frac{2}{3} \Delta_{so} \Delta_{cr} \right]^{1/2}, \qquad (5.6a)$$

The reverse relation gives Δ_{so} and Δ_{cr}:

$$\left. \begin{matrix} \Delta_{so} \\ \Delta_{cr} \end{matrix} \right\} = \frac{1}{2} \left[\Delta_{C,A} + \Delta_{B,A} \pm \left(2\Delta_{C,B}^2 - \Delta_{B,A}^2 - \Delta_{C,A}^2 \right)^{1/2} \right] \qquad (5.6b)$$

where $\Delta_{X,Y} = E_X - E_Y$.

Thus, in order to find energies Δ_{cr} and Δ_{so} from Eq. (5.6), we need only the differences of energies between excitons. Thus, we can calculate $\Delta_{so} = 17.9 \pm 1.2$ meV, and $\Delta_{cr} = 8.8 \pm 0.3$ meV. Since the value of the spin-orbit parameter of cubic GaN is $\Delta_{so} = 17 \pm 1$ meV [30], we can conclude that the spin-orbit splitting in wurtzite and zinc blende GaN is very similar. The values of crystal-field and spin-orbit splitting parameters obtained on heteroepitaxial layers by different authors [13,15,16,27,31] are collected in Table 5.2 and compared with our appropriate results [17] for homoepitaxial GaN. Although in the case of the spin-orbit splitting there is good agreement among the results, there is wide variation in the values obtained for the crystal-field splitting, since this parameter value is strongly affected by strain. This observation is in agreement with theoretical calculations of the elastic properties of GaN [16,32] (see Section 5.2.4). The data reported here obtained for strain-free homoepitaxial GaN might be used as a reference to determine the shift of excitonic resonances in heteroepitaxial layers.

Table 5.2 Comparison of valence band splitting parameters of hexagonal GaN obtained by various authors.

Δ_{cr} *(meV)*	Δ_{so} *(meV)*	*Source*
22	11	Experiment, [13]
72.9	15.6	Calculation, [31]
35	18	Calculation, [16]
11	17	Experiment, [16]
10.0	17.6	Experiment, [15]
22	15	Experiment, [27]
8.8 ± 0.3	**17.9 ± 1.2**	**Experiment, [17]** (this work)

Appropriate analysis of the reflectance spectrum permits reconstruction of the corresponding absorption spectrum. We transformed the measured reflectance spectrum into the absorption one using the Kramers–Kronig relations [33]. Typically, for Kramers–Kronig analysis the transformed spectra have to be known in a wide frequency range. Our data cover a rather narrow spectral range (3.43 eV–3.55 eV). For energies above the measured range (up to 30 eV) the data available in the literature [34] were used in order to reconstruct the corresponding spectrum. Additionally, we assumed that for the energy range 0–3.43 eV there is no absorption present in GaN (we neglect the absorption processes due to phonons that exists below 0.1 eV). In this region the reflectivity was extrapolated numerically [35].

The absorption coefficient derived from Kramers–Kronig analysis of the reflectance is shown in Figure 5.5b with a solid line. The dotted curve in the same figure represents the "external" absorption coefficient $\alpha_s = (2\omega/c) \cdot \mathrm{Im}(n_s)$, calculated directly from the polariton model, including the dead-layer effect Eq. (5.5), with parameters previously used to reproduce the reflectance spectra. As can be seen in Figure 5.5b, both methods applied to reconstruct the fundamental absorption edge in the investigated sample give very similar results. They are slightly different from the expected "bulk" absorption coefficient $\alpha_{eff} = (2\omega/c) \cdot \mathrm{Im}(n_{eff})$ Eq. (5.4), which should be applied to reconstruct the fundamental absorption edge in GaN (Figure 5.5b, solid thin line). This spectrum of the fundamental absorption in GaN has been successfully reproduced with theory without any additional fitting parameters [36].

As shown in Figure 5.5b, the absorption spectrum of the investigated GaN layer shows three dominant structures, A, B, and C, which respectively peak at 3.476 eV, 3.481 eV, and 3.498 eV. One can also distinguish the steplike absorption band which begins around 3.49 eV and which is ascribed to an absorption continuum corresponding to valence band–conduction band transitions. It is interesting to note that in addition to the strong lines due to free excitons there is also a weak structure at 3.471 eV attributed to a donor-bound exciton (note the corresponding feature in the reflectance spectrum). This observation indicates strong donor-bound exciton oscillator strength in GaN. Transitions related to bound excitons are rarely observed in reflectance spectra, though such an observation has been reported for high-quality CdS [37].

The amplitude of the excitonic absorption coefficient α of about $3 \times 10^5\,\text{cm}^{-1}$ (see Figure 5.5b) is similar to previously reported values for heteroepitaxial GaN: $1 \times 10^5\,\text{cm}^{-1}$ [38], $2.5 \times 10^5\,\text{cm}^{-1}$ [39], and $3 \times 10^5\,\text{cm}^{-1}$ [40].

5.2.3 Temperature Dependence

The evolution of the GaN band structure has been studied in the temperature range 4–400 K [41]. The reflection spectra of a homoepitaxial GaN layer were measured at various temperatures using a continuous-flow helium cryostat with a tungsten lamp as the light, source and an incident angle of 45° Spectra were analyzed with a SPEX 500M monochromator.

In order to determine the energetic positions of the excitons, we used a simplified procedure for reflectance spectra analysis. Assuming a Lorentzian shape for the excitonic contribution to the dielectric constant [22] and the dead-layer effect Eq. (5.5), we approximated the spectral shape due to an exciton transition at energy $h\nu$ by:

$$R(\nu) = R_0 + \sum_x A_x \,\mathrm{Re}\left(\frac{h\nu_x - h\nu + i\Gamma_x}{(h\nu_x - h\nu)^2 + \Gamma_x^2} e^{i\Theta_x}\right) \tag{5.7}$$

where A_X is amplitude, $h\nu_X$ is energy, and Γ_X is a broadening parameter of exciton X, R_0 is background reflectivity, and Θ_X is a phase. In order to reduce number of fitting parameters, it was assumed that $\Theta_A = \Theta_B$. We obtained $\Theta_A = (0.03 \pm 0.1)\pi$, $\Theta_C = (0.25 \pm 0.1)\pi$. The energies obtained for excitons at $T = 4.2$ K were $E_A = 3.4776(3)$ eV, $E_B = 3.4827(3)$ eV, $E_C = 3.5015(20)$ eV, which is in good agreement with the results of more precise analysis (see Table 5.1).

The reflectance spectra are shown in Figure 5.6. At low temperature, two dispersion lines, A and B, are clearly visible. Line C is visible only up to 170 K, but the remains of lines A and B exist even at 390 K. In order to obtain values of excitonic energies at different temperatures, the curves defined by Eq. (5.7) were fitted to spectra. The number of adjustable parameters was reduced to five (E_A, E_B, Γ_X, R_X, R_0) by assuming that $\Theta_A = \Theta_B = 0$, $R_A = R_B$, and $\Gamma_A = \Gamma_B$. Up to 170 K it is possible to separate lines A and B, and it was found that the energy distance between them is constant: $\delta E = 5.1 \pm 0.2$ meV. At temperatures above 170 K the lines are too broad to

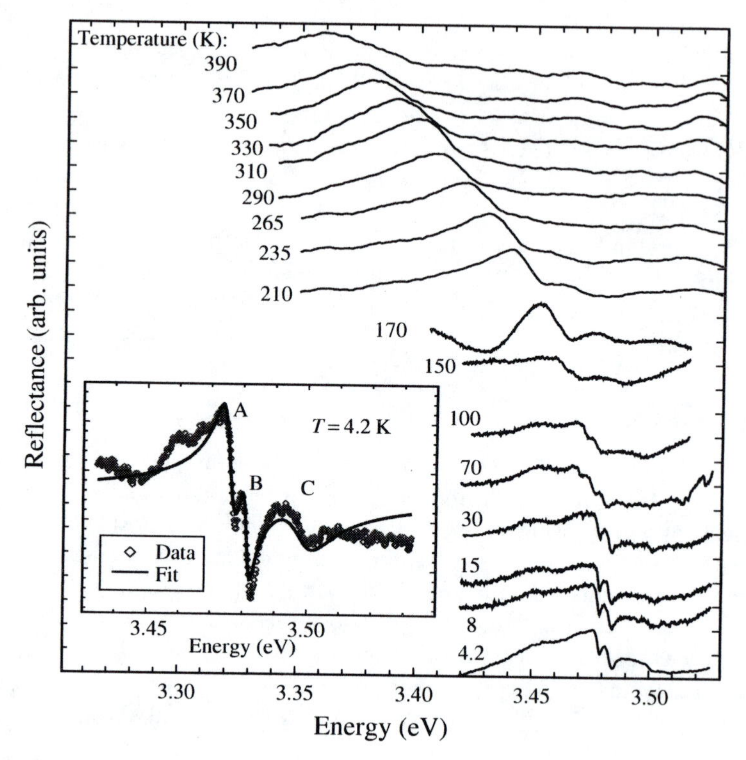

Figure 5.6 Reflection spectra of GaN at various temperatures. Inset: curve given by Eq. (5.7) fitted to experimental data for $T = 4.2$ K.

separate them, so it was assumed that $E_B = E_A + 5.1$ meV, which reduced the number of fitting parameters to four (E_A, Γ_X, R_X, R_0). This procedure gives reliable results even for spectra taken at 390 K.

The temperature dependence of the energy gap determined in this way is shown in Figure 5.7a. The change in the energy gap of a semiconductor caused by temperature should be proportional to the number of phonons, $n_q \sim 1/(\exp(\theta/T) - 1)$ [42], where θ is the Debye temperature. Thus, we used a similar formula to describe the experimental variation of the energy gap with temperature. It was found that in the temperature range 4.2–390 K the following formula gave the best fit to the experimental data (Figure 5.7):

$$E(T) = E_0 - \lambda/(\exp(\beta/T) - 1), \qquad (5.8)$$

where λ and β are fitting parameters: $\lambda = 0.121 \pm 0.005$ eV, and $\beta = 316 \pm 8$ K. E_0 represents low-temperature energies for the excitons (described previously).

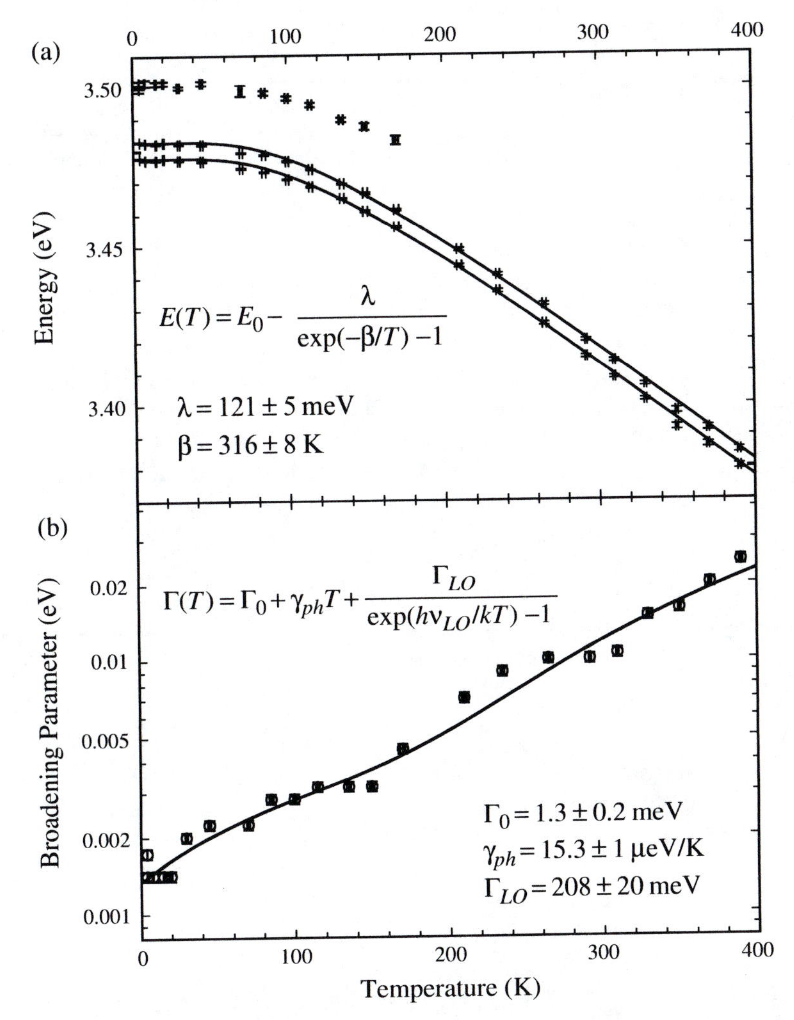

Figure 5.7 (a) Points, temperature dependence of free-exciton energies; solid lines, fit according to Eq. (5.8) (for the low-temperature energies E_0 of excitons see Table 5.1). (b) Points, temperature dependence of the broadening parameter of the exciton lines, solid line, fit according to Eq. (5.10).

The experimental data can also be fitted with the Varshni formula [43]. We found that the following curve fit the experimental data:

$$E(T) = \frac{E_0 - 0.00181T^2}{T + 2300\ K}.$$

However, using Eq. (5.8) fit gives about half the χ^2 value than that obtained by applying the Varshni formula.

Table 5.3 Pressure coefficients for donor bound (D^0X) and free excitons and the energy gap in GaN.

	E(0) (eV)	*α (meV/GPa)*	*β (meV/GPa²)*	*Source*
D^0X	**3.469(2)**	**42.9 ± 0.6**	**-0.45 ± 0.08**	**Homoepitaxial, PL, [57]** (this work)
D^0X	3.472(2)	40 ± 1		Homoepitaxial, PL, [56]
D^0X	3.4808	38.6(5)	-0.08	Heteroepitaxial, PL, [53]
D^0X	3.473(2)	41.5 ± 1.5	-0.30 ± 0.2	Heteroepitaxial, PL, [51]
D^0X	3.472	44 ± 1	-0.11 ± 0.02	Heteroepitaxial, PL, [52]
Exciton A	**3.4762(20)**	**42.7 ± 0.6**	**-0.39 ± 0.08**	**Homoepitaxial, R, [57]** (this work)
Exciton A	3.479(2)	41.4 ± 1.0	-0.30 ± 0.1	Heteroepitaxial, R, [51]
Exciton A	3.4875	39.0 ± 0.5	-0.18	Heteroepitaxial, PL, [53]
Exciton B	**3.4813(20)**	**42.9 ± 0.6**	**-0.39 ± 0.08**	**Homoepitaxial, R, [57]** (this work)
Exciton C	**3.498(2)**	**43.0 ± 0.7**	**-0.40 ± 0.1**	**Homoepitaxial, R, [57]** (this work)
E_g	**3.497(3)**	**43.0 ± 0.6**	**-0.4 ± 0.1**	**Homoepitaxial, R, [57]** (this work)
E_g	3.503(2)	43.8 ± 1.4	-0.50 ± 0.2	Heteroepitaxial, R, [51]
E_g	3.457	47	-0.18	Bulk, A, [50]
E_g	3.5	42 ± 4		Heteroepitaxial, A, [54]
E_g	2.64	39	-0.32	Calculations (LDA), [55]

Note: Coefficients α and β were found by parabolic fit: $E = E(0) + \alpha P + \beta P^2$. The data are from photoluminescence (PL), reflectance (R), or absorption (A) measurements, and from theoretical (LDA) calculations.

The temperature coefficient of the energy gap at $T = 300$ K was $dE_g/dT = -0.36 \pm 0.02$ meV/K. The E_g change due to the temperature-dependent dilatation of the lattice is described by the formula

$$\left(\frac{dE_g}{dT}\right)_{dil} = -3\alpha B \frac{dE_g}{dp}, \tag{5.9}$$

where $\alpha = 3.5 \times 10^{-6}\,\mathrm{K}^{-1}$ [44] is the linear thermal expansion coefficient, and $B = 245$ GPa [45] is the bulk modulus. Taking $dE_g/dp = 43.0 \pm 0.7$ meV/GPa (see Table 5.3) and using Eq. (5.9) we obtained $(dE_g/dT)_{dil} = -0.11 \pm 0.01$ meV/K. The remaining part of the energy gap change is the contribution from electron–phonon interaction $(dE_g/dT)_{ph} = -0.26 \pm 0.04$ meV/K, which means that at room temperature the phonon mechanism is the dominant one (about 70%). This situation is typical for many semiconductors [43].

Figure 5.7b shows the broadening parameter Γ of excitons A and B obtained by fitting of Eq. (5.7) to the reflection spectra. The temperature dependence of Γ can be interpreted as a broadening due to exciton–phonon scattering [46,47]. It has been found that at low temperature ($T < 120$ K) Γ increases linearly with the temperature, which indicates the influence of acoustic phonon scattering; however, at temperatures above 130 K Γ increases nearly exponentially, which suggests that LO phonon scattering also should be included. The data may be fitted by the following expression [41]:

$$\Gamma(T) = \Gamma_0 + \frac{\gamma_{ph} T + \Gamma_{LO}}{\exp(h\nu_{LO}/kT) - 1} \tag{5.10}$$

where $h\nu_{LO} = 92.3\,\text{meV}$ is the energy of an LO phonon in homoepitaxial GaN [48], Γ_0 is a zero-temperature broadening parameter, γ_{ph} is the coupling strength of exciton–acoustic phonon interaction [46], and Γ_{LO} is a parameter describing the exciton–LO phonon interaction [47]. It was found that a reasonably good fit could be obtained for $\Gamma_0 = 1.3 \pm 0.2\,\text{meV}$, $\gamma_{ph} = 15.3 \pm 1\,\mu\text{eV/K}$, and $\Gamma_{LO} = 208 \pm 20\,\text{meV}$. The obtained value of γ_{ph} is similar to values reported for other semiconductors [46], whereas Γ_{LO} is very high, probably due to a very high Frölich constant [49] and high LO phonon energy in GaN.

5.2.4 Pressure Dependence

Early experimental reports on high-pressure properties of GaN described measurements on GaN polycrystals [44], GaN single crystals [45,50] and heteroepitaxial layers [51–54]. Some theoretical calculations concerning the band structure of GaN also have been published [31,32,55]. Homoepitaxial layers, however, are the best samples for pressure measurements due to their high optical quality and because these layers have the same values of elastic constants as the substrates. Some high-pressure measurements on a homoepitaxial GaN have recently been reported [56,57].

The high-pressure reflectivity and luminescence measurements presented in Ref. 57 were performed in a diamond anvil cell filled with helium and, cooled down to 10 K in a continuous-flow helium cryostat. In order to fit into a diamond anvil cell the sample was cleaved to about 100 μm in diameter. Applied pressures were determined by monitoring the shift of the ruby line [58]. A xenon arc lamp was used for the reflectance measurements. The light was focused on a pinhole of 100 μm diameter, which was imaged with about unity magnification onto the sample surface. The spectra were analyzed by a 0.6-m single-grating spectrometer. Within experimental resolution all pressure effects on the optical spectra were fully reversible.

The reflectance spectra of GaN under pressure are shown in Figure 5.8a. The exciton energies obtained by fitting of Eq. (5.7) to the reflectance spectra are plotted in Figure 5.9. The energy-versus-pressure dependence was fitted by a second-order polynomial: $E = E(0) + \alpha P + \beta P^2$. The pressure coefficients α and β for all excitons are summarized in Table 5.3 and

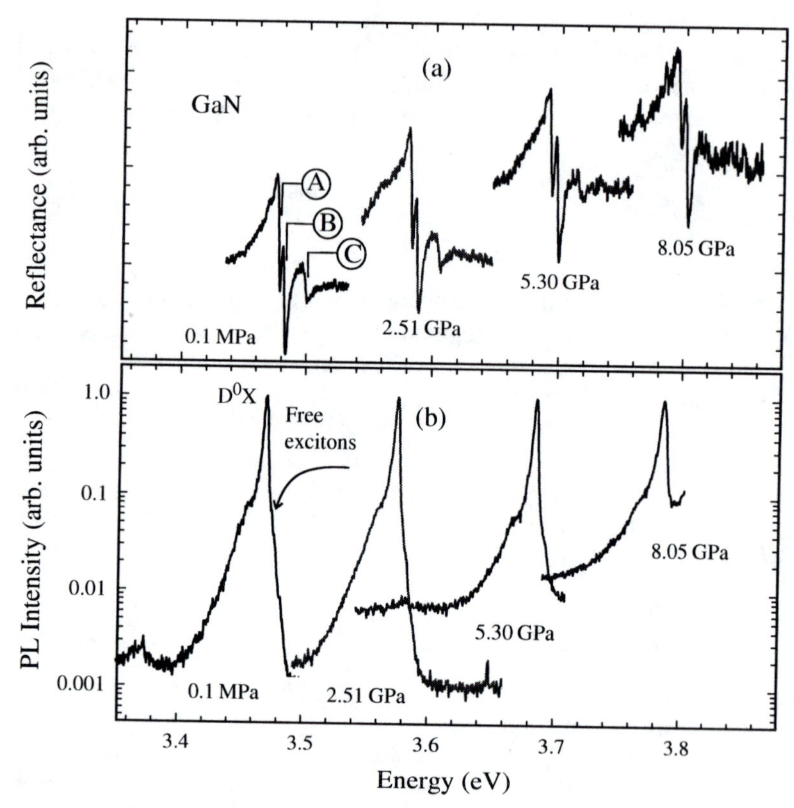

Figure 5.8 Optical spectra of homoepitaxial GaN under pressure. (a) Reflectance (spectra are shifted up proportionally to the pressure). (b) PL spectra of GaN (spectra are normalized to 1).

compared with previously published data. It can be seen that the data for the homoepitaxial sample [57] lie in the middle of the previously published data, but the standard deviations of these data are among the smallest. It is worth noting that both the first- and the second-order pressure coefficients (α and β) are in good agreement with theoretical predictions [55].

In order to calculate the energy gap of GaN, one should take into account the binding energies G of the excitons. We took the data for the ambient pressure from experiment [59]: $G_A = G_B = (21 \pm 1)$ meV, and $G_C = (23 \pm 1)$ meV. The pressure dependence of the binding energies is due to the change in the exciton masses and the decrease in the dielectric constant with pressure. Taking into account Luttinger parameters [31], we can estimate that the relative change in the effective exciton masses under hydrostatic pressure $\mu^{-1} * d\mu/dp$ is of the order of $10^{-4}\,\mathrm{GPa}^{-1}$, so it is negligible. The change in the dielectric constant under pressure $dln(n)/dp = -0.0030(4)\,\mathrm{GPa}^{-1}$ [49] causes the increase in the binding energy $dG/dp = 0.25(4)$ meV/GPa, which is similar for all free excitons.

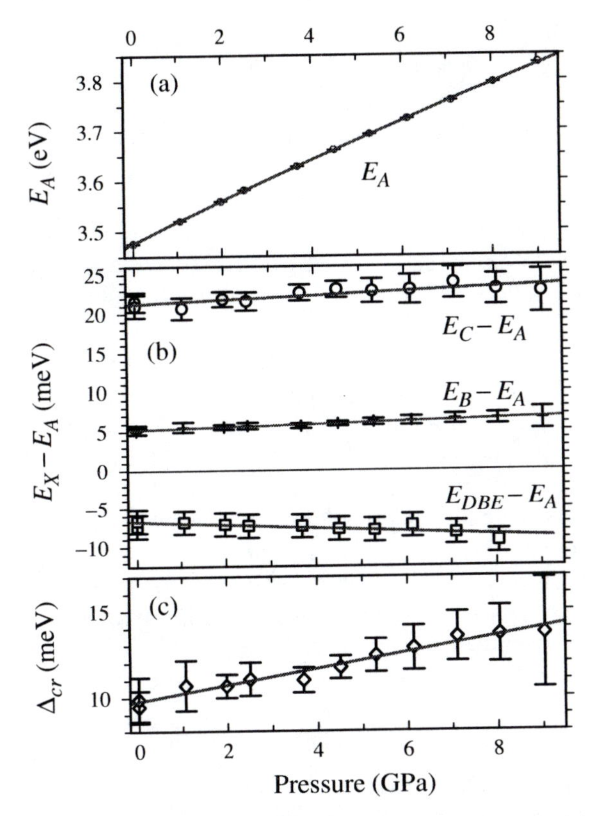

Figure 5.9 Dependence of exciton energies versus pressure for homoepitaxial GaN. (a) The free exciton A; experimental points are fitted with a parabolic curve (for parameters see Table 5.3). (b) Differences in energy between exciton A and free excitons B and C. (c) Crystal-field splitting.

By adding this value to the pressure coefficient of the A exciton, we obtain the bandgap pressure coefficient $dE_g/dp = 43.0$ meV/GPa (at zero pressure). The $E_g(0)$ value and pressure coefficients α and β for the energy gap are shown in Table 5.3 and compared with previously published data.

There is much scatter in the published data [44,45,53] for the bulk modulus of GaN. We adopted the value of 207 GPa, which was obtained in recent single-crystal X-ray diffraction, studies [60]. Based on this number, we calculated the GaN band gap deformation potential: $D_{E_g} = -27 \pm 3$ eV.

Figure 5.9b shows the differences in energies of the free excitons. The change in the E_{AB} energy is visible, but very small. The observed energy versus pressure dependence is described by the relation $E_{AB} = 5.2 \pm 0.2\,\text{eV} + 0.15 \pm 0.03\,\text{meV/GPa} \times P$. (Note that it is possible to determine the relative positions even better than the absolute values.) The situation is similar in the case of the C exciton. The pressure coefficient obtained for the

C exciton is only 0.3 ± 0.1 meV/GPa. In the cubic approximation the energy differences among the A, B, and C excitons caused by crystal-field Δ_{cr} and spin-orbit Δ_{so} splittings are described by Eq. (5.6). Since the changes in binding energies are similar for all free excitons, we can neglect their influence and calculate the energies of the splitting parameters using Eq. (5.6). The spin-orbit splitting was found to be independent (within experimental error) of pressure, which is in agreement with theoretical assumptions [29]. Its zero-pressure value, $\Delta_{so} = 17 \pm 2$ meV, is similar to previously reported numbers (see Table 5.2). For further calculations we took the constant value $\Delta_{so} = 18$ meV, deduced as the best from Table 5.2. Taking into account the changes in relative positions of the A, B lines allows a pressure dependence of Δ_{cr} to be calculated (see Figure 5.9c). We estimated the pressure coefficient of the crystal-field splitting to be $\alpha_{\Delta cr} = 0.4 \pm 0.1$ meV/GPa.

It is a well-known fact [15,16,27] that the energies of the A, B, and C excitons in GaN are very sensitive to biaxial strain. This sensitivity was first observed experimentally [15] and then explained by theoretical calculations [32]. The energy splitting between the A and the B excitons E_{AB} in homoepitaxial layers is 5 meV [17,41] (see also Section 5.2.2). In thick heteroepitaxial layers, $E_{AB} = 6$ meV has been reported [13]; however, in heteroepitaxial layers strain can increase the splitting even up to $E_{AB} = 12$ meV [27], which is in contrast with our results that under high hydrostatic pressure the distance between the excitonic lines is nearly constant. The following relation describes the pressure dependence of the crystal-field splitting [29]:

$$\Delta_{cr} = \Delta_1 + D_3\varepsilon_{zz} + D_4(\varepsilon_{xx} + \varepsilon_{yy}), \tag{5.11}$$

where D_3, D_4 are deformation potentials, and ε_{xx}, ε_{yy}, ε_{zz} are components of the strain tensor ε.

On the basis of the elastic constants of GaN, $c_{11} = (296 \pm 18)$ GPa, $c_{12} = (130 \pm 11)$ GPa, $c_{13} = (158 \pm 6)$ GPa, $c_{33} = (267 \pm 18)$ GPa [44], it is possible to calculate a strain caused by hydrostatic pressure [61]: $\varepsilon_{xx} = \varepsilon_{yy} = -0.00170\,\text{GPa}^{-1} \times P$, $\varepsilon_{zz} = -0.00172\,\text{GPa}^{-1} \times P$. The deformation potentials for GaN are known from theoretical calculations: $D_3 = 6.61$ eV, $D_4 = -3.55$ eV [32]. Results from experimental analysis of biaxial strain are in agreement with these values: $D_3 = 8.82$ eV ($D_4 = -D_3/2$ assumed) [27]. From theoretically calculated values of the deformation potentials [32] one can obtain $\alpha_{\Delta cr} = 0.7$ meV/GPa which is of the same order of magnitude as the experimental result $\alpha_{\Delta cr} = 0.4$ meV/GPa.

It is worth noting that in the case of a uniaxial stress along the z-axis, the resulting strains, $\varepsilon_{xx} = \varepsilon_{yy} = 0.00248\,\text{GPa}^{-1} \times P_z$, $\varepsilon_{zz} = -0.00668\,\text{GPa}^{-1} \times P_z$, are of opposite sign. Thus, according to Eq. (5.11), the obtained pressure coefficient $\alpha^u_{\Delta cr} = 62$ meV/GPa for the uniaxial strain, is very strong, which causes the large changes in the exciton energy splittings in the strained GaN layers.

In the cubic approximation [29] a sum of the deformation potentials $D_3 + 2D_4$ should be equal to zero for the ideal quasi-cubic crystal, but this is not true in real crystals. For example, the parameter $\gamma = (D_3 + 2D_4)/D_4$ is equal to 1.4, 1.5, 0.18, and −0.24 for ZnO [62], CdS [62], CdSe [62], and AlN [32], respectively. The value obtained on homoepitaxial GaN is $\gamma = 0.07$, so we can conclude that GaN is a nearly ideal example of the quasi-cubic model of wurtzite-structured crystals. This is probably due to nearly ideal tetrahedral coordinates of the first neighbors in GaN: the c/a ratio for GaN is equal to 1.6262 [3] (it should be 1.6330 for the ideal structure) and the u parameter is equal to 0.377 [63] (it should be 0.375). For example, for AlN the values are $c/a = 1.601$, and $u = 0.382$ [63].

The PL spectra, measured at different values of hydrostatic pressure, are shown in the Figure 5.8b. The energy shift caused by pressure is very fast (~40 meV/GPa). Since the energy of the exciting laser line was 3.815 eV, it was impossible to observe excitonic luminescence above 8.1 GPa. The pressure dependence of the neutral-donor-bound exciton (D^0X) energy could be very accurately fitted with a parabolic curve: $E_{D^0X} = 3.469 \pm 0.002\,\text{eV} + 42.9 \pm 0.6\,\text{meV/GPa} \times P - 0.45 \pm 0.08\,\text{meV/GPa}^2 \times P^2$. The fit quality was very good; the maximum deviation between experimental data and the fitted curve was 0.001 eV. The obtained pressure coefficients were similar for the free excitons and the bandgap (see Table 5.3), so we concluded that changes of the exciton-to-donor binding energy under pressure are small.

5.2.5 Free-Exciton Magneto-Optics

In this section we discuss the symmetry of excitons in wurtzite-structure GaN crystals on the basis of high-field magnetoreflectance experiments [64]. The magnetic-field pattern of the resonances is confirmed with calculations based on the Luttinger model [65] completed by introducing the crystal-field and electron–hole exchange interactions, as proposed in Ref. [64]. We have shown that symmetries of the observed magneto-excitonic levels can be satisfactorily parameterized using only two coupling constants, i.e., the g-factor for the conduction-band electrons and a single "kappa" parameter for all three valence subbands. The interplay between the crystal-field and spin-orbit interactions leads to the unexpected spin symmetries of the resulting exciton states.

In GaN, as in most semiconductors, free-exciton resonances involve a coupled pair of an s-type conduction-band electron and a p-type valence-band hole. As discussed in previous sections, this 12-fold degenerate exciton ground state is split in the presence of spin-orbit interaction and crystal field, and three structures related to excitons A, B, and C can be observed in reflectivity measurements. In the external magnetic field the subsequent splitting caused by the magnetic field coupling to the angular momentum and spin allows detailed studies of the conduction- and valence-band structure.

In the past, the magnetic field pattern of different excitonic states for wurtzite-type semiconductors such as CdS, has been discussed only in terms of the effective g-factors, separately for each valence subband [21,66,67]. Such a many-parameter description offers a rather limited understanding of the origin of the splitting pattern observed for the excitonic resonances, particularly when the amplitudes of spin-orbit and crystal-field interactions are of the same order. In the case of GaN, spin-orbit and crystal-field interactions are of comparable strength, and both are smaller than exciton binding energy (Section 5.2.2). Thus, A, B, C exciton components are close in energy and most likely form a coupled state and therefore need to be understood within the most general model possible with a minimum number of parameters.

The experiments were performed using homoepitaxial GaN layers grown by MOCVD on GaN bulk crystals. A detailed description of the samples and the experimental setup was given in Section 5.2.2. Low-temperature reflectivity spectra were measured with the magnetic field **B** parallel and perpendicular to the **c**-axis of the GaN crystal lattice. The representative raw data measured in the magnetic field configuration $\mathbf{B}\|\mathbf{c}$, in the range $\mathbf{B} = 0$ to 27 T, are presented in Figure 5.10. The spectrum obtained at $\mathbf{B} = 0$ shows three characteristic features, which correspond to the optically active excitons A, B, and C. The origin of this structure, which is typical for wurtzite-type semiconductors, is described in detail in Section 5.2.2. The magnetic field evolution of the reflectivity spectra shows the Zeeman effect, which is essentially different for different exciton components and also depends on the magnetic field orientation. For example, in the $\mathbf{B}\|\mathbf{c}$ configuration (see Figure 5.10) we clearly observe the spin splitting of B and C excitons, whereas exciton A does not split in this configuration, although a tiny feature appears on its low-energy wing [68].

The energy positions of the excitonic resonances were defined from the measurements by the standard method of calculating the derivatives dR/dE of the reflectivity spectra and assigning the maxima of $-dR/dE$ to the optically active transverse excitons. This method can introduce some systematic error in the absolute value of the exciton energy but is quite accurate for describing the relative displacement of the Zeeman components. The energy positions of the exciton Zeeman components, measured for different magnetic fields and for two experimental configurations, are shown in Figure 5.11. Representative derivatives of the reflectance spectra measured at $\mathbf{B} = 27$ T are also presented in this figure. Using the "derivative" method we can very easily follow the magnetic field evolution of excitons A and B. This procedure is less effective in the region of exciton C because of the significant broadening of this resonance and because of the overlap of exciton C with the excited states of excitons A and B. Nevertheless, the effect of magnetic

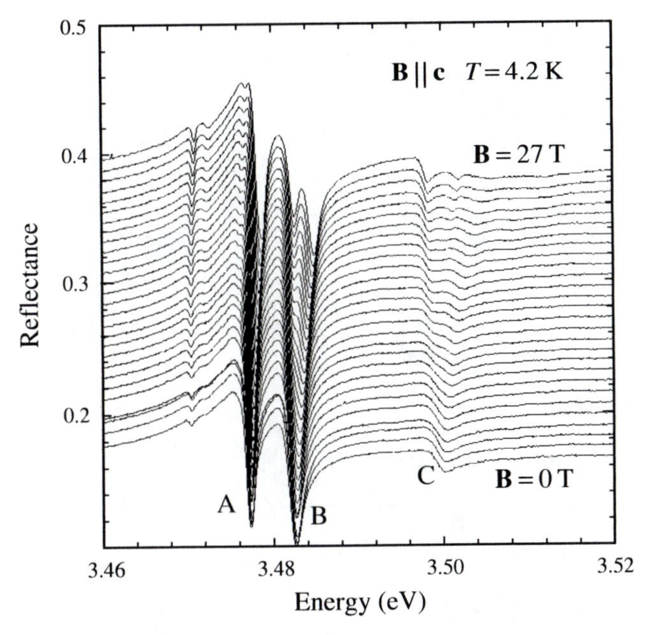

Figure 5.10 Magneto reflectance spectra of GaN measured in magnetic fields up to 27 T applied along the **c**-axis of the wurtzite crystal. For clarity, subsequent spectra are shifted vertically.

fields on exciton C can be quite clearly seen in the raw data (see Figure 5.10). Note, for example, that exciton C splits into only two components in the **B**||**c** configuration, whereas the feature appearing between the two spin-split components (see Figure 5.11a) is related to the 2*s* state of exciton A.

We note the following characteristic features of splitting patterns presented in Figure 5.11a and b:

(1) **B**||**c** configuration: The spin splitting of the exciton A is negligibly small, and in this case proper analysis of the observed structures requires complex analysis using the same method as discussed for the zero-field spectrum (Section 5.2.2.); exciton B splits into two components of comparable intensities; the largest splitting is observed for exciton C.

(2) **B**⊥**c** configuration: Exciton A clearly splits into two components; exciton B shows up to four spin-split components at highest fields; exciton C splits into more than two optically active components. (Due to a broadening and the appearance of the excited states of exciton A, the splitting pattern of exciton C is less precisely determined.)

We first describe these observations qualitatively, introducing the simplified approach to excitonic transitions in terms of band-to-band transitions,

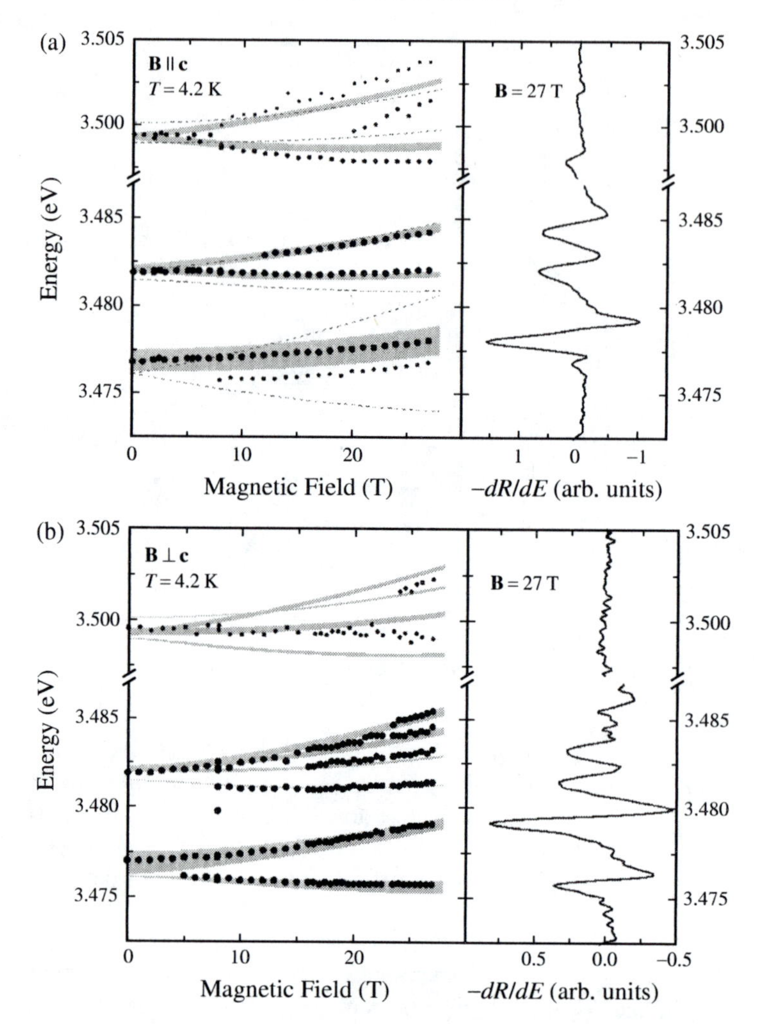

Figure 5.11 (a) Left side: energies of the Zeeman components for the exciton in GaN as a function of the magnetic field for $\mathbf{B}\|\mathbf{c}$; circles, experiment; shadow and broken lines, calculated energies of optically active and "dark" excitons, respectively. The width of the shadowed area is proportional to the calculated oscillator strength of a given Zeeman component. Right side: derivative of the reflectance spectrum at $\mathbf{B} = 27$ T. (b) The same as (a) for the magnetic field direction $\mathbf{B} \perp \mathbf{c}$.

as schematically shown in Figure 5.12. In this approach, the symmetry of each exciton component is characterized by two quantum numbers (s_e, j_z) that correspond to the electron and hole involved in an excitonic transition, where $s_e = \pm 1/2$ and $j_z = \pm 3/2, \pm 1/2$ denote, respectively, the projections of the conduction-band spin and of the total angular momentum of the valence-band states onto the direction parallel to the **c**-axis of the crystal.

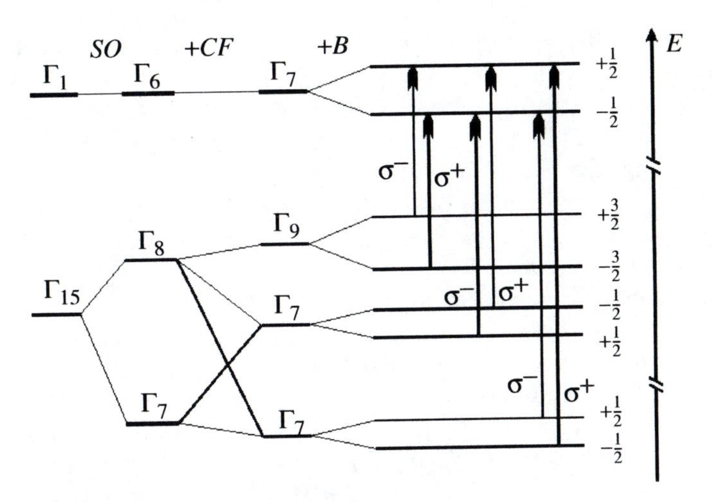

Figure 5.12 Breaking-symmetry scheme of the conduction and valence bands in GaN. The influence of the subsequent interactions: spin-orbit, crystal-field, and magnetic field (**B**||**c**) is shown from left to right. The total angular momentum of the resulting states is given on the right side. The optically active excitons correspond to the allowed transitions, which are marked by arrows.

As shown in Figure 5.12, the spin-orbit and crystal-field interactions reduce the high symmetry of initial states to those with $\Gamma_9(j_z = \pm 3/2)$ and $\Gamma_7(s_e = \pm 1/2$ or $j_z = \pm 1/2)$ symmetries. The orientation of the magnetic field with respect to the crystal **c**-axis is an important factor in the determination of a transition's symmetry: states with Γ_7 symmetry split in both configurations of the magnetic field, whereas states with Γ_9 symmetry split in the **B**||**c** configuration but not when $\mathbf{B} \perp \mathbf{c}$ [69].

Our results confirm the $\Gamma_9 \rightarrow \Gamma_7$ character of the transition associated with exciton A. As expected, this transition shows only two Zeeman components of equal intensity in the $\mathbf{B} \perp \mathbf{c}$ configuration. The splitting of exciton A in this configuration is determined solely by the spin splitting in the conduction band, allowing a direct measure of the electron g_e-factor. In the $\mathbf{B} \perp \mathbf{c}$ configuration up to four Zeeman components are expected for each $\Gamma_7 \rightarrow \Gamma_7$ interband transition, as indeed is observed for excitons B and C.

The case of the **B**||**c** configuration is relatively simple due to the clearly defined $s_e - j_z = \pm 1$ selection rules for optically allowed transitions (see Figure 5.12). In this configuration, each exciton A, B, and C is expected to split into a doublet. Apparently, the spin splitting of the $\Gamma_9 \rightarrow \Gamma_7$ transition is not experimentally resolved. This is because spin splittings in the conduction and in the valence band compete with one another, leading to an almost exact cancellation of the Zeeman effect for exciton A in this configuration. We assumed that $g_e > 0$ and that its value is not sensitive to the field orientation [70], therefore concluding that the g-factor value for the Γ_9 valence band states is positive: $g_{vA\|} \approx g_e > 0$.

Although exciton A remains accidentally degenerate when **B**‖**c**, the doublet splitting is indeed observed for excitons B and C. We note that the spin splitting of exciton B in the **B**‖**c** configuration is smaller than the spin splitting of exciton A in the **B** ⊥ **c** configuration, the latter splitting being determined solely by the spin splitting in the conduction band. According to the selection rules shown in Figure 5.12, we concluded that the g-factor of Γ_7 valence states associated with exciton B is negative ($g_{vB\|} < 0$). In contrast, the spin splitting of exciton C in the **B**‖**c** configuration is larger than the spin splitting in the conduction band, and therefore the g-factor of Γ_7 valence states associated with exciton C is positive ($g_{vC\|} > 0$). The resulting spin order of Zeeman components of the valence states is shown in Figure 5.12.

It is interesting to note that in the limit of finite spin-orbit splitting ($\Delta_{so} > 0$) but vanishing crystal-field effect ($\Delta_{cf} = 0$), one would expect that $g^0_{vB\|} = (1/3)g^0_{vA\|} > 0$ and $g^0_{vC\|} = ((1/3)g_{vA\|} - 1) < 0$ [65] in contrast with our observation, $g_{vB\|} < 0$, $g_{vC\|} > 0$. Although the crystal-field effects are relatively weak in GaN ($\Delta_{so} > \Delta_{cf}$) [17], our results clearly show that mixing of valence-band states in this compound is sufficiently strong to reverse the spin order for the two Γ_7 valence subbands.

The preceding qualitative considerations can be justified by comparing the experimental data with the following, more rigorous model proposed in Ref. 64. We assumed that the exciton is composed of an s-like conduction-band electron and a p-type valence-band hole, which can be described using a $|s_e, l_z, s>$ basis, where the **z**-direction is chosen along the **c**-axis. These three quantum numbers correspond to the spin operator of the electron $\mathbf{s}_e(s_e = \pm 1/2)$, orbital angular momentum, and spin operators of the hole $\mathbf{l}(l_z = 1, 0, -1)$ and $\mathbf{s}$ ($s = \pm 1/2$), respectively. The Hamiltonian of this 12-fold degenerate exciton family can be stated in the following form:

$$H = H_0 + H_B + H_D \tag{5.12}$$

where H_0 is the zero-field Hamiltonian, and H_B and H_D describe the linear and quadratic magnetic field effects, respectively.

H_0 is a 1s exciton Hamiltonian for the wurtzite structure, in the simplest, quasi-cubic approximation [29,71]:

$$H_0 = E_0 - \Delta_{cf} l_z^2 - \tfrac{2}{3}\Delta_{so}\vec{\mathbf{l}}\vec{\mathbf{s}} + 2\Delta_{ex}\vec{\mathbf{s}}_e\vec{\mathbf{s}} \tag{5.13}$$

where E_0 corresponds to the energy gap; Δ_{cf} and Δ_{so} denote, respectively, the crystal-field and spin-orbit splittings, and Δ_{ex} is the isotropic exchange splitting. The values of the crystal-field and spin-orbit interaction parameters determine the Γ_9, Γ_7, and Γ_7 energies of the split valence bands (Section 5.2.2).

According to Luttinger [65], the linear magnetic field effects can be included as follows:

$$H_B = g_e \mu_B \vec{\mathbf{B}}\vec{\mathbf{s}}_e + 2\mu_B \vec{\mathbf{B}}\vec{\mathbf{s}} - \mu_B(3\tilde{\kappa} + 1)\vec{\mathbf{B}}\vec{\mathbf{I}} \qquad (5.14)$$

Here, the first term corresponds to the Zeeman splitting of the conduction band and is described by the effective g-factor of the electron g_e. The next two terms account for the valence subband splitting and are related to the spin and orbital angular momentum of the hole, respectively. The parameter $\tilde{\kappa}$ gives directly the Zeeman splitting of the hole bound in exciton A for the $\mathbf{B}\|\mathbf{c}$ configuration: $g_{hA\|} = 6\tilde{\kappa}$. Its value may be directly expressed by the κ parameter of the free hole, and other valence-band parameters [72]. Note that g-factors of the electron and of the corresponding hole are of opposite sign: $g_{vA\|} = -g_{hA\|}$, etc.

The diamagnetic shift, clearly visible in the experimental results (Figure 5.11), was introduced in the simplest way, through $H_D = d\mathbf{B}^2$.

When the direction of the incoming light is normal to the crystal surface, the following four states among exciton basis are optically active: $| + 1/2, \pm 1, -1/2 >$ and $|-1/2, \pm 1, +1/2 >$. For unpolarized light their oscillator strengths are equal.

The diagonalization of the Hamiltonian gives the energies and eigenfunctions of the exciton Zeeman components, and their oscillator strengths can be calculated. The resulting exciton spectrum is presented in Figure 5.11 in comparison with the experimental results. Four eigenstates of the Hamiltonian with the lowest energy correspond to the components of exciton A. For the magnetic field $\mathbf{B}\|\mathbf{c}$ two of them, with Γ_5 symmetry, are almost pure basis states, namely, $|+1/2, -1, -1/2>$ and $|-1/2, +1, +1/2>$, and are optically active [67]. For this configuration, the magnetic field–induced splitting between these states is equal to $\Delta_{A\|}(B) = |(g_e + 6\tilde{\kappa})\mu_B B|$. Since this splitting is not resolved, one can estimate $\tilde{\kappa} \approx -g_e/6$.

Two other eigenstates: $|-1/2, -1, -1/2>$ and $|+1/2, +1, +1/2>$ belong to "dark excitons" with Γ_6 symmetry. A magnetic field applied perpendicular to the **c**-axis introduces coupling between the optically active and "dark" A-excitons. In the first approximation, neglecting the interactions with excitons from the B and C groups, the observed splitting is described by [69]: $\Delta_{A\perp}(B) = \sqrt{\Delta_{ex}^2 + (g_e \mu_B B)^2}$.

Thus, the zero-field exciton energies, A-exciton diamagnetic shift, and its splitting measured for two configurations $\mathbf{B}\|\mathbf{c}$ and $\mathbf{B} \perp \mathbf{c}$, give sufficient information to estimate all parameters involved in the Hamiltonian. Finally, the values were corrected using a least-squares-fit procedure, taking into account all transitions resolved in the experiments (see Figure 5.11). The obtained parameter values are listed in Table 5.4.

Table 5.4 Band structure parameters of GaN obtained from analysis of the near-band-edge magnetoreflectance, and the resulting energies of the optically active excitons for $B = 0$.

Δ_{cf} (meV)	Δ_{so} (meV)	Δ_{ex} (meV)	g_{el}	κ	d ($\mu eV/T^2$)	E_A (meV)	E_B (meV)	E_C (meV)
10.2(1)	18.1(2)	−0.91(5)	1.94(2)	−0.36(1)	1.84(10)	3476.9(3)	3482.1(3)	3499.3(8)

Our model is in fair agreement with the experimental results presented in Figure 5.11a and b. Some disagreement in the values observed for exciton C can be related to its interferences with the excited states of excitons A and B. The resulting spin-splitting pattern of excitons A, B, and C (see Figure 5.12) allows us to determine the symmetry-breaking scheme of the corresponding valence states. The nonintuitive spin ordering for the two Γ_7 valence states is an interesting consequence of the valence-band mixing. The electron–hole exchange interaction is quite important in this compound due to a strong localization of the exciton wave functions. The g-factor value of the electron involved in the free exciton is, within experimental error, the same as obtained for the shallow donor [70] $g_D = 1.95$.

The crystal-field parameter $\Delta_{cf} = 10.2$ meV presented here differs from the value of 8.8 meV reported in Table 5.2, since in Section 5.2.2 the exchange interaction was not taken into account and since exciton energies differ slightly among different samples due to a small residual strain.

Concluding the reflectivity measurements in a high magnetic field, we have shown that a complicated Zeeman splitting pattern, observed for free-exciton states in a wurtzite-symmetry semiconductor, can be well described using the classical Luttinger model with a single coupling constant kappa for all three valence-band components. The competition between the spin-orbit and crystal-field interactions leads to an unexpected sequence of the g-factor signs for the B and C valence subbbands.

5.3 BOUND EXCITONS

Low-temperature PL spectra of a high-quality undoped homoepitaxial layer show several sharp peaks related to free and bound excitons (Figure 5.13). The absolute energies of this transition are slightly sample dependent and change within the range of 1 meV; however, separations between particular lines remain constant at this energy scale. The line observed at 3.472 eV [(Figure 5.13a] usually dominates the photoluminescence of the nominally undoped GaN and is assigned to the exciton bound to the neutral donor (D^0X) [4,9,16,19,73,74]. High-resolution measurements [(Figure 5.13b] show that in the range of the D^0X transition up to five lines should be resolved (see also

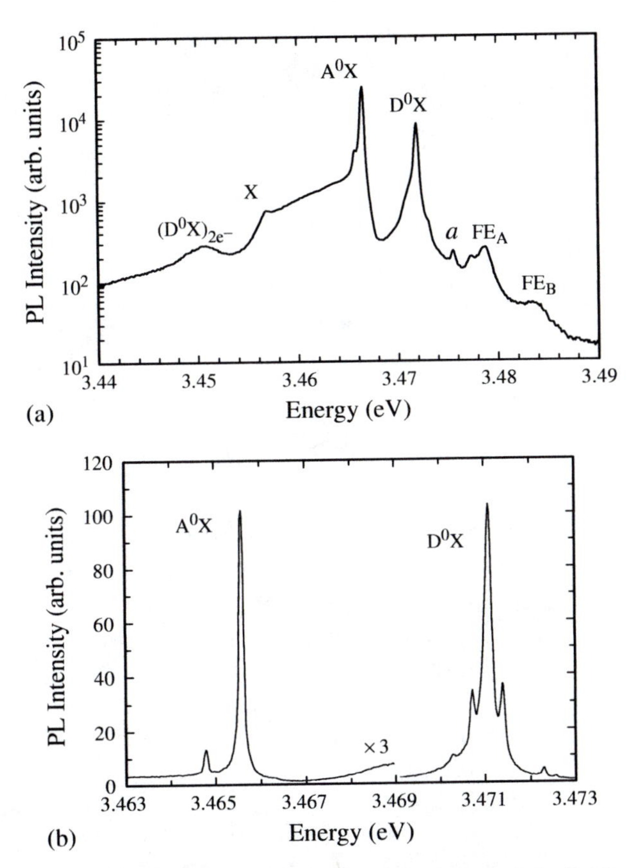

Figure 5.13 (a) Low-temperature photoluminescence spectra of the lightly Mg-doped homoepitaxial GaN layers. (b) Details of the luminescence spectrum in the range of bound exciton recombination measured on the nominally undoped sample.

Ref. 75). They could result from the different donor-bound excitons [4,7] as well as from the excited states of these complexes. The line at 3.467 eV has been assigned to the recombination of the exciton bound to a neutral acceptor (A^0X) [4,9,19,76] or to the exciton bound to an ionized donor (D^+X) [77,78]; however, the last interpretation has not been confirmed by magneto-optical studies [9,76]. The fine structure of A^0X emission, especially the satellite lying 0.8 meV below the main line, has been attributed to the internal structure of the acceptor state in GaN [9,76]. Its origin will be discussed later. The characteristic low-energy wing of the A^0X transition is probably related to its acoustic phonon replica. The peak labeled X (see Figure 5.13) observed around 3.455 eV does not correlate with the A^0X line (see Figure 5.14). It could tentatively be related to the deeper acceptor-bound exciton.

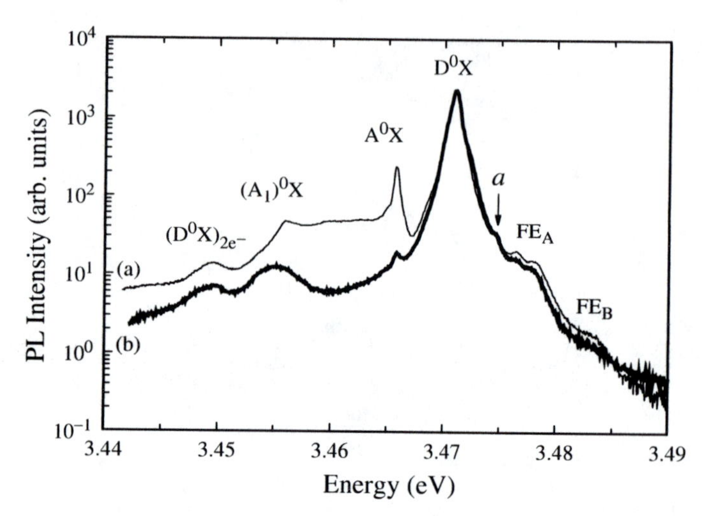

Figure 5.14 Comparison of luminescence spectra of nominally undoped layers grown at different conditions.

The high-energy part of the spectrum shows pronounced emission lines at 3.478 and 3.484 eV. They have been attributed to the recombination of free A-, and B-excitons (FE_A and FE_B lines) involving holes from Γ_9 and Γ_7 valence bands, respectively [17,64]. The double structure on the FE_A line has been related to polariton scattering processes with contributions from both lower and upper polariton branches (Section 5.2.2). The a-line observed at 3.476 eV (see Figure 5.13) has been attributed to the neutral-donor-bound exciton formed with the participation of a hole from the Γ_7 valence band [4] or to the exciton bound to an intrinsic donor (N vacancy or Ga interstitial) characterized by a positive central cell shift [16]. Another possibility is to relate the a-line to the excited state of a neutral-donor-bound exciton [4,76]. The emission peak at 3.450 eV has been attributed to a two-electron transition $(D^0X)_{2e^-}$ of a donor-bound exciton recombination in which the neutral donor is left in the excited 2s state. The 22-meV energy difference between the D^0X emission and its two-electron replica gives within the framework of effective-mass theory a donor binding energy of 29 meV [79]. Recent high-resolution measurements have confirmed this result and provided a more accurate value for 1s–2s in the donor of 21.6 meV, which implies a value of 28.8 meV for the donor binding energy [80].

The localization energies of neutral-acceptor- and neutral-donor-bound excitons, namely, $E^b_{D^0X} = E_{FE_A} - E_{D^0X} = 6\,\text{meV}$ and $E^b_{A^0X} = E_{FE_A} - E_{A^0X} = 11.4(5)\,\text{meV}$ are in reasonable agreement with the theoretical estimates giving the ranges 3.8–7.8 meV and 6.6–12.6 meV for D^0X and A^0X, respectively [6]. One could expect that these values should describe the temperature dependencies of excitonic emission.

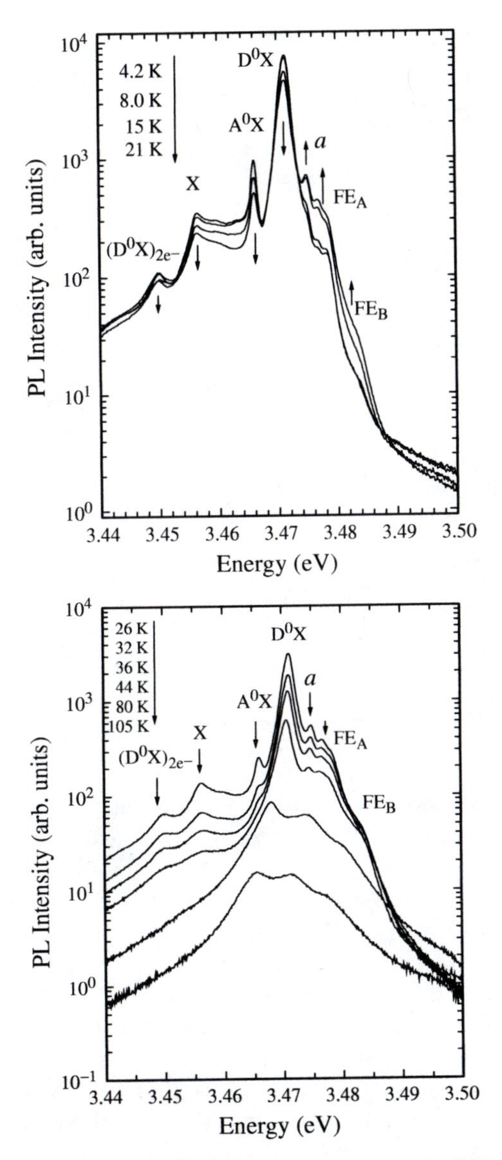

Figure 5.15 Temperature evolution of excitonic luminescence in GaN. The arrows show direction of intensity change with rising temperature.

The temperature dependence of the luminescence spectrum in the exciton region is shown in Figure 5.15. Increasing temperature above 4.2 K initially causes an increase and then a decrease in the intensity of free exciton recombination and line *a* (see Figure 5.15). Despite the fact that the A^0X line is at a greater distance from the free exciton than is D^0X, this spectrum disappears

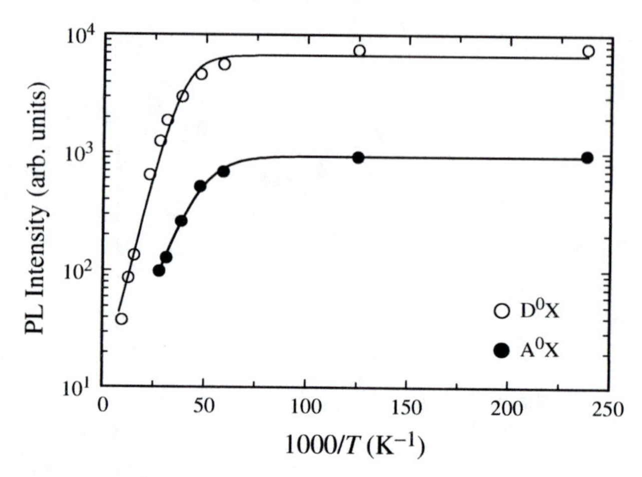

Figure 5.16 Temperature dependence of D^0X (open circles) and A^0X (solid circles); solid lines represent fit to the experimental data (see text).

first (above 40 K). The intensity of bound excitons decreases monotonically with rising temperature (see Figure 5.16). Such behavior can be represented using the simple formula

$$I(T) = \frac{I_0}{1 + C\exp(-\Delta E/kT)}, \tag{5.15}$$

where ΔE is activation energy, and C is a constant.

The activation energies obtained for the D^0X and A^0X lines are about 14 and 9 meV, respectively. These values, however, do not reflect the localization energies of these complexes. Such behavior can be explained by exciton kinetic processes [4]. Time-resolved PL experiments performed on GaN homoepitaxial layers [81] showed that D^0X decay time can be almost an order of magnitude longer than for A^0X. At higher temperature (35 K) the whole edge part of the photoluminescence has the same decay time as the free-exciton emission. It has been postulated that temperature-induced tunneling from a neutral-acceptor-bound exciton to a donor-bound one, as well as to a free exciton, takes place and allows much faster exciton decay. As a consequence of this process, the DC PL intensity of the A^0X line rapidly decreases with increasing temperature.

5.3.1 Bound-Exciton Magneto-Optics

Low-temperature luminescence spectra were measured with the magnetic field parallel and perpendicular to the **c**-axis of the GaN layer. The zero-field PL spectrum of the lightly Mg doped sample is shown in Figure 5.13a.

When a sufficiently high magnetic field is applied, all the lines previously discussed split. The behavior of the split components depends on the magnetic field orientation with respect to the **c**-axis of the sample (wurtzite structure). In the case of the D^0X line, the splitting is quite pronounced for a configuration with the magnetic field perpendicular to the **c**-axis (Figure 5.17, $\mathbf{B} \perp \mathbf{c}$). This characteristic splitting into two components of the same intensity has been reported previously, in Refs. 19 and 74.

The energies of these components in the magnetic field can be calculated using a model with two phenomenological parameters characterizing the D^0X line: the effective g-factor and the diamagnetic shift coefficient D:

$$E_{\perp\pm} = E_0 \pm \tfrac{1}{2} g_{\perp} \mu_B B + D_{\perp} B^2 \tag{5.16}$$

Fitting this formula to the experimental data for magnetic fields up to 27 T allowed us to obtain the following values for the D^0X line: $E_0^{D^0X} = 3.4724\,\text{eV}$, $g_{\perp}^{D^0X} = 1.87$, and $D_{\perp}^{D^0X} = 7.3 \times 10^{-7}\text{eV/T}^2$.

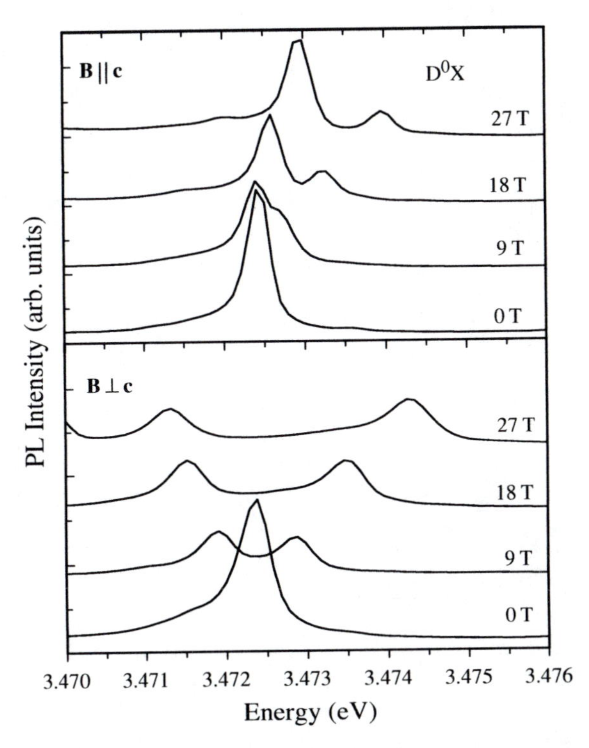

Figure 5.17 Photoluminescence spectra of D^0X emission for different magnetic fields.

For the magnetic field parallel to the **c**-axis the D^0X line splitting is significantly smaller (Figure 5.17, $\mathbf{B}\|\mathbf{c}$), and for our samples only well resolved above 10 T. This is one of the reasons why D^0X splitting for this configuration was not observed initially [19,74]. One should notice that for **B** parallel to the **c**-axis the higher-energy D^0X spin-split component is less intense. For this configuration the fitting of Eq. (5.16) to the experimental data gives $g_{\|}^{D^0X} = 0.64$, and $D_{\|}^{D^0X} = 1.4 \times 10^{-6}$eV/T^2.

In the case of the A^0X line the magnetic field splitting pattern is different from those observed for D^0X. The zero-field spectrum shows an additional, less intense emission structure, at an energy 0.7 meV below the main line [Figure 5.13b]. For the magnetic field parallel to the **c**-axis, a splitting can be resolved into two components only for the dominant line at a magnetic field above 20 T (Figure 5.18, $\mathbf{B}\|\mathbf{c}$—the small shoulder visible for 27 T on the lower-energy side of the spectrum); however, these lines show clearly resolved selection rules, and their splitting can be

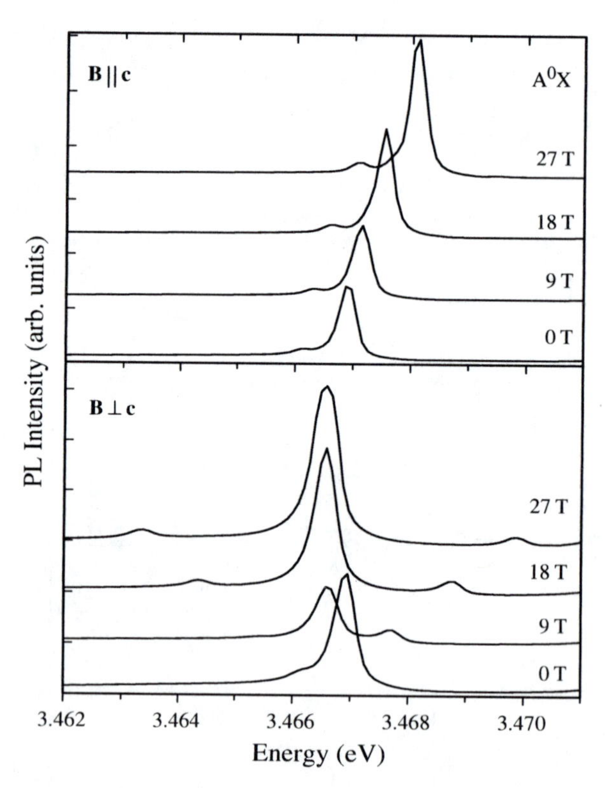

Figure 5.18 Photoluminescence spectra of A^0X emission for different magnetic fields.

clearly resolved in magnetic field studies of the circularly polarized luminescence [76]. The corresponding effective g-factor and diamagnetic shift parameter obtained from the fitting procedure are equal to $g_{\parallel}^{A^0X} = 0.22$ and $D_{\parallel}^{A^0X} = 1.3 \times 10^{-6}\,\text{eV/T}^2$, respectively. For this configuration, the lower-energy component of the dominant A^0X line is less intense.

For the magnetic field perpendicular to the **c**-axis, a clear splitting of the dominant A^0X line is resolved into two components (Figure 5.18, $\mathbf{B} \perp \mathbf{c}$). At low magnetic field these components have comparable intensities, but with increasing magnetic field, the intensity of the higher one decreases, showing a thermalization dependence (Figure 5.19) [19].

$$\frac{I_{up}}{I_{down}} = \exp\left(-\frac{g_{XA}^{*}\mu_B \cdot B}{kT_{eff}}\right) \tag{5.17}$$

with an effective temperature $T_{eff} = 9.3\,\text{K}$.

Note that at the same time the intensities of the D^0X-line components remain almost unchanged.

The A^0X line splitting can be described by the effective g-factor $g_{\perp}^{A^0X} = 2.1$ and diamagnetic shift by $D_{\perp}^{A^0X} = 1.6 \times 10^{-6}\text{eV/T}^2$. For sufficiently high magnetic fields, a third, less intense component is observed at lower energy.

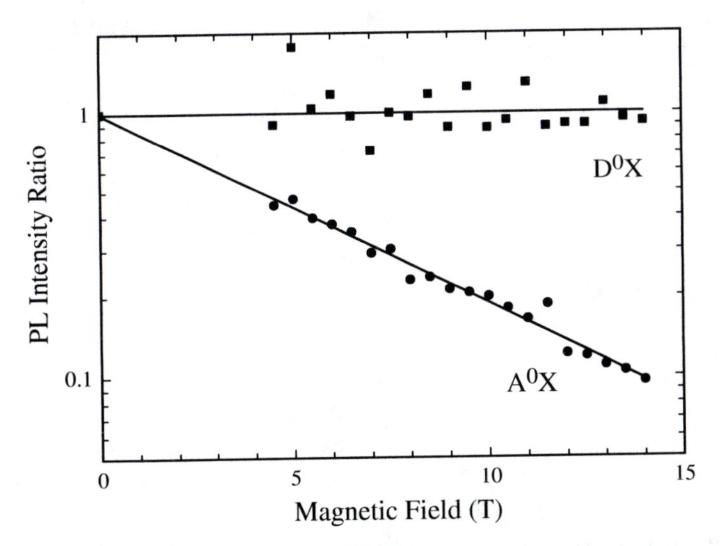

Figure 5.19 Magnetic field dependence of the intensity ratio of the dominant components (higher energy to lower energy) of the spin-split exciton bound to the neutral acceptor A^0X and to the neutral donor D^0X, respectively. Solid line for A^0X represents a fit according to Eq. (5.17) with an effective temperature $T_{eff} = 9.3\,\text{K}$.

5.3.2 Discussion of Bound-Exciton Magneto-Optics

The shallow impurity ground state within the effective-mass approximation should reflect the symmetry of the top of the corresponding conduction or valence band [82]. As discussed in Section 5.2, it is generally accepted that the conduction band has Γ_7 symmetry, and that the top of the valence band has Γ_9 symmetry.

In the case of the shallow donor ground state in GaN one can expect that the ground state has the Γ_7, almost spherical symmetry. Electron spin resonance (ESR) studies [70,83] and far-infrared intradonor magneto-optics [84] have confirmed this prediction. From effective-mass theory one expects the Γ_9 symmetry of the acceptor ground state, which should lead to a strong anisotropic variation in the external magnetic field. This is a consequence of the crystal-field interaction, which for Γ_9 states fixes the angular momentum direction parallel to the **c**-axis. Such a doubly degenerate state should not split in a magnetic field perpendicular to the **c**-axis. On the contrary, in ESR studies of Mg-doped samples, a relatively small g-factor anisotropy ($g_{\parallel} = 2.073$ and $g_{\perp} = 1.989$) was found for the acceptor ground state [83].

Impurity-related luminescence for the classical wurtzite-type semiconductor (for example, CdS) has been intensively studied in the past. A basic discussion of the Zeeman effect for impurity-bound excitons has been provided by Thomas and Hopfield [69]. According to this discussion we can conclude that the observed behavior of the D^0X line (Section 5.3.1) is in full agreement with the properties of the neutral-donor-bound exciton recombination [9,19,74]. For this transition the initial state is composed of two electrons of opposite spin and one hole. The final state of the D^0X recombination contains one electron. In a magnetic field the initial state splits as the Γ_9 hole state, while the final state splitting is just the Zeeman splitting of the neutral donor. As a consequence, for the magnetic field direction parallel to the **c**-axis, both the final and initial states split. The observed luminescence splitting, described by $g_{\parallel}^{D^0X} = 0.64$, corresponds to the difference between the effective g-factors for the electron and the hole. Different intensities of the split components reflect the thermalization process in the initial state. Since the higher-energy component is less intense, one can conclude that the initial state splitting (hole state splitting) is larger than the final (electron) one. For the magnetic field direction perpendicular to the **c**-axis the Γ_9 hole state does not split, so there is no splitting in the initial state of the neutral-donor-bound exciton recombination. As a result, in this configuration the D^0X line splits into two components of the same intensity (no thermalization), reflecting the final state splitting. The obtained g-factor value $g_e^{D^0X} = 1.87$ is slightly smaller than $g_{el} = 1.95$ derived from low-field ESR studies [70,83]. Assuming that the effective g-factor for electrons is isotropic, one can add the effective g-factor $g_{\perp}^{D^0X} = 1.87$ and $g_{\parallel}^{D^0X} = 0.64$, obtaining an estimate of the hole effective g-factor $g_{h\|}^{D^0X} = 2.5$ for the magnetic field parallel to the **c**-axis.

The behavior of the A^0X structure in a magnetic field is not so clear as in the case of D^0X. If we take into account the splitting pattern of the dominant component of this line, we can easily conclude [19] that the observed behavior in a magnetic field is typical for the neutral-acceptor-bound exciton recombination. For the shallow acceptor state, as discussed in Ref. 69, the initial state for this transition is composed of two holes of opposite spin and one electron. The final state of the A^0X recombination contains one hole. As a consequence, for the magnetic field direction parallel to the **c**-axis the luminescence line splits, and the observed splitting ($g_{\|}^{A^0X} = 0.22$) corresponds to the difference between initial (electron) and final (hole) state splittings. The lower intensity of the low-energy component confirms the smaller splitting of the electron (initial) state. On the other hand, for the magnetic field direction perpendicular to the **c**-axis the final state (hole) does not split, and the observed splitting corresponds to the Zeeman splitting of the electron involved in the neutral-acceptor-bound exciton complex. The effective g-factor of such localized electrons is equal to $g_{e\perp}^{A^0X} = 2.1$. With this information we can calculate the effective g-factor for a hole (involved in the A^0X complex) in a magnetic field parallel to the **c**-axis $g_{h\|}^{A^0X} = 2.3$, which is very close to the value $g_{h\|}^{D^0X} = 2.5$ obtained for the D^0X complex. The obtained parameters, which describe the behavior of the bound excitons, are collected in Table 5.5. At this level the interpretation, based on the effective-mass approximation model of Thomas and Hopfield [69] provides a consistent explanation of the behavior of D^0X and A^0X in a magnetic field; however, it does not explain the origin of the weak, lower-energy satellite of A^0X that is clearly visible in the spectrum (Figures 5.13 and 5.18). At present, we can only speculate on the origin of this line. It could be a manifestation of the internal structure of the acceptor state, and the line could correspond to the transition at which the final state (neutral acceptor) is left in the excited state. This excited state energy would increase with increasing magnetic field, leading to a decrease in the transition energy. We must point out that the assignment proposed in Refs. 77 and 78 for the A^0X line to the ionized donor-bound exciton in incomplete disagreement with our experimental data. For such a complex a distinct splitting pattern in a magnetic field should be observed [69].

5.4 COUPLING OF LO PHONONS TO EXCITONS IN GaN

The LO phonon replica of the exciton region is shown in Figure 5.20. Despite the fact that the emission due to D^0X is about one order of magnitude

Table 5.5 Effective g-factors and diamagnetic shift parameters for D^0X and A^0X luminescence lines.

D^0X		A^0X	
Effective g-factors	*Diamagnetic shift*	*Effective g-factors*	*Diamagnetic shift*
$g_{\parallel}^{D^0X} = 0.64$	$D_{\parallel}^{D^0X} = 1.4 \times 10^{-6}$ eV/T^2	$g_{\parallel}^{A^0X} = 0.22$	$D_{\parallel}^{A^0X} = 1.3 \times 10^{-6}$ eV/T^2
High-energy line weak; therefore $g_u > g_l$		High-energy line strong; therefore $g_u < g_l$	
$g_{\perp}^{D^0X} = 1.87$	$D_{\perp}^{D^0X} = 7.3 \times 10^{-7}$ eV/T^2	$g_{\perp}^{A^0X} = 2.1$	$D_{\perp}^{A^0X} = 1.6 \times 10^{-6}$ eV/T^2
Both lines equal in strength; therefore $g_u = 0$		High-energy line weak; therefore $g_u > g_l$	
$g_e^{D^0X} = 1.87$ $g_{h\parallel}^{D^0X} = 2.5$		$g_e^{A^0X} = 2.1$ $g_{h\parallel}^{A^0X} = 2.3$	

Note: g_u and g_l refer to the values of the upper (initial) and lower (final) states, respectively. g-Factors of electrons and holes localized in these complexes were obtained as described in the text.

stronger than the A^0X luminescence, the predominant structure observed in the LO phonon sideband of the excitonic spectrum is dominated by optical transitions involving A^0X [85]. The intensity ratio of the first LO phonon to no-phonon emission of A^0X and D^0X is about 10^{-2} and 10^{-4}, respectively. The characteristic shape of the A^0X emission is well reproduced in the phonon replica (Figure 5.20). Also, the temperature dependence of the replica reproduces the intensity changes of no-phonon emission (Figure 5.21). At 4.2 K the A^0X-related structure dominates the spectrum. Its intensity decreases faster with rising temperature than that of D^0X (as in the case of the no-phonon line in Figure 5.15). Finally, above 35 K the A^0X-related emission disappears from the LO replica spectrum.

The energy shift of the first LO phonon replica is slightly different for different excitons (see Figure 5.20), perhaps due to coupling to different local modes with slightly less energy than that of the lattice LO phonon [86].

The result that the LO phonons are so strongly coupled to A^0X can be understood as the manifestation of the neutral acceptor–LO phonon bound-state formation [86]. Such a state can be formed in GaN because the ionization energy of a shallow acceptor being about 255 meV (see Section 5.5), is much larger than the LO phonon energy (close to 92 meV). In the case of D^0X the bound state with an LO phonon is much less probable because the energy of the LO phonon is larger than the donor binding energy of

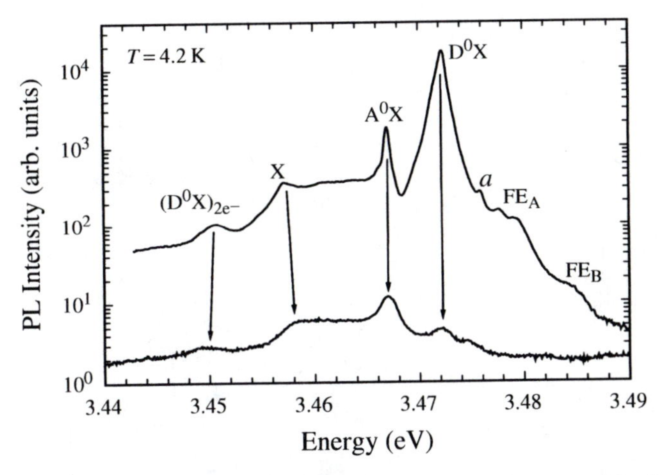

Figure 5.20 Comparison of no-phonon excitonic luminescence (upper curve) with its LO phonon replica (lower curve). The LO replica is shifted by 91.8 meV for a comparison with the no-phonon spectrum.

32 meV [84]. Thus, a neutral donor–LO phonon bound state is degenerate with the free electron continuum and decays rapidly by donor ionization [87]. As a result the LO phonon replica of the D^0X should show much shorter decay time, which implies a broadening of the D^0X replica, making it harder to observe. In the case of CdS and CdSe D^0X replica manifest a broad structure with a deep in the middle, which has been interpreted as due to resonant behavior of local and lattice phonon modes with the donor continuum [87]. The D^0X replica in homoepitaxial GaN also shows a double structure (Figure 5.20), which could be interpreted in the same way; however time-resolved luminescence studies have not confirmed such an interpretation. They showed that the decay times of the first and second LO phonon replicas of D^0X and A^0X structures are similar to those of the zero-phonon lines [88]. Moreover, the second LO exciton phonon replicas of D^0X and A^0X excitons show proportions characteristic of zero-phonon lines and do not show substantial broadening induced by decay time shortening (see Figure 5.22) [88]. The double structure could be the result of coupling of two donor species to local modes with slightly different energies.

A weak coupling between an LO phonon and D^0X is more probably caused by a delocalization of its wave function in real space [89]. The larger D^0X radius implies that the volume of the D^0X in the k-vector space is much smaller than that of the A^0X complex. It also implies that the total number of LO phonons that could interact with the D^0X exciton is smaller than the respective number expected for the A^0X exciton. The effect can be particularly strong when a donor-bound exciton interacts with a single phonon, leading to the small relative intensity of 1-LO D^0X replica. For the

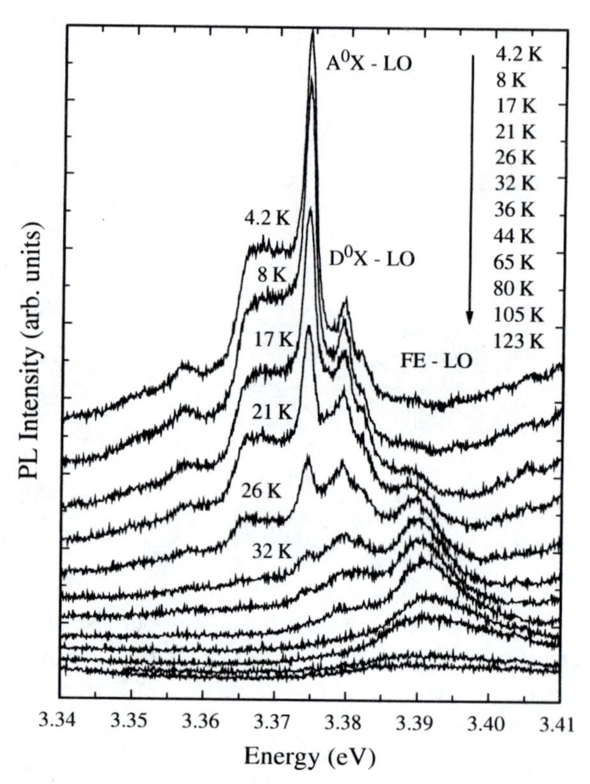

Figure 5.21 One-LO phonon replica of the exciton emission measured at various temperatures.

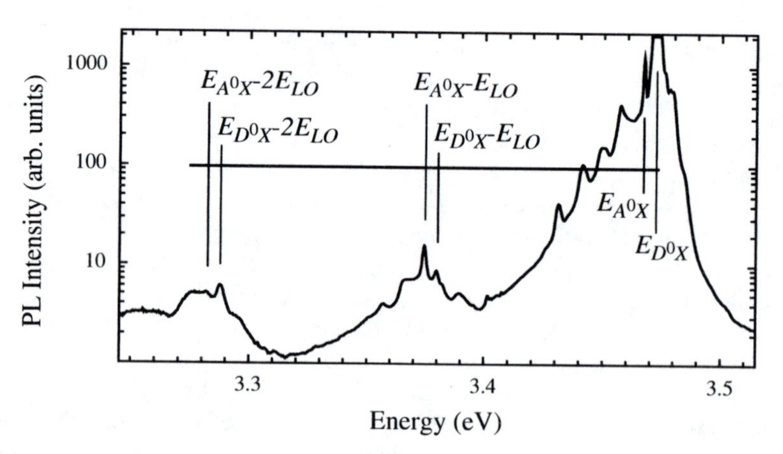

Figure 5.22 First and second phonon replicas of excitons in homoepitaxial GaN. Vertical lines separated by the energy of the LO phonon (ELO = 92 meV) indicate expected positions of the replicas.

second phonon replica many different combinations of the two-phonon wave vectors are possible for a given k-vector of the exciton, and the exciton–phonon interaction may be efficient even in the case of D^0X.

With rising temperature, LO phonon replica of free excitons start to dominate the spectrum (Figure 5.21). The spectral shape of this emission can be described by the following formula, which assumes a Maxwell distribution of exciton kinetic energy [90]:

$$J_N(\varepsilon) = \varepsilon^{1/2} \exp\left(\frac{-\varepsilon}{kT}\right) W_N(\varepsilon), \tag{5.18}$$

where ε is the exciton kinetic energy, $W_{N(\varepsilon)}$ is the probability of radiative exciton recombination with emission of N phonons, and T is the temperature. Assuming the power dependence of $W_{N(\varepsilon)} \sim \varepsilon^L$, Eq. (5.18) can be rewritten in a simple form:

$$J_{N(\varepsilon)} \sim \varepsilon^{L+1/2} \exp(-\varepsilon/kt) \tag{5.19}$$

In this case the energy position of the intensity maximum with respect to the low energy of emission limit ($\varepsilon = 0$) is proportional to the temperature:

$$\Delta = \left(L + \tfrac{1}{2}\right) kT \tag{5.20}$$

The temperature dependency is well reproduced by experimental data (Figure 5.23), giving $L = 0.2 \pm 0.1$, however, in GaN the energetic positions

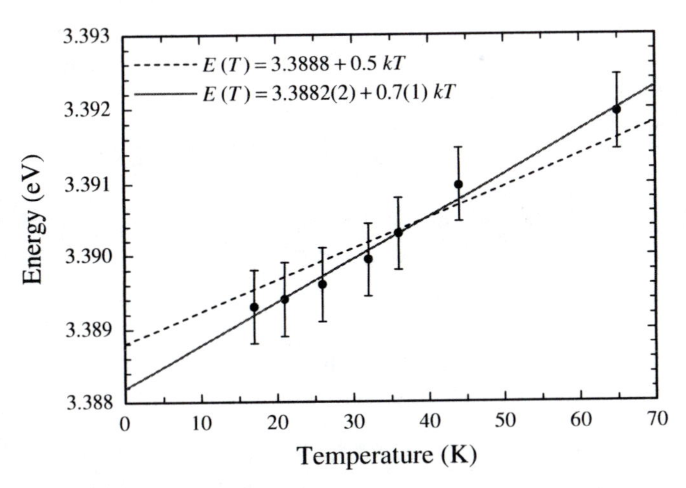

Figure 5.23 Temperature variation of the intensity maximum energy of LO phonon replica of free excitons in undoped homoepitaxial GaN.

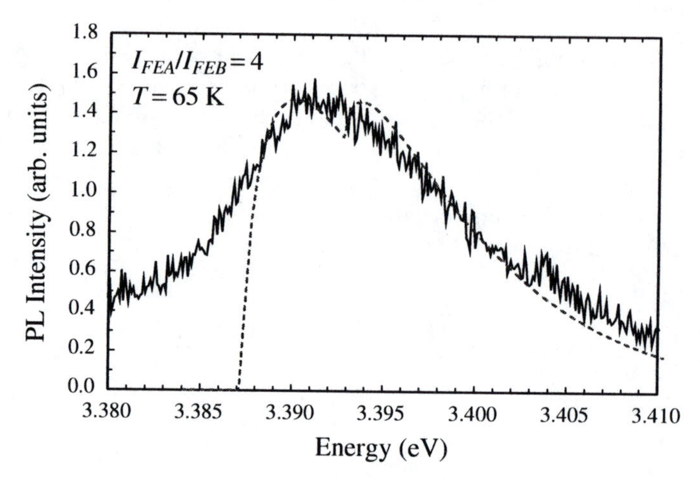

Figure 5.24 Comparison the LO phonon replica of free excitons measured at 65 K in undoped homoepitaxial GaN layer (solid line) with theory, Eq. (5.18) (dashed line).

of free excitons A and B, so one can expect that at elevated temperature one can see participation of both free excitons in the LO phonon replica. This can artificially increase the value of L obtained experimentally.

The theoretical shape according Eq. (5.18), assuming $W_{N(\varepsilon)} = \text{constant}$, is compared with experimental data in Figure 5.24. The theoretical curve includes two components related to excitons A and B with the intensity ratio observed in the no-phonon emission. The agreement between the theoretical curve and experimental results is quite good.

The result that recombination probability for the first LO replica does not depend on the exciton kinetic energy is quite different from results observed for high-quality GaN grown on sapphire [91,92] as well as for II–VI compounds [90], perhaps as a result of exciton scattering by point defects in the material [93]. Without such scattering processes the recombination of free excitons with the emission of one phonon has a forbidden character, resulting from the k conservation law. Such a result is surprising, since it should be expected that in homoepitaxial GaN that shows a few orders lower dislocation density [8], the point defect concentration should also be much lower.

5.5 DONOR–ACCEPTOR PAIR RECOMBINATION

Emission spectra of slightly p-type–doped homoepitaxial layers show more than 30 sharp lines in the energy range below excitonic transitions (see Figure 5.25). The low-energy lines are superimposed on the slope of the broad peak at 3.266 eV (not shown in Figure 5.25). The middle part of the spectrum is related to the 92.3-meV LO phonon replica of the A^0X exciton luminescence [85]. These lines (excluding the LO phonon replica of A^0X) have

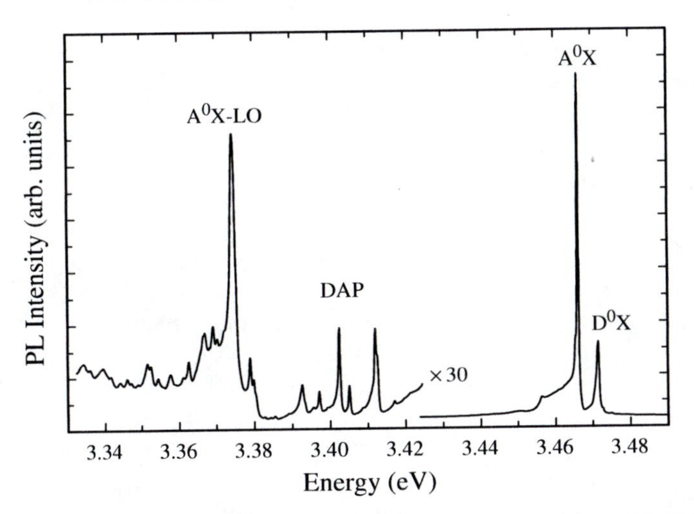

Figure 5.25 Low-temperature luminescence spectrum of Mg-doped homoepitaxial GaN layer.

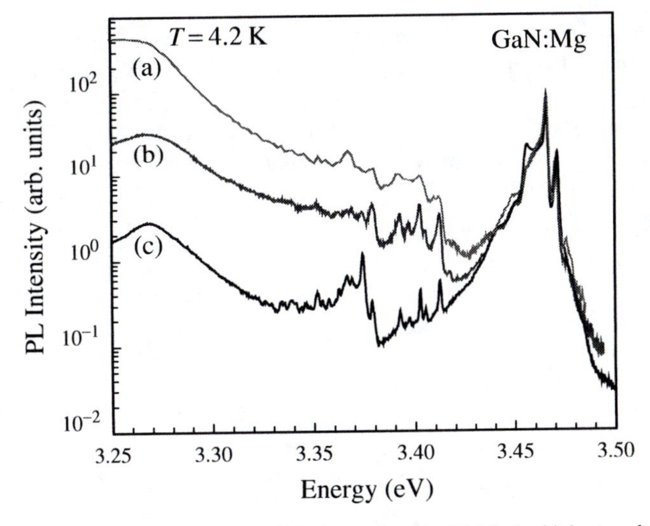

Figure 5.26 Luminescence spectra of GaN layers doped with Mg (a, highest; c, lowest Mg concentration). Spectra were normalized at A^0X line.

been interpreted as a recombination of differently separated donor–acceptor pairs [48,94]. Identical pair spectra have been observed for several different Mg-doped crystals (see Figure 5.26). The positions of pair lines and their relative intensities remain unchanged with increasing Mg concentration. The transition energy for a given line depends on the donor–acceptor separation R_n. Taking into account the Coulomb interaction between ionized centers existing in the final state of the recombination process, the energy emitted as a photon can be calculated using the simplified formula

$$E_n = E_g - E_D - E_A + \frac{e^2}{\varepsilon R_n} - \frac{\alpha_w}{R_n^6} \tag{5.21}$$

where E_g, E_D, and E_A are the energy gap, the donor, and the acceptor ionization energies respectively; ε is the dielectric constant R_n is the donor–acceptor separation; and α_w is the van der Waals interaction coefficient, describing the interaction of neutral donor and acceptor centers.

The analysis performed for the wurtzite GaN lattice shows that donor and acceptors involved in the pair spectra are located in different sublattices (so-called type II spectra) [48,94]. The characteristic feature observed as an "energy gap" in the spectrum close to 3.385 eV is well reproduced by the calculations. The result obtained is in agreement with the assumption that the main native shallow donor in GaN is located at the nitrogen site, perhaps a nitrogen vacancy or oxygen. The plot of the donor–acceptor transition energy E_n versus $1/R_n$ gives approximately a straight line (Figure 5.27). From the slope of this line, the static dielectric constant $\varepsilon = 9.6(1)$ and the limiting value of $E_\infty = E_n(R_n \to \infty) = 3.210\,\text{eV}$ were determined. Taking into account that $E_\infty = E_g - E_D - E_A$, where the bandgap energy $E_g = 3.497(3)\,\text{eV}$ (Table 5.3), the sum of the isolated donor and acceptor binding energies of 287 meV was calculated. Assuming a shallow donor ionization energy in GaN of 32 meV [84], an acceptor ionization energy of about 255 meV can be deduced.

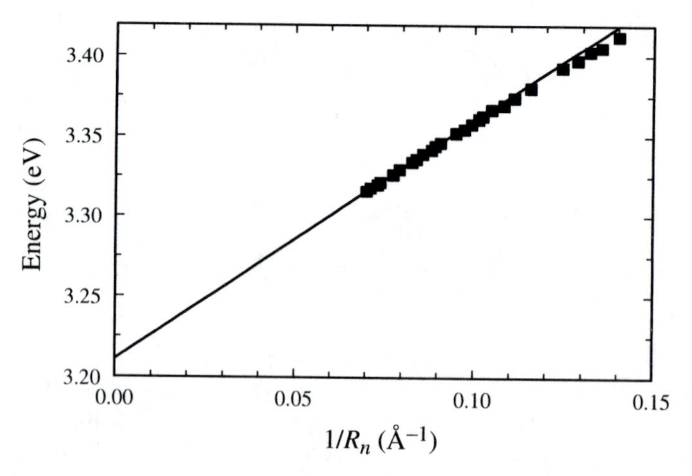

Figure 5.27 Variation of the energy E_n of donor–acceptor transitions versus the calculated reciprocal donor–acceptor separation $1/R_n$.

Intensity calculations of discrete D–A transitions have been performed assuming that luminescence intensity is proportional to the number of pairs at separation R_n multiplied by the square of the optical matrix element for a product of donor and acceptor wave functions [95]. The transition energy E_n is described by Eq. (5.21). The resulting theoretical spectrum shown in Figure 5.28 was obtained for an effective donor radius of 23 Å and minimum separation between donor and acceptor of 7.13 Å. The van der Waals term with $\alpha_w = 1000\,\text{eVÅ}^6$ is important only for the closest pairs (Figure 5.28). The minimum separation of 7.13 Å, which is several times greater than the distance between nearest lattice sites, seems to be reasonable; a similar result was obtained for type-II spectra such as O–Cd or O–Zn in GaP [96]. The minimum separation is smaller when one of the particles is more tightly bound. When the donor–acceptor core separation is smaller than the donor radius, the electron sees the attractive D^+ and neutral A^0 states; however, when the core separation approaches a critical distance close to the acceptor radius, the hole will not be able to be bound. In this case the extended electron will see D^+ and A^- core potentials and will not be bound either. According to Hopfield [97] this critical separation can be estimated as $1.75R_A$, where R_A is the radius of the acceptor. Taking into account $R_0 = 7.13$ Å, one get the acceptor Bohr radius of 4.1 Å. Much more advanced calculations [98], taking into account the interaction between neutral donor and acceptor, show that critical distance R_0 depends on the electron-to-hole mass ratio $\sigma = m_e/m_h$

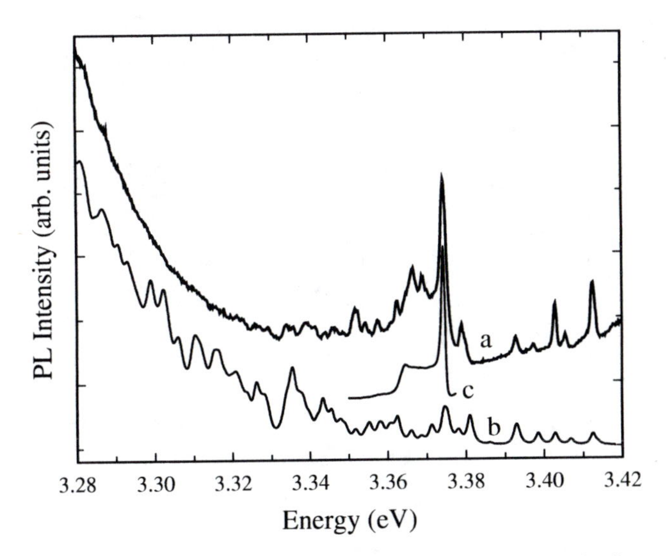

Figure 5.28 Comparison of experimental (a) and simulated (b) spectrum for discrete donor–acceptor pair luminescence. The middle part of the experimental spectrum is affected by the LO phonon replica of A^0X luminescence (c).

and donor Bohr radius a_D:

$$0.64\sigma a_D < R_0(\sigma) < \frac{2\sigma(1+\sigma)}{1+\sigma+\sigma^2} a_D$$

Assuming that for GaN $\sigma \cong 0.29$ and $a_D = 23$ Å, one gets the estimation $4.5\,\text{Å} < R_0 < 12.7\,\text{Å}$, which is in good agreement with the experimental result of 7.13 Å.

The temperature dependence of donor–acceptor pair recombination has been discussed in Ref. 94. Despite the fact that the LO phonon replica of the A^0X line as well as the zero-phonon line disappear above 40 K [4], the donor–acceptor pair spectrum does not change significantly (see Figure 5.29), confirming that under the laser illumination at 40 K neutral donors and acceptors are present in the sample. The energy variation of the sharp D–A lines caused by temperature is the same as the temperature dependence of the GaN energy gap; however, the temperature dependence of the 3.268-eV peak reveals that its energy increases with temperature (about 0.18 meV/K) with respect to the energy gap [94]. Such temperature behavior is characteristic of direct conduction band–acceptor (e–A) recombination. It means that emission from distant D–A pairs is hidden under strong e–A luminescence. Assuming that the 3.268-eV energy corresponds to the difference between the energy gap and the acceptor binding energy $E_g - E_A$, one

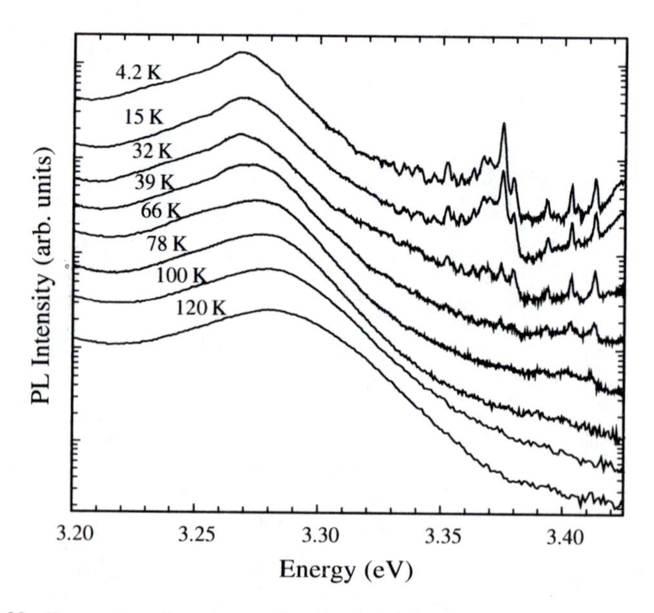

Figure 5.29 Temperature dependence of e–A and D–A luminescence in GaN.

can calculate the acceptor energy $E_A = 255 \pm 15$ meV, which is in good agreement with the value provided by the analysis of discrete D–A pairs.

5.5.1 Donor–Acceptor Pair Magneto-Optics

As was shown in Figure 5.25, the emission spectrum in the energy range below the excitonic transitions, close to 3.4 eV, consists of more than 30 sharp lines of halfwidth close to 0.5 meV. These lines (excluding LO phonon replica of A^0X) have been interpreted as recombination of differently separated donor–acceptor pairs [48,94].

In contrast to the behavior observed for bound excitons (see Section 5.3.1), in the case of donor–acceptor pair spectra, surprisingly, the line splitting pattern is very similar for both $\mathbf{B} \| \mathbf{c}$ and $\mathbf{B} \perp \mathbf{c}$ magnetic field configurations (Figure 5.30); however, the oscillator strengths of the split components strongly depend on the magnetic field orientation. In addition to the structures observed at zero field, new lines, forbidden without a magnetic field, appear on the lower-energy side of the allowed components. The intensity ratios of split components at a given magnetic field are different for different pair lines. Some lines show a doublet structure that can be related to donor–acceptor pairs of slightly different separation or to different

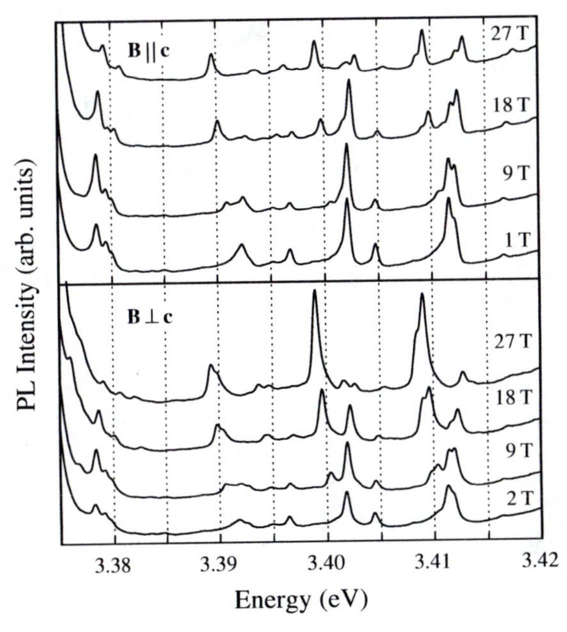

Figure 5.30 Photoluminescence spectra of donor–acceptor pair emission for different magnetic fields.

alignments with respect to the **c**-axis. The difference in energy between doublet constituents, well resolved for the line at 3.4118 eV, is about 0.5 meV, irrespectively of magnetic field strength.

The observed splitting between allowed and forbidden components can be described using the phenomenological formula

$$\Delta_E = \sqrt{\Delta_0^2 + (g_{eff}\mu_B B)^2} \tag{5.22}$$

where Δ_0 is the zero-field splitting and g_{eff} denotes the effective g-factor. Within experimental error, the same value $g_{eff} = 2.2(1)$ was obtained for different pairs. The term Δ_0 is close to 1.1 meV and seems to be different for different pairs, slightly decreasing when the donor–acceptor pair distance increases.

The behavior of donor–acceptor pair spectra in a magnetic field displays properties that are difficult to explain with the simple effective-mass approximation. Our results are different from those observed for donor–acceptor pair spectra in CdS, where both the splitting pattern and the intensity changes of split components are very sensitive to the orientation of the magnetic field [99,100].

In CdS the splittings of pair lines have been described using two parameters: the electron–hole exchange coupling constant A, and a local crystal-field parameter $|D|$, caused by the presence of the donor. This additional crystal-field interaction induces a mixing between the Γ_9 and Γ_7 hole states. The value of the A parameter was assumed either as constant for different pairs [99] or dependent on the donor–acceptor separation [100]. A decrease in the A value with increasing donor–acceptor pair separation is expected as a result of the reduction of the overlap of the electron and hole wave functions. The value of the crystal-field splitting parameter $|D|$ varied from one pair to the next. In the case of a magnetic field parallel to the **c**-axis, the splitting was described using the difference of the electron g_{e^-} and Γ_9 hole effective $g_{h\|}$-factor for the allowed doublet, and the sum of these g-factors for the forbidden doublet. Thus, the splitting of the forbidden doublet was about twice as larger as the splitting of the allowed one [99,100]. For a magnetic field perpendicular to the **c**-axis the observed splitting was fitted using only the electron effective g-factor. Thus, in CdS the splitting observed in magnetic fields parallel and perpendicular to the **c**-axis are different. Generally speaking, the main source of such behavior is that the Γ_9 hole, as discussed in Section 5.3.2, does not split in a magnetic field perpendicular to the **c**-axis but only in one parallel to the **c**-axis.

In the case of donor–acceptor pairs in GaN the observed splittings do not depend on the orientation of the magnetic field with respect to the **c**-axis. Moreover, the zero-field splitting does not change much from one pair to the next. It seems that the main part of the zero-field splitting, which comes either from the exchange interaction or the crystal-field splitting, has almost the same value for all observed pairs. If we assume that the crystal-field parameter $|D|$ is equal to zero, the magnetic field effect may be explained by the charged-donor- or charged-acceptor-bound exciton energy relations of Thomas and Hopfield [69]. The splitting of these excitons in the magnetic field configuration perpendicular to the **c**-axis can be described by Eq. (5.22). The zero-field splitting parameter Δ_0 is equivalent to the exchange interaction parameter A, and the g_{eff} parameter corresponds to the effective g-factor of the electron. Surprisingly enough, this formula fits well the splittings of donor–acceptor pairs in GaN for both magnetic field configurations. The value of Δ_0 starting from 1.1 meV obtained for the highest-energy pair decreases slightly when the donor–acceptor separation is increased. This behavior is in agreement with the results obtained for CdS, where the exchange interaction becomes smaller at larger donor–acceptor separations. The value $g_{eff} = 2.2(1)$ is slightly larger than the electron g-factors obtained for D^0X ($g_{\perp}^{D^0X} = 1.87$) and A^0X ($g_{\perp}^{A^0X} = 2.1$), but it resembles the g-factor of the electron rather than that of holes involved in bound excitons ($g_{h\|}^{A^0X} = 2.3$, $g_{h\|}^{D^0X} = 2.5$).

The preceding results, especially the splitting isotropy, suggest that the properties of the hole involved in donor–acceptor pair recombination are quite different from the Γ_9 valence hole, which has to show highly anisotropic behavior. The localized nature of the bound hole can make it very sensitive to the symmetry of the local defect potential, which may change the magnetic properties of this hole [101]. One suggestion is that, in contrast with the CdS case, some local fields play the dominant role in the Mg acceptor center in GaN, and the magnetic field "activates" optical transitions for pairs oriented perpendicular to the field direction.

On the other hand, in recent ESR studies of Mg-doped samples only a small g-factor anisotropy was reported [83]. Such anisotropic behavior is expected rather for the Γ_7 valence band hole and may suggest that the wave function of the hole in the donor–acceptor complex does not reflect the symmetry of the highest valence band (Γ_9). One possible reason is that the highly localized nature and the complicated internal structure of the acceptor state in GaN make the effective-mass approximation insufficient in this case. The existing ab initio calculations [102] also confirm the strong p_{xy}-like-character anisotropy of the shallow Mg acceptor state.

5.6 CONCLUSIONS

Extensive studies of the optical properties of GaN have been performed. Because homoepitaxial GaN layers that are strain free were used, we believe that the observed results are directly related to the internal properties of this material. Basic band structure parameters of GaN were estimated from free-exciton reflectivity studies performed under different conditions.

As discussed in Sections 5.3–5.5, different processes related to neutral impurities dominate low-temperature luminescence in GaN, as is typical for wide-gap semiconductors. Although the nature of neutral donors that are involved in bound-exciton recombination is well understood, the case of the acceptor center in GaN seems to be less clear. In particular, the splitting pattern obtained for neutral-donor-bound excitons is in good agreement with the model (effective-mass theory) developed by Thomas and Hopfield for bound excitons in wurtzite CdS; however, the behavior of A^0X in a magnetic field is not so clear as in the case of D^0X. We observed some discrepancies between the description obtained from the effective-mass approximation and experimental data obtained for transitions in which acceptor centers are involved, namely, (1) the additional weak line on the lower-energy side of the A^0X complex recombination and (2) the almost spherical behavior of the donor–acceptor pair spectra in magnetic field. Both observations can be tentatively attributed to the internal structure of the acceptor state in GaN. A strong localization, local fields induced by donors, and/or a possible pairing of hydrogen with acceptor centers can play an important role in the explanation of the observed properties.

Despite the problems encountered with the microscopic structure of the acceptor center, the effective-mass theory can be successively applied to an exciton bound to a shallow neutral impurity. Effective g-factors have been obtained for electrons and holes weakly bound in D^0X and A^0X complexes, as well as diamagnetic shifts for D^0X and A^0X complex recombination lines, in different field configurations (see Table 5.5).

The parameters collected in this chapter have substantially improved the state of knowledge about this important wide-gap semiconductor.

Acknowledgments

The authors would like to thank to all the coauthors of our common papers, which have been essential in the writing of this chapter. In particular, we would like to thank Dr I. Grzegory from the High Pressure Research Center, Polish Academy of Sciences, for growing GaN single-crystal substrates, K. Pakuła from the Institute of Experimental Physics, Warsaw University, for growing homoepitaxial layers, and to Dr M. Potemski from the Grenoble High Magnetic Field Laboratory for significant contribution to the results obtained in high magnetic fields.

A. W. and K. P. K. gratefully acknowledge support (Forschungs-stipendium) by the Alexander von Humboldt Foundation during their work at the Max Planck Institut, Stuttgart.

References

1. K. Pakuła, J. M. Baranowski, R. Stępniewski, A. Wysmołek, I. Grzegory, J. Jun, M. Sawicki, S. Porowski, and K. Starowieyski, *Acta Phys. Pol.*, **88**, 861 (1995).
2. S. Porowski, J. Jun, M. Boćkowski, M. Leszczyński, S. Krukowski, M. Wróblewski, B. Łucznik, and I. Grzegory, *Proc. of the 8th Conf. on Semiinsulating III–V Materials*, edited by M. Godlewski (World Scientific, New York, 1994) p. 61.
3. M. Leszczyński, H. Teisseyre, T. Suski, I. Grzegory, M. Boćkowski, J. Jun, S. Porowski, K. Pakuła, J. M. Baranowski, C. T. Foxon, and T. S. Cheng, *Appl. Phys. Lett.*, **69**, 73 (1996).
4. K. Pakuła, A. Wysmołek, K. P. Korona, J. M. Baranowski, R. Stępniewski, I. Grzegory, M. Boćkowski, J. Jun, S. Krukowski, M. Wróblewski, and S. Porowski, *Solid State Commun.*, **97**, 919 (1996).
5. S. Porowski, J. M. Baranowski, M. Leszczyński, J. Jun, M. Boćkowski, I. Grzegory, S. Krukowski, M. Wróblewski, B. Łucznik, G. Nowak, K. Pakuła, A. Wysmołek, K. P. Korona, and R. Stępniewski, *Proc. of the Int. Symp. on Blue Laser and Light Emiting Diodes*, Chiba, 1996, edited by A. Yoshikawa, K. Kishino, M. Kobayashi, and T. Yasuda (Ohmsha, Tokyo, 1996) p. 38.
6. R. Stępniewski and A. Wysmołek, *Proc. of the XXV Int. School on Physics of Semiconducting Compounds*, Jaszowiec, 1996, *Acta Phys. Pol.*, **90**, 681 (1996).
7. J. M. Baranowski and S. Porowski, *Proc. of the 23rd Int. Conf. on Physics of Semiconductors*, Berlin, 1996, edited by M. Sheffler and R. Zimmermann (World Scientific, Singapore, 1996) p. 497.
8. J. M. Baranowski, Z. Liliental-Weber, K. Korona, K. Pakuła, R. Stępniewski, A. Wysmołek, I. Grzegory, G. Nowak, S. Porowski, B. Monemar, and J. P. Bergman, *MRS Proc.*, **449**, 393–405 (1997).
9. R. Stępniewski, A. Wysmołek, M. Potemski, J. Łusakowski, K. Korona, K. Pakuła, J. M. Baranowski, G. Martinez, P. Wyder, I. Grzegory, and S. Porowski, *Proc. of the 8th SLCS*, Montpellier, 1998, *Phys. Status Solidi B*, **210**, 373–383 (1998).
10. T. Steiner, M. L. W. Thewalt, E. S. Coteles, and J. P. Salerno, *Phys. Rev. B*, **34**, 1006 (1986).
11. D. Sell, R. Dingle, S. E. Stokowski, and J. V. Di Lorenzo, *Phys. Rev. B*, **7**, 4568 (1973).
12. J. J. Hopfield and D. G. Thomas, *Phys. Rev.*, **132**, 563 (1963).
13. R. Dingle, D. D. Sell, S. E. Stokowski, and M. Ilegems, *Phys. Rev. B*, **4**, 1211 (1971).
14. B. Monemar, *Phys. Rev. B*, **10**, 676 (1974).
15. B. Gil, O. Briot, and R. L. Aulombard, *Phys. Rev. B*, **52**, R17028 (1995).
16. D. Volm, K. Oettinger, T. Streibl, D. Kovalev, M. Ben-Chorin, J. Diener, B. K. Meyer, J. Majewski, L. Eckey, A. Hoffman, H. Amano, and I. Akasaki, *Phys. Rev. B*, **53**, 16543 (1996).
17. R. Stępniewski, K. P. Korona, A. Wysmołek, J. M. Baranowski, K. Pakuła, M. Potemski, G. Martinez, I. Grzegory, and S. Porowski, *Phys. Rev. B*, **56**, 15151 (1997).
18. K. Torii, T. Deguchi, T. Sota, K. Suzuki, S. Chichibu, and S. Nakamura, *Phys. Rev. B*, **60**, 4723 (1999).
19. R. Stępniewski, M. Potemski, A. Wysmołek, K. Pakuła, J. M. Baranowski, G. Martinez, I. Grzegory, M. Wróblewski, and S. Porowski, *Proc. of the 23rd Int. Conf. on Physics of Semiconductors*, Berlin, 1996, edited by M. Sheffler and R. Zimmermann (World Scientific, Singapore, 1996) p. 549.
20. I. Broser, M. Rosenzweig, R. Broser, M. Richard, and E. Birkicht, *Phys. Status Solidi B*, **90**, 77 (1978).
21. M. Rosenzweig, *Phys. Status Solidi B*, **129**, 187 (1985).
22. S. Nakajima, Y. Toyozawa, and R. Abe, in *The Physics of Elementary Excitations* (Springer-Verlag, New York, 1980).
23. R. J. Elliott, *Phys. Rev.*, **108**, 1384 (1957).
24. S. I. Pekar, *J. Phys. Chem. Solids*, **5**, 11 (1958).
25. J. Lagois, *Phys. Rev. B*, **16**, 1699 (1977).
26. J. C. Powell, in *Numerical Data and Functional Relationships in Science and Technology*, edited by K.-H. Hellwege, and J.-L. Olsen, Landolt-Börnstein, New Series, Group III (Springer-Verlag, Berlin, 1985) Vol. 15b, 228.

27. A. Shikanai, T. Azuhata, T. Sota, S. Chichibu, A. Kuramata, K. Horino, and S. Nakamura, *J. Appl. Phys.*, **81**, 417 (1997).
28. S. Pau, J. Kuhl, F. Scholz, V. Haerle, M. A. Khan, and C. J. Sun, *Phys. Rev. B*, **56**, R12718 (1997).
29. G. L. Bir and G. E. Pikus, in *Symmetry and Strain-Induced Effects in Semiconductors* (Halsted, London, 1974).
30. G. Ramirez-Flores, H. Navarro-Contreras, A. Lastras-Martinez, R. C. Powell, and J. E. Greene, *Phys. Rev. B*, **50**, 8430 (1994).
31. M. Suzuki, T. Uenoyama, and A. Yanase, *Phys. Rev. B*, **52**, 8132 (1995).
32. J. A. Majewski, M. Städele, and P. Vogl, *MRS Internet J. Nitride Semicond. Res.*, **1**, art. 30 (1996).
33. F. Stern, in *Solid State Physics*, edited by F. Seitz and D. Turnbull (Academic, New York, 1963) Vol. 15, 341.
34. S. Logethetidis, J. Petlas, M. Cardona, and T. D. Moustakas, *Phys. Rev. B*, **50**, 18017 (1994).
35. K. Jezierski, *J. Phys C: Solid State Phys.*, **17**, 475 (1984).
36. M. Sobol and W. Bardyszewski, *Acta Phys. Pol. A*, **94**, 534 (1998).
37. K. Bohnert, G. Schmieder, S. El-Dessouki, and C. Klingshirn, *Solid State Commun.*, **27**, 295 (1978).
38. B. B. Kosicki, R. J. Powell, and J. C. Burgiel, *Phys. Rev. Lett.*, **24**, 1421 (1970).
39. S. Logothetidis, J. Petelas, M. Cardona, and T. D. Moustacas, *Mater. Sci. Eng., B*, **29**, 65 (1995).
40. J. I. Pankove, H. P. Maruska, and J. E. Berkeyheiser, *Appl. Phys. Lett.*, **17**, 197 (1970).
41. K. P. Korona, A. Wysmołek, K. Pakuła, R. Stępniewski, J. M. Baranowski, I. Grzegory, B. Łucznik, M. Wróblewski, and S. Porowski, *Appl. Phys. Lett.*, **69**, 788, (1996).
42. H. Y. Fan, *Phys. Rev.*, **82**, 900 (1951).
43. Y. P. Varshni, *Physica*, **34**, 149 (1967).
44. A. U. Sheleg and V. A. Savastenko, *Vesti Akad. Nauk BSSR, Ser. Fiz. Mat. Nauk.*, **3**, 126 (1976); V. A. Savastenko, A. U. Sheleg, *Phys. Status Solidi A*, **48**, K135 (1978).
45. P. Perlin, C. Jauberthie-Carillon, J. P. Itie, A. San Miquel, I. Grzegory, and A. Polian, *Phys. Rev. B*, **45**, 83 (1992).
46. R. Hellmann, M. Koch, J. Feldmann, S. T. Cundiff, E. O. Gobel, D. R. Yakovlev, A. Waag, and G. Landwehr, *Phys. Rev. B*, **48**, 2847 (1993).
47. M. O'Neill, M. Oestreich, W. W. Ruhle, and D. E. Ashenford, *Phys. Rev. B*, **48**, 8980 (1993).
48. A. Wysmołek, J. M. Baranowski, K. Pakuła, K. P. Korona, I. Grzegory, M. Wróblewski, and S. Porowski, *Proc. of Int. Symp. on Blue Laser and Light Emitting Diodes*, Chiba, 1996, edited by A. Yoshikawa, K. Kishino, M. Kobayashi, and T. Yasuda (Ohmsha, Tokyo, 1996) p. 492.
49. A. S. Baker and M. Illegens, *Phys. Rev. B*, **7**, 743 (1973).
50. P. Perlin, I. Gorczyca, N. E. Christensen, I. Grzegory, H. Teisseyre, and T. Suski, *Phys. Rev. B*, **45**, 13307 (1992).
51. Z. X. Liu, S. Pau, K. Syassen, J. Kuhl, W. Kim, H. Morkoç, M. A. Khan and C. J. Sun, *Phys. Rev. B*, **58**, 6696 (1998).
52. S. Kim, I. P. Herman, J. A. Tuchman, K. Doverspike, L. B. Rowland, and D. K. Gaskill, *Appl. Phys. Lett.*, **67**, 380 (1995).
53. W. Shan, T. J. Schmidt, R. J. Hauenstein, J. J. Song, and B. Goldenberg, *Appl. Phys. Lett.*, **66**, 3492 (1995).
54. D. L. Camphausen and G. A. N. Connell, *J. Appl. Phys.*, **42**, 4438 (1971).
55. N. E. Christensen and I. Gorczyca, *Phys. Rev. B*, **50**, 4397 (1994).
56. H. Teissyre, B. Kozankiewicz, M. Leszczyński, I. Grzegory, T. Suski, M. Boćkowski, S. Porowski, K. Pakuła, P. M. Mensz, and I. B. Bhat, *Phys. Status Solidi B*, **198**, 235 (1996).
57. Z. X. Liu, K. P. Korona, K. Syassen, J. Kuhl, K. Pakuła, J. M. Baranowski, I. Grzegory, and S. Porowski, *Solid State Commun.*, **108**, 433–438 (1998).
58. G. J. Piermarini, S. Block, J. D. Bernett, and R. A. Forman, *J. Appl. Phys.*, **46**, 2774 (1975).

59. W. Shan, B. D. Little, A. J. Fischer, J. J. Song, B. Goldenberg, W. G. Perry, M. D. Bremster, and R. F. Davis, *Phys. Rev. B*, **54**, 16369 (1996).
60. M. Leszczyński, T. Suski, P. Perlin, H. Teissyre, I. Grzegory, M. Boćkowski, J. Jun, S. Porowski, and J. Major, *J. Phys. D*, **28**, A146 (1995).
61. J. F. Nye, in *Physical Properties of Crystals* (Oxford University Press, Oxford, 1976).
62. D. W. Langer, R. N. Euwema, K. Era, and T. Koda, *Phys. Rev. B*, **2**, 4005 (1970).
63. H. Schulz and K. H. Thiemann, *Solid State Commun.*, **23**, 815 (1977).
64. R. Stępniewski, M. Potemski, A. Wysmołek, K. Pakuła, J. M. Baranowski, J. Łusakowski, G. Martinez, P. Wyder, I. Grzegory, and S. Porowski, *Phys. Rev. B*, **60**, 4438 (1999).
65. J. M. Luttinger, *Phys. Rev.*, **102**, 1030 (1956).
66. I. Broser and M. Rosenzweig, *Phys. Rev. B*, **22**, 2000 (1980).
67. G. Blattner, G. Kurtze, G. Schmieder, and C. Klingshirn, *Phys. Rev. B*, **25**, 7413 (1982).
68. We do not discuss in this paper the weak structure at 3.471 eV, attributed to a donor bound exciton (see Section 5.3).
69. D. G. Thomas and J. J. Hopfield, *Phys. Rev.*, **128**, 2135 (1962).
70. W. E. Carlos, J. A. Freitas Jr., M. Asif Khan, D. T Olson, and J. N. Kuznia, *Phys. Rev. B*, **48**, 17878 (1993).
71. J. Wrzesinski and D. Fröhlich, *Solid State Commun.*, **105**, 301 (1998).
72. A. B. Malyshev, Merkulov, and Rodina, *Solid State Physics*, **40**, 917 (1998) [*Russ. Fiz. Tver. Tela*, **40**, 1002 (1998)].
73. T. Matsumoto and M. Aoki, *Jap. J. Appl. Phys.*, **13**, 1804 (1974).
74. D. Volm, T. Streibl, B. K. Meyer, T. Detchprohm, H. Amano, and I. Akasaki, *Solid State Commun.*, **96**, 53 (1995).
75. K. Kornitzer, T. Ebner, K. Thonke, R. Sauer, C. Kirchner, V. Schwegler, M. Kamp, M. Leszczyński, I. Grzegory, and S. Porowski, *Phys. Rev. B*, **60**, 1471 (1999).
76. A. Wysmołek, M. Potemski, R. Stępniewski, J. Łusakowski, K. Pakuła, J. M. Baranowski, G. Martinez, P. Wyder, I. Grzegory, and S. Porowski, *Proc. of the ICNS3*, Montpellier, 1999, *Phys. Status Solidi B*, **216**, 11 (1999).
77. D. C. Reynolds, D. C. Look, B. Jogai, V. M. Phanse and R. P. Vaudo, *Solid State Commun.*, **103**, 533 (1997).
78. B. Santic, C. Merc, U. Kaufman, R. Niebuhr, H. Obloh, and K. Buchen, *Appl. Phys. Lett.*, **71**, 1837 (1997).
79. A. Fiorek, J. M. Baranowski, A. Wysmołek, K. Pakuła, and M. Wojdak, *Acta Phys. Pol., A*, **92**, 742–745 (1997).
80. K. Kornitzer, T. Ebner, M. Grehl, K. Thonke, R. Sauer, C. Kirchner, V. Schwegler, M. Kamp, M. Leszczyński, I. Grzegory, and S. Porowski, *Proc. of the ICNS3*, Montpellier 1999, *Phys. Status Solidi B*, **216**, 5 (1999).
81. M. Godlewski, A. Wysmołek, K. Pakuła, J. M. Baranowski, I. Grzegory, J. Jun, S. Porowski, J. P. Bergman, and B. Monemar, *Proc. of the Int. Symp. on Blue Laser and Light Emiting Diodes*, Chiba, 1996, edited by A. Yoshikawa, K. Kishino, M. Kobayashi, and T. Yasuda, (Ohmsha, Tokyo, 1996) p. 356.
82. J. Bernholc and S. T. Pantelides, *Phys. Rev. B*, **15**, 4935 (1977).
83. M. Palczewska, B. Suchanek, R. Dwilinski, K. Pakuła, A. Wagner, and M. Kaminska, *3rd European Gallium Nitride Workshop*, Warsaw, *MRS Internet J. Nitride Semicond. Res.*, **3** (1998).
84. A. M. Witowski, M. L. Sadowski, K. Pakuła, and P. Wyder, *Phys. Status Solidi B*, **210**, 385 (1998).
85. A. Wysmołek, P. Łomiak, J. M. Baranowski, K. Pakuła, R. Stępniewski , K. P. Korona, I. Grzegory, M. Boćkowski, and S. Porowski, *Proc. of the XXV Inter. School on Physics of Semiconducting Compounds*, Jaszowiec, 1996, *Acta Phys. Pol.*, **90**, 981–984 (1996).
86. P. J. Dean, D. D. Manchon, and J. J. Hopfield, *Phys. Rev. Lett.*, **25**, 1027 (1970).
87. C. H. Henry and J. J. Hopfield, *Phys. Rev. B*, **6**, 2233 (1972).
88. K. P. Korona, A. Wysmołek, J. M. Baranowski, K. Pakuła, I. Grzegory, and S. Porowski, *Proc. of Materials Research Society Fall Meeting*, Boston, 1997, *Mat. Res. Soc.*, **482**, 501 (1998).
89. M. Sufczynski and L. Wolniewicz, *Phys. Rev. B*, **40**, 6250 (1989).

90. S. Permogorov, in *Excitons*, edited by E. I. Rashba and M. D. Sturge (North Holland, Amsterdam, 1982), p. 177.
91. D. Kovalev, B. Averboukh, D. Volm, B. K. Meyer, H. Amno, and I. Akasaki, *Phys. Rev. B*, **54**, 2518 (1996).
92. M. Wojdak, A. Wysmołek, K. Pakuła, and J. M. Baranowski, *Proceedings of the ICNS3*, Montpellier, 1999, *Phys. Status Solidi B*, **216**, 95 (1999).
93. I. A. Buyanova, J. P. Bergmann, B. Monemar, H. Amano, I. Akasaki, A. Wysmołek, P. Łomiak, J. M. Baranowski, K. Pakuła, R. Stępniewski, K. P. Korona, I. Grzegory, M. Boćkowski, and S. Porowski, *Solid State Commun.*, **105**, 497 (1998).
94. A. Wysmołek, K. P. Korona, K. Pakuła, J. M. Baranowski, R. Stępniewski, I. Grzegory, M. Wróblewski, S. Porowski, *Proc. of the 23rd Int. Conf. Phys. Semicond.*, Berlin, 1996, edited by M. Sheffler and R. Zimmermann (World Scientific, Singapore, 1996) p. 2925.
95. D. G. Thomas, J. J. Hopfield, and W. M. Augustyniak, *Phys. Rev.*, **140**, A202 (1965).
96. P. J. Dean, C. H. Henry, and C. J. Frosch, *Phys. Rev.*, **168**, 812 (1968).
97. J. J. Hopfield, *Proc. 7th International Conf. Phys. Semicond.*, Paris, 1964 (Dunod, Paris, 1965).
98. G. Munschy and B. Stébé, *Phys. Status Solidi B*, **59**, 525 (1973).
99. C. H. Henry, R. A. Faulkner, and K. Nassau, *Phys. Rev.*, **183**, 798 (1969).
100. D. C. Reynolds and T. C. Collins, *Phys. Rev.*, **188**, 1267 (1969).
101. B. Monemar, U. Lindenfelt, and W. M. Chen, *Physica B and C*, **146**, 256 (1987).
102. V. Fiorentini, F. Bernardini, A. Bosin, and D. Vanderbilt, *Proc. of the 23rd Int. Conf. on Physics of Semiconductors*, Berlin, 1996, edited by M. Sheffler and R. Zimmermann (World Scientific, Singapore, 1996) p. 2877.

CHAPTER 6

Physics and Optical Properties of GaN–AlGaN Quantum Wells

P. LEFEBVRE[1], B. GIL[1], J. MASSIES[2], N. GRANDJEAN[2], M. LEROUX[2], P. BIGENWALD[3] and H. MORKOÇ[4]

[1]*Groupe d' Etude des Semiconducteurs—Centre National de la Recherche Scientifique, Université Montpellier II—Case courrier 074, 34095 Montpellier cedex 5, France*
[2]*Centre de Recherche sur l'Hétéro-Epitaxie et Applications, Centre National de la Recherche Scientifique, Rue Bernard Gregory, Sophia-Antipolis, 06560 Valbonne, France*
[3]*Université d'Avignon et des Pays de Vaucluse, 33, Rue Pasteur, 84000 Avignon, France*
[4]*Virginia Commonwealth University, College of Engineering and Department of Physics, P. O. Box 843072, Richmond VA, 23284-3072, USA*

6.1 INTRODUCTION

Group III nitride–based quantum wells (QWs) are currently the central element in the fabrication of efficient light emitting diodes (LEDs) and laser diodes (LDs) operating in the blue–ultraviolet range of the spectrum. The recent success obtained in this field has motivated numerous theoretical and experimental studies aimed at a detailed understanding of radiative recombination processes in these QWs and of their possible improvement [1–3]. Up to now, the greatest interest has been in InGaN/GaN QWs, or multi-QWs (MQWs). Recent breakthroughs [4–15] have emphasized the major role played by the localization of electrons and holes on deep potential fluctuations in the plane of InGaN layers. This localization seems to allow the carriers to avoid the nonradiative recombination channels, such as structural defects, which are numerous in group III nitride epitaxial layers. These fluctuations result from very large variations in the indium composition, with typical extensions of 2–3 nm [13–15], which suggested to some workers that the recombinations occurred in self-organized "quantum dots." However, the local distribution of compositions, which is related to demixion mechanisms, and the size of the corresponding indium-rich dots seem to depend strongly on the whole growth process and up to now hardly have been controlled [16]. In fact, the average composition itself seems really difficult to monitor during growth, which explains in part why the dependence of the bandgap of InGaN on the In composition is not yet fully understood. Now, these parameters are very crucial if one intends to perform bandgap engineering to optimize light-emitting devices. Moreover, recent theoretical and experimental studies [17–34] have shown that huge built-in electric fields are present in III-nitride QW systems, originating from the large piezoelectric coefficients and spontaneous polarization of these natural wurtzite materials. These fields of several hundred kV/cm necessarily have an important impact on the optical processes and on their dynamics [24,27,29,31,33], yet the respective roles played by carrier localizations and by these fields still have to be unraveled.

We have chosen to approach this problem by focusing our investigations on GaN/AlGaN QWs, which also are of technological interest for UV light emission and detection. Indeed, by taking a binary compound as the confining layer, we avoid dealing with alloy demixion issues. In fact, we use GaN/AlGaN QWs as model systems that, much like GaAs/AlGaAs QWs, allow us to validate useful parameters for bandgap engineering, including the built-in electric fields. In Section 6.2, we briefly recall the principal characteristics of the optical properties of excitons in QWs based on zinc-blende crystals. This allows us to emphasize, in Section 6.3, the specific properties of GaN QWs due to their wurtzite form, namely, those related to their particular band structure and those related to the piezoelectric and pyroelectric effects. In Section 6.4, we present and discuss the results of continuous-wave (CW) optical spectroscopy. Section 6.5 is devoted to the modeling of excitons in GaN/AlGaN QWs, including electric-field effects. The results of Sections 6.4 and 6.5 serve as a basis for commenting on the time-resolved photoluminescence (PL) results described in Section 6.6, allowing us to discuss unexpected dynamical behaviors. A summary and concluding remarks are presented in Section 6.7.

6.2 QUANTUM WELLS BASED ON CUBIC SEMICONDUCTORS

Group III nitrides constitute an exception among III–V semiconductors because they generally crystallize in the hexagonal (wurtzite) phase. They can be forced to grow in the cubic (zinc-blende) phase, by heteroepitaxy on proper substrates, but this has been done, up to now, to the detriment of structural and electronic properties [1–3]. Nevertheless, the principal subjects of interest in the last decades were strained layers of cubic semiconductors of group IV elements, or of III–V and II–VI compounds, grown by various kinds of heteroepitaxy.

As an extreme case of such strained layers, semiconductor QWs and superlattices (SLs) were first grown at the end of the 1970s. A QW is formed by a thin slice (typically a few nanometers) of a "low"-bandgap semiconductor cladded between thick slabs of a wider-bandgap compound [35], which constitute the barriers. The motion of carriers confined in such QWs is called *bidimensional* because the dispersion relation is suppressed for carriers moving along the growth axis, being replaced by quantized "levels." In fact, these levels are the extrema of $E(\mathbf{k})$ curves (or "subbands") that still exist for carriers propagating in the plane of the QW. Superlattices are periodic structures in which the basic building block—a QW with thin barriers—is repeated a large number of times. If thin enough barriers are used, coupling of adjacent QWs is induced by a tunnel effect that one can control by adjusting the barrier width and height. Then, one ends up with a totally artificial, three-dimensional material with largely anisotropic properties.

To calculate the subband extrema of QWs it is rather convenient, and still really accurate, to use the effective-mass approximation within the so-called envelope function approach. The simplest expression of this model consists in taking a square potential well of width L with infinitely high barriers.

This straightforward approach yields a series of quantized energies, obtained analytically as a function of an integer quantum number α for conduction-band states, and β for valence-band states. These energies depend only on the QW width and on the effective masses: m_c for the conduction band and m_v for the valence band. What is really obtained by spectroscopy is the transition energy between the αth conduction subband and the βth valence subband. This transition energy is given by

$$E^{\alpha,\beta} = E_0 + \alpha^2 \frac{\hbar^2 \pi^2}{2m_c L^2} + \beta^2 \frac{\hbar^2 \pi^2}{2m_v L^2}. \tag{6.1}$$

In this simplistic model, the envelope functions vanish at the well boundaries, and the oscillator strengths of such subband-to-subband transitions strictly vanish if $\alpha \neq \beta$. Beyond this idealistic picture, several characteristics of "real" systems induce interesting effects.

1. The potential wells have finite depths: the sum of these depths for conduction and valence bands is simply the bandgap difference between the well and barrier materials. The envelope functions of the carriers then leak into the barrier material, and the effective masses in this material then have to be included in a nonanalytic calculation of subband energies. The electronic structure of multiple quantum wells is obtained from the single quantum well one just by using a transfer matrix formalism. The superlattice case is treated by taking into account the prescriptions of the Born–Von Karman cyclic conditions. The transition energies differ from their analogues in the infinite well model. The selection rules also are different: a transition is allowed if α and β have the same parity. This more sophisticated (although still simple) model introduces the necessity of a crucial parameter: the valence- (or conduction-) band offset, i.e., the distribution of the bandgap difference between conduction and valence bands. For all material systems investigated, the value of this parameter has been subject to debates. It seems that consensual values can be reached by fitting theoretical results to those of optical spectroscopy, but only after accumulation of numerous experimental results and only after all other parameters have been independently determined, which is seldom the case.

2. An important approximation, sometimes forgotten, of the preceding models lies in the assumption that Bloch functions with the same **J** and J_z values are strictly identical in the two materials. This approximation is too drastic, in fact, for QWs grown from two compounds with neither a common anion nor cation [36], or for QWs grown along low-symmetry directions, such as the (311) direction [37] in zinc-blende crystals. In such cases, the complexity of the band structure away from the Γ-point needs to be included, especially for valence-band states, by a multiband treatment in the framework of Kane's description [38]. In particular, only such a treatment

can help explain the optical anisotropy along the growth plane of these specific QWs [37].

3. Under most usual experimental conditions, the transitions observed by low-temperature optical spectroscopy are excitonic transitions rather than band-to-band transitions. Practically, one should substract the exciton binding energy from the band-to-band energy to obtain the excitonic transition energy. It is well known that the binding energy of the exciton in a QW is larger than its value in the corresponding bulk crystal, provided that the well width is commensurable with the Bohr diameter. The treatment of this famous quasi-2D exciton problem was pioneered by Bastard [35,39] and developed by many others [40–45]. For a general modeling of all quasi-2D systems, including the most exotic ones, it is important to include the electrostatic attraction between the electron and the hole as part of the mechanisms of carrier confinement along the quantization axis. This has led to the development of methods in the "effective potential" context [43–45], which must be used for systems like type II QWs, where the two types of carriers lie in separate layers, or for QWs exposed to electric fields. It should be noted, also, that one can use a mathematical approach in metric spaces with noninteger dimensionalities [46,47] to describe quite accurately some characteristics of QW excitons, which are never exactly 2D or 3D. For wells wider than the exciton Bohr diameter, but still narrower than the carriers coherence length, it is more convenient to consider the confinement of the exciton center of mass, taking the exciton as a whole particle [48–50]. To be more precise, it is better to consider, in fact, that we have a thin-slab interference pattern involving exciton–polaritons rather than photons.

4. Quantum wells and superlattices are generally based on lattice-mismatched materials, including the GaAs–AlAs combination, for which the small lattice mismatch between AlAs and GaAs explains some symmetry properties of the conduction band in short-period superlattices [51]. For cubic materials grown on (001) planes, the resulting internal strains provoke particular fine-structure effects, mainly on the energies and ordering of states of different symmetries like, e.g., heavy-hole and light-hole excitons. For QWs and SLs grown away from the (001) direction, internal piezoelectric fields can appear, resulting from the asymmetrical distortion of the elementary tetrahedra of the fourfold chemical coordination. Such fields separate the electron from the hole, so that the oscillator strength of the transition may become very small [52], as sketched in Figure 6.1. By applying a reverse bias, or under sufficient photoinjection of electron–hole pairs, a quasi-flat-band situation can be recovered, giving an efficient modulation of the optical response.

5. Last, there is an important consideration in strained-layer epitaxy, namely, the critical thickness for coherent growth. Below such thickness the lattice mismatch is accommodated by elastic deformations. Above this

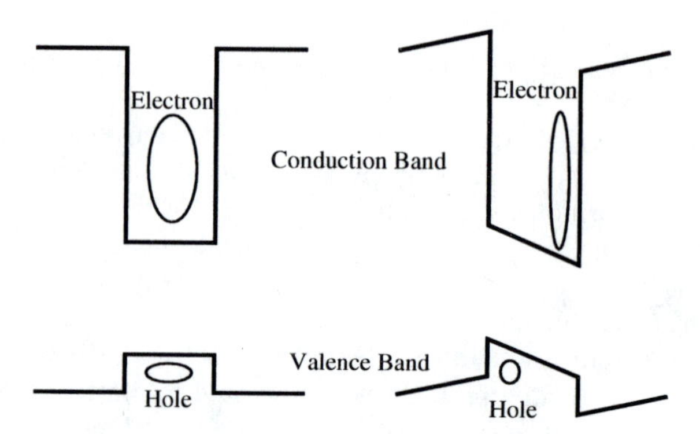

Figure 6.1 Qualitative sketch of the electron and hole spatial distributions for the first quantized states. In the square-well context, on the left, the probability densities peak in the middle of the well. In the case of an internal electric field, these densities are shifted in opposite directions, and an electron–hole dipole is created.

value dislocations are created, with deleterious influence on the optical and structural properties. If some growth parameters are properly set, the plastic deformation regime can be replaced by a coherent epitaxy of three-dimensional self-organized nano-islands [53,54].

6.3 SPECIFIC PROPERTIES OF GaN-BASED QUANTUM WELLS

6.3.1 Band-Structure and Strain Effects

In their wurtzite form, group III nitrides are characterized by two lattice parameters. The lattice parameter a, in the $(10\bar{1}0)$ directions, equals 3.189, 3.112, and 3.53–3.60 Å for GaN, AlN, and InN, respectively, whereas lattice parameter c, in the (0001) directions equals 5.185, 4.982, and 5.96–5.74 Å [55]. These materials are often grown on lattice-mismatched substrates by using modern epitaxy techniques that produce large areas of crystals with reasonable structural qualities. A variety of lattice-mismatched substrates exist, among which are sapphire, 6H–SiC, Si (111), and ZnO [56]. There is also a mismatch of thermal expansion coefficients between these substrates and nitrides. When the samples are cooled to room temperature after growth, this mismatch results in residual internal strain, a different c/a ratio, and thus a different optical response. The best-known effect of such strains is to alter the eigenenergies of electrons in the crystal.

The conduction and valence bands of III–V semiconductors are principally built up from s-like and p-like orbitals, respectively. The latter should be threefold degenerate in cubic symmetry and in a spinless description. In hexagonal crystals, this threefold degeneracy is lifted into a singlet state and a doublet one. Including spin effects the sixfold valence band yields a fourfold Γ_8 state and a twofold Γ_7 one for cubic crystals, whereas the wurtzite symmetry yields a twofold Γ_9 and two split Γ_7 states. In cubic crystals any anisotropic strain field will split the Γ_8 fourfold state. Therefore, the bandgap of the unstrained semiconductor can be determined from the measurement (or extrapolation) of the transition energy when there is no splitting of the fourfold Γ_8. In wurtzite semiconductors, the crystal fields splitting exists even in the absence of strain [56]. This is why the "unstrained" value of the bandgap has been difficult to determine, since all heteroepitaxial layers sustain some residual strain, as stated previously. Nevertheless, this determination has been achieved, but the connection of the actual strain with growth parameters is complicated and still not fully understood. The available values for the excitonic gap of unstrained GaN are either close to 3.470 eV or close to 3.477 eV [56]. The first value corresponds to thick GaN layers grown on sapphire, supposed to be relaxed or under slight compression, or was based on measurements of sample curvature. The second value was determined from homoepitaxial samples.

Most of the GaN–AlGaN QW systems are grown on top of a GaN buffer layer, with a typical thickness of 1 to 3 μm, meant to optimize the crystalline quality of the QWs. This high-temperature GaN layer should be distinguished from the GaN or AlN buffer layer, grown at low-temperature, a few tens of nanometers thick, and directly deposited on the substrate surface to facilitate the nucleation of the high-temperature nitride layers. Under so-optimized growth conditions, typical values of the excitonic bandgap of GaN at $T = 2$ K range from 3.45 eV (silicon substrate) to 3.50 eV (on A-plane sapphire [57]). Such large variations demonstrate how crucial it is to know the strain state of each layer if one aims to master the optical properties of GaN–AlGaN QWs. Fortunately, optical spectroscopy can help us evaluate the strain in the underlying GaN buffer and then deduce the strain of individual layers in the QW system from simple elasticity, since the critical thickness is seldom reached.

Figure 6.2 displays the 2 K reflectance features from GaN layers buried below more or less complicated MQW structures (optical properties are shown in Figure 6.5). The samples were grown by molecular beam epitaxy (MBE) on a series of C-plane sapphire substrates cut from the same batch. The values of the A-excitonic transitions are 3.480, 3.477, and 3.482 eV from the top to the bottom of the figure; these GaN layers, although grown in identical conditions, undergo slightly different strains. The point we wish to emphasize here is the importance of performing this measurement, which is not necessary for conventional QWs grown on unstrained semiconductor substrates. This measurement yields the "real" valence-to–conduction-band energies in the GaN that constitutes the QWs, once the exciton binding energy of 26 meV [57] is properly included.

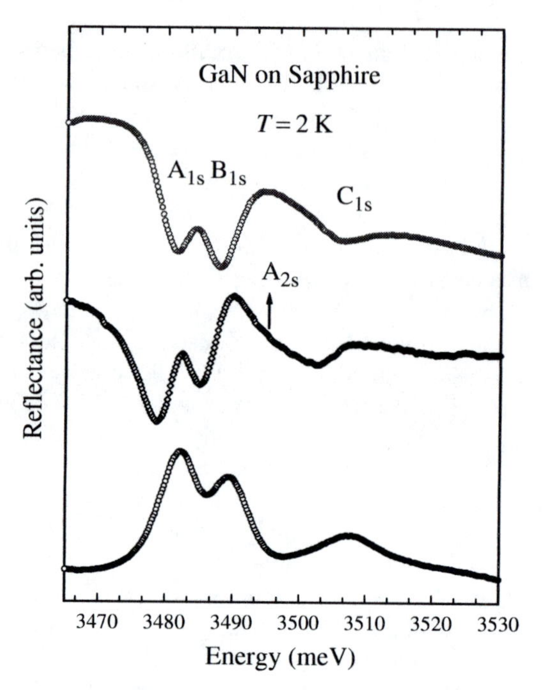

Figure 6.2 Typical reflectance spectra corresponding to GaN buffer layers grown on C-plane sapphire before the growth of GaN/AlGaN QWs. The MBE growth conditions were kept identical. The phases of reflectance features result from interferences between the beams reflected at the surface and at the upper interface of the GaN buffer. Because thicknesses of the MQW structures are different for these three samples, so are the phases of the reflectance features.

Such optical measurements can be usefully complemented by structural characterization methods, such as X-ray diffraction reciprocal space mapping. An example concerning the AlGaN barriers in an MQW sample is shown in Figure 6.3. The sample consists of 10 periods of GaN (6 MLs)/AlGaN (19ML) deposited on a 2 μm-thick GaN layer [25,27]. Note that due to specificities of MBE growth [54], we frequently use monolayers (MLs) as thickness units. One GaN monolayer is 2.59 Å. Figure 6.3 shows that $Al_{0.11}Ga_{0.89}N$ barriers are pseudomorphically strained onto the GaN template. Both values of **a** and **c** lattice constants (3.189 Å and 5.185 Å, respectively) and the energy of the A free exciton (3.478 eV) indicate that the GaN thick buffer layer is nearly relaxed. The AlGaN barriers are thus under tensile stress.

6.3.2 Internal Electric Field Effects

For QWs based on wurtzite III-nitrides, the electric fields in the well (e.g., GaN) and barrier material (e.g., AlGaN) result from the difference in their

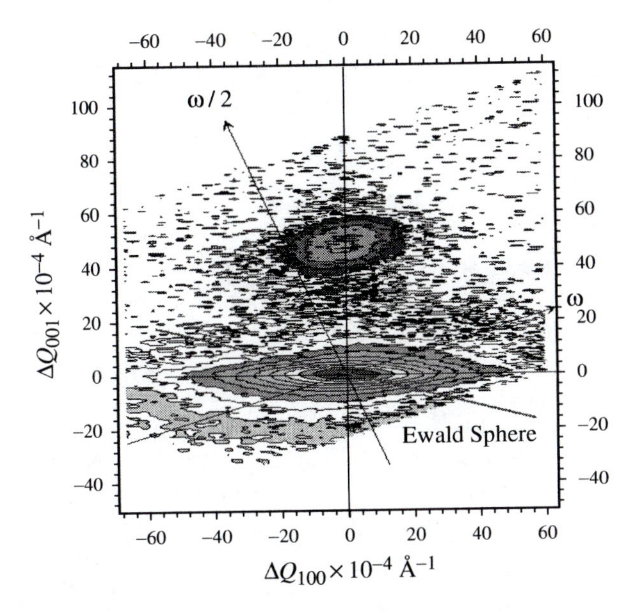

Figure 6.3 X-ray diffraction reciprocal space mapping of a 10-period $GaN/Al_{0.11}Ga_{0.89}N$ MQW (6 ML/19 ML). The abscissa is proportional to the inverse of the *a* lattice parameter, the ordinate to the inverse of the *c* parameter. The alignment of both Bragg spots on a vertical axis shows that the layers are pseudomorphically strained (AlGaN strained to GaN).

respective electrostatic polarizations [20]. For each material, this polarization results from the superposition of the piezoelectric effect, which depends on the strain state of the material, and the pyroelectric effect (also referred to as the spontaneous or equilibrium polarization). These effects strongly influence the optical properties of heterostructures. They are also of prime importance for determining the two-dimensional charge densities in heterostructures for field-effect transistors [18,19,21–23,34].

The effects of piezoelectric polarization are already known—and utilized—in some strained-layer heterostructures based on zinc-blende materials grown along the (111) axis, such as (Ga,In)As/GaAs [52,58,59] or CdTe/CdZnTe [60] QWs. In such systems, the piezoelectric effect appears because of the strain and of the absence of inversion symmetry: when the elementary tetrahedron is elongated or compressed along the (111) axis, the average positions of negative and positive charges are no longer the same. A dipole is created due to the difference in the barycenters of positives charges (all the ions) and negative charges (valence electrons). This dipole divided by the volume of the unit cell is the piezoelectric polarization P^{piezo} (in C/m^2). A simplified picture of the phenomenon consists of considering an elongated tetrahedron of group III cations and group V anions, taken as effective positive and negative charges, respectively; but this picture is misleading, from the quantitative viewpoint. For instance, the piezoelectric coefficients have *opposite* signs in III–V and II–VI zinc-blende compounds [20], i.e., a similar tetrahedron deformation yields opposite effects in these two families.

Turning to wurtzite nitride QWs, there are large differences in lattice constants among AlN, GaN, and InN: the heterostructures made of these materials are strained. This strain induces a piezoelectric polarization, which, in the usual case of biaxial (0001) strain, is along the c-axis and given by $P_z^{piezo} = 2[e_{31}-(C_{31}/C_{33})e_{33}]\varepsilon_{xx}$, where ε_{xx} is the in-plane strain, C_{31} and C_{33} are elastic constants, and e_{31} and e_{33} are piezoelectric coefficients (C/m^2) in Voigt notations (actually a third-order tensor). It turns out that piezoelectric coefficients are much larger in group III nitrides than in conventional semiconductors [20].

Keeping in mind that the piezoelectric effect is in fact a strain-induced modification of the unit cell polarization, it is sufficient to say that, even in the absence of strain, the wurtzite symmetry (contrary to the zinc-blende symmetry) does not require that the barycenters of positive and negative charges be at the same position. In other words, the spontaneous polarization P^{spont} is a natural consequence of the wurtzite symmetry, and there is no reason why it should not exist. The total polarization of the material is simply $P = P^{spont} + P^{piezo}$. Nevertheless, the question remains, what are the respective magnitudes of the two effects? The answer obviously depends on the materials.

For III-nitrides, the spontaneous polarization has been adressed recently from the theoretical viewpoint [20,21]. Very large values have been found, and, in particular, there is a large difference in spontaneous polarization between AlN and GaN [20]. This difference is connected to the polar nature of III-nitrides. For instance, Bernardini et al. [20] calculated the spontaneous polarization of AlN using the ideal wurtzite parameters, i.e., the ideal c/a ratio, and the resulting polarization is 2.5 times smaller than what is found for the actual structure. The polarization (spontaneous and piezoelectric) depends on the polarity of the crystal, namely, whether the bonds along the c-direction are from cation sites to anion sites or vice versa. By Ga (resp. N) polarity one means that if one were to cut the perfect solid along the c-plane where one breaks only one bond, one would end up with a Ga (resp. N) terminated surface. By convention, the Ga face corresponds to the (0001) direction, the N face to the (000-1) one.

As for the effects of this total polarization, standard electrostatics [61] teaches us that an infinite polar material is neutral. In fact, measurable effects appear as soon as interfaces are present. When a slab of polar material is in vacuum or cladded by two semi-infinite layers of nonpolar material, the polarization of the slab results in the presence of two charged planes, of opposite sign, at the interfaces. The charge density is given by $\sigma = \mathbf{P}\cdot\mathbf{n}$, where $\mathbf{n}$ is the unit vector normal to the interface. For (0001) wurtzite slabs, $\sigma = \pm P$.

When the slab (e.g., a well) is cladded between two layers of another polarized material (e.g., the barriers), then $\sigma = (\mathbf{P_b} - \mathbf{P_w}) \cdot \mathbf{n}$, where $\mathbf{P_w}$ and $\mathbf{P_b}$ are the polarization of the well and of the barrier, respectively. Due to the conservation of the displacement vector, one gets a discontinuity of the electric field at each interface, resulting from [62]: $\varepsilon_w \varepsilon_0 E_w - \varepsilon_b \varepsilon_0 E_b = P_b - P_w$ (in MKSA units), where ε_0 is the permeability of vacuum, $\varepsilon_w(\varepsilon_b)$ is the well (barrier) relative dielectric constant, and $E_w(E_b)$ is the field in the well (barrier); however, the actual value of the field depends on the sample geometry. In the simple case of infinite barriers, the contributions to the electric field created by each charge plane ($E = \pm\sigma/2\varepsilon\varepsilon_0$) add to each other in the well and cancel in the barriers; then, $E_b = 0$ and $E_w = (P_b - P_w)/\varepsilon_w \varepsilon_0$. Of course, as discussed by Fiorentini et al. [28], if $E_w L_w \geq eE_g$, where L_w is the slab thickness and E_g the bandgap, the sample should turn metallic. In bulk samples or micrometric epilayers, charge transfers occur, possibly connected to the presence of adsorbates at the surface [63], which counterbalance the polarization charges [28]. Therefore, the built-in electric field is usually neglected in such samples. This is not the case for QWs.

For "real" samples, such as those discussed in this chapter, embedding several layers with different widths and strain state, it is really difficult, at present, to draw a definitive picture of the overall band profile. A simple approximation is to assume a global flat-band situation, i.e., $\Sigma_{w,b} E_b L_b + E_w L_w = 0$ throughout the structure. For infinite, periodic MQWs (or superlattices), this approximation results in cyclic boundary conditions leading to:

$$E_w = \frac{L_b(P_b - P_w)}{(L_w \varepsilon_b \varepsilon_0 + L_b \varepsilon_w \varepsilon_0)} \tag{6.2}$$

and

$$E_b = \frac{L_w(P_w - P_b)}{(L_w \varepsilon_b \varepsilon_0 + L_b \varepsilon_w \varepsilon_0)}. \tag{6.3}$$

These equations show that the larger the barriers, the higher the field in the wells, and vice versa. As commented later, this is precisely what is observed. However, the flat-band approximation is probably not exact, due to complicated effects at the sample surface and at the interface between the GaN buffer and the QW structure, and due to free carriers resulting from the residual doping. Band-curvature effects are likely to be present, which can be modeled only via a self-consistent resolution of Schrödinger's and Poisson's equations [28], provided that all pertinent parameters are known (not the case yet) and provided that the correct boundary conditions are assessed. At the time this chapter was written, much work remained necessary to meet these requirements.

We now look at some practical cases. For GaN/AlGaN QW samples with the Ga polarity, which corresponds for instance to growth on nitridated sapphire, the electric field in the well points toward the substrate. In the wells, electrons are pushed toward the surface, and holes toward the substrate. The situation is the same for biaxially compressed (Ga,In)N/GaN QWs [26,31]. For coherently strained $Al_yGa_{1-y}N$ on a nearly relaxed GaN template, corresponding to the samples discussed in this chapter [25], the strain ε_{xx} is tensile (positive) and given by $\varepsilon_{xx} = y(a_{GaN} - a_{AlN})/a_{AlGaN}$. Using c-plane lattice constants of 3.112 Å and 3.189 Å for AlN and GaN, respectively, we end up with $\varepsilon_{xx} \approx 0.024y$. The piezoelectric polarization is then $P^{piezo} \approx y \cdot 2.66 \times 10^{13}\,\mathrm{ecm}^{-2}$, using the values given by Bernardini et al. [20] (neglecting the bowing induced by the composition dependence of elastic constants and piezoelectric coefficients). The difference in spontaneous polarization between $Al_yGa_{1-y}N$ and GaN is given by $y \cdot 3.25 \times 10^{13}\,\mathrm{ecm}^{-2}$. Such values are really huge, much larger than those produced in piezoelectric zinc-blende QWs [58–60], and they yield internal fields of several hundred kV/cm.

Experimentally, it is not straightforward to separate the piezo- and pyroelectric effects. Consider, for instance, GaN QWs grown on a relaxed GaN, with strained $Al_yGa_{1-y}N$ barriers. In these barriers, due to the tensile strain, the piezoelectric and spontaneous polarizations have the same sign. If the same structure is then grown on relaxed $Al_yGa_{1-y}N$, the piezoelectric effect will take place only in the biaxially compressed GaN well, thus with an opposite sign. Because the electric fields are proportional to polarization *differences*, E_w should be nearly the same in both cases. In fact, exchanging the strained materials in this way should result only in deformation potential effects, changes in some material constants, and band offsets [20,21].

In fact, the list of practical obstacles for determining electric fields is long. First, the exact respective strain states of the GaN and AlGaN layers are not known perfectly. Generally, the GaN template is biaxially compressed in an uncontrolled way by the sapphire substrate. One may assume that this strain is transmitted to all GaN layers. The strain of AlGaN barriers, also, may not be pseudomorphic but partially relaxed. Moreover, the clear definition of a polarization-induced built-in electric field requires that no inversion domains exist in the structure.

6.4 CONTINUOUS-WAVE SPECTROSCOPY OF GaN/AlGaN QWs

Low-temperature photoluminescence (PL) is used more often than reflectance or absorption to characterize low-dimensional heterostructures, mainly because fair-quality samples may show weak and damped reflectance features but always do provide some PL signal. This technique provides the spectral distribution of carriers over their allowed states in the pseudo-thermal equilibrium induced by continuous photoexcitation. As far as QWs are concerned, this thermalization is the reason why PL spectra are generally dominated by the contribution from localized excitons rather than from free excitons, i.e., excitons freely propagating in the well plane. Localized excitons occupy lower-energy states and are thus more populated at low temperatures. Moreover, the localization breaks the **k**-selection rule, which normally restricts *radiative* free excitons to those having the photon wavenumber. For undoped QWs this localization takes place in regions of the sample where the confining layer is wider than its average width [64–69]. To measure the localization energy, it is really useful to combine PL and reflectance results, when possible: this localization energy is simply the difference (Stokes shift) between the reflectance and the PL transition energies.

In some cases, reflectance features cannot be observed, though the quality of the samples would normally permit it. For example, wide-enough GaN/AlGaN QWs emit below the GaN excitonic gap, due to the quantum-confined Stark effect induced by internal fields. Then, the reflectance features characteristic of the QW are often unobservable, because they are weak (reduction of oscillator strength by electron–hole separation [24]) and because they are superimposed on the strong interference fringes of the GaN/sapphire cavity. In such cases, a convenient estimation of localization energies can be obtained from temperature-dependent PL experiments. If $k_B T$ becomes large enough compared with the localization energy, a detrapping of excitons occurs, giving back a **k**-space–extended Maxwellian distribution of the exciton population. The PL becomes dominated by free excitons with a long radiative lifetime. In fact, nonradiative recombination channels become more efficient at high temperatures, strongly reducing the radiative efficiency and thus the PL intensity. Nevertheless, the PL signal often remains sufficient to measure the temperature-dependent PL energy. The Stokes shift is deduced as the difference between the low-temperature PL maximum and the low-temperature extrapolation of the high-temperature variation of this maximum.

In Figure 6.4 are shown typical reflectance (top) and PL (bottom) spectra of GaN/$Ga_{0.93}Al_{0.07}N$ MQWs. The thickness of GaN wells is 2.5 nm; the thickness of the barriers is 10 nm. The PL was excited here by the 325-nm line of a He–Cd laser. The result of a simple envelope-function calculation (detailed later) of transition energies for the fundamental E_1–Γ_1^9 and E_1–Γ_1^7 confined excitons is shown by arrows. The high-energy shoulder of the PL spectrum of the QW, at 3.550 eV, corresponds to the principal minimum of reflectivity and to the calculated ground exciton energy. This result indicates that the dominant line of the cw PL, at 3.535 eV, corresponds to localized excitons, induced by well-width and depth fluctuations, that are

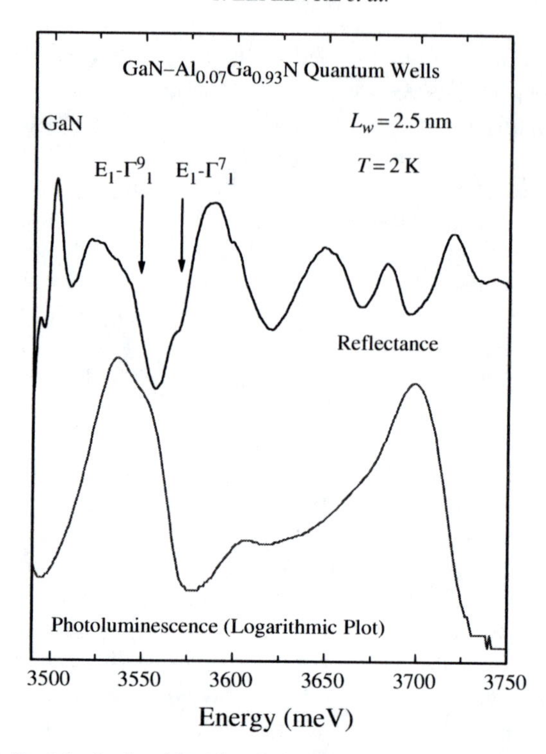

Figure 6.4 Reflectivity (top) and log plot of PL (bottom) spectra from a GaN/GaAlN MQW, under standard cw conditions. Note that both localized excitons and free excitons associated with the MQW structure are resolved.

saturated under the present excitation conditions [70]. The PL line (B) at 3.70 eV arises from the $Ga_{0.93}Al_{0.07}N$ barriers, with a LO-phonon replica at 3.61 eV. The intensity of the B line indicates the poor efficiency of charge transfers between the barriers and the QWs at low temperature, despite the small barrier width (10 nm). This is attributed to carrier localization due to fluctuations of the Al composition in the barriers.

Increasing the Al composition leads to a larger compositional disorder in the barrier layers and increases the localization energy of the exciton. This is shown in Figure 6.5, which is the analogue of Figure 6.4 for three samples. The Al composition is $x = 0.11$ for the two samples containing several QWs [Figures 6.5a and b] and $x = 0.09$ for the single QW [Figure 6.5c], respectively. The barriers are 5 nm thick for the two multiple QWs and 30 nm for the single QW. It is worth noticing that all samples were grown in the same setup under identical growth conditions and as part of the same sequence of growths [29]. Thus, there is no reason to expect important differences among them, in terms of bulk or interface nonradiative processes, for instance. Figure 6.5a shows the cw PL and reflectivity spectra of a sample

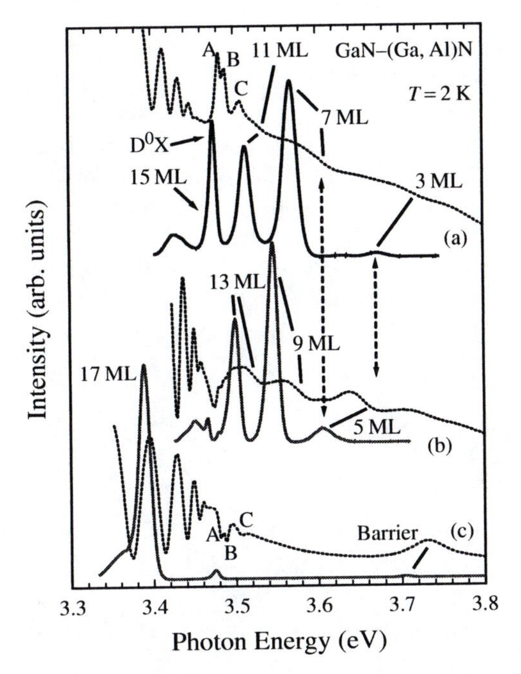

Figure 6.5 Reflectance (dotted line) and PL (solid line) spectra taken from three GaN/AlGaN QW samples. The first two samples (a and b) contain four QWs of different widths, labeled in the figure. The third sample contains a single 17-ML-wide QW. Thick lines connect reflectance features to the corresponding PL peaks (Stokes shift). Dashed arrows on two examples match PL energies for a given nominal well width with the reflectance features for widths 2 ML larger.

that contains four $GaN/Ga_{0.89}Al_{0.11}N$ QWs with respective thicknesses of 3, 7, 11, and 15 MLs from the surface toward the GaN buffer layer. The Stokes shifts are in the range of 15 to 20 meV between PL lines and reflectance features (larger for the narrower wells), indicating that the PL corresponds to the recombination of excitons localized on areas where the QWs are wider by 2 MLs (1 ML = 2.59 Å) than their nominal thickness. These Stokes shifts also correspond very well to the energy difference between PL peaks from QWs that differ by 2 MLs in width. This can be checked in Figure 6.5b, which is the equivalent but for a sample with three $GaN/Ga_{0.89}Al_{0.11}N$ QWs of 5, 9, and 13 MLs respective thicknesses. These lines nicely intercalate between those of the previous sample. The PL line from the 15-ML-wide QW in sample 1 overlaps the line of D^0X complexes in the GaN buffer. A deconvolution with Gaussian lineshapes allows us to separate an intense PL from the D^0X, at 3.475 eV, and a weaker PL from the 15-ML-wide QW, at 3.462 eV. The third sample in Figure 6.5 gives PL clearly below that of the GaN template. This behavior is typical of either the existence of an electric field in GaN well and the subsequent quantum-confined Stark effect (QCSE) or a type-II configuration of the band lineups. As mentioned previously, it has been understood recently [24,25] that a strong QCSE exists, due to the presence of huge internal electric fields. This is confirmed by the absence of any marked feature in the reflectance spectrum of Figure 6.5c, which is, rather, dominated by Fabry–Pérot interferences below the bandgap of the GaN template. This is evidence of the very small exciton oscillator strength in this wide QW.

Before proceeding to time-resolved PL results, let us see how one can model excitonic energies and oscillator strengths, in the particular context of GaN/AlGaN QWs with huge built-in electric fields.

6.5 ENVELOPE-FUNCTION CALCULATIONS

6.5.1 Basic Parameters

The microscopic mirror parity in real space is an important symmetry operator in familiar heterostructures such as GaAs–GaAlAs QWs grown along their (001) axis. In the case of QWs based on group III nitrides, this symmetry is destroyed by the presence of the internal electric field. Nevertheless, the basic elements of the theoretical treatment in the effective-mass approximation are the same.

The eigenstates of the quantum structure are built from Bloch states of the binary and ternary materials multiplied with envelope functions. The decoupled electrons and heavy (Γ_9) hole states (labeled hh_i hereafter) are then calculated within the context of the one-band envelope-function approach, since we place ourselves at $\mathbf{k}_{//} = \mathbf{0}$. For these particles, the global wave functions are $\Psi_e(\pm1/2) = F_e|1/2, \pm1/2\rangle$, and $\Psi_{hh}(\pm3/2) = F_{hh}|3/2, \pm3/2\rangle$, where the Bloch functions $|1/2, \pm1/2\rangle$ and $|3/2, \pm3/2\rangle$ are $S(^{\uparrow}{}_{\downarrow})$ and $\mp1/\sqrt{2}.|[X \pm iY](^{\uparrow}{}_{\downarrow})\rangle$, respectively. As usual, *S* represents an s-like function in real space, and *X*, *Y*, and *Z* are three p-like functions. The light-hole (lh_i) and spin-orbit split-off-hole (so_i) states both transform like Γ_7, and a two-band envelope-function calculation is required to find the corresponding eigenstates. In spherical space, in terms of the projections of the angular momenta, we take

$$|3/2, \pm1/2\rangle = 2/\sqrt{6}|Z(^{\uparrow}{}_{\downarrow})\rangle \mp 1/\sqrt{6}|[X \pm iY](^{\uparrow}{}_{\downarrow})\rangle, \tag{6.4}$$

$$|1/2, \pm1/2\rangle = \pm1/\sqrt{3}|Z(^{\uparrow}{}_{\downarrow})\rangle + 1/\sqrt{3}|[X \pm iY](^{\uparrow}{}_{\downarrow})\rangle. \tag{6.5}$$

In the representation where the good quantum numbers are the components of the total hole angular momentum, the corresponding wave functions are written as follows:

$$\Psi_{\Gamma 7}(\pm 1/2) = \Phi_{LH}|3/2, \pm 1/2\rangle + \Phi_{SO}|1/2, \pm 1/2\rangle, \tag{6.6}$$

which we first rearrange into

$$\Psi_{\Gamma 7}(\pm 1/2) = \Phi_1|[X(\pm)iY](^{\uparrow}{}_{\downarrow})\rangle + \Phi_2|Z(^{\uparrow}{}_{\downarrow})\rangle, \tag{6.7}$$

with $\Phi_1 = -1/\sqrt{6}\Phi_{LH} + 1/\sqrt{3}\Phi_{SO}$ and $\Phi_2 = 2/\sqrt{6}\Phi_{LH} - 1/\sqrt{3}\Phi_{SO}$.

Neither of these two expansions is more appropriate than the other one for calculating the two-component envelope functions. Details about the procedure can be found in the paper by Suzuki and Uenoyama [71].

In order to include the proper the potential well depths, we need:

(1) the composition dependence of the bandgap of the alloy. There is large scatter among the reported values of the bowing parameter in (Al,Ga)N [72,73, and references therein]. Lee et al. [73] discuss possible origins for this scatter. The calculations shown in this chapter use a bowing parameter of 0.6–0.7 eV, in agreement with this reference and our PL and reflectivity results on (Al,Ga)N barriers.

(2) the deformation potentials of the alloy. We take the same values for GaN and AlGaN.

(3) the effective masses, the crystal-field splitting parameters, and the spin-orbit interaction parameters. They are taken from Kumagai et al. [74]. It is important to note that the "light hole" is as heavy as the "heavy hole" along the (0001) direction in these materials. In addition, the A–B splitting is of the order of 8–10 meV in bulk GaN. Therefore, the splitting between the first confined heavy-hole and light-hole states is quite small, and the envelope functions will have comparable extensions in the barrier layers.

(4) the valence-band offset. A type I, 70/30 distribution was retained between the conduction band and heavy-hole band lineups in our calculations; however, we wish to point out that, once again, this value is far from being definitely assessed, in particular regarding its dependence on strain.

6.5.2 Results and Comparison with Experiments

By using the preceding parameters we can fit fairly well the transition energies of the first MQW presented in Figure 6.4, with $x = 0.07$ as the Al composition in the barriers. The accuracy of this fitting, however, is largely improved by including a field of 200–250 kV/cm in the wells. For the other

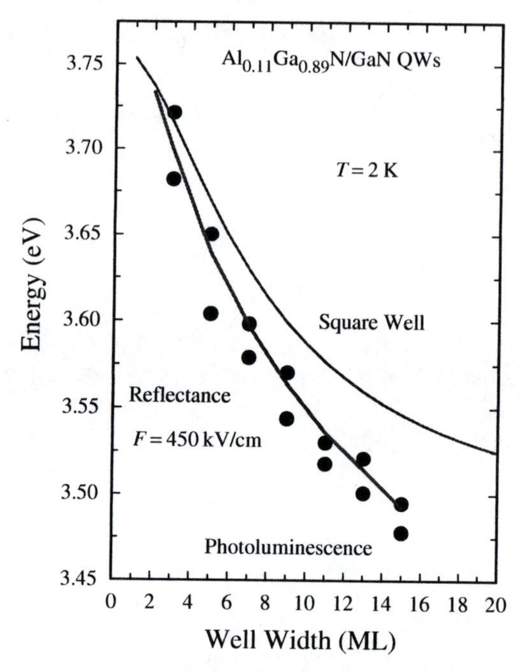

Figure 6.6 Experimental transition energies in GaN/$Ga_{0.83}Al_{0.11}N$ QWs with 5-nm-thick barriers. For each well width, the high (low) energy data correspond to reflectance (PL) measurements. The result of a "square-well" calculation is shown by the thin line. The fit to reflectance data was obtained by including an internal electric field of 450 kV/cm in the wells.

samples presented in Figure 6.5, including an electric field is absolutely necessary. For the MQW samples with $x = 0.11$ [Figures 6.5a and b], the best fit is obtained with an electric field of 450 kV/cm, which we compensate for with an opposite field in the barriers, chosen so that there is no potential drop through the whole structure [25,27]. Figure 6.6 illustrates the agreement between the experimental data and the calculation for the series of QWs with $x = 0.11$. The result of the square-well calculation (the exciton binding energy is included) is shown together with the result of a triangular-well calculation with an electric field of 450 kV/cm in the wells. The energies of reflectance features are shown (high-energy set) as well as those of PL maxima (low-energy set), illustrating the Stokes shift induced by exciton localization. Of course, the correct evaluation of the electric field can be made only by fitting reflectance data and *not* PL data. Alternatively, PL data can be exploited if an assumption is made on the "effective" well width corresponding to localized excitons: Figure 6.6 clearly shows that a nice agreement would be reached between calculations and PL data if all well widths were uniformly increased by 2 MLs.

One puzzling result, at first sight, was that a field of 600 kV/cm was necessary to fit the data for the single QW of Figure 6.5c, although the Al composition was lower than in the previous samples ($x = 0.09$ instead of 0.11). Careful examination of the signature of the GaN templates revealed no drastic variation of the strain in these layers. Thus, the electric field does not depend only on the Al composition. For a given x, the electric field in single or multiple QWs does not have the same magnitude; there is a redistribution of the electric field in MQW samples. To show this, we grew a series of samples including single and multiple QWs in which we varied the barrier width between 5 and 60 nm. The main result was that the transition energy for an MQW with thin barriers (say, 5 nm) is blueshifted relative to a similar, *single* QW. Moreover, as expected from the equations in Section 6.3.2., if the barriers are wide enough (say, 30 nm), the multiple and single QWs again show the same transition energy. For GaN/$Ga_{0.83}Al_{0.17}N$ QWs, we find that the electric field *in the well* increases from $\sim$600 kV/cm to $\sim$800 kV/cm when the barrier width increases from 5 to 60 nm. This purely electrostatic is nicely compatible with the equations in Section 6.3.2. [30], although strictly speaking, they should be applied to ideal, periodic structures.

This type of comparison between theory and experiment is thus a good way to estimate the built-in field. For instance, Grandjean et al. evaluated the dependence of this field on the Al composition of the barrier. To this aim, they grew a series of MQW samples with GaN well widths of 4, 8, 12, and 16 MLs and varied the Al composition up to 27% in the barriers but kept constant (10 nm) the thickness of these barriers. The electric field in the well was found to increase linearly at a rate of 40–50 kV/cm per %Al [32]. Typical PL spectra are given in Figure 6.7, and the variation of the electric field in the QW is plotted in Figure 6.8 versus Al composition.

To close this theoretical discussion we wish to address the modeling of low-dimensional exciton GaN–AlGaN QWs. As a matter of fact, a correct estimation of electric fields in these QWs is often a matter of a few meV. It is thus crucial that the exciton binding energies be properly calculated. Moreover, Coulomb effects must be properly included for a correct estimation of oscillator strengths, especially when they are made small by internal fields.

Calculating the binding energy of excitons in QWs requires a variational procedure. In the simplest approach (hereafter named 2D), one can use a two-dimensional trial function proportional to $\exp(-\rho/\lambda)$ that accounts only for the relative in-plane motion of the electron–hole pair. The total exciton energy has to be minimized versus the in-plane extension parameter (pseudo Bohr radius), λ. A more sophisticated approach (3D hereafter) consists in improving the choice of the trial function by using, e.g., the two-parameter ansatz $\exp(-\sqrt{\rho^2 + \alpha^2 z^2}/\lambda)$, although setting the anisotropy parameter, α, to 1 generally yields satisfactory results. It is important, also, to perform a self-consistent calculation that includes *the alterations of the potential wells*

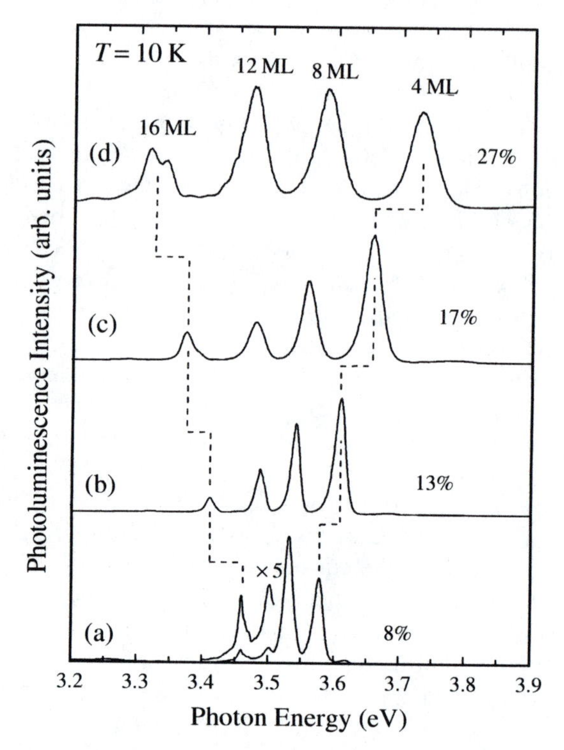

Figure 6.7 PL spectra of two GaN/$Ga_{1-x}Al_xN$ MQWs with $x = 0.08, 0.13, 0.17$, and 0.27, respectively, and with the same 10-nm barrier width.

produced by the electron–hole mutual attraction. In nitrides, the valence bands are split by the wurtzite symmetry, and the choice of 3D trial function asymptotically provides the values of the bulk exciton binding energy (26 meV) for infinitely wide wells. In Figure 6.9 are shown the predictions of 2D and 3D calculations for GaN/$Ga_{1-x}Al_xN$ QWs with x ranging up to 0.27. The results of square-well calculations (zero electric field) allow us to recover the well-known trends of the physics of GaAs–GaAlAs QWs but with the corresponding orders of magnitude for nitrides. As expected, the 2D calculation underestimates the binding energy down to values below the bulk GaN value for wide wells. This unrealistic behavior is not found with the 3D calculation. For more realistic calculations, we included the internal electric field of 1.1 MV/cm, corresponding to $x = 0.27$ in the barriers. The 2D calculation is plotted for the whole series of compositions, but the 3D computation is restricted to the highest composition in order to keep the figure legible.

In the range of well widths investigated, the exciton binding energy collapses due to the electric field–induced separation of the electron and hole wave functions. The 3D calculation leads to a reduction of the binding energy

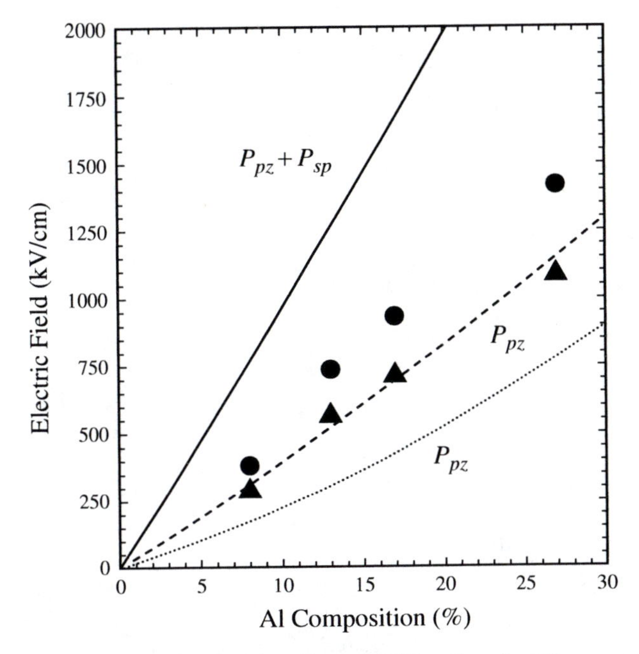

Figure 6.8 The internal electric field found in our MQWs with well-width sequences of 4, 8, 12, and 16 ML, plotted versus the Al content in the barriers. The width of the AlGaN barriers is 10 nm in all cases. The triangles are the measured values; circles are values corrected for finite barrier width effects.

by a factor of 4. The corresponding decrease of the oscillator strength is much more dramatic, as shown in Figure 6.10. The overlap integrals of electron and hole envelope functions have been calculated versus well widths for the square-well case (open diamonds) and for the real triangular-well situation (solid diamonds), for $x = 0.27$. This nonexcitonic calculation gives a first insight into the effect of the electric field: when the well width increases from 5 to 30 MLs and if no electric field is considered, the oscillator strength varies by a factor of 2, whereas it decreases by as much as *five orders of magnitude* when the electric field is included. The same trends are found by calculation of the excitonic oscillator strengths, i.e., a moderate variation with no electric field, and a decrease over *six orders of magnitude* when the field is included. In Figure 6.11 are plotted some typical probability densities in these quantum wells. In the top and middle of the figure are plotted the probability densities for two wells of 10-ML width. The Al compositions are 0.08 and 0.27 for the top and middle parts of the figure, respectively, and the electric fields in the wells are 380 kV/cm and 1.1 MV/cm. The separation between the electron and the hole increases when the Al composition (the electric field) increases, as expected, but the overlap of their probability distributions remains significant. Use of the self-consistent calculation

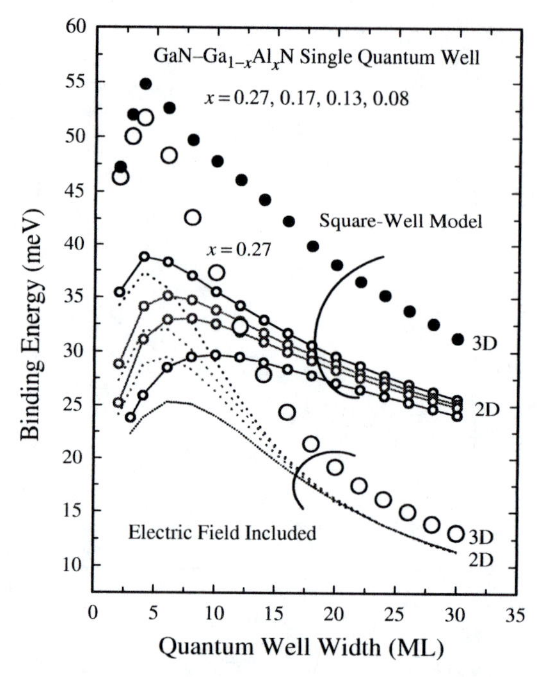

Figure 6.9 QW width dependence of the exciton binding energies calculated for GaN–$Ga_{1-x}Al_xN$ QWs, including either zero electric field ("square-well model") or the x-dependent field determined from Figure 6.8 ("electric field included"). The results of the 2D calculation are given for four values of x. The results of a 3D calculation are restricted to $x = 0.27$.

(dotted line), which accounts for the mutual attraction between the two carriers, shifts the electron probability density toward that of the hole more efficiently when the Al composition is low than when it is high. This is due to the competition between the Stark effect and the Coulomb attraction. When the Stark effect increases, it becomes more difficult for the Coulomb interaction to compensate for it. We thus expect that this attraction will have a small influence on oscillator strengths in the case of wide wells, as shown at the bottom of the figure for a 30-ML-wide QW ($x = 0.27$). In this case we no longer distinguish the influence of the self-consistent treatment on the electron probability density. The excitonic oscillator strength is now obviously very small, although the exciton binding energy remains of the order of 12 meV (see Figure 6.9).

6.6 TIME-RESOLVED SPECTROSCOPY IN GaN-AlGaN QUANTUM WELLS

The preceding calculations show that the clearest manifestations of strong internal fields should be (1) a significant redshift of transitions due to the

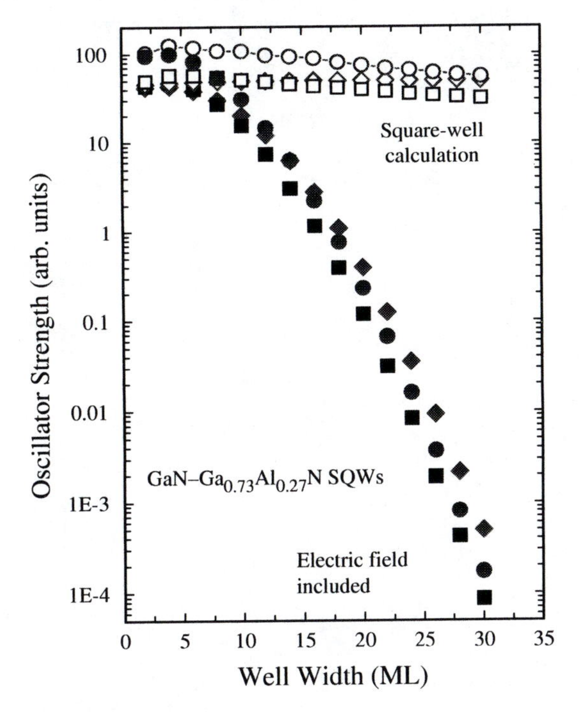

Figure 6.10 Evolution of the excitonic oscillator strength versus well width for $x = 0.27$ in the barriers. Squares and circles correspond to the 2D and 3D calculations, respectively. The overlap between electron and heavy-hole envelope functions is also plotted for comparison, using diamonds. Open symbols show the result of assuming zero electric field. Solid symbols show the result of including a realistic field of 1.1 MV/cm.

quantum confined Stark effect and (2) a drastic decrease of the excitonic oscillator strength versus well width. These points were first verified by Im et al. [24]: they observed a strong redshift, even capable of pushing the transition energy below the GaN bandgap, accompanied by a drastic increase in the PL decay time. These first experiments corresponded in fact to excitons localized at well-width fluctuations, as mentioned above. From previous studies made on GaAs–AlGaAs QWs it is now well established that the recombination dynamics of localized excitons are really different from those of excitons that are free to propagate along the well plane [70]. For instance, the decay time of bound excitons is almost temperature independent at moderate temperatures. For free excitons, a good quantum number is the wavenumber **K** of their center of mass. Their kinetic energy is a parabolic (in the simplest case) function of **K**. Therefore, given a temperature above a quantum limit, and provided that interaction times of carriers with the lattice are much faster than radiative recombination times, one can assume for those free excitons a Maxwellian (thermalized) distribution in the reciprocal space.

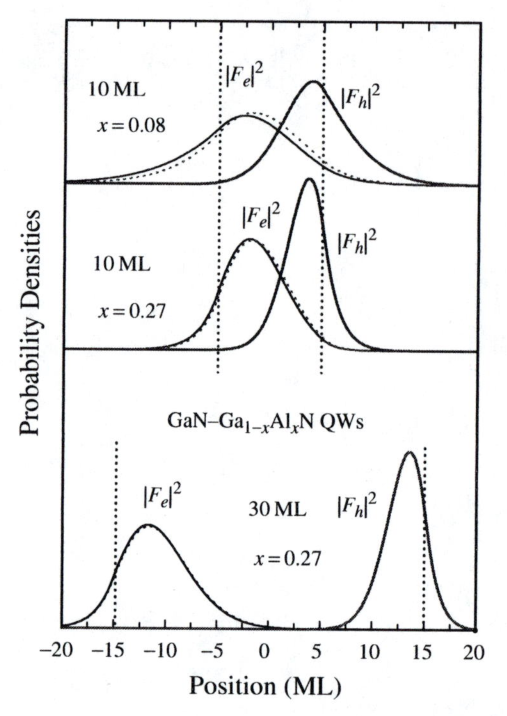

Figure 6.11 Probability densities for electrons and heavy holes in GaN/Ga$_{1-x}$Al$_x$N QWs with various widths and x values. The correction brought to the electron envelope function by the Coulomb attraction of the hole is shown by dashed lines. Clearly, in the case of wide wells with large x (thus large field) the spatial separation of the carriers is so large that the oscillator strength is extremely small.

The larger the lattice temperature, the larger the average value of **K**. Now, excitons capable of coupling efficiently with the radiation field are those having a wavenumber **K** as close as possible to that of the photon, equal to $\mathbf{k}_0$ in the vacuum. The photon is emitted in the growth direction: the exciton propagates in the growth plane. In fact, the radiative recombination rate of excitons has a pole near $\sqrt{\varepsilon}\mathbf{k}_0$. Then, the larger the temperature, the larger the average **K** assumed by excitons and the larger the average exciton lifetime. In 2D systems, one predicts a linear dependence of this exciton lifetime versus T [75]. Now, in practical cases, the effective recombination time τ depends on the radiative lifetime τ_{rad} and on the nonradiative lifetime τ_{nr} via $1/\tau = 1/\tau_{rad} + 1/\tau_{nr}$. The nonradiative time accounts for all scattering mechanisms reducing the radiative efficiency η, which can be expressed as $\eta = \tau_{nr}/(\tau_{rad} + \tau_{nr})$. In other words, the dependence of the PL intensity versus T is essentially the same as the dependence of η. In high-quality QWs, it is well known that radiative recombinations dominate the PL dynamics at $T = 2\,\mathrm{K}$, and nonradiative recombinations dominate at high temperatures.

From measurements of both the PL decay time and the PL intensity versus T, one can extract the variation of the *radiative* decay time as $\tau_{rad}(T) = \tau(T)/\eta(T)$.

Quantum mechanics allows us to relate the exciton radiative lifetimes to the oscillator strengths in QWs. The reference quantity is the radiative lifetime at the in-plane exciton–polariton bottleneck ($\mathbf{k} = \sqrt{\varepsilon}\mathbf{k}_0$): τ_0. If we use a 2D trial function for the QW exciton, from Fermi's golden rule the lifetime τ_0 is *analytically* connected to the oscillator strength by [70]: $\tau_0 = \lambda^2/(2 \cdot I_{eh}^2\sqrt{\varepsilon}k_0\omega_{LT}a_B^3)$, where ω_{LT} is the longitudinal transverse splitting of the exciton–polariton in the bulk, I_{eh} is the overlap integral of the electron and hole envelope functions, and a_B is the 3D Bohr radius. Concerning the temperature dependence of the average (thermalized) radiative lifetime, we end up with $\partial\tau/\partial T = \tau_0 \cdot (\xi M k_B/\hbar^2 k_0^2)$ with $\xi = 3$ if heavy-hole and light-hole excitonic transition energies are well split, and $\xi \cong 2$ when they are nearly degenerate. M is the total exciton mass, and k_B is Boltzmann's constant.

The experiment we show here was performed on the GaN/$Ga_{0.93}Al_{0.07}N$ MQW shown in Figure 6.4. The 2 K PL spectrum of this sample exhibits lines from both free and localized excitons. The changes versus T of the PL intensity and of the experimental decay time are shown in Figure 6.12. This offers us the oportunity to extract the temperature dependence of the radiative lifetime, as shown in Figure 6.13. From the experimental value $\partial\tau/\partial T = 20.5 \pm 0.7$ ps/K and assuming $k_0 = 1.80 \times 10^7$ m^{-1}, $n = 3$, and $M = 1.2m_0$, we obtain $\tau_0 = 2.4$ ps. This gives $\hbar\omega_{LT} = 0.6$ meV, which is in fairly good agreement with data obtained from lineshape fitting

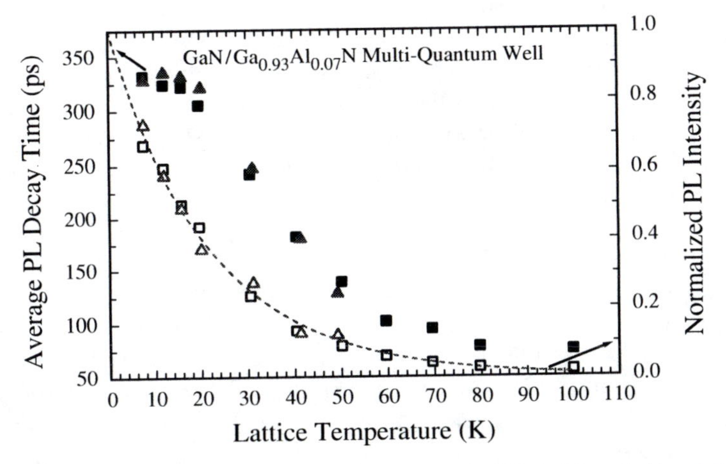

Figure 6.12 Variation of the PL decay time of excitons in the QWs (solid symbols) and of their spectrally and temporally integrated PL intensity (open symbols) versus temperature. Squares and triangles correspond to the probing of two differents spots on the sample surface.

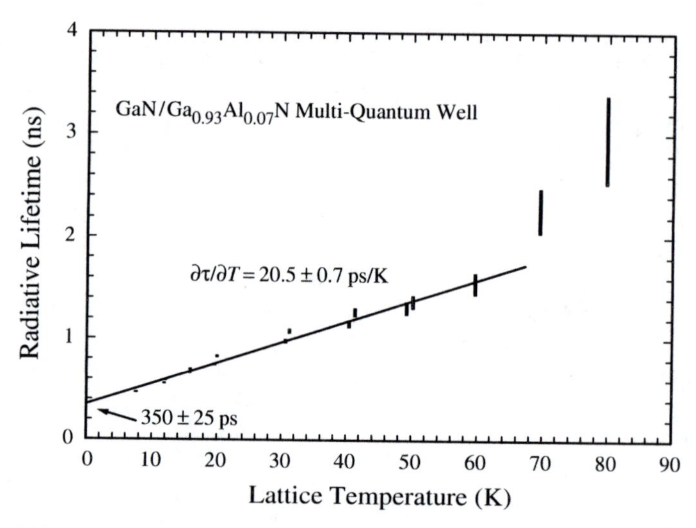

Figure 6.13 Radiative lifetime of excitons in $GaN/Ga_{0.93}Al_{0.07}N$ MQWs, deduced from the results of Figure 6.12. The line shows the linear fitting to the data. Error bars show the total inaccuracy due to both measurements of the decay time and of the time-integrated PL intensity.

of reflectance on high-quality epilayers [76]. Thus, we deduce the radiative lifetime at the bottleneck from the variation of this lifetime with temperature. One could argue that the absolute value of the decay time results from an overlap of free-exciton and bound-exciton contributions. In fact, at low temperature we have a spectral dependence of the decay time (Figure 6.14), and the asymptotic value at low temperatures in Figure 6.13 corresponds to the decay time on the high-energy wing of the PL band. Figure 6.14 shows that increasing the localization energy increases the decay time. A thermal detrapping of localized excitons can be produced experimentally by increasing the temperature, which increases the population of free excitons and decreases the population of localized excitons. Another example of this condition is shown in Figure 6.15 for several QWs for which the localization energies are different. Thermal detrapping occurs with increasing temperature, and the energy of the PL band follows the trend expected for the reflectance feature. The results in Figure 6.15 show that the PL lines from this series of samples are strictly due to localized excitons at low temperature.

Now, from the preceding equations it is clear that the slope of the radiative lifetime should vary with the thickness of the quantum well, via the evolution of the overlap integral I_{eh} and of the 2D variational Bohr radius λ. It would be interesting to check this effect, but this would require other samples showing free-exciton features.

Nonetheless, time-resolved PL measurements on the three samples in Figure 6.5 are interesting all the same. The PL decays measured at $T = 8$ K

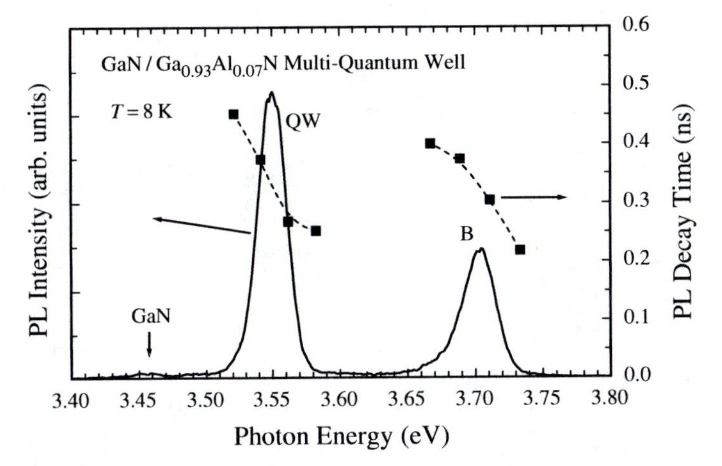

Figure 6.14 Time-integrated PL spectrum of the sample discussed in Figures 6.12 and 6.13. The spectral dependence of the decay time for the PL from the QW and from the barrier (B) is also shown by squares.

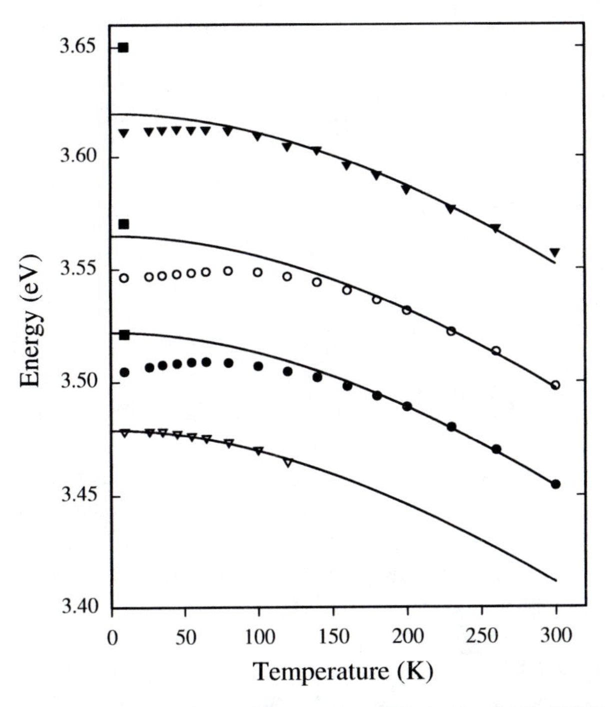

Figure 6.15 Temperature dependence of the energies of PL maxima for the MQW also shown in the middle of Figure 6.5. The solid squares are free exciton energies deduced from reflectivity measurements. The lowest curve corresponds to the GaN buffer.

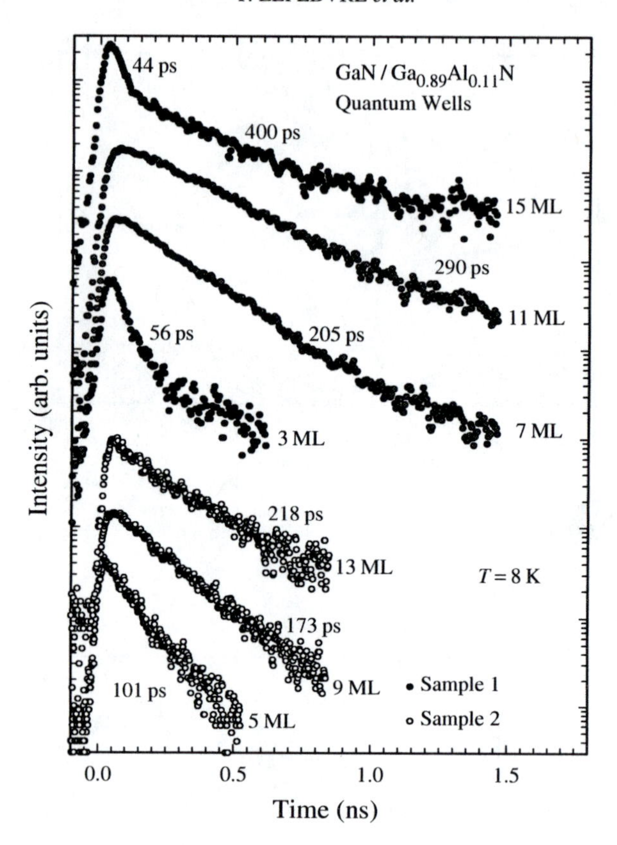

Figure 6.16 PL intensity decays for the two MQWs also shown in Figures 6.5a and b.

are displayed in Figure 6.16 for the different well widths. Although the energies of PL maxima nicely intercalate for the first two samples in Figures 6.5a and b, the PL decay times do not. There seems to be a linear dependence of these decay times versus well width, but the slopes are different for the two samples, as shown in Figure 6.17. Lefebvre et al. [33] calculated the well width dependence of the radiative lifetimes for localized excitons in these samples (see Figure 6.18) by including the electric field of 450 kV/cm, which yielded a correct fit to the transition energies. However, there is a strong discrepancy between the calculation in Figure 6.18 and the experimental trend of Figure 6.17. In fact, the experimental decay times are shorter than expected, and their increase with well width is much slower than the increase in the radiative lifetime. It seems that some efficient nonradiative process takes place *in these particular samples* embedding sequences of QWs with different widths, separated by 5-nm-wide barriers.

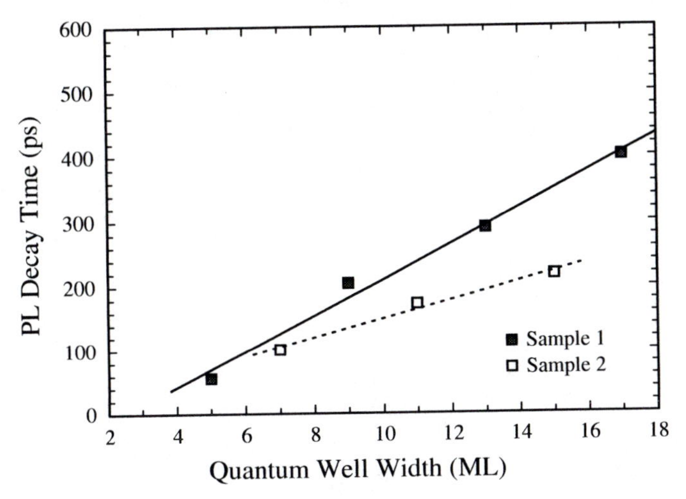

Figure 6.17 Plot of the PL decay times versus "real well widths", i.e., the nominal widths plus 2 ML, for the two samples also shown in Figures 6.5a and b.

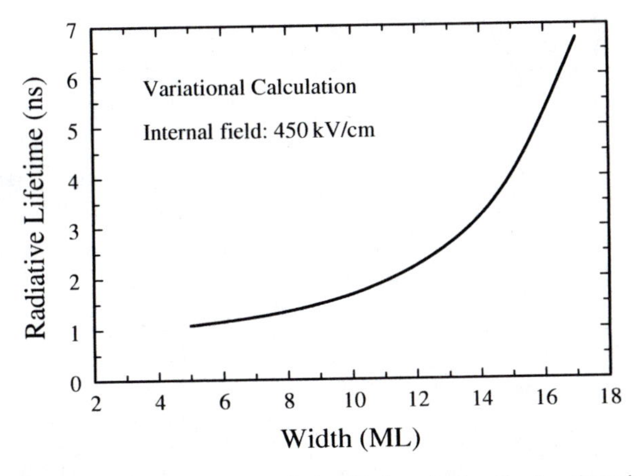

Figure 6.18 Theoretical calculation of radiative lifetimes for localized excitons in GaN QWs, with a longitudinal electric field of 450 kV/cm in the GaN layer.

We eventually understood that the decay times measured in these cases were also dependent on the escape of the carriers out of the well toward their neighbors across the barriers, in these MQW samples. The corresponding transfer time depends in a complex way on the electric field in the barrier layers and on the distance of the adjacent wider well. In particular, the values of electric fields in the barriers and in the wells depend on their respective thicknesses. Moreover, it is not clear at this stage whether the interwell transfer is merely due to the tunnel effect or if other mechanisms should be considered.

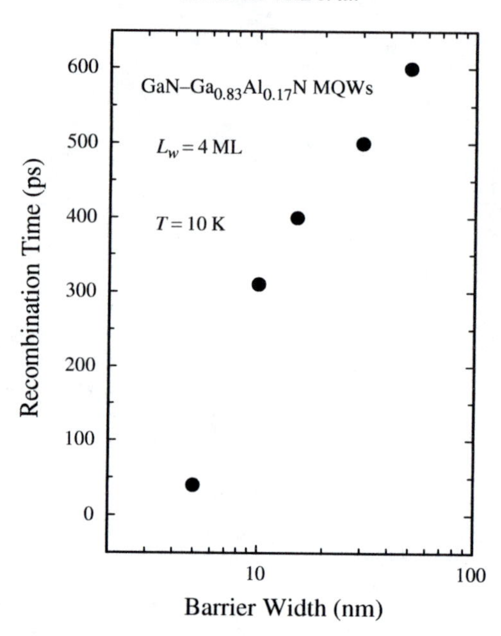

Figure 6.19 Barrier width dependence of the decay times measured for a 4-ML-wide QW in a series of samples, each of which embeds four QWs of widths 4, 8, 12, and 16 ML, respectively.

To demonstrate the existence of this effect and to study the transfer mechanisms, we measured the PL decay times for a series of samples containing four GaN/$Ga_{0.83}Al_{0.17}N$ QWs of respective nominal widths of 4, 8, 12, and 16 ML. For the different samples, only the thickness of the barrier layer was changed between 5 nm and 50 nm, thus changing the electric field in the wells from 0.6 MV/cm to 0.8 MV/cm (see Section 6.4). The general trend is increasing decay time versus barrier width, for a given well width. For example, we show in Figure 6.19 that the decay time measured at 10 K for the 4-ML-wide QWs increases strongly with barrier width. Now, for such a narrow QW, the calculation predicts almost no variation (5% at most) of the excitonic oscillator strength versus barrier width, i.e., versus the electric field in the well. Therefore, the radiative lifetime is almost the same for all the 4-ML-wide QWs presented here. Moreover, all samples were grown in the same MBE chamber and as part of the same growth series; only the barrier thicknesses were varied. Thus there is no reason to invoke any change of structural quality or defect density between the different samples. The fast increase in decay time with barrier width indicates that there is indeed a transfer of carriers toward the adjacent 8-ML-wide QW. This transfer rapidly becomes inefficient with increasing barrier width. The decay time measured for 50-nm-wide barriers should be quite close to the "true" radiative lifetime for localized excitons in such a 4-ML-wide QW.

Calculations of the escape times are really difficult, since many mechanisms certainly contribute, in addition to the tunnel effect, such as impurity-assisted transfers or accidental interwell resonances, which were impossible to assign quantitatively at the time this chapter was written. The interwell tunnel effect has been confirmed by previous obervations of inter-well optical recombinations [27,77]. Our most recent results [33] seem to indicate, however, that the barrier width dependence of the decay time is much too steep to be caused by only the tunnel effect.

6.7 CONCLUSIONS

The physics of GaN–AlGaN QWs is rapidly reaching maturity, mainly due to recent progress in growth methods and to the intensive investigations of past years on other QW systems. We have used this model system to investigate the most striking—and hopefully useful—feature of these systems, i.e., the huge built-in electric fields, and their effect on the dynamics of optical recombinations. We have listed the theoretical and numerical parameters to be included in modeling excitonic transitions in such QWs submitted to piezo- and pyroelectric effects and have discussed the origin and consequences of these effects. Nevertheless, cw and time-resolved optical spectroscopy on purposely designed samples has allowed us to check the magnitude of internal fields versus Al composition and versus well/barrier widths. Particular temporal behaviors have evidenced efficient interwell carrier transfers, even for relatively thick and high barriers. These transfers are probably related to extrinsic scattering mechanisms involving defects. Indeed, the great challenge in the near future will be the control of the intrinsic quality of the group III nitrides. Although considerable progress has been made during the last two years, nonintentional doping as well as the dislocation density remains high. One way to obviate this problem might be the intentional growth of planes of GaN quantum dots, as recently demonstrated [78], with a larger surface density than the dislocations, thus realizing, in a controlled way, what is hardly controlled for InGaN QWs.

Acknowledgments

We thank A. Hangleiter (Universität Stuttgart) for his critical reading of this manuscript and his constructive comments.

References

1. S. Nakamura and G. Fasol, eds., *The Blue Laser Diode* (Springer, Berlin, 1997).
2. B. Gil, ed., *Group III Nitride Semiconductor Compounds* (Clarendon Press, Oxford, 1998).
3. *Proc. of the 3rd Int. Conf. on Nitride Semiconductors*, Montpellier, France, July 1999, Edited by A. Hoffmann and P. Lefebvre, *Phys. Status Solidi*, **216**, 1 (1999).

4. J. S. Im, V. Härle, F. Scholz, and A. Hangleiter, *MRS Internet J. Nitride Semicond. Res.*, **1**, 37 (1996).
5. C.-K. Sun, S. Keller, G. Wang, M. S. Minsky, J. E. Bowers, and S. P. DenBaars, *Appl. Phys. Lett.*, **69**, 1936 (1996).
6. E. S. Jeon, V. Kozlov, Y.-K. Song, A. Vertikov, M. Kuball, A. V. Nurmikko, H. Liu, C. Chen, R. S. Kern, C. P. Kuo, and M. G. Craford, *Appl. Phys. Lett.*, **69**, 4194 (1996).
7. Wei Li, P. Bergman, B. Monemar, H. Amano, and I. Akasaki, *J. Appl. Phys.*, **81**, 1005 (1997).
8. C.-K. Sun, T.-L. Chiu, S. Keller, G. Wang, M. S. Minsky, S. P. DenBaars, and J. E. Bowers, *Appl. Phys. Lett.*, **71**, 425 (1997).
9. J. Allègre, P. Lefebvre, W. Knap, J. Camassel, Q. Chen, and A. Khan, *MRS Internet J. Nitride Semicond. Res.*, **2**, no. 34. (http://nsr.mij.mrs.org/2/34/); *Materials Science Forum*, **264–268**, 1295 (1998).
10. Y. Narukawa, Y. Kawakami, S. Fujita, S. Fujita, and S. Nakamura, *Phys. Rev. B*, **55**, R1938 (1997); Y. Kawakami, Y. Narukawa, K. Sawada, S. Saijyo, S. Fujita, S. Fujita, and S. Nakamura, *Mater. Sci. Eng., B*, **50**, 256 (1997).
11. M. S. Minsky, S. B. Fleischer, A. C. Abare, J. E. Bowers, E. L. Hu, S. Keller, S. P. Denbaars, *Appl. Phys. Lett.*, **72**, 1066 (1998).
12. A. Sohmer, J. Off, H. Bolay, V. Härle, V. Syganow, J. S. Im, V. Wagner, F. Adler, A. Hangleiter, A. Dörnen, F. Scholtz, D. Brunner, O. Ambacher, and H. Lakner, *MRS Internet J. Nitride Semicond. Res.*, **2**, 14 (1997) (http://nsr.mij.mrs.org/2/14/).
13. S. Chichibu, T. Azuhata, T. Sota, and S. Nakamura, *Appl. Phys. Lett.*, **69**, 4188 (1996).
14. Y. Narukawa, Y. Kawakami, M. Funato, S. Fujita, and S. Nakamura, *Appl. Phys. Lett.*, **70**, 981 (1997).
15. K. P. O'Donnell, R. W. Martin, and P. G. Middleton, *Phys. Rev. Lett.*, **82**, 237 (1999).
16. S. Nakamura, *Semicond. Sci. Technol.*, **14**, R27 (1999).
17. K. P. O'Donnell, T. Breitkopf, H. Kalt, W. Van der Stricht, I. Moerman, P. Demeester, and P. G. Middleton, *Appl. Phys. Lett.*, **70**, 1843 (1997).
18. A. Bykhovsky, B. L. Gelmont, and M. Shur, *J. Appl. Phys.*, **81**, 6332 (1997).
19. T. Takeuchi, H. Takeuchi, S. Sota, H. Sakai, H. Amano, and I. Akasaki, *Jpn. J. Appl. Phys., Part 2*, **36**, L177 (1997).
20. F. Bernardini, V. Fiorentini, and D. Vanderbilt, *Phys. Rev. B*, **56**, R10026 (1997).
21. M. B. Nardelli, K. Rapcewicz, and J. Berholc, *Phys. Rev. B*, **55**, R7323 (1997); *Appl. Phys. Lett.*, **71**, 3135 (1997).
22. P. M. Asbeck, E. T. Yu, S. S. Lau, G. J. Sullivan, J. Van Hove, and J. M. Redwing, *Electron. Lett.*, **33**, 1230 (1997).
23. E. T. Yu, G. J. Sullivan, P. M. Asbeck, C. D. Wang, D. Qiao, and S. S. Lau, *Appl. Phys. Lett.*, **71**, 2794 (1997).
24. J. S. Im, H. Kollmer, J. Off, A. Sohmer, F. Scholz, and A. Hangleiter, *Phys. Rev. B*, **57**, R9435 (1998).
25. M. Leroux, N. Grandjean, M. Laügt, J. Massies, B. Gil, P. Lefebvre, and P. Bigenwald, *Phys. Rev. B*, **58**, R13371 (1998).
26. T. Takeuchi, C. Wetzel, S. Yamaguchi, H. Sakai, H. Amano, I. Akasaki, Y. Kaneko, S. Nakagawa, Y. Yamaoka, and N. Yamada, *Appl. Phys. Lett.*, **73**, 1691 (1998).
27. B. Gil, P. Lefebvre, J. Allègre, H. Mathieu, N. Grandjean, M. Leroux, J. Massies, P. Bigenwald, and P. Christol, *Phys. Rev. B*, **59**, 10246 (1999).
28. V. Fiorentini, F. Bernardini, F. Della Sala, A. Di Carlo, and P. Lugli, *Phys. Rev. B, to be published (15 September 1999).*
29. P. Lefebvre, J. Allègre, B. Gil, H. Mathieu, N. Grandjean, M. Leroux, J. Massies, and P. Bigenwald, *Phys. Rev. B*, **59**, 15563 (1999).
30. M. Leroux, N. Grandjean, J. Massies, B. Gil, P. Lefebvre, and P. Bigenwald, *Phys. Rev. B*, **60**, 1496 (1999).
31. A. Hangleiter, Jin Seo Im, H. Kollmer, S. Heppel, J. Off, and F. Scholz, *MRS Internet J. Nitride Semicond. Res.*, **3**, no. 15. (http://nsr.mij.mrs.org/3/15/).
32. N. Grandjean, J. Massies, and M. Leroux, *Appl. Phys. Lett.*, **74**, 2361 (1999).
33. P. Lefebvre, M. Gallart, T. Taliercio, B. Gi, J. Allègre, H. Mathieu, N. Grandjean, M. Leroux, J. Massies, and P. Bigenwald, in Ref. 3, page ??

34. O. Ambacher, J. Smart, J. R. Shealy, N. G. Weimann, K. Chu, M. Murphy, W. J. Schaff, L. F. Eastman, R. Dimitrov, L. Wittmer, M. Stutzmann, W. Rieger, and J. Hilsenbeck, *J. Appl. Phys.*, **85**, 3222 (1999).
35. G. Bastard, *Wave Mechanics Applied to Semiconductor Heterostructures* (Les Editions de Physique, Paris 1988).
36. O. Krebs and P. Voisin, *Phys. Rev. Lett.*, **77**, 1829 (1996).
37. Y. El Khalifi, P. Lefebvre, J. Allègre, B. Gil, H. Mathieu, and T. Fukunaga, *Solid State Commun.*, **75**, 677 (1990).
38. E. O. Kane, *J. Phys. Chem. Phys. Solids*, **1**, 249 (1957).
39. G. Bastard, E. E. Mendez, L. L. Chang, and L. Esaki, *Phys. Rev. B*, **26**, 1974 (1982).
40. R. L. Greene, K. K. Bajaj, and D. E. Phelps, *Phys. Rev. B*, **29**, 1807 (1984).
41. G. D. Sanders and Y. C. Chang, *Phys. Rev. B*, **32**, 5517 (1985).
42. G. Peter, E. Deleporte, G. Bastard, J. M. Berroir, C. Delalande, B. Gil, J. M. Hong, and L. L. Chang, *J. Lumin.*, **52**, 147 (1992).
43. A. Bellabchara, P. Lefebvre, P. Christol, and H. Mathieu, *Phys. Rev. B*, **50**, 11840 (1994).
44. P. Bigenwald and B. Gil, *Solid State Commun.*, **91**, 33 (1994); *Phys. Rev. B*, **51**, 9780 (1995).
45. B. Gil and P. Bigenwald, *Solid State Commun.*, **94**, 883 (1995).
46. P. Lefebvre, P. Christol, and H. Mathieu, *Phys. Rev. B*, **48**, 17308 (1993).
47. C. Tanguy, P. Lefebvre, H. Mathieu, and R.J. Elliott, *J. Appl. Phys.*, **82**, 798 (1997) and references therein.
48. L. Schulteis and K. Ploog, *Phys. Rev. B*, **29**, 7058 (1984).
49. H. Tuffigo, R. T. Cox, G. Lentz, N. Magnea, and H. Mariette, *J. Crystal Growth*, **101**, 778 (1990).
50. P. Lefebvre, V. Calvo, N. Magnea, T. Taliercio, J. Allègre, and H. Mathieu, *Phys. Rev. B*, **56**, R10040 (1997).
51. P. Lefebvre, B. Gil, H. Mathieu, and R. Planel, *Phys. Rev. B*, **39**, 5550 (1989).
52. D. L. Smith and C. Mailhiot, *Rev. Mod. Phys.*, **62**, 173 (1990).
53. J. Y. Marzin, J. M. Gérard, A. Izrael, D. Barnier and G. Bastard, *Phys. Rev. Lett.*, **73**, 716 (1994).
54. N. Grandjean et al. This book.
55. H. Morkoç, *Nitride Semiconductors and Devices* (Springer-Verlag, Heidelberg, 1999).
56. B. Gil, *Semiconductors and Semi-metals*, edited by J. I. Pankove and T. D. Moutsakas (Wiley, New York), 1999 Vol. 57, Chap. 6, p. 209.
57. A. Alemu, B. Gil, M. Julier, and S. Nakamura, *Phys. Rev. B*, **57**, 3761 (1998).
58. P. Boring, K. J. Moore, P. Bigenwald, B. Gil, and K. Woodbridge, *J. Phys.* (France) IV, Vol. 3, colloque **C5**, p. 249 (1993).
59. P. Boring, B. Gil, and K. J. Moore, *Phys. Rev. Lett.*, **71**, 1875 (1993).
60. R. André, J. Cibert, and Le Si Dang, *Phys. Rev. B*, **52**, 12013 (1995).
61. M. Ravaille, *Electrostatique Electrocinétique* (Baillière, Paris, 1964).
62. D. Vanderbilt and R. D. King-Smith, *Phys. Rev. B*, **48**, 4442 (1993).
63. O. Gfrörer, J. Off, F. Scholz, and A. Hangleiter, in Ref. 3 page ??
64. G. Bastard, C. Delalande, M. H. Meynadier, P. M. Frijlink, and M. Voos, *Phys. Rev. B*, **29**, 7042 (1984).
65. C. Delalande, M. H. Meynadier, and M. Voos, *Phys. Rev. B*, **31**, 2497 (1985).
66. M. Colocci, M. Gurioli, and J. Martinez-Pastor, *J. Phys.*, (France) **C5**, Vol. 3, 3 (1993).
67. D. S. Citrin, *Phys. Rev. B*, **47**, 3832 (1993).
68. A. V. Kavokin, *Phys. Rev. B*, **50**, 8000 (1994).
69. E. L. Ivchenko, A. V. Kavokin, *Sov. Phys. Semicond.*, **25**, 1070 (1991).
70. P. Lefebvre, J. Allègre, B. Gil, A. Kavokin, H. Mathieu, H. Morkoç, W. Kim, A. Salvador, A. Botchkarev, *Phys. Rev. B*, **57**, R9447 (1998).
71. M. Suzuki and T. Uenoyama, in Ref. 2, p. 307.
72. T. Ochalski, B. Gil, P. Lefebvre, N. Grandjean, M. Leroux, J. Massies, S. Nakamura, and H. Morkoç, *Appl. Phys. Lett.*, **74**, 3353 (1999).
73. S. R. Lee, A. F. Wright, M. H. Crawford, G. A. Petersen, J. Han, and R. M. Biefeld, *Appl. Phys. Lett.*, **74**, 3344 (1999).
74. M. Kumagai, S. L. Chuang, and H. Ando, *Phys. Rev. B*, **57**, 15302 (1998).

75. L. C. Andreani, *Solid State Commun.*, **77**, 641 (1991).
76. A. Hoffmann, *Mater. Sci. Eng. B*, **43**, 185 (1997).
77. H. Kollmer, Jin Seo Im, S. Heppel, J. Off, F. Scholz, and A. Hangleiter, *Appl. Phys. Lett.*, **74**, 82 (1999).
78. B. Damilano, N. Grandjean, F. Semond, J. Massies, and M. Leroux, *Appl. Phys. Lett.*, **75**, 962 (1999).

CHAPTER 7

Characterization of GaN and Related Nitrides by Raman Scattering

H. HARIMA

Department of Applied Physics, Osaka University, Suita, Osaka 565-0871, Japan
harima@ap.eng.osaka-u.ac.jp

7.1 INTRODUCTION

Raman scattering is one of the standard optical characterization techniques for both lattice and electronic properties of semiconductor materials. Compared with other standard spectroscopic techniques, Raman scattering characterization has various merits; it is in principle nondestructive, and contactless, and it requires no special sample preparation such as thinning or polishing. When a standard Raman microscope is employed with common visible lasers, the probe laser beam is focused to a diameter of $\sim 1\,\mu m$, which roughly gives the lateral resolution of the measurement. Since GaN and related compounds including their alloys (hereafter referred to as the nitrides) are in many cases transparent to common visible lasers, the depth resolution is somewhat worse than the lateral one; however, a depth resolution similar to the lateral one can be obtained if a suitable optical setup such as confocal arrangement is used.

Although material transparency lowers depth resolution in general, it does produce a large scattering volume. In other words, an intense Raman signal is easily obtained. As is now well recognized by the recent rapid development in blue/green lasers, the nitrides are relatively rugged compared with other III–V compounds or II–VI compounds. Such inherent material ruggedness is advantageous in sample treatment. For example, it is possible to use intense laser irradiation in Raman microprobe excitation without causing serious thermal damage.

If the incident photon energy exceeds the bandgap of the nitrides, e.g., if hexagonal GaN is excited with UV lasers with photon energy above 3.4 eV ($\sim$ 360 nm wavelength), a very different situation results: because of efficient creation of photocarriers by interband transition, a resonant enhancement of the Raman signal is induced. This is an advantage for local analysis because the light scattering volume is limited. For example, a resonant Raman study introduced in this chapter probed an InGaN QW layer only a few nanometers wide sandwiched by much thicker barrier layers.

If the surface layer undergoes resonant excitation, the near-surface layer can easily be probed compared with visible laser excitation because the laser penetration depth is greatly reduced. A disadvantage of resonant excitation is that because the nitrides are generally very luminescent materials, as already well known, weak Raman signals are hindered by the luminescence. This difficulty can be overcome however, by carefully selecting the incident laser energy so as to observe Raman signals in the frequency range far from the luminescence peak.

Owing to these advantages and recent rapid progress in sample preparation technology for high-quality materials, characterization of basic physical parameters in the nitrides is now steadily progressing, especially for the lattice or phonon properties; most of the Raman studies reported so far on the nitrides belong to this category. Typical applications of phonon studies include characterization of strain in epilayers, and of alloy composition and homogeneity. Phonons coupled with collective oscillation of free carriers have also been intensively studied and widely used for evaluation of free-carrier density and mobility.

In this chapter we review the present status of Raman scattering experiments on the nitrides by classifying them into those for the lattice and the electronic properties. Important parameters such as phonon frequencies reported in recent papers are tabulated. In particular, characteristic behavior of phonon frequencies in ternary compounds is described. As we shall see, however, Raman studies on the electronic properties including band structures, impurity levels, and carrier scattering processes are now at the starting point. Rapid development in such work is anticipated for characterization of future nitride devices.

7.2 LATTICE PROPERTIES

7.2.1 Introduction

Raman scattering provides various types of information on the lattice properties of crystalline solids: bond strength from the phonon frequency, lattice strain from the phonon-frequency shift, and structural disorder from the broadening of phonon peaks, and so on. In ternary compounds, phonon frequency is also a good measure of atomic composition. Structural defects or impurity atoms may yield weak extra signals in Raman spectra called local vibrational modes. Most of these subjects have already been examined in GaN, AlN, InN and related alloys, as described in the following sections.

Section 7.2.2 describes observations of phonon-mode spectra by Raman scattering for the basic binary compounds GaN, AlN, and InN. As for the most common material, hexagonal GaN, precise phonon frequencies at the Γ-point in the Brillouin zone are known to a precision of $1\,\mathrm{cm}^{-1}$, and the values seem to have gained a consensus. Other compounds have been less intensively investigated. For cubic AlN, and hexagonal and cubic InN, the data are especially limited; thus, reported phonon frequencies for these nitrides have larger experimental uncertainties than those for hexagonal GaN. Nevertheless, important phonon frequencies for these basic binary compounds, both in hexagonal and cubic phases, have been lined up very recently. Precise phonon spectra have been measured for these binary compounds, which give second-order phonon Raman scattering. The data present a critical test for calculated phonon density of states or phonon dispersion curves.

Section 7.2.3 describes phonon-frequency measurements in alloys. If the frequencies are plotted against the atomic composition, some phonon branches show complex behavior: they may show bowing like the dependence of bandgap on atomic composition, or they may show splitting into different branches, called two-mode behavior. It is shown that such behavior depends critically on cation mass and size and on the ionicity of the atomic bonds. The results give interesting bases for considering phonon dynamics of alloys.

Section 7.2.4 describes stress-induced effects on phonon spectra. In most cases the nitride samples are grown by heteroepitaxy; therefore, stress effects induced by the difference in lattice constants and/or thermal expansion coefficients between the layer in question and the neighboring layers or substrates are inevitable. For example, a hexagonal GaN layer grown by MOCVD on a sapphire (0001) plane generally has large compressive strains, whereas a layer grown on a SiC substrate may have small tensile strains. Stress effect is an important issue that critically affects, e.g., photoemission spectra of LEDs with QW structures. In terms of alloys, if the atomic composition in a ternary compound is estimated from a phonon frequency, stress effects that induce additional frequency shift will have to be considered. Depending on the direction of stresses, phonon peaks may shift and undergo splitting. This section treats only stress-induced shifts, since no experimental reports on the splitting have been reported for the nitrides.

Section 7.2.5 describes Raman studies at high temperatures. Heating induces effects similar to those produced by stress; namely, phonon peaks shift and show spectral broadening. High-temperature experiments clearly reveal anharmonic terms involved in the phonon Hamiltonian. The data are also conveniently used for practical applications such as monitoring of temperature in nitride devices under operation.

Section 7.2.6 describes Raman studies on local vibrational modes. When light impurities occupy a host lattice site in a crystal, lattice vibrations localized to the vicinity of impurity atoms may be induced. Defects, and interstitial atoms and their complexes may also produce additional phonon Raman peaks. In this section, Raman spectra from Mg-doped GaN layers are highlighted, because the roles of hydrogen impurities in the activation process of acceptors are of interest both from a physical and an application standpoint. Ion implantation will be a standard doping technique in future nitride devices, so characterization of implantation damage is also important. Raman scattering can contribute such information.

7.2.2 Phonon Modes of AlN, GaN, and InN

7.2.2.1 First-Order Spectra

The group III nitrides crystallize in two structural modifications; 2H- (or hexagonal in wurtzite) structure with space group C_{6v}^4, and $3C$- (or cubic in zinc blende) structure with space group T_d^2. In the case of GaN, for example, its primitive unit cell in the wurtzite structure consists of two Ga–N atom pairs, whereas that in the zinc-blende type consists of one Ga–N atom pair. In both cases, an atom of one kind is tetrahedrally surrounded by atoms of the other species, so that they have the same nearest-neighbor surroundings. The 3C and 2H modifications differ only in the stacking sequence of Ga–N bilayers along the cubic [111] or the hexagonal [0001] direction, i.e., the stacking order is ABCABC... in the cubic phase, and ABAB... in the hexagonal phase. Here, A, B, and C denote the allowed sites in the close-packed layer of spheres as usually employed [1]. The unit cell length of the cubic phase along [111] is given by the width of one unit bilayer, whereas that of the hexagonal structure along [0001] is twice as large. Therefore, as shown in Figure 7.1, the phonon dispersion in the 2H phase along [0001] ($\Gamma \rightarrow$ A) in the Brillouin zone is approximated by folding back the phonon dispersion along [111] ($\Gamma \rightarrow$ L) in the 3C phase. By this folding, the TO phonon mode at the L-point of the cubic Brillouin zone reduces to the wurtzite E_2 mode at the Γ-point. This mode is denoted here as E_2^H, the higher-frequency E_2 mode. There is another lower-frequency E_2 mode, labeled E_2^L. In the wurtzite structure, due to the anisotropy in the macroscopic electric field induced by the polar phonons, the TO mode splits into the A_1(TO) and E_1(TO) modes, and the LO mode also into the A_l(LO) and E_l(LO) modes. This splitting is not shown in Figure 7.1.

The first-order phonon Raman spectra present signals of phonon modes near $k = 0$ (Γ-point) because of the momentum conservation rule in the light scattering process. In the wurtzite structure, group theory predicts eight sets of phonon normal modes at the Γ-point, $2A_1 + 2E_1 + 2B_1 + 2E_2$. Among them, one set of A_1 and E_1 modes are acoustic, and the remaining six modes, $A_1 + E_1 + 2B_1 + 2E_2$, are optic. As shown in Figure 7.1, one A_1 and one B_1 mode (denoted B_1^H) derive from a singly degenerate LO-mode branch in the cubic system by zone folding, and one E_1 and one E_2 mode (E_2^H) derive from a doubly degenerate TO mode in the cubic system.

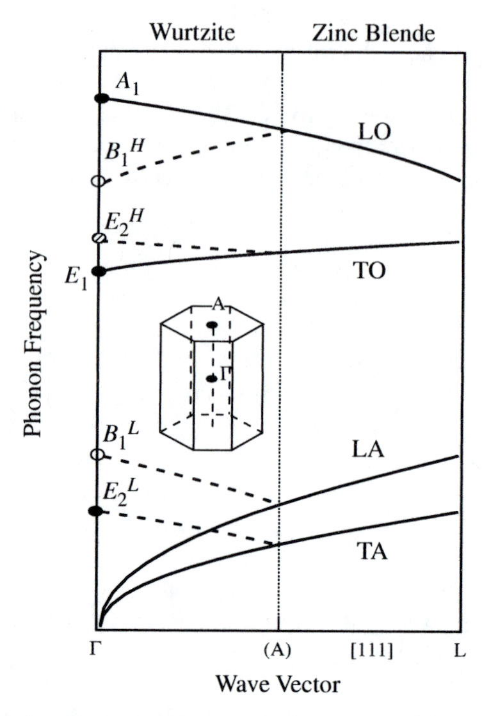

Figure 7.1 Schematic representation of phonon dispersion. Phonon branches along [111] in the zinc-blende structure are folded back, approximating those of the wurtzite structure along [0001].

Vibrational schemes for these optic modes in hexagonal systems are shown in Figure 7.2. Here, the A_1 and E_1 modes are both Raman and IR active, whereas the two E_2 modes are only Raman active, and the two B_1 modes are both Raman and IR inactive (silent modes) [2].

Since the Raman tensor for the A_1 mode has only diagonal components [3], it can be observed when incident and scattered light is parallel polarized. The A_1(LO) phonon mode has atomic displacements parallel to the c-axis and propagates along the c-axis. Therefore, if the z-direction is taken along the c-axis and x- and y-directions vertical to the c-axis, the A_1(LO) mode can be observed by backscattering from the c-plane of wurtzite samples with $z(x, x)-z$ scattering configuration. Here, conventional notation is employed for the scattering geometry; outside the parentheses, the symbols show, from left to right, the direction of incident and scattered light, respectively, and inside the parentheses, from left to right, they give the polarization direction of the incident and scattered light, respectively. Table 7.1 summarizes the scattering geometries in which Raman-active modes are observed in wurtzite systems. In contrast, Raman selection rules for zinc-blende–type crystals are relatively simple; if backscattering geometry is employed with

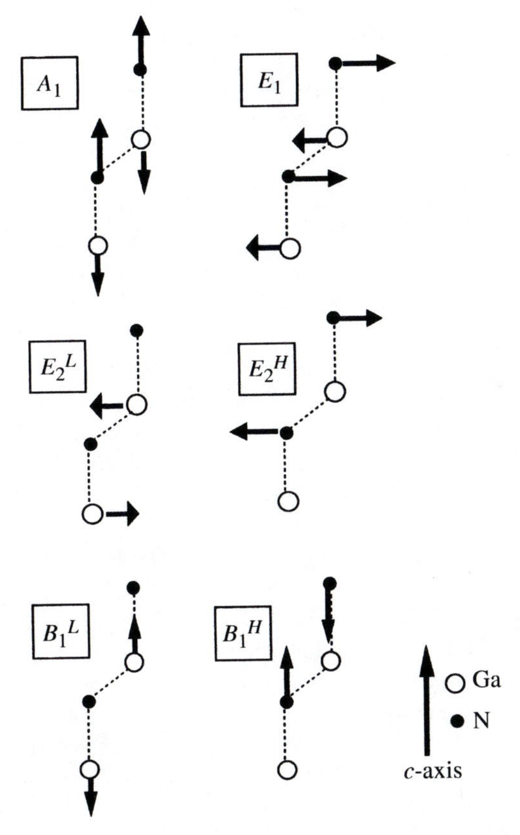

Figure 7.2 Optical phonon modes in the wurtzite structure. There are two types of E_2 and B_1 modes, which are distinguished in this chapter by superscripts L and H.

no polarization detection, only the LO phonon is observable from the (100) plane, and only the TO phonon is observable from the (110) plane. From the (111) plane, TO and LO phonons can be observed simultaneously [2].

Figure 7.3 shows typical phonon spectra of hexagonal GaN observed by a Raman microscope at room temperature [4]. The samples were grown a few micrometers thick on sapphire (0001) substrate by MOCVD. Different scattering geometries were tested, as shown schematically in the inset; the upper figure shows backscattering geometry from the c-plane, where two E_2 modes (E_2^L, E_2^H) and one A_1(LO) mode are observed. Among the phonon modes, the higher-frequency E_2 mode (E_2^H) gives the strongest signal, which is common to the hexagonal compounds InN and AlN. The lower figure shows backscattering spectra from the cross section of the epilayer, where A_1(TO) and E_1(TO) modes produce dominant features. The observed phonon frequencies are listed in Table 7.2 together with the

Table 7.1 Raman configurations for the allowed modes in wurtzite nitrides.

Mode	*Configuration*			
A_1(TO)	$x(y, y) - x$	$x(z, z) - x$		
A_1(LO)	$z(x, x) - z$			
E_1(TO)	$x(y, z) - x$	$x(y, z)y$		
E_1(LO)	$x(y, z)y$			
E_2	$z(y, y) - z$	$z(x, y) - z$	$x(y, y) - x$	$x(y, y)z$

values for other nitrides. It is readily noticed that the lower-frequency E_2 modes (E_2^L) give a fairly narrow linewidth. The intrinsic width would be less than $\sim 0.3\,\text{cm}^{-1}$ in good-quality crystals. This is common to the hexagonal compounds AlN and InN. The narrow linewidth means a longer lifetime for this phonon compared with other higher-frequency optic modes [5]. Such a long lifetime suggests small anharmonic decay channels for low-frequency phonons. (See Section 7.2.5 for anharmonic interactions in the phonon Hamiltonian).

Although the A_1(TO) and E_2^H modes should be forbidden in the scattering geometry in the bottom half of Figure 7.3, $x(z, y) - x$, they are actually observed. The E_1(LO) mode is also weakly observed in a forbidden geometry $x(z, z) - x$ (denoted here as quasi-E_1(LO) mode). There are two possible factors contributing to this "leakage effect"; first, the objective lens of a Raman microscope allows a large solid angle of radiation for excitation and signal collection. Such an experimental setup usually relaxes the Raman selection rule. Second, hexagonal GaN belongs to a system in which long-range electrostatic force predominates over the crystalline anisotropy [3], thus, mixing of the A_1 and E_1 modes easily occurs for probe laser incidence that is not strictly parallel to the optical axis or to the basal plane. These mixed modes are called quasi-TO or -LO modes [6–8]. The predominance of long-range electrostatic force is easily confirmed by comparing in Table 7.2 the TO–LO splitting (about 180–200 cm^{-1}) with the separation between the A_1 and E_1 modes (7–30 cm^{-1}). Since wurtzite nitride is a uniaxial crystal, purely transverse or longitudinal phonons with A_1 or E_1 symmetry can be observed only when the phonon propagation direction is parallel to the c-axis or lies in the c-plane. For other propagation directions, the quasi-TO and -LO modes take some intermediate frequency between the A_1 and E_1 mode because of the mixing [3];

$$\omega_Q^2(\text{TO}) = \omega^2(E_1[\text{TO}])\cos^2\theta + \omega^2(A_1[\text{TO}])\sin^2\theta \tag{7.1}$$

$$\omega_Q^2(\text{LO}) = \omega^2(A_1[\text{LO}])\cos^2\theta + \omega^2(E_1[\text{LO}])\sin^2\theta \tag{7.2}$$

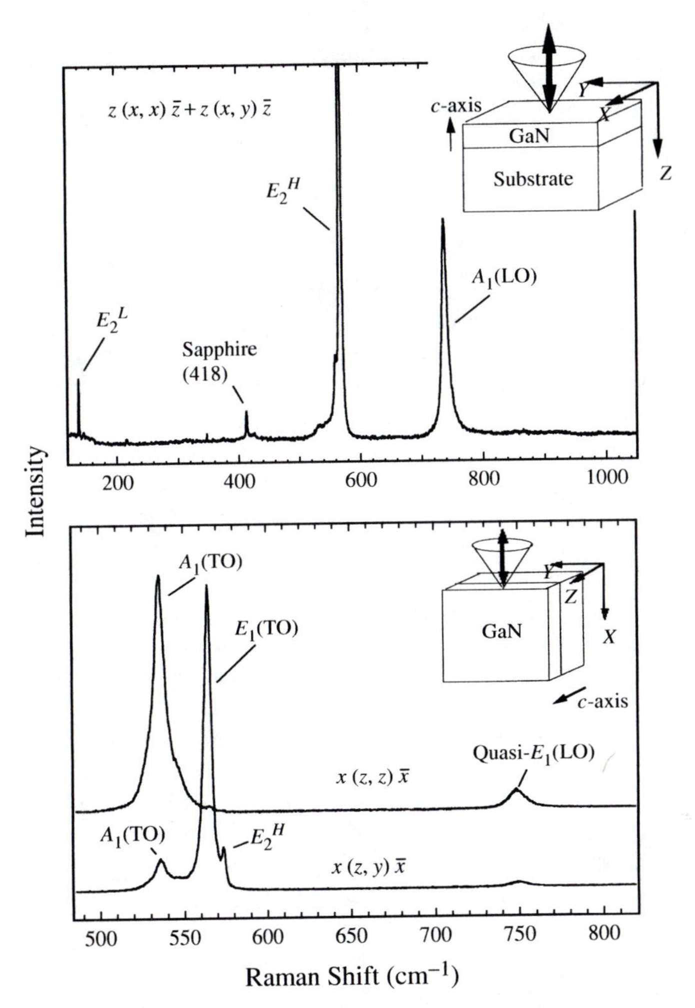

Figure 7.3 Typical Raman spectra in hexagonal GaN observed at different scattering geometries [4]. The inset shows schematically the direction of incident and scattered light for GaN epitaxial layers grown on a substrate with the c-axis perpendicular to the growing surface.

where θ is the angle between the phonon propagation direction and the c-axis. This relation was confirmed in nitrides by a Raman study on hexagonal AlN [9]. The quasi-E_1 (LO) mode seen in Figure 7.3 almost agrees in frequency with that of pure E_1(LO) mode (741 cm^{-1}). This means that θ is effectively $\sim 90°$ in Eq. (7.2), as expected from the employed geometry.

The Γ-point phonon frequencies have been intensively studied by Raman scattering on GaN and AlN [6,7,10–23], but to a limited extent on InN [24,25]. Table 7.2 lists some representative values selected from recent

Table 7.2 Observed phonon frequencies at 300 K (cm^{-1}).

	AlN		*GaN*			*InN*
Hexagonal						
E_2^L	248.6	247.9	144 ± 1	144.0	143.5 ± 0.5	87
A_1(TO)	611.0	610.0	531 ± 2	531.8	531.0 ± 0.5	447
E_1(TO)	670.8	670.0	557 ± 1	558.8	558.6 ± 0.5	476
E_2^H	657.4	656.3	567.3 ± 0.1	567.6	567.7 ± 0.5	488
A_1(LO)	890.0	889.1	735 ± 1	734.0	734.2 ± 0.5	586
E_1(LO)	912.0	912.3	742 ± 3	741.0	741.0 ± 0.5	593
	Ref. 23	Ref. 26	Ref. 47	Ref. 23	Ref. 4	Ref. 25
Cubic						
TO	655		555		555	457
LO	902		740		741	588
	Ref. 27		Ref. 21		Ref. 22	Ref. 25

Raman experiments at 300 K. It should be noted that some phonon frequencies in the table were obtained from thin epitaxial layers, which means that the values are more or less affected by residual strains (see Section 7.2.4). We briefly describe here the sample preparation; as for hexagonal GaN, relatively thick samples were selected; Raman data from bulk single crystals grown by a high-pressure technique [17], from bulk single crystals grown on quartz by a sublimation method [4], and from 50–70 μm-thick layers grown by hydride vapor-phase epitaxy (HVPE) on sapphire substrates [23], are listed here. All the data show good agreement. The peak position frequencies of the LO-phonon modes are sensitive to the presence of free carriers, which cause a blue-shift to the phonon-modes frequencies due to plasmon; however, the listed LO frequencies [4] were obtained for pure samples with a carrier density of $<1 \times 10^{17}$ cm^{-3}, and the frequency shift due to plasmon coupling was considered to be less than 1 cm^{-1}. Since the high-frequency E_2 mode (E_2^H) in hexagonal GaN is often used to calibrate residual stresses in epitaxial layers (see Section 7.2.4), its precise value in stress-free samples is expected. Careful inspection of the published data suggests that the value 567.5 ± 0.5 cm^{-1} is most reasonable and can be recommended at present. The data on cubic GaN were obtained independently by Siegle et al. [21] and Tabata et al. [22] from epitaxial layers grown by MBE on GaAs (100). Siegle et al. [21] tested a 0.6-μm-thick epilayer. The results show good agreement. For Hexagonal AlN, data from 5–7-μm-thick layers grown by HVPE on sapphire substrates [23], and those from bulk needle crystals [26] are cited. Both sets of data agree within 1 cm^{-1}. The data for cubic (c-) AlN were obtained from a sandwiched layer grown by MBE on Si (001) substrate; c-GaN (3 nm)/c-AlN (0.6 μm)/c-GaN (150 nm)/3C–SiC (10–20 μm)/Si (100) [27]. Precise InN-phonon frequencies were recently reported by Raman scattering: Davydov et al. measured hexagonal InN layer grown by MBE on sapphire (0.1–0.7 μm thick) [24], and Tabata et al. measured cubic InN grown by MBE on GaAs [25].

Table 7.2 shows a systematic decrease in phonon frequency from AlN to InN for all modes. In a simple model assuming a common force constant for the oscillating atomic pairs, phonon frequencies should be proportional to the inverse square root of the reduced mass (9.2, 11.7, and 12.5 amu for AlN, GaN, and InN, respectively). Thus, phonon frequencies should have the ratio 1.13:1:0.97 for AlN/GaN/InN. All the phonon modes in Table 7.2 give ratios close to this: (1.15–1.22):1:(0.8–0.85) for both hexagonal and cubic systems (only the E_2^L mode gives exceptional value). The small deviation between these ratios may be attributed to the difference in the force constant, or in other word, the bonding nature or polarity.

7.2.2.2 *Second-Order Spectra*

To extract information on phonon energies at non-Γ- ($k \neq 0$) points in the Brillouin zone, we usually rely on neutron-scattering experiments; however, there are no such reports on the nitrides probably because large enough single crystals are rarely available. A less direct, but a very convenient, method for probing such phonon frequencies is to analyze the second-order phonon Raman spectra. In that case, the momentum conservation rule in the light-scattering process demands that summation of the phonon wave vectors involved in the scattering process be nearly zero; $k_1 + k_2 = 0$. Thus, all phonons throughout the Brillouin zone can be Raman active in principle. Actually, however, phonons at k-vectors with large density of states can give notable contributions to the spectra. Therefore, the experiment can be a critical test for calculated phonon density of states at high symmetry points in the Brillouin zone. If highly disordered crystals are treated, the momentum conservation rules is greatly relaxed. Then, the spectra can be a direct measure of phonon density of states [2,28]. Thus, they can also be used as a critical test of calculated phonon dispersion curves [23,24,29–31].

Typical second-order phonon spectra from hexagonal (or α-) and cubic (β-) GaN are shown in Figure 7.4 [30]. The spectra were recorded at room temperature by backscattering from the grown surface. Besides the strong first-order modes at 569 cm^{-1} (E_2^H) and 735 cm^{-1} [A_1(LO)] in the hexagonal sample and those at 555 cm^{-1} (TO) and 742 cm^{-1} (LO) in the cubic material, rich structures due to second-order Raman scattering process can be seen. Siegle et al. made a group-theoretical analysis for the second-order Raman selection rule in hexagonal systems and assigned the observed

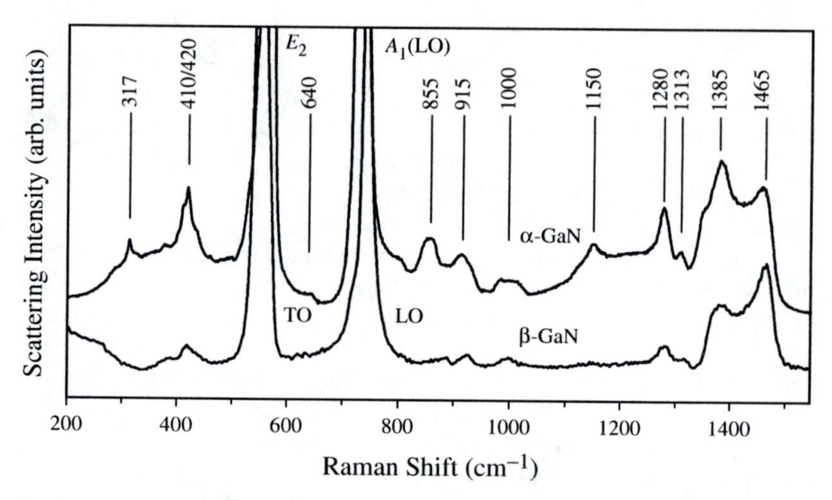

Figure 7.4 Second-order Raman spectra of hexagonal (upper) and cubic (lower) GaN grown on (0001) sapphire and (001) GaAs. The spectra were recorded at room temperature in backscattering from the grown surface [30].

structures [30]. The results are summarized as follows: (i) Overtones always contain the representation A_1, whereas combinations of phonon states belonging to different irreducible representations never contain A_1. (ii) More overtones or combinations become allowed with decreasing symmetry of the point in the Brillouin zone considered. (iii) A consequence of the different packing sequence of hexagonal GaN compared with cubic material is halving of the Brillouin-zone dimension in the cubic [111] direction, as already described. Siegle et al. made the following assignments based on this rule: The modes in hexagonal GaN at 317–420 cm^{-1} are acoustic overtones, the modes at 855–1000 cm^{-1} are combinations of optic and acoustic branches, and other higher-frequency modes at 1150–1465 cm^{-1} are optic–phonon combinations or overtones. The highest-energy distribution of the second-order signal cutoff at 1495 cm^{-1} is attributed to an overtone of the zone-center E_1(LO) mode.

Murugkar et al. also observed second-order phonon Raman spectra for hexagonal GaN [32], and more recently, Davydov et al. reported precise second-order spectra for hexagonal GaN and AlN [23].

7.2.3 Phonon Modes of Alloys

7.2.3.1 $Al_xGa_{1-x}N$

We look first at typical Raman spectra from a hexagonal $Al_xGa_{1-x}N$ alloy. Figure 7.5 shows the variation in the phonon spectra for Al content $x = 0$

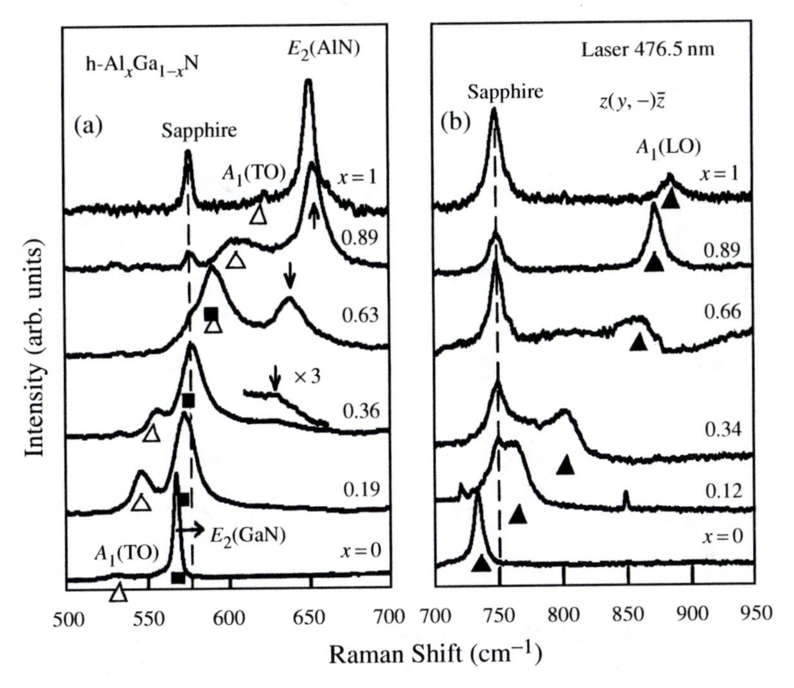

Figure 7.5 Raman spectra of hexagonal $Al_xGa_{1-x}N$ alloy for the (a) TO- and (b) LO-phonon region [33]. Phonon modes of A_1(TO) (open triangle), E_2^H (square and arrow), and A_1(LO) (closed triangle) shift to higher frequency as the Al content x increases from bottom (GaN) to top (AlN). Dashed lines denote the phonon frequencies of the sapphire substrate.

to 1 [33]. The spectra were obtained in the backscattering geometry $z(x, x) - z + z(x, y) - z$, and the TO- and LO-phonon regions are depicted in Figure 7.5a and b, respectively. The A_1(TO) mode (open triangle), which should be Raman forbidden in this scattering geometry, was observed because of a leakage effect in Raman microprobing. This figure shows that the A_1(TO) (open triangle), E_2^H (square and arrow), and A_1(LO) (closed triangle) modes shift to higher frequency as x increases from $x = 0$ (GaN, bottom) to $x = 1$ (AlN, top). The intensity of the GaN-type E_2^H mode (square) gradually decreases with increasing x, and a new mode appears at $x = 0.36$ as a broad peak at 629 cm^{-1} [see the arrow in the third spectrum from the bottom in Figure 7.5a]. The E_2^H-mode signal is enhanced when x is further increased, and finally is linked to the AlN-type E_2^H mode. This shows coexistence of two kinds of E_2^H mode at an intermediate range of alloy composition—one characteristic of GaN and the other characteristic of AlN. Thus, the E_2^H mode exhibits a two-mode–type behavior. In contrast, the A_1(LO)-phonon mode (closed triangle) shifts monotonously from GaN to AlN with increasing x, exhibiting a one-mode–type behavior. No AlN localized mode was observed in this spectral range.

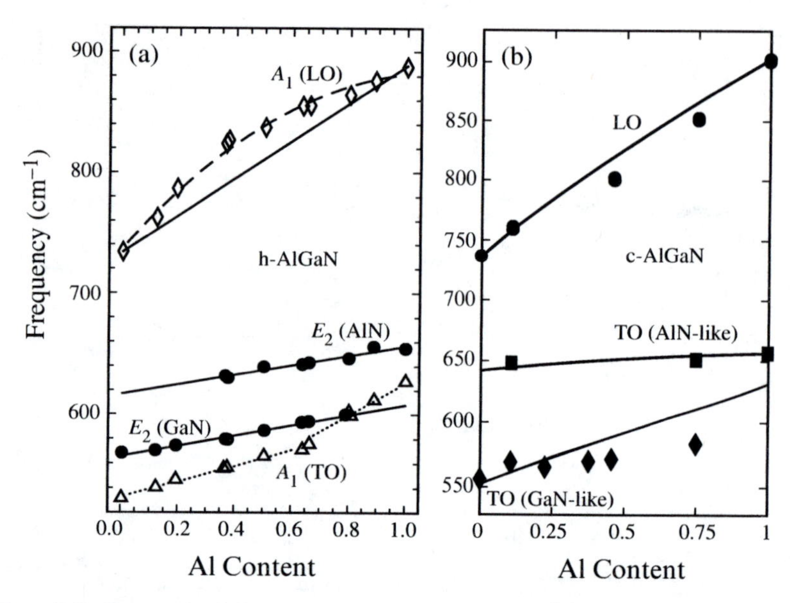

Figure 7.6 Compositional dependence of phonon frequencies in (a) hexagonal $Al_xGa_{1-x}N$ [33], and (b) cubic $Al_xGa_{1-x}N$ [36]. The thick dashed curve in (a) for A_1(LO) denotes fit to the Raman data by the quadratic $736.5 + 268.3x - 125.1x^2$ (cm^{-1}). Solid lines in (b) show theoretical calculations by the REI model [34,35]. The E_2 mode in (a) and the TO mode in (b) show two-mode–type behavior, whereas the LO modes for both systems show one-mode–type behavior.

Observed phonon frequencies are plotted against x in Figure 7.6a [33]. It is clearly seen that the E_2^H mode shows two-mode–type behavior, whereas the A_1(LO) mode gives one-mode–type behavior. As for the A_1(TO) mode, there is a small argument, as will be described later. The A_1(LO)-phonon mode does not follow a straight line connecting the phonon frequency of GaN and AlN (solid line) but bends up toward higher frequency, as seen by the dashed curve. This is a fit to the observed frequency by a quadratic function $A_1(\text{LO})(x) = 736.5 + 268.3x - 125.1x^2$ (cm^{-1}). The one- or two-mode behavior of phonons in cubic $Al_xGa_{1-x}N$ was theoretically predicted by Grille et al. by using a random-element isodisplacement (REI) model [34,35]. According to their calculation, one-mode–type behavior is expected for the LO phonon, whereas two-mode–type behavior is expected for the TO mode. If this is true, the E_2^H mode in hexagonal $Al_xGa_{1-x}N$ should also have two-mode–type character, because it is linked to the cubic TO-phonon branch by zone folding, as we saw in Section 7.2.1. This preference for one-mode–type behavior in the LO mode for both cubic and hexagonal systems was attributed to the strong ionicity in the cation–nitrogen bond and the fact that the LO mode is more strongly affected by the surrounding electric field than the TO mode. The theoretical prediction by Grille et al. [34] on cubic $Al_xGa_{1-x}N$ has been experimentally verified by Harima et al. [36], as shown in Figure 7.6b, where is clearly demonstrated that the LO mode gives a one-mode–type variation, whereas the TO mode gives a two-mode–type behavior.

In a recent paper, Grille et al. [35] compiled recent experimental data on the nitride alloys for both hexagonal and cubic systems and compared them with their theoretical calculations based on a modified REI model. There is a consensus on hexagonal $Al_xGa_{1-x}N$ between experiments [16,33,37,38] and theory [35] that the A_1(LO) and the E_1(LO) modes show one-mode–type behavior, whereas the E_2^H mode shows two-mode–type behavior. There is disagreement, however, in the TO phonon region 500–700 cm^{-1} on mode symmetry and on classification as one- or two-mode behavior: Demangeot et al. [39] obtained a result similar to Figure 7.6a. However, according to their mode assignment, no mode crossing occurs between the A_1(TO) and GaN-like E_2^H mode at an Al content of $x \sim 0.8$, as seen in Figure 7.6a. This result is in line with the theoretical prediction by Grille et al. [34,35]. Wiesniewski reported a two-mode behavior for the E_1(TO) mode by IR reflection [40]. It is clear that this frequency range shows complex features in this composition range, because mode overlapping occurs between the A_1(TO), E_2 and E_1(TO) modes, and therefore we obviously need more precise experimental data.

7.2.3.2 $In_xGa_{1-x}N$

Because of difficulties in obtaining good-quality single crystalline layers for a wide range of In content, only limited Raman scattering data are available at present for In content, $x < \sim 0.3$, or $x \sim 1$ for both hexagonal and cubic $In_xGa_{1-x}N$.

For hexagonal $In_xGa_{1-x}N$ alloys, phonon Raman spectra were observed by Behr et al. for $x = 0.019$–0.113 [41], Harima et al. for $x = 0$–0.7 [42], and by Wieser et al. for $x = 0$–0.33 [43]. The samples were grown on sapphire substrates by MOCVD with thick buffer layers [41,43], or by hot-wall epitaxy without thick buffer layers [42]. Behr et al. [41] employed a UV-laser excitation (3.00 eV) to induce resonant-Raman enhancement in the LO-phonon signal (see Section 7.4.2 for resonant Raman scattering). Resonant enhancement can be more easily observed with conventional blue/green lasers at high In content [43]. This is a technical advantage of observing $In_xGa_{1-x}N$ layers compared with $Al_xGa_{1-x}N$.

For cubic $In_xGa_{1-x}N$, phonon modes were observed by Tabata et al. for $x = 0$–0.31 in MBE films grown on GaAs (001) substrate [44]. They found that both the TO and LO phonons exhibit a one-mode–type behavior, and their frequencies display a linear dependence on the In composition.

The observed phonon frequencies are plotted in Figure 7.7 for (a) hexagonal [24,41–43,45] and (b) cubic $In_xGa_{1-x}N$ alloys [25,44]. The solid and

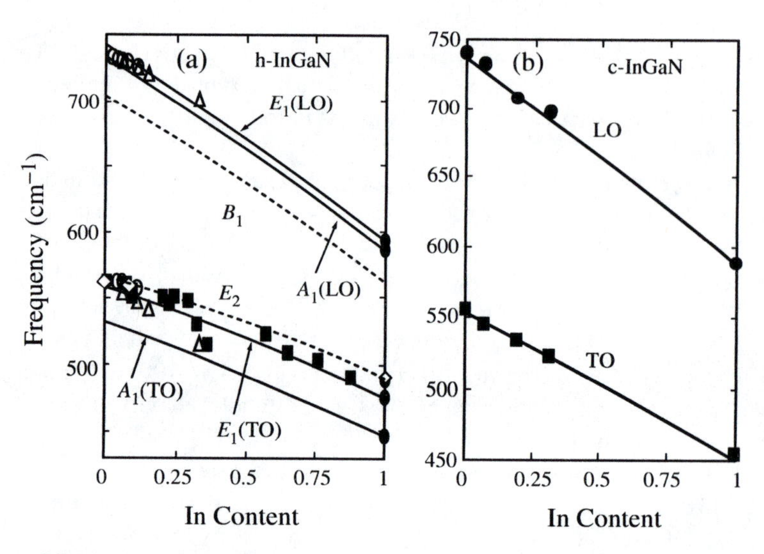

Figure 7.7 Compositional dependence of phonon frequencies in (a) hexagonal $In_xGa_{1-x}N$ [24,41–43,45,46], and (b) cubic $In_xGa_{1-x}N$ [25,44]. In (a), Raman data for A_1(LO) and E_2^H are plotted with open triangles, solid circles [24,45], open circles [41], and open squares [42]. Solid squares for E_1(TO) were obtained from IR reflection [46]. In both figures solid and dashed lines show theoretical calculations by the REI model [35].

dashed lines are theoretical calculations by the REI model [35]. This figure also includes IR data by Osamura et al. (solid squares) [46]. Their results should be recognized as a pioneering work on hexagonal $In_xGa_{1-x}N$ that covers the whole In composition range: in their experiment. 0.5-mm-thick polycrystalline alloy films were grown on sapphire or quartz substrate by an electron-beam plasma technique, and IR reflection was measured in the Reststrahlen region. They obtained a linear relation connecting the phonon frequency of GaN (558 cm^{-1}) and InN (478 cm^{-1}). Although Osamura et al. did not make the claim explicitly, this mode corresponds to the E_1(TO) mode.

We close this section by comparing the phonon-mode type of $Al_xGa_{1-x}N$ and $In_xGa_{1-x}N$. The LO mode in $Al_xGa_{1-x}N$ shows one-mode–type behavior for both cubic and hexagonal systems. Although the $In_xGa_{1-x}N$ data are not yet sufficient, one-mode–type is suggested for the LO mode in both cubic and hexagonal alloys, and the bowing is small. A difference appears in the TO phonon region: cubic $Al_xGa_{1-x}N$ clearly shows two-mode–type behavior, as seen in Figure 7.6b, whereas one-mode–type is suggested for cubic $In_xGa_{1-x}N$ alloys, as seen in Figure 7.7b. Similarly, two-mode–type character is clear for the E_2^H mode in hexagonal $Al_xGa_{1-x}N$ [Figure 7.6a], whereas one-mode–type is suggested for the E_1(TO) and the E_2^H modes in hexagonal $In_xGa_{1-x}N$ [Figure 7.7a]. This difference can be interpreted as follows [35]. TO phonons in cubic systems exhibit either one- or two-mode–type behavior depending on the cation size: since the atomic mass of nitrogen is much smaller than that of Al, Ga, and In, the reduced masses of AlN, GaN, and InN are almost the same as that of nitrogen. Thus, the difference in reduced mass has to be studied in detail: the reduced mass of InN (12.5 amu) comes closer to that of GaN (11.7) than that of AlN (9.2). Consequently, the In–N atomic unit can follow the frequency of the Ga–N unit vibration more easily than can the Al–N unit. Therefore, TO phonons in $Al_xGa_{1-x}N$ tend to show independent modes in the intermediate composition range. The difference in phonon kinetics between the cubic (zinc-blende) and hexagonal (wurtzite) system is not essential, because phonon branches of both systems are connected by zone folding. Therefore, whether the alloy system is cubic or hexagonal is not the key for choosing one- or two-mode–type. This simple picture does not hold for the LO mode, where the long-range electrostatic force that derives from the strong ionicity of the cation–nitrogen bond dominates the reduced-mass difference. Thus, one-mode–type behavior is preferred for LO phonons in both $Al_xGa_{1-x}N$ and $In_xGa_{1-x}N$ alloys, regardless of the cubic or hexagonal structures.

7.2.4 Stress Effects

As is well known, residual stress or strain in nitride layers is an important issue. For example, stress and strain often play key roles in deforming electronic band structures. There are several origins for the lattice strain. First, the lattice constant is affected by the existence of native defects or by the incorporation of various impurities. Such point defects will induce hydrostatic strain components when they are uniformly distributed in the crystal. Second, in many cases thin nitride layers are grown on foreign substrates that have different lattice constants and thermal expansion coefficients. Such differences will induce mainly biaxial strains. Generally, hydrostatic and biaxial strain components can coexist in nitride layers.

Probing of phonon frequency by Raman scattering is one of the most convenient methods for characterizing residual stress. Raman microscopy, especially, is a powerful tool for probing stress distribution with micrometer-order spatial resolution, and stress imaging for various layered structures is now intensively conducted. For example, the depth distribution of stress in epilayers may be obtained by observing the cross section. It is also well accepted that ELOG (epitaxially laterally overgrown) samples have a unique lateral distribution of stress depending on the location (window or mask). These topical applications will be described in Section 7.4.2.

Here we present only Raman studies on hexagonal GaN. There are no systematic Raman studies on the pressure effect on other nitrides or alloys.

For a more general aspect of strain effects on the nitrides, see recent good review articles [47–49].

7.2.4.1 Hydrostatic Pressure

Application of hydrostatic pressure usually increases the phonon frequency because the interatomic force constant increases with decreasing interatomic distance. Precise Raman studies on bulk crystals using a diamond anvil cell for applying hydrostatic pressure have been conducted for hexagonal AlN [13,19] and hexagonal GaN [17]. No such experiments have been reported on InN. Perlin et al. [17] tested a needle-type bulk hexagonal GaN crystal grown by a high-pressure method [50], up to 47 GPa, as shown in Figure 7.8a. The strongest signal in the bottom spectrum at 2 GPa was assigned to the A_1(TO) mode (531 cm^{-1} at ambient pressure). The faint structures L_2–L_5 were not clearly assigned in their first report [17], but L_5 was later assigned to the A_1(LO)-phonon mode. Perlin et al. explained the advent of this mode at above 20 GPa as follows [51]: The samples were heavily unintentionally doped at ambient pressure with $n = 10^{19}$–10^{20} cm^{-3}, thus the LO-peak intensity was severely weakened because of the plasmon–LO–phonon coupling (see Section 7.3.2), however, the A_1(LO) mode appeared at high pressure because of a carrier localization effect (see discussions in Section 7.3.5).

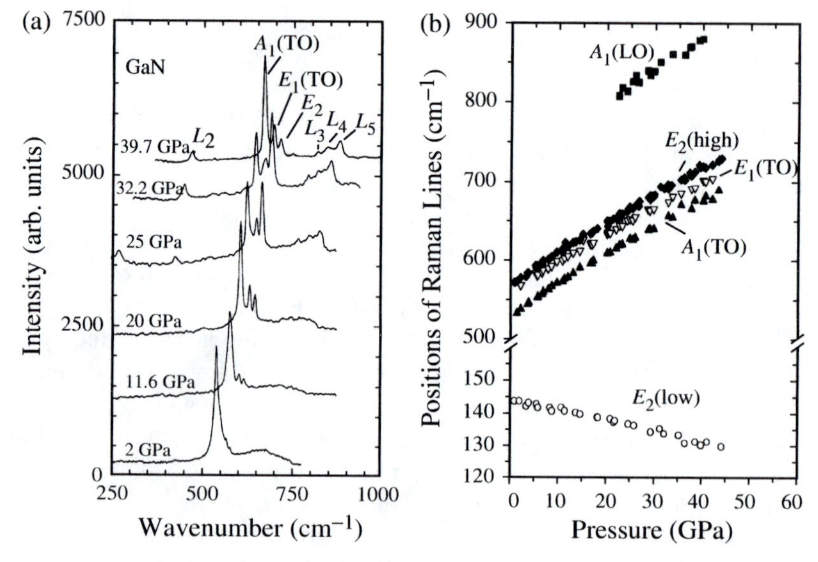

Figure 7.8 Raman spectra of hexagonal GaN under hydrostatic pressure (a) [17], and the phonon frequencies plotted against the pressure (b) [17,47]. The weak signal L_5, which appeared at above 20 GPa in (a), was attributed to the A_1(LO) mode (see Section 7.3.5 for carrier localization effects at high pressure). For the other faint structures L_2–L_4, see Ref. 17.

The phonon frequency is plotted in Figure 7.8b against the pressure [47]. All the modes showed nearly linear pressure dependence: the A_1(TO), E_1(TO) and higher-frequency E_2 mode (E_2^H) shifted to higher frequency with increasing pressure, whereas only the lower-frequency E_2 mode (E_2^L) shifted to lower frequency. In hexagonal AlN, the A_1(TO), E_2^H and E_1(LO) mode showed the same tendency up to 13 GPa [19]. Table 7.3 lists the linear stress shift coefficient K for each phonon mode. It is defined as $\omega = \omega_0 + Kp$, where p is the pressure in GPa, and ω_0 is the phonon frequency at zero pressure in cm^{-1}. This table also includes the mode Grüneisen parameter γ, which is defined as the frequency shift relative to the volume compression, and is rewritten by using the bulk modulus B_0 as

$$\gamma = \frac{d\omega/\omega_0}{dV/V} = \frac{B_0}{\omega_0}\frac{d\omega}{dp}. \tag{7.3}$$

Perlin and coworkers obtained γ by substituting the observed K values for $d\omega/dp$ and using $B_0 = 210$ and 207.9 GPa for GaN and AlN, respectively [19,47].

The peculiar behavior of the E_2^L mode in GaN, i.e., small negative pressure dependence and a negative Grüneisen parameter, should be noted. Perlin and coworkers discussed this so-called soft-mode behavior in conjunction with the mode softening of cubic (zinc blende) lattices in the transverse acoustic (TA) phonon modes at the L-point in the Brillouin zone [17,19,47]. As pointed out by Weinstein [52], some zinc-blende–type crystals such as ZnTe undergo softening of zone-boundary TA phonons under hydrostatic pressure. Therefore, the Grüneisen parameter is correlated with the pressure at which a structural phase transition occurs. As we saw in Figure 7.1, lattice vibration of the E_2^L mode at the Γ-point in the wurtzite structure is linked to the zone-boundary TA phonon in the zinc-blende structure by zone folding.

Table 7.3 Hydrostatic-pressure shift of phonon modes in hexagonal GaN.

Mode	E_2^L	$A_1(TO)$	$E_1(TO)$	E_2^H	$A_1(LO)$	$E_1(LO)$	
h-GaN							Ref. 47
K (cm^{-1}/GPa)	−0.32	3.8	3.3	3.6	3.8	—	
γ	−0.46	1.50	1.24	1.33	1.09	—	
h-AlN							Ref. 19
K (cm^{-1}/GPa)	—	4.63	—	3.99	—	1.67	
γ	—	1.58	—	1.26	—	0.38	

Note: The linear shift coefficient K is defined as $\omega = \omega_0 + Kp$, where p is the pressure in GPa, and ω (ω_0) is the phonon frequency under (without) hydrostatic pressure. The mode Grüneisen parameter γ is also shown.

According to Perlin and coworkers, a phase transition from wurtzite to rock-salt structure occurred at 47 and 16.6 GPa of hydrostatic pressure in GaN and AlN, respectively [17,53].

7.2.4.2 Biaxial Strain

In the general case, stresses can induce shift and splitting of phonon modes; however, the most common example in nitride systems is the growth of hexagonal layers on a foreign hexagonal (0001) or cubic (111) substrate. In thin films, we have only to consider uniform biaxial strain in the c-plane of epilayers with strain tensor components $e_{xy} = e_{yz} = e_{zx} = 0$ and $e_{xx} = e_{yy}$. Then, the layer may be compressed (or expanded) in the c-plane, and in the reverse manner along the c-axis. It means that hexagonal symmetry is maintained under stress, thus, only a phonon frequency shift is expected.

Let us consider as an example hexagonal GaN layers grown on sapphire or 6H–SiC (0001) substrate and consider only the difference in the thermal expansion coefficient. The larger thermal expansion coefficient of sapphire ($\alpha_{\perp} = 7.5 \times 10^{-6}$K^{-1}) than that of GaN (5.59×10^{-6}K^{-1}) [54] will induce compressive stress in the GaN c-plane in the cooling process from the growth temperature. For example, GaN layer will acquire $\sim$0.2% tensile strain when grown at 1000 °C. In contrast, 6H–SiC is used as the substrate, the GaN layer will have tensile strain, because $\alpha_{\perp} = 4.2 \times 10^{-6}$K^{-1} for 6H–SiC [55]. The residual strain in hexagonal GaN epitaxial layers, i.e., its magnitude as well as its polarity (compressive or tensile), is easily judged by looking at the stress shift of the E_2^H-phonon peak. This mode is readily observed in the backscattering geometry from the c-plane (see Figure 7.3) and, furthermore, is very sensitive to biaxial strains in the c-plane, as expected from the atomic displacement scheme (see Figure 7.2). Figure 7.9 shows typical variation in the E_2^H-mode spectra for different substrates [56]. Here, the middle spectrum peak at 566.2 cm^{-1} was observed in a "stress-free bulk" sample for comparison. (Note that this frequency is slightly downshifted compared with the suggested "stress-free" value of $\sim$568 cm^{-1} discussed in Section 7.2.2). According to the authors this is because the lattice constant of bulk GaN varies locally [56]. It is clear anyway that the GaN layer on SiC acquires a tensile stress because the frequency is downshifted from the stress-free value, whereas the GaN layer on sapphire (Al_2O_3) acquires a large compressive stress. For comparison, let us look at a recent Raman experiment on a hexagonal GaN layer grown on the (111) plane of a spinel ($MgAl_2O_4$) substrate [57]. The $MgAl_2O_4$ substrate has a smaller lattice mismatch with GaN (effective value, 9.5%) than does sapphire (13%) [58] and a smaller mismatch in the thermal expansion coefficient [54], thus it is used as a substrate for a GaN laser diode [59]. Li et al. [57] tested 1.3-μm-thick GaN epilayers grown by MOCVD at 1050 °C on $MgAl_2O_4$ (111) with a 20-nm-GaN buffer layer and observed the E_2^H-phonon peak at 568.0–568.2 cm^{-1}.

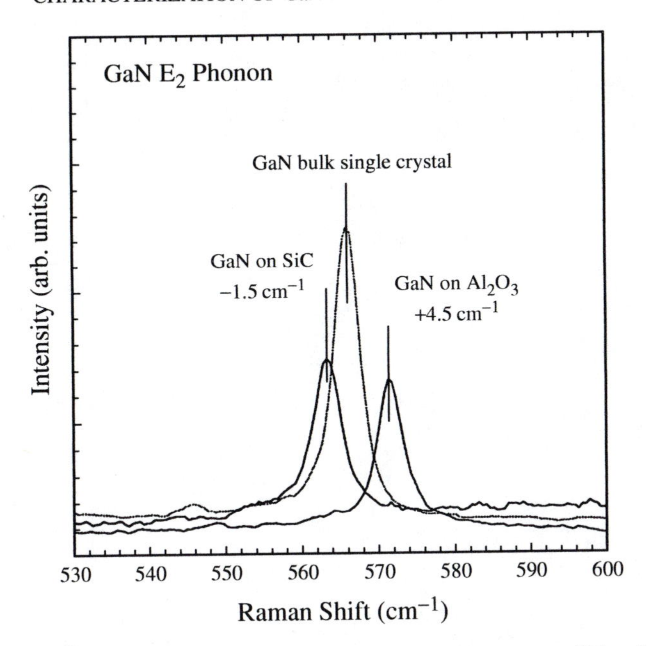

Figure 7.9 E_2^H-phonon spectra observed in hexagonal GaN grown on SiC and sapphire substrate [56].

This value is in good agreement with the standard stress-free value, showing that $MgAl_2O_4$ substrate is a good candidate for growing hexagonal GaN from the viewpoint of residual stresses.

Intensive Raman studies have been done to measure precisely the frequency shift of the E_2^H mode against the compressive strain [8,50,60–63]. Demangeot et al. observed stress shifts for the A_1(LO) and E_2^H modes, as shown in Figure 7.10 [62], and deduced stress shift rates of 0.8 and 2.9 cm^{-1}/GPa for the A_1(LO) and the E_2^H modes, respectively. This means that the A_1(LO) mode is insensitive to strains and thus is not suitable for evaluating stress; however, this becomes an advantage when we analyze the A_1(LO)-mode profile to evaluate free-carrier density: this peak frequency is sensitive to the free-carrier density through LO-phonon–plasmon coupling, and the peak frequency can be used as a measure of free-carrier density (see Section 7.3.2).

Table 7.4 lists reported biaxial–stress shift coefficients for the E_2^H mode. There is no good agreement at present, which may be attributed to the different methods used in calibrating the biaxial stress; photoreflection [62], photoluminescence [56], X-ray diffraction [8], and wafer bending [60]. Demangeot et al. deduced deformation potential constants for the A_1(LO) and E_2^H modes from the biaxial-stress shift coefficients as follows [62]: within the context of a perturbation approach and in the linear

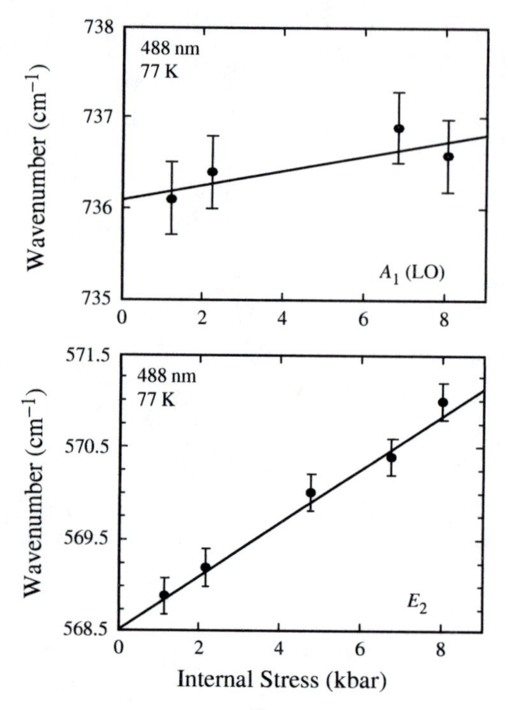

Figure 7.10 Stress shift of the A_1(LO)- and E_2^H-phonon peaks in hexagonal GaN on sapphire substrate [62]. The lines denote the best linear fits to the data.

Table 7.4 Biaxial-stress shift coefficient of the E_2^H-phonon mode in hexagonal GaN (cm^{-1}/GPa).

6.2	2.9 ± 0.3	4.2 ± 0.3	2.7 ± 0.3
Ref. 60	Ref. 62	Ref. 56	Ref. 8

approximation on a wurtzite crystal [64], the stress shifts of the phonon frequency under hydrostatic or biaxial stress can be expressed as

$$\Delta \nu_J = 2a'_J \sigma_{xx} + b'_J \sigma_{zz} \qquad (J = A_1, E_1, \text{ and } E_2), \tag{7.4}$$

where the σ_{JJ}'s are the diagonal components of the stress tensor, with $\sigma_{xx} = \sigma_{zz}$ for hydrostatic pressure and $\sigma_{zz} = 0$ for biaxial stress in the c-plane. The coefficients a'_J and b'_J are expressed in terms of elastic constants S_{ij} and deformation potentials a_J and b_J as

$$a'_J = a_J(S_{11} + S_{12}) + b_J S_{13} \quad \text{and} \quad b'_J = 2a_J S_{13} + b_J S_{33}. \tag{7.5}$$

The biaxial stress coefficient $d\nu_J/d\sigma$ with $\sigma = \sigma_{xx}$ ($\sigma > 0$ for compression) can be written as

$$2a_J(S_{11} + S_{12}) + 2b_J S_{13} = -K_J^B. \tag{7.6}$$

Given the elastic constants and the biaxial stress coefficient, one more relation is needed for the determination of the set of $\{a_J, b_J\}$ deformation potentials. This relation is obtained from the hydrostatic-pressure-induced stress shift, which is written from Eqs. (7.4) and (7.5) as

$$2a_J(S_{11} + S_{12} + S_{13}) + (2S_{13} + S_{33})b_J = -K_J^H. \tag{7.7}$$

Demangeot et al. obtained $\{a_J, b_J\} = \{-818, -797\}$ for the E_2^H mode and $\{-685, -997\}$ in cm^{-1}/unit strain for the A_1(LO) mode. Davydov et al. also obtained deformation potential constants on hexagonal GaN based on Raman and X-ray diffraction studies [8]. Their result for the E_2^H mode shows reasonable agreement with the preceding values: $\{a_J, b_J\} = \{-850, -920\}$ cm^{-1}/unit strain.

7.2.5 Temperature Effects

We have seen that uniform or hydrostatic pressure induces lattice compression, in general shifting the phonon frequency toward higher frequency. Rising temperature induces the opposite effect: namely, phonon frequency usually decreases with the dilatation of the lattice. At high temperatures, furthermore, anharmonic terms in the phonon Hamiltonian rapidly increase, so that phonon frequency further decreases, and the peak broadens [65]. This process may qualitatively be understood as follows: with rising temperature, channels of phonons are decayed by dissipating the phonon energy to lower-frequency phonons, thus, the phonon self-energy is suppressed. Link et al. recently reported using Raman scattering to observe the variation of the frequency and width with temperature in the E_2^H and A_1(LO) modes for hexagonal AlN and GaN at 85–760 K [66]. Figure 7.11 shows their result for GaN. The tested GaN layers were grown on sapphire substrates by MOCVD to a thickness of 2.2 μm. The lines are theoretical fits to the data considering anharmonic decay terms up to the fourth order of phonon. Effects of lattice expansion of the GaN layer and strains induced by the difference in thermal expansion coefficients between the substrate and the GaN layer were also considered in the fitting. The observed temperature dependence is well explained by the theoretical treatment. As an application of this data on temperature variation, Link et al. measured a local temperature in a GaN pn-diode during operation [66].

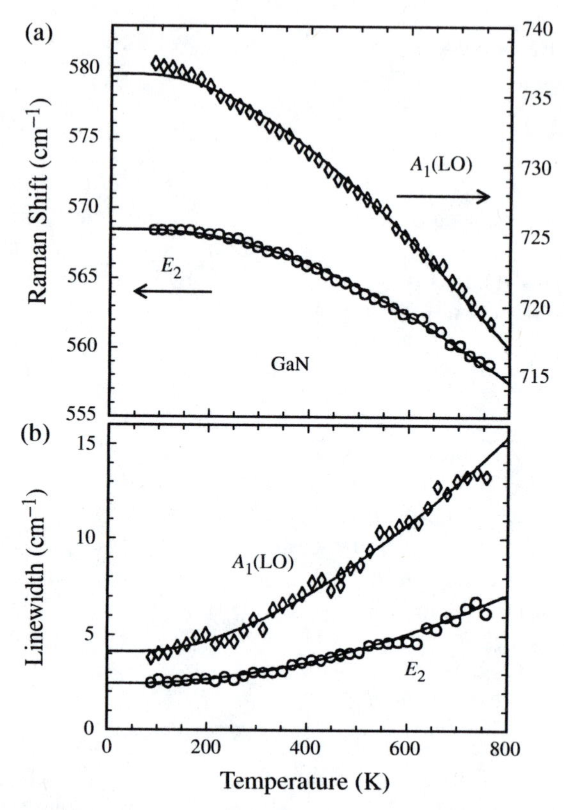

Figure 7.11 Temperature variation of E_2^H- and A_1(LO)-phonon modes in hexagonal GaN; (a) and (b) denote peak frequency and linewidth (FWHM), respectively. The lines show theoretical fits to the data considering anharmonic terms in the phonon Hamiltonian and effects of lattice expansion and strain [66].

In the case of hexagonal GaN, only the E_2^L-phonon mode at 144 cm^{-1} showed anomalous behavior like a "soft mode" against hydrostatic pressure, as we discussed in the preceding section. A similar anomaly is observed for this mode at high temperature, i.e., the frequency increases with rising temperature [47].

7.2.6 Defect and Impurity Modes

7.2.6.1 Local Vibrational Mode in Mg-Doped GaN

It is well known that impurity atoms in a crystal lighter than the host atoms can give rise to a local vibrational mode (LVM). Atomic motions in LVM are confined primarily to the impurity itself and to its nearest neighbors, with rapidly decaying vibrational amplitude for more distant host atoms. Usually, the lighter the impurity, the higher the vibrational frequency and more localized [67]. Raman scattering, like IR absorption, is a powerful technique for observing LVM in semiconductors. For example, Si and Mg impurities that occupy the cation sites are commonly used for n- and p-type doping in GaN, respectively. Although LVMs can be expected in both cases, experimental data are limited so far to the case of p-type doping. A systematic observation of LVM to study the Mg-activation process is reviewed in this section.

It is well known that low-energy electron beam irradiation [68] or thermal annealing in N_2 atmosphere [69] is necessary to obtain p-type conductivity in Mg-doped GaN films grown by MOCVD using H_2 as the carrier gas of precursors, and NH_3 as the nitridation source, as commonly employed. Although the underlying mechanism for the Mg-activation process is not yet fully understood, it is widely believed, as first suggested by Nakamura et al. [69], that H impurities incorporated in the growth and/or annealing process passivate Mg acceptors by forming Mg–H neutral complexes. Observation of LVM for Mg–N–H complex by IR absorption [70] in as-grown MOCVD samples gives convincing evidence for this hypothesis.

Harima and coworkers recently reported a systematic Raman study on the thermal annealing process from the viewpoint of LVM observation [71,72]. They observed $\sim$2-μm-thick Mg-doped hexagonal GaN layers grown on sapphire substrate. The samples were annealed at different temperatures in the range 500–1000 °C in a N_2 atmosphere. Figure 7.12 shows the Raman spectra observed at room temperature by backscattering from the c-plane [71]. Figures 7.12a, b, and c show different frequency regions, and the spectra are compared with each other after being normalized to the peak height of the E_2^H mode at 568 cm^{-1}. In Figure 7.12a all spectra are greatly expanded vertically to show small LVM signals. Sharp phonon peaks of the E_2^H and the A_1(LO) modes (735 cm^{-1}) are therefore scaled out, and the strong sapphire signal (418 cm^{-1}) is observed. The bottom spectrum, derived from the as-grown, highly resistive sample, coincides with the standard spectrum shown in Figure 7.3 (upper). The spectra showed different features at various annealing temperatures, as seen in the upper traces: At an annealing temperature, above 600 °C LVM appeared at 657 and 260 cm^{-1}, and a continuum band appeared in the low-frequency region below $\sim$400 cm^{-1}. These structures are most evident at annealing temperatures of 800–900 °C, but suddenly weaken at 1000 °C.

This continuum band was assigned to the inter-valence-band transition of holes (See Section 7.3.4). Thus, the advent and growth of the continuum band with rising annealing temperature was interpreted as an indication of the activation process of Mg impurities [27]. The 657 cm^{-1} mode was assigned to an LVM of the (activated) Mg–N bond because (i) the frequency agrees with an estimate calculated from the optical mode frequency of GaN at $\sim$560 cm^{-1} considering the difference in the reduced mass between Mg–N

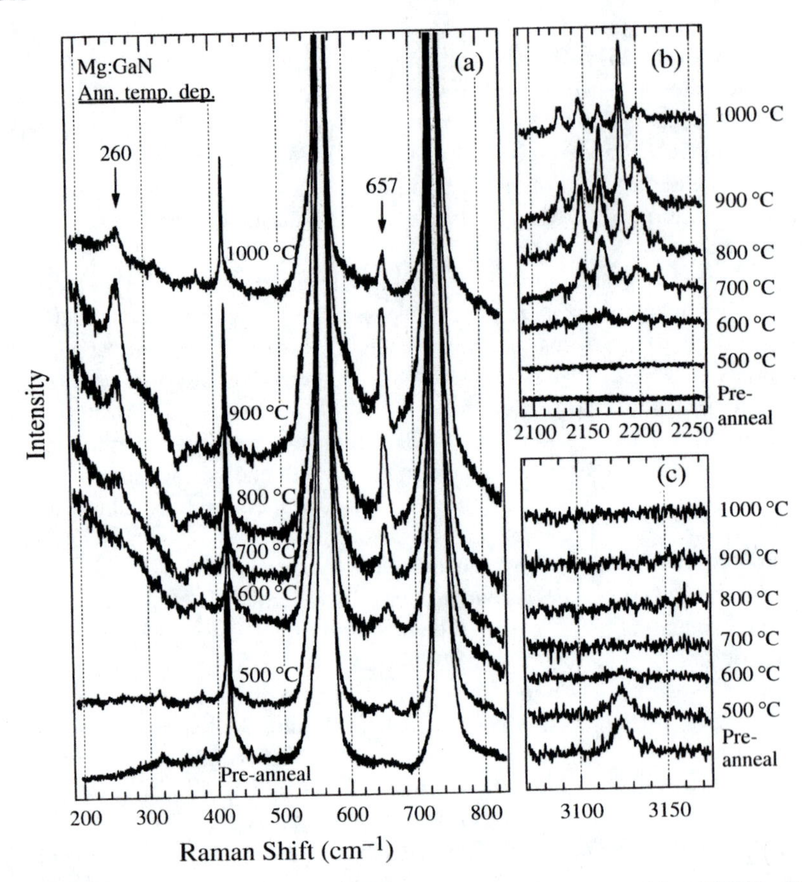

Figure 7.12 Raman spectra of Mg-doped hexagonal GaN films prepared by MOCVD with postannealing at various temperatures in a N_2 atmosphere [71]. The frequency region is separated into three parts: (a), (b), and (c). The spectra are normalized to the intensity of the E_2^H-mode at 568 cm^{-1}. The bottom spectrum in each figure was obtained from the as-grown samples.

and Ga–N pairs, and (ii) the mode intensity is proportional to the hole density. This mode was found also in as-grown MBE films [73]. It is known that when a Mg-doped GaN layer is grown by the MBE process, it has native p-type character [73]. Therefore, it is not surprising that the 657 cm^{-1} mode was observed in the as-grown MBE films. The 260 cm^{-1} mode in Figure 7.12 may also be related to the (activated) Mg–N bond. Thus, Figure 7.12a tells us that p-type conductivity is obtained by annealing at 600–900 °C but is lost when the annealing temperature is too high, as at 1000 °C. The p-type conductivity was lost at such temperatures probably because carrier compensating centers such as nitrogen vacancies (V_N) increased rapidly [71].

Raman spectra in the higher-frequency regions shown in Figures 7.12b and c give a hint of the role of H atoms in the Mg-passivation process [72]. An LVM is observed at 3123 cm^{-1}, as seen in Figure 7.12c in the pre-anneal sample (bottom spectrum). It disappears by annealing at above 600 °C, but at the same time, several new LVMs appear at 2000–2200 cm^{-1}, as observed in Figure 7.12b. The 3123 cm^{-1} mode was first observed by Götz et al. [70] by IR absorption in MOCVD films, and the mode was confirmed to be H-related by a deuteration experiment. Götz et al. [70] considered that this frequency, which is close to that of N–H stretching in the NH_3 molecule, suggests that H atoms are bonded to N atoms and form a Mg–N–H complex. This view is supported by recent theoretical studies: hydrogen ions bonded to nitrogen, and located at the antibonding (AB_N) [74,75] or body-centered site (BC_N) [76] of Mg–N bonds are energetically most stable. LVM signals similar to Figure 7.12b were previously observed by Brandt et al. [77] and more recently by Kaschner et al. [73,78] in as-grown p-type conductive MBE films. These authors attributed these LVMs mainly to H-decorated defects such as nitrogen vacancies (V_N) and extended defects. In addition to such candidates, other possibilities such as Ga–H and Mg (neutral) –H pairs may be included, because they have been observed in common semiconductors such as Ga–H in Si (2171 cm^{-1}) and Mg–H in GaAs (2144 cm^{-1}) [79]. Complex centers including interstitial Mg have also recently been proposed as a candidate [80]. Although the new bonding partner of hydrogen is not well known at present, there are probably multiple origins for the 2100–2200 cm^{-1} modes, because peak intensities of the LVMs in Figure 7.12b show different temperature variations.

To summarize, Figure 7.12 shows that H atoms are released from Mg by thermal annealing in N_2 at above 600 °C and diffuse in the host lattice; they partially remain in the material to form new bondings with some partner, and show LVMs at 2000–2200 cm^{-1}. The absence of the Mg–N mode at 657 cm^{-1} in Figure 7.12 in the pre-annealed sample can be understood if H^+ ions bonded to N significantly disturb the Mg–N vibration. It was further shown in another experiment [72] that when N_2 gas was used instead of H_2 as the carrier gas in the MOCVD process, then the 657 cm^{-1} mode appeared even in the as-grown films, and H-related modes were not observed at around 2100 or 3100 cm^{-1} eiher before or after annealing. This means that the sample has native p-type character, which is consistent with the result of Hall measurements [81].

7.2.6.2 Related Modes

LVM studies are also convenient for finding other light impurities or defects in GaN. Structural damage in GaN by ion implantation has been investigated by Raman scattering [82] and IR absorption [83–85]. As a typical example, Figure 7.13 shows Raman spectra of hexagonal GaN grown by MBE and

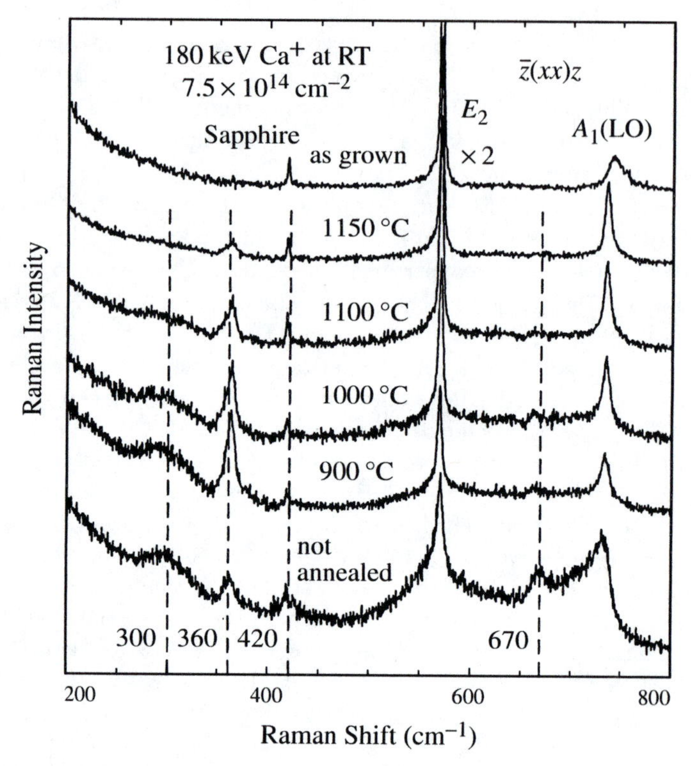

Figure 7.13 Raman spectra of hexagonal GaN with Ca^+ implantation and subsequent annealing at various temperatures [82]. The bottom spectrum was derived from the as-implanted sample. The dashed lines denote defect modes.

implanted with Ca^+, with subsequent annealing at various temperatures [82]. The pre-annealed sample (bottom) shows that the A_1(LO) and E_2^H modes are heavily broadened with strong asymmetric tails, and characteristic modes appear at 300, 360, 420, and 670 cm^{-1}, as indicated by the dashed lines. Because implantation of other ionic species such as Ar, Mg, and P yielded the same spectra [82], the observed features were assigned to LVM not by substitutional impurities but by vacancy-related defects: the strong tails to lower frequency for the A_1(LO) and E_2 modes and the 300 cm^{-1} peak were attributed to defect-activated Raman scattering (DARS). It appeared as a result of breakdown of Raman selection rules in a disordered system: In defective crystals, phonons not limited to the Γ-point in the Brillouin zone can contribute to the Raman spectra. Therefore, the spectrum reflects the phonon density of states to some extent [23]. Seager et al. reported IR absorption for N–H bonds at 3183 and 3219 cm^{-1} in H-implanted GaN [83]. Weinstein et al. [85] also observed many sharp LVMs at 3000–3150 cm^{-1} in H-implanted GaN, which were attributed to LVM by Ga vacancy decorated with hydrogen (V_{Ga}–H_n) [86]. Duan et al. reported LVM by IR absorption at 1730 and 2960 cm^{-1} for Ga–H and N–H complexes, respectively [84].

Kuball et al. [87] studied the thermal stability of the GaN host lattice in the process of high-temperature annealing. No degradation was observed below 900 °C in a N_2 gas atmosphere, whereas a broad peak centered at 610 cm^{-1} appeared between 900 and 1000 °C, and other distinct peaks appeared at 630, 656, and 770 cm^{-1} at above 1000 °C. These modes were assigned to DARS or LVM at the GaN/substrate (sapphire) interface.

LVMs related to CH, CH_2, and CH_3 defect complexes have also been observed by IR absorption at 2850, 2923, and 2956 cm^{-1} [88,89].

7.2.7 Isotope Effects

Zhang et al. performed a unique study on the effect of isotope replacement on phonon dynamics: they replaced natural nitrogen atoms (99.6% ^{14}N) isotopically pure ^{15}N in hexagonal GaN [90]. The Raman spectra showed that frequencies of the polar optical phonons, A_1(TO,LO) and E_1(TO,LO), shifted following the inverse-square-root dependence of the reduced masses, as expected: but the two nonpolar modes, E_2^H and E_2^L, showed isotope shifts that were significantly different from the reduced-mass dependence and from the dependence of pure N or Ga vibrations (see the atomic displacements shown in Figure 7.2). It was concluded that the E_2-phonon modes involve mixing of Ga and N vibrations. It was further shown that in alloys in which the same amount of different N isotopes is included, Ga^{14}N$_{0.5}$^{15}N$_{0.5}$, the A_1 and E_1 mode frequencies showed an isotopic disorder effect.

7.3 ELECTRONIC PROPERTIES

7.3.1 Introduction

Raman spectroscopy is a convenient and nondestructive tool for characterizing the electronic as well as the crystalline properties of semiconductors. Raman scattering related to electronic transitions is generally called *electronic Raman scattering*. In principle, free carriers or bound electrons (or holes) can participate in electronic Raman scattering. Electronic Raman transitions that are coupled with phonon excitations may also be classified in this category. A variety of such examples have been reported in common semiconductors such as GaAs, GaP, Si, and SiC [91]. Related papers for the nitrides have appeared quite recently, yet the number is still very limited when compared with those on phonon Raman scattering.

Free-carrier transitions observed in Raman scattering are classified into single-particle and collective-mode excitations [91,92]. In polar semiconductors, collective excitation of free carriers (plasmons) can couple with LO phonons to form an LO–phonon–plasmon coupled mode. This mode has been observed in hexagonal n-type GaN [20,51]. Profiles of LO–phonon–plasmon coupled modes can be analyzed to evaluate electrical transport parameters such as carrier density and mobility [20,27]. In contrast, there are very few Raman studies on the p-type nitrides. This is due partly to experimental difficulties in growing good-quality p-type crystals, especially with high carrier densities. Furthermore, the inactive nature of p-type material has showed electronic Raman studies. For example, LO–phonon–plasmon coupled modes in the p-type show much less change with carrier density than the n-type.

Raman scattering by single-particle excitation of electrons has been studied by time-resolved technique using fs lasers, and relaxation times of nonequilibrium electrons and LO phonons have been reported [93–96]. This work will be described in detail in Section 7.4.3 for time-resolved Raman scattering.

Electronic Raman scattering related to shallow donor or acceptor states has scarcely been reported on the nitrides. As an exceptional case, there is a report on the electronic transition between shallow donor states in n-type GaN, but there is still controversy over the spectral interpretation.

7.3.2 LO–Phonon–Plasmon Coupled Modes

In polar semiconductors such as GaN, collective oscillation of free carriers can interact with LO phonons via longitudinal electric fields and form LO–phonon–plasmon coupled (LOPC) modes. Raman scattering from the LOPC mode has been extensively studied in III–V, II–VI, and IV–IV compound semiconductors [91]. LOPC modes in the nitride were first reported in hexagonal n-type GaN [20,51]. As first demonstrated in GaAs in LOPC mode [97], the mode generally comprises upper- and lower-frequency branches (referred to as the L^+ and L^- mode here, respectively); however, in some wide-bandgap semiconductors such as SiC [98], the L^- mode cannot be observed clearly. This is because the plasmon is overdamped ($\omega_P \tau < 1$, where ω_P is the plasmon frequency, and τ is the plasmon scattering time). Figure 7.14 shows the variation of LOPC modes in n-type hexagonal GaN with the free-carrier density [99]. This figure shows that the sample belongs to the former category (small plasmon damping) because the L^- mode is observed clearly, changing in lineshape with the carrier density n. The spectra were observed here by backscattering from the c-plane, $z(x, x+y)-z$, thus these LOPC modes correspond to the axial type that propagates along the c-axis. As n is increased, the L^+ peak broadens, shifts to higher frequency, and becomes asymmetric with a longer tail at higher frequency. The solid and dashed lines are theoretical fits, as explained below. Figure 7.15 shows the observed variation of the L^+-peak parameters: from top to bottom, circles give the peak-frequency shift, peak width (FWHM),

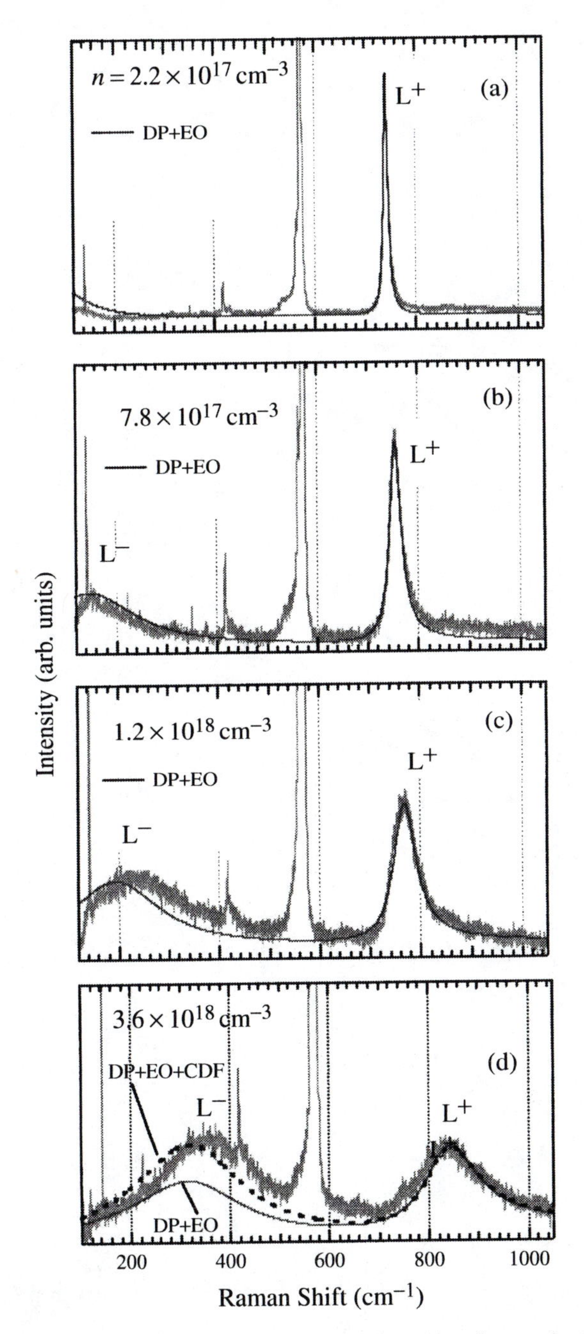

Figure 7.14 Raman spectra of n-type hexagonal GaN films with various carrier densities [99]. Smooth solid lines denote theoretical fits to L^+ mode considering DP and EO mechanisms. Dashed line in (d) includes CDF contribution.

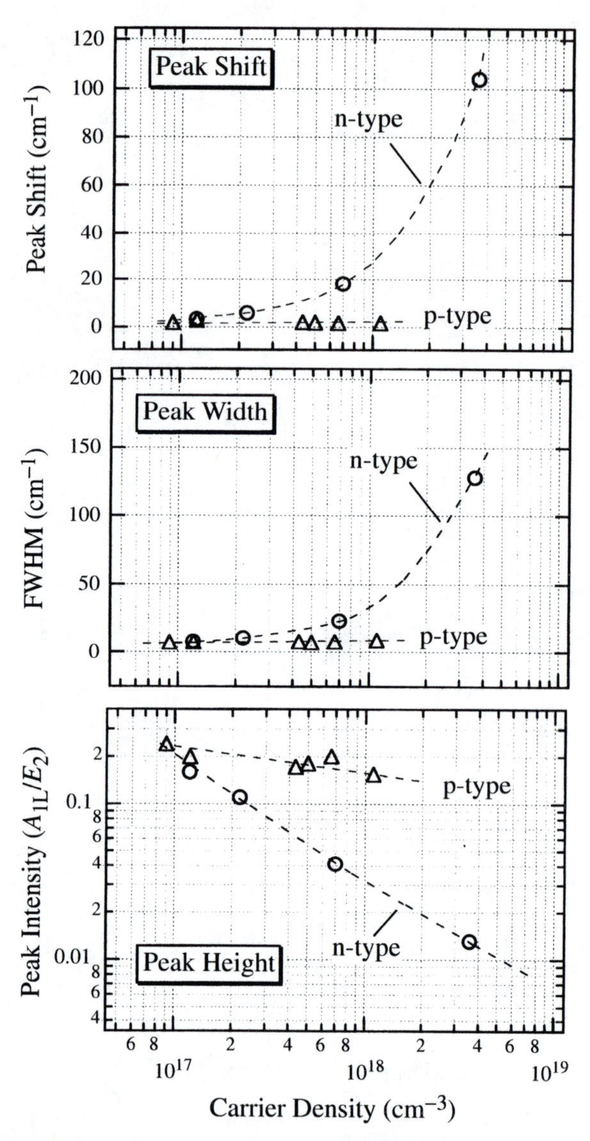

Figure 7.15 Peak-frequency shift, width (FWHM), and peak height (normalized to the E_2^H-mode peak at 568 cm^{-1}) of the L^+ mode in hexagonal GaN plotted against the free-electron density (circles). Results for the p-type are also shown for comparison (triangles).

and peak height normalized to the E_2^H-phonon peak height at 568 cm^{-1} (independent of the carrier density). The triangles represent the results of p-type hexagonal GaN. These results are briefly discussed at the end of the next section.

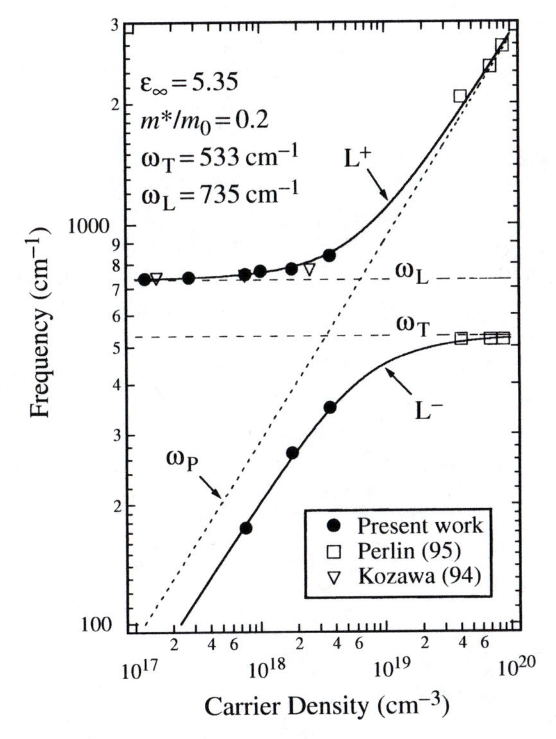

Figure 7.16 Carrier-density dependence of the axial-type LOPC-mode frequency in n-type hexagonal GaN [99]. Solid curves denote theoretical values obtained by solving $\varepsilon = 0$. Raman data by Perlin et al. (squares) [51] and Kozawa et al. (triangles) [20] are also plotted. Dashed lines show plasmon frequency and the A_1(TO, LO) phonon frequencies in the undoped sample.

The peak-frequency variation of the LOPC mode in n-type hexagonal GaN is plotted in Figure 7.16 against the carrier density with filled circles [99], open triangles [20], and open squares [51]. The dashed line shows plasmon frequency calculated by using $\varepsilon_\infty = 5.35$ for the optical dielectric constant and m^* $(=0.2m_0)$ for the effective mass of the electron [100]. The solid curves denote LOPC mode frequencies obtained by solving the equation for longitudinal excitations, $\varepsilon(\omega) = 0$. Here $\varepsilon(\omega)$ is the dielectric function consisting of phonon and plasmon contributions:

$$\varepsilon = \varepsilon_\infty \left[1 + \frac{\omega_L^2 - \omega_T^2}{\omega_T^2 - \omega^2 - i\omega\Gamma} - \frac{\omega_P^2}{\omega^2 - i\omega\gamma} \right], \tag{7.8}$$

where $\omega_T = 531\,\text{cm}^{-1}$ $(\omega_L = 734\,\text{cm}^{-1})$ is the uncoupled TO (LO) phonon frequency of the A_1-type mode (see Table 7.2), ω_P is the plasmon frequency given by

$$\omega_P = \sqrt{\frac{4\pi\, ne^2}{\varepsilon_\infty\, m^*}}. \tag{7.9}$$

In Eq. (7.8) Γ and $\gamma(=1/\tau)$ are the damping rate of phonons and plasmons, respectively. They were both set to zero in solving $\varepsilon(\omega) = 0$ to approximate LOPC mode frequencies. In the spectral lineshape fitting analysis described in the next section, these factors are treated as adjustable parameters. It is seen in Figure 7.16 that the observed peak frequencies almost agree with the calculated curves.

Free-electron density n in GaN can be estimated by simply comparing the observed L^+-mode parameters, peak shifts, FWHM, or peak intensity with the results of Figure 7.15. Convenient empirical formulas to deduce carrier density n (cm^{-3}) from the L^+-mode peak frequency ω_M (cm^{-1}) have been proposed, e.g., [101]

$$n = 1.1 \times 10^{17}\, (\omega_M - 736)\,; \tag{7.10}$$

however, we can make more precise estimates of on n by a lineshape-fitting analysis described in the next section. A further advantage of the fitting analysis is that since the plasmon damping rate γ can be obtained from the fitting analysis, we can also evaluate the drift mobility by

$$\mu = \frac{e}{m^*\gamma}. \tag{7.11}$$

The axial-type LOPC modes presented in Figure 7.14 yield carrier mobility along the c-axis ($\mu_\parallel$), whereas if the planar-type mode is observed by a right-angle scattering geometry, $x(y, z)y$, we can probe mobility perpendicular to the c-axis ($\mu_\perp$).

This method of evaluating carrier transport parameters without electrodes is very convenient when microscopic distributions are needed. For example, we can analyze carrier transport parameters layer by layer by observing the cross sections. There are many reports on microscopic characterization of carrier densities in various types of epitaxial layers including ELOG samples [27,102–105]. This topic will be treated in Section 7.4.2 on micro-Raman imaging.

7.3.3 Analysis of Coupled Modes

The lineshape-fitting analysis goes as follows: In a semiclassical approach considering the contribution of deformation-potential (DP) and electro-optic

(EO) mechanisms, the Raman scattering efficiency is expressed as [106,107]

$$I(\omega) = SA(\omega)\mathrm{Im}[-1/\varepsilon(\omega)] \tag{7.12}$$

where ω is the Raman shift, S is a proportionality constant, $\varepsilon(\omega)$ is the dielectric function, and $A(\omega)$ is a coefficient given by

$$\begin{aligned} A(\omega) = {} & 1 + 2C\omega_T^2\left[\omega_P^2\gamma(\omega_T^2 - \omega^2)\right. \\ & \left.-\omega^2\Gamma(\omega^2 + \gamma^2 - \omega_P^2)\right]/\Delta + C^2(\omega_T^4/\Delta) \\ & \times\left[\omega_P^2\{\gamma(\omega_L^2 - \omega_T^2) + \Gamma(\omega_P^2 - 2\omega^2)\}\right. \\ & \left.+\omega^2\Gamma(\omega^2 + \gamma^2)\right]/(\omega_L^2 - \omega_T^2). \end{aligned} \tag{7.13}$$

Here, C is a dimensionless parameter called the Faust–Henry coefficient [108], and we have defined

$$\Delta \equiv \omega_P^2\gamma\left[(\omega_T^2 - \omega^2)^2 + (\omega\Gamma)^2\right] + \omega^2\Gamma(\omega_L^2 - \omega_T^2)(\omega^2 + \gamma^2). \tag{7.14}$$

The factor C is rewritten as a measure of the relative strength of deformation-potential to electro–optic coupling in the electron–phonon interaction:

$$C = \frac{e^*}{\omega_T^2 M}\left[\frac{(\partial\alpha/\partial u)}{(\partial\alpha/\partial E^L)}\right], \tag{7.15}$$

where e^* is the effective charge associated with the optical phonons, M is the reduced mass of the oscillating atomic pair, u is the atomic displacement, E^L is the macroscopic electric field, and α is the polarizability. The factor C is experimentally deduced using the following formula if the Raman intensity ratio of the LO- and TO-phonon bands is measured in undoped samples:

$$\frac{I_{LO}}{I_{TO}} = \left(\frac{\omega_1 - \omega_L}{\omega_1 - \omega_T}\right)^4 \frac{\omega_T}{\omega_L}\left(1 + \frac{\omega_T^2 - \omega_L^2}{C\omega_T^2}\right)^2. \tag{7.16}$$

where, ω_1 is the incident photon frequency. Values that have been proposed for the C coefficient of the axial mode in GaN include 0.4 [20], 0.48 [99], and −3.8 [109].

The smooth solid curves in Figure 7.14 are the best fit to the L^+-mode profiles obtained by adjusting the parameters n, γ, and Γ as listed in Table 7.5 [99]. Samples 2–5 in this table gave Raman spectra (a)–(d), respectively, in Figure 7.14. Table 7.5 also lists the plasmon frequency ω_P and the mobility μ evaluated by Eqs. (7.9) and (7.11), respectively. For comparison, the values

Table 7.5 Best-fit parameters n (carrier density), γ (plasmon damping), and Γ (phonon damping) used for fitting the L^+ modes shown in Figure 7.14.

Sample	n *(cm^{-3})*	ω_P *(cm^{-1})*	γ *(cm^{-1})*	Γ *(cm^{-1})*	μ *(cm^2/Vs)*
1	1.2×10^{17} (1.2×10^{17})	140	100	6	460 (470)
2 (a)	2.2×10^{17} (3.1×10^{17})	140	120	8	390 (430)
3 (b)	7.8×10^{17} (1.1×10^{18})	260	200	14	230 (290)
4 (c)	1.2×10^{18} (1.4×10^{18})	320	220	40	210 (310)
5 (d)	3.6×10^{18} (5.2×10^{18})	550	360	50	130 (170)

Source: H. Harima, H. Sakashita, and S. Nakashima, *Mater. Sci. Forum*, **264–268**, 1363 (1998).
Note: Samples 2–5 correspond to spectra (a)–(d) in Figure 7.14. Values in parentheses for n and μ (mobility) were obtained from Hall measurements.

from Hall measurements are shown in the parentheses. There is reasonable agreement between the Hall and Raman measurements. This table shows that $\omega_P \geq \gamma$, or $\omega_P \tau \geq 1$; namely, the plasmon is not overdamped, as discussed in the introduction of Section 7.3.2.

Figure 7.14 shows that the best-fit curves to the L^+ mode (solid lines) also explain the L^--mode profiles to a reasonable extent. There, Harima et al. included [99] the contribution of the charge-density-fluctuation (CDF) mechanism [106,107] for Raman scattering to improve the fitting. If the CDF contribution is included, Eq. (7.12) can be modified as follows:

$$I(\omega) = S[A(\omega) + KB(\omega)]Im[-1/\varepsilon(\omega)], \tag{7.17}$$

where K is a mixing constant adjusted by fitting, and $B(\omega)$ gives the CDF contribution,

$$B(\omega) = \left[\omega_P^2 \gamma(\omega_L^2 - \omega^2)^2 + \omega_P^4 \gamma(\omega_L^2 - \omega_T^2) + \gamma\Gamma^2\omega_P^2\omega^2\right]/\Delta. \tag{7.18}$$

The dashed curve in Figure 7.14d includes the CDF contribution. It gives a better fit to the L^- mode but gives no effective contribution to the L^+-mode profile [99]. This means that the CDF mechanism contributes to the L^- mode. A similar situation occurs in n-type GaAs [110], where the CDF gives a comparable contribution to the DP and EO mechanisms. The importance of the CDF contribution to the L^- mode, especially in heavily doped GaN, has also been reported by Demangeot et al. [111].

As the triangle plots in Figure 7.15 show, the L^+ mode in p-type GaN shows a much smaller variation than in the n-type; therefore, precise line-shape analysis as done in the n-type to obtain carrier density and mobility is difficult. A numerical simulation suggests that this is due to heavy damping of hole plasmons [112]. Such a small frequency shift has also been confirmed experimentally by Popovici et al. [113] up to $p = 7 \times 10^{17}\,\mathrm{cm}^{-3}$ and Demangeot et al. [109] up to $3 \times 10^{18}\,\mathrm{cm}^{-3}$ carrier concentration.

LOPC modes in cubic nitride are hardly reported. Ramsteiner et al. examined cubic n-type GaN layers on GaAs and found the overdamped nature of plasmons [114]. They explained the spectral features by considering the CDF mechanism and also a contribution from an impurity-induced Frölich mechanism.

7.3.4 Inter-Valence-Band Transitions

It was shown in Section 7.2.6 that a continuum band appeared in the low-frequency region in Mg-doped GaN when the acceptor impurities were activated (see Figure 7.12). Figure 7.17 compares Raman spectra of Mg-doped samples grown by MOCVD with postannealing in a N_2 gas atmosphere at 800 °C [112]. The samples had different hole densities p, as denoted in the figure. The bottom spectrum was obtained from an as-grown pre-annealed sample, which is highly resistive. Here, the spectra are normalized to the peak height of the strongest phonon peak at $568\,\mathrm{cm}^{-1}$ (E_2^H). This figure shows that the continuum band at $< \sim 400\,\mathrm{cm}^{-1}$ grows in intensity with increasing p. This monotonous change is clearly depicted in the lower figure. Here, the continuum band intensity at $144\,\mathrm{cm}^{-1}$ was normalized to the E_2^L-phonon band intensity (assumed independent of p in the tested range), and plotted. Harima et al. assigned this continuum band to inter-valence-band transition of holes (heavy-hole-to light-hole-band transition), and claimed that the band intensity could be used as a measure of p [112]. This type of transition has been reported before in heavily doped p-type Si and GaAs [115,116]. In these cases, the continuum band in the Raman spectra extended up to the optic–phonon region and induced asymmetric distortion called *Fano interference profiles* [92,117]. This feature may appear for the E_2^L-phonon mode if tested samples have higher hole concentrations.

7.3.5 Shallow Impurity Levels

Electronic Raman transitions involving shallow donor or acceptor levels have been reported in common semiconductors [91]. However there have been no such reports on GaN or related nitrides, with the exception of, Ramsteiner et al. [118], who reported electronic Raman transitions from 1s to 2s, and 1s to 3s donor levels in a hexagonal- and cubic-phase mixture of nominally undoped GaN grown on GaAs by MBE. The authors reported that these donor levels lie within 32 meV below the conduction band edge; however, there are different opinions about assigning the observed bands to

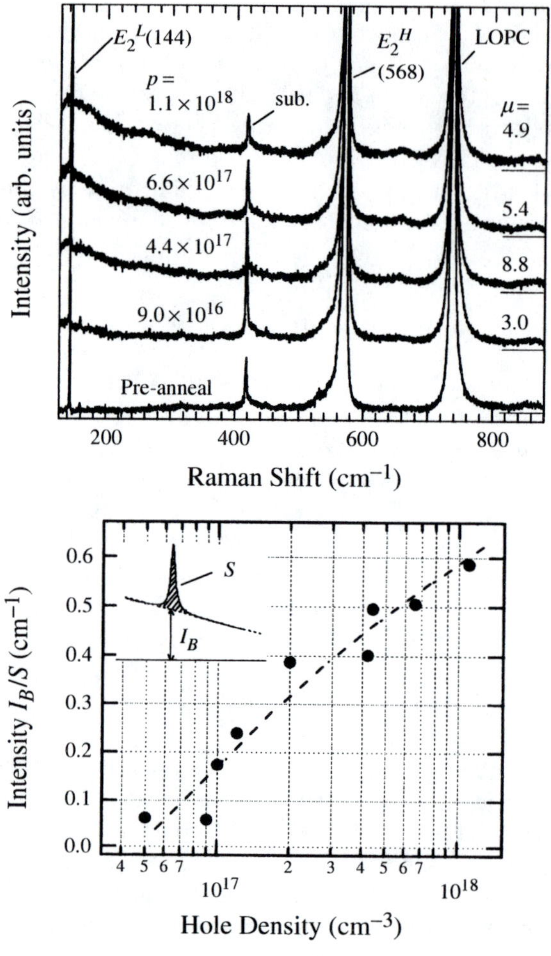

Figure 7.17 Raman spectra of Mg-doped p-type hexagonal GaN with various hole densities. (upper figure) [112]. The spectral intensities are normalized to the E_2^H-mode peak intensity at 568 cm^{-1} and shifted vertically for comparison. The lower figure shows the variation of the low-frequency continuum band at 144 cm^{-1} (normalized to the E_2^L-mode peak). The dashed line is drawn as a guide.

As impurities incorporated in the GaN lattice, and this argument is not yet settled [119–121].

A unique study on shallow donor levels was conducted using heavily doped GaN samples by applying high hydrostatic pressure [101,122,123]. Figure 7.18 shows the result for a highly oxygen-doped layer grown by HVPE (upper) and an unintentionally doped bulk sample grown by a high-pressure technique (lower) [123]. At low hydrostatic pressure the LO phonon is not observed in either figure because it couples with a plasmon and is

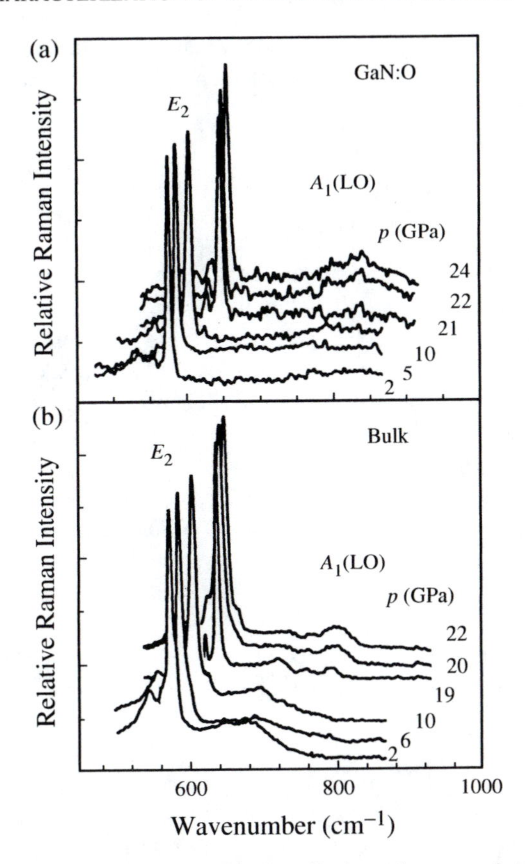

Figure 7.18 Raman spectra plotted as a function of hydrostatic pressure for a highly O-doped hexagonal GaN layer (2×10^{19} cm^{-3}) grown by HVPE (a), and an unintentionally doped bulk sample grown by a high-pressure technique (b) [123]. The A_1(LO) mode appears at above 20 GPa for both samples.

severely damped out due to the contained high carrier density ($\sim 10^{19}$ cm^{-3}), however, when hydrostatic pressure greater than 20 GPa is applied, the phonon appears, grows in intensity, and shifts to higher frequency with increasing pressure. This behavior was described in Section 7.2.4 for pressure effects on phonon lines (see Figure 7.8b). Wetzel et al. [123] considered that a carrier freezing occurs at high pressure, and concluded that the electronic ground state of O_N undergoes a transition from a shallow hydrogenic level (dilute doping case) or degenerate level (high doping case) below the critical pressure (20 GPa) to a strongly localized gap state above it [123]. According to the authors, no such carrier freezing occurs up to 25 GPa in Si-doped GaN; thus, Si is a good hydrogenic donor to cover a wide pressure range. They further claimed that since alloying of GaN with AlN induces effects similar to applying high pressure, namely, reduction of interatomic distance, Si should also be a good hydrogenic donor in AlGaN alloys.

7.4 TOPICAL CASES

7.4.1 Resonant Raman Scattering

A resonant enhancement in the first-order Raman scattering efficiency for optical phonons occurs when the incident photon energy is sufficiently close to that of the fundamental or higher-energy gap of a semiconductor crystal [124]. Resonant enhancement helps us observe weak Raman signals from a limited scattering volume, which are barely detectable under off-resonant conditions. Although there are number of publications on Raman scattering by optical phonons in GaN and related nitrides, most of them were observed under off-resonance conditions using visible lasers at around 500 nm. Resonant Raman scattering using UV lasers has been reported relatively recently, and the reports are still limited. InN or In-rich InGaN alloys present exceptional cases where resonant enhancement is expected with common visible lasers [24].

Wagner and coworkers reported a series of resonant Raman spectra on hexagonal GaN layers on sapphire [125], single QWs of GaN/AlGaN [38,41], and InGaN/GaN [126], and heterostructures of InGaN/GaN/AlGaN [41] and InGaN/GaN [127]. The merits of resonant Raman scattering are depicted in Figure 7.19. Here the spectra for a $Al_{0.15}Ga_{0.85}N$ (0.1 μm)/GaN (3 nm)/$Al_{0.15}Ga_{0.85}N$ (1.5 μm) QW are compared for on-resonance excitation (top) and off-resonance excitation (middle), and a nonresonant excitation for a thick GaN sample (bottom) [38]. The excitation energy at 3.54 eV is close to resonance with the interband transitions in the GaN well layer between the topmost valence-band state and the lowest quantized electronic level. It is clearly seen in the middle spectrum that the LO-phonon region is dominated by the signal of AlGaN barrier layers, whereas in the top spectrum the GaN well-layer signal is strongly enhanced and becomes the dominant peak. Recalling that the incident photon well penetrates the whole thickness of the QW structure, and that the well layer is much thinner (3 nm) than the barrier layers (1.5 μm), the resonance enhancement effect is remarkable. Behr et al. [38] concluded that there was no cation intermixing at the interface. The authors also observed resonant enhancement in $In_xGa_{1-x}N$/GaN/AlGaN heterostructures [41]. By the resonant Raman effect, they further demonstrated compositional inhomogenity in InGaN layers on GaN [127], and the presence of strain-induced piezoelectric fields in InGaN/GaN QWs [126]. It should be emphasized that resonant Raman scattering provides a powerful method for characterizing structural or compositional properties as well as electronic properties for QW layers that are

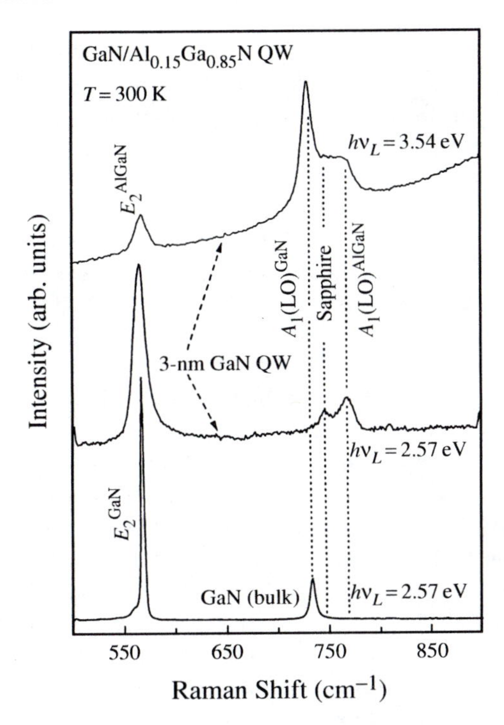

Figure 7.19 Raman spectra of hexagonal $Al_{0.15}Ga_{0.85}N$/GaN (3 nm)/$Al_{0.15}Ga_{0.85}N$ single QW (middle and top) and of a hexagonal bulklike GaN reference sample (bottom) [38]. The top spectrum was recorded with excitation at 3.54 eV, which is close to resonance with the interband transition in the GaN layer; the others were recorded with off-resonance excitation at 2.57 eV.

a few nanometers thick. This method is therefore very suitable for examining future device structures. Compositional inhomogenity or phase separation has also been investigated by resonant Raman scattering in hexagonal GaN [128,129] and cubic [130] GaN layers. A resonant Raman study on quantum dot structures of hexagonal GaN embedded in AlN has also recently been reported by Gleize et al. [131]. The authors observed resonance enhancement of LO-phonon signals at 3.8 eV excitation, which was nearly resonant with the fundamental transition energy of the quantum dots.

7.4.2 Micro-Raman Imaging

Micro-Raman imaging, or mapping of Raman microprobe signals or their derivatives, is now a standard spectroscopic technique for getting an overview of the spatial variation of physical properties in various materials [132]. This technique is now frequently employed for the nitrides to characterize inhomogenities in strain, carrier concentration, polytype (cubic or hexagonal), and the like, for a variety of epitaxial layers [103,133–135]. ELOG layers also have been intensively studied by Raman imaging [104,105,136]. Since only an outline is presented here, the readers are strongly recommended to see the original reports with their beautiful colorful mappings.

Ponce et al. [102] observed a lateral distribution of the A_1(LO)-phonon-mode intensity with a spatial resolution of 0.5 μm in an n-type GaN layer grown by MOCVD on sapphire. The authors obtained the Si-dopant distribution in the GaN layer, assuming that the LO phonon quenches in the region and including high donor density. Using the same analytical technique, Harima et al. [27] studied the distribution of carrier density and mobility in a GaN MOCVD layer and discussed anisotropy between the directions parallel and perpendicular to the source-gas flow.

Siegle et al. [103] observed the cross section of an undoped 220-μm-thick GaN layer grown on sapphire by HVPE. They obtained a carrier-density variation from 10^{17} cm^{-3} (near surface) to 10^{20} cm^{-3} (substrate interface) by analyzing the LOPC mode. The latter large value was attributed to the high density of defects near the interface with the substrate. Peak-frequency variation of the E_2^H-phonon mode revealed relaxation of biaxial compressive strain toward the surface. The authors studied the spatial correlation between Raman scattering and microscopic photoluminescence data; the peak position of the near-bandgap luminescence shifted to lower energies with decreasing distance from the substrate, but a strong blueshift occurred at the interface. The authors correlated these effects with the inhomogeneous free-carrier distribution and strain gradient found by Raman scattering.

A similar result, i.e., high carrier density near the substrate interface, was observed by Goldys et al. in a thick HVPE film [135]. Siegle et al. [133] observed the cross section of a 400-μm-thick HVPE film to see the variation in the phonon intensity ratio between the E_2^H-mode at 568 cm^{-1} and the A_1(TO) mode at 531 cm^{-1}. They found a reorientation of the c-axis near the substrate interface.

Papers related to Raman imaging for ELOG–GaN samples are rapidly appearing. For example, Pophristic et al. [136] used confocal Raman microscopy [132] to improve spatial resolution and found that the GaN layer is unintentionally doped with Si deriving from the SiN mask. Holtz et al. [105] observed isolated hexagonal islands grown on a square-patterned mask. They reported larger biaxial compressive strain and smaller free-carrier density in the window region than in the overgrowth region. Bertram et al. [104] also characterized the distributions of free-carrier density and compressive strains in ELOG–GaN samples. The results were combined with the cathode luminescence imaging. They reported qualitatively the same results as Holtz et al. [105] for the strain and the free-carrier density. The authors further clarified that local luminescence from the coherent-growth region at the window is dominated by narrow excitonic emission, showing superior crystalline quality.

7.4.3 Time-Resolved Raman Scattering

Tsen and coworkers reported a series of time-resolved Raman studies on hexagonal GaN. [93–96] In the early experiments they used 600-fs laser pulse trains at a photon energy of 4.36 eV [94]. The pulses were split into two equally intense but perpendicularly polarized beams: one was used to excite electron–hole pairs, and the other to probe nonequilibrium LO phonon populations by Raman scattering with variable delay times. The density of photoexcited electron–hole pairs was obtained by fitting the time-integrated luminescence. The authors observed transient Raman spectra due to single-particle scattering (SPS) by electrons with a crossed polarization geometry, $z(x, y) - z$, which probes contributions by a spin-density fluctuation mechanism [91,92]. The SPS profiles were theoretically fitted [137] using the electron collision time τ_e as an adjustable parameter. The temporal variation of the A_1(LO)-mode intensity was also measured at up to 5 ps of delay time to deduce the population decay time τ_p of nonequilibrium LO phonons. The authors obtained $\tau_e = 15$–20 fs at an electron density of 10^{17}–$10^{18}\,\mathrm{cm^{-3}}$, and $\tau_p = 5$ ps at an electron–hole pair density of $10^{17}\,\mathrm{cm^{-3}}$. They also showed that at an electron–hole pair density greater than $5 \times 10^{17}\,\mathrm{cm^{-3}}$, the electron distribution can be well described by Fermi–Dirac distribution functions with the temperature of the electron substantially higher than that of the lattice.

In later experiments, Tsen et al. [95] used 50-fs UV-laser pulses for the time-resolved Raman scattering and obtained an electron–LO phonon scattering rate of $4 \times 10^{13}\,\mathrm{s^{-1}}$. The large scattering rate, thus strong electron–phonon interaction, was attributed to the strong ionicity of Ga–N bonds in hexagonal GaN. Using the same experimental technique, the authors further investigated the decay process of LO phonons: they concluded that the zone-center LO phonons in hexagonal GaN decay creating primarily large-wave-vector TO phonons and large-wave-vector LA or TA phonons.

7.5 CONCLUSIONS

Owing to the rapid progress in the growth technology and growing need for characterization of optoelectronic devices, there has recently been dramatic development in Raman studies on GaN and related compounds, especially in the studies on the lattice properties of binary compounds: Γ-point phonon frequencies of GaN, AlN, and InN are all established. The data for hexagonal GaN are precisely known, including stress and temperature dependence.

As for the ternary compounds, phonon frequencies in the hexagonal AlGaN system are relatively well studied over the entire compositional range. The nature of one- or two-mode–type behavior is almost elucidated for most of the phonon modes, and the physical background has been intensively studied. For InGaN alloy data, on the contrary, there a large range of atomic composition is missing, especially at high In content, where phase separation problems degrade sample quality.

Raman studies on electronic properties are still in the beginning stage and are far behind other spectroscopic techniques such as photoluminescence and photoreflection. Even bound-electronic transitions have not yet been experimentally confirmed by Raman scattering. Purely electronic signals are generally weak compared with phonon signals; therefore, precise measurements are usually conducted at low temperature. Since impurity vibrations may hide weak electronic signals even at low temperatures, high-purity samples must be prepared. Time-resolved Raman experiments using femtosecond laser pulse trains show a new direction toward understanding the dynamic properties of free carriers.

Raman scattering due to the LO–phonon–plasmon coupled mode in hexagonal n-type GaN has been well studied experimentally and has been intensively applied to the characterization of electric transport parameters such as carrier density and mobility. Raman scattering obviously provides a convenient electrodeless technique for characterizing future device structures. Development of a similar characterization technique for p-type material is an urgent issue.

Raman scattering will doubtless continue to find wide use in GaN and related nitrides as a basic spectroscopic testing tool for electronic and lattice properties, for at least two reasons: first, the growth technology of nitrides is still steadily developing, so the need for contactless, nondestructive characterization in advanced structures such as quantum wells and dots will surely increase. Second, rapid progress is also being made on the instrument side. The recent increase in UV–Raman studies is a good example. Because of the large-bandgap nature of nitrides, Raman experiments are naturally oriented to short-wavelength excitation, as evidenced by the recent remarkable success in resonant Raman studies.

Acknowledgments

I would like to thank the many people involved in this work through collaboration and discussion, including H. Sakashita, T. Inoue, H. Sone, E. Kurimoto, T. Hosoda, S. Nakashima, K. Kisoda, K. Mizoguchi, K. Furukawa, M. Ishida, H. Mouri, M. Taneya, H. Grille, F. Bechstedt, V. Yu. Davydov, T. Inushima, N. Wieser, H. Okumura, Y. Ishida, S. Yoshida,

T. Koizumi, S. Chu, A. Ishida, and H. Fujiyasu. I am also very grateful to S. Sakai for supplying a bulk GaN sample.

References

1. A. Trampart, O. Brandt, and K. H. Ploog, in *Semiconductors and Semimetals*, edited by J. I. Pankove, and T. D. Moustakas (Academic Press, San Diego, 1998) Vol. 50, p. 167.
2. M. Cardona, in *Light Scattering in Solids II*, edited by M. Cardona, and G. Günterrodt, Topics in Applied Physics (Springer-Verlag, Berlin, 1982) Vol. 50, p. 19.
3. R. Loudon, *Adv. Phys.*, **13**, 423 (1964).
4. H. Harima, H. Sakashita, and S. Nakashima, unpublished.
5. L. Bergman, D. Alexson, P. L. Murphy, and R. J. Nemanich, *Phys.Rev. B*, **59**, 12977 (1999).
6. T. Azuhata, T. Sota, K. Suzuki, and S. Nakamura, *J. Phys.: Condensed Matter*, **7**, L129 (1995).
7. C. A. Argüello, D. L. Rousseau, and S. P. S. Porto, *Phys. Rev.*, **181**, 1351 (1969).
8. V. Yu. Davydov, N. S. Averkiev, I. N. Goncharuk, D. K. Nelson, I. P. Nikitina, A. S. Polkovnikov, A. N. Smirnov, M. A. Jacobson, and O. K. Semchinova, *J. Appl. Phys.*, **82**, 5097 (1997).
9. L. Bergman, D. Alexson, R. J. Nemanich, M. Dutta, M. A. Atroscio, C. Balkas, and R. Davis, *MRS Internet J. Nitride Semicond. Res.*, **4S1**, G6.65 (1997).
10. D. D. Manchon, A. S. Barker, P. J. Dean, and R. B. Zetterstrom, *Solid State Commun.*, **8**, 1227 (1970).
11. G. Burns, F. Dacol, J. C. Marinace, B. A. Scott, and E. Burstein, *Appl. Phys. Lett.*, **22**, 356 (1973).
12. Brafman, G. Lengyel, S. S. Mitra, J. N. Gielisse, J. N. Plendl, and L. C. Mansur, *Solid State Commun.*, **6**, 523 (1968).
13. J. A. Sanjuro, E. L.-Cruz, P. Vogl, and M. Cardona, *Phys. Rev. B*, **28**, 4579 (1983).
14. C. Carlone, K. M. Lankin, and H. R. Shanks, *J. Appl. Phys.*, **55**, 4010 (1984).
15. A. Cingolani, M. Ferrara, M. Lugara, and G. Scamarcio, *Solid State Commun.*, **58**, 823 (1986).
16. K. Hayashi, K. Itoh, N. Sawaki, and I. Akasaki, *Solid State Commun.*, **77**, 115 (1991).
17. P. Perlin, C. J. Carillon, J. P. Itie, A. S. Miguel, I. Grzegory and A. Polian, *Phys. Rev. B*, **45**, 83 (1992).
18. L. E. McNeil, M. Grimsditch, and R. H. French, *J. Am. Ceram. Soc.*, **76**, 1132 (1993).
19. P. Perlin, A. Polian, and T. Suski, *Phys. Rev. B*, **47**, 2874 (1993).
20. T. Kozawa, T. Kachi, H. Kano, Y. Taga, and M. Hashimoto, *J. Appl. Phys.*, **75**, 1098 (1994).
21. H. Siegle, L. Eckey, A. Hoffmann, C. Thomsen, B. K. Meyer, D. Schikora, M. Hankeln, and K. Lischka, *Solid State Commun.*, **96**, 943 (1995).
22. A. Tabata, R. Enderlein, J. R. Leite, S. W. da Silva, J. C. Galzerani, D. Schkora, M. Kloidt, and K. Lischka, *J. Appl. Phys.*, **79**, 4137 (1996).
23. V. Yu. Davydov, Yu. E. Kitaev, I. N. Goncharuk, A. N. Smirnov, J. Graul, O. Semchinova, D. Uffmann, M. B. Smirnov, A. P. Mirgorodsky, and R. A. Evarestov, *Phys Rev. B*, **58**, 12899 (1998).
24. V. Yu. Davydov, V. V. Emtsev, I. N. Goncharuk, A. N. Smirnov, V. D. Petrikov, V. V. Mamutin, V. A. Vekshin, S. V. Ivanov, M. B. Smirnov, and T. Inushima, *Appl. Phys. Lett.*, **75**, 3297 (1999).
25. A. Tabata, A. P. Lima, L. K. Teles, L. M. R. Scolfaro, J. R. Leite, V. Lemos , B. Schöttker, T. Frey, D. Schikora, and K. Lischka, *Appl. Phys. Lett.*, **74**, 362 (1999).
26. H. Harima, T. Inoue, and S. Nakashima, unpublished.
27. H. Harima, H. Sakashita, T. Inoue, and S. Nakashima, *J. Cryst. Growth*, **189/190**, 672 (1998).
28. M. H. Brodsky, in *Light Scattering in Solids*, edited by M. Cardona, Topics in Applied Physics (Springer, Berlin, 1975). Vol. 8, p. 205.

29. T. Azuhata, T. Matsunaga, K. Shimada, K. Yoshida, T. Sota, K. Suzuki, and S. Nakamura, *Physica B*, **219/220**, 493 (1996).
30. H. Siegle, G. Kaczmarczyk, L. Filippidis, A. P. Litvinchuk, A. Hoffmann, and C. Thomsen, *Phys. Rev. B*, **55**, 7000 (1997).
31. K. Karch, J.-M. Wagner, and F. Bechstedt, *Phys. Rev. B*, **57**, 7043 (1998).
32. S. Murugkar, R. Merlin, A. Botchkarev, A. Salvador, and H. Morkoç, *J. Appl. Phys.*, **77**, 6042 (1995).
33. A. Cros, H. Angerer, R. Handschuh, O. Ambacher, M. Stutzmann, R. Höpler, and T. Metzger, *Solid State Commun.*, **104**, 35 (1997).
34. H. Grille and F. Bechstedt, *J. Raman Spec.*, **27**, 201 (1996).
35. H. Grille, Ch. Schnittler, and F. Bechstedt, to be published in *Phys. Rev. B*.
36. H. Harima, T. Inoue, S. Nakashima, H. Okumura, Y. Ishida, S. Yoshida, T. Koizumi, H. Grille, and F. Bechstedt, *Appl. Phys. Lett.*, **74**, 191 (1999).
37. S. Clur, O. Briot, J.-L. Rouviere, A. Andonet, Y.-M. Le Vaillant, B. Gil, R. L. Aulombard, J. F. Demangeot, J. Frandon, and M. Renucci, *Mat. Res. Soc. Symp. Proc.,* **462**, 23 (1997).
38. D. Behr, R. Niebuhr, J. Wagner, K.-H. Bachem, and U. Kaufmann, *Appl. Phys. Lett.*, **70**, 363 (1997).
39. F. Demangeot, J. Groenen, J. Frandon, M. A. Renucci, O. Briot, S. Clur, and R. L. Aulombard, *Appl. Phys. Lett.*, **72**, 2674 (1998).
40. P. Wiesniewski, W. Knap, J. P. Malzac, J. Camassel, M. D. Bremser, R. F. Davis, and T. Suski, *Appl. Phys. Lett.*, **73**, 1760 (1998).
41. D. Behr, R. Niebuhr, H. Obloh, J. Wagner, K. H. Bachem, and U. Kaufmann, *Mat. Res. Soc. Symp. Proc.*, **468**, 213 (1997).
42. H. Harima, E. K urimoto, Y. Sone, S. Nakashima, S. Chu, A. Ishida, and H. Fujiyasu, *Phys. Status Solidi A*, **216**, 785 (1999)
43. N. Wieser et al., private communication.
44. A. Tabata, J. R. Leite, A. P. Lima, E. Silveira, V. Lemos, T. Frey, D. J. As, D. Schikora, and K. Lischka, *Appl. Phys. Lett.*, **75**, 1095 (1995).
45. H. Harima, Y. Nanishi, N. Teraguchi, and A. Suzuki, unpublished.
46. K. Osamura, S. Naka, and Y. Murakami, *J. Appl. Phys.*, **46**, 3432 (1975).
47. N. E. Christensen and P. Perlin, in *Gallium Nitride* I, edited by J. I. Pankove, and T. D. Moustakas, Semiconductors and Semimetals, (Academic Press, London, 1998) Vol. 50, p. 409.
48. B. Gil, in *Gallium Nitride* II, edited by J. I. Pankove, and T. D. Moustakas, Semiconductors and Semimetals, (Academic Press, London, 1999) Vol. 57, p. 209.
49. C. Kisielowski, in *Gallium Nitride* II, edited by J. I. Pankove, and T. D. Moustakas, Semiconductors and Semimetals, (Academic Press, London, 1999) Vol. 57, p. 275.
50. I. Grzegory and S. Krukowski, *Phys. Scr.,* T**36**, 242 (1991).
51. P. Perlin, J. Camassel, W. Knap, T. Taliercio, J. C. Chervin, T. Suski, I. Grzegory, and S. Porowski, *Appl. Phys. Lett.*, **67**, 2524 (1995).
52. B. A. Weinstein, *Solid State Commun.*, **24**, 595 (1977).
53. I. Gorczyca and N. E. Christensen, *Solid State Commun.*, **79**, 1033 (1991).
54. *Numerical Data and Functions in Science and Technology*/Landolt–Börnstein, (Springer-Verlag, Berlin, 1982) Vol. 3/17a.
55. A. Taylor and R. M. Jones, in *Silicon Carbide—A High Temperature Semiconductor*, edited by J. R. O'Conner, and J. Smiltens (Pergamon Press, Oxford, 1960) p. 147.
56. C. Kisielowski, J. Krüger, S. Ruvimov, T. Suski, J. W. Ager III, E. Jones, Z. L. Weber, M. Rubin, E. R. Weber, M. D. Bremser, and R. F. Davis, *Phys. Rev. B*, **54**, 17745 (1996).
57. G. H. Li, W. Zhang, H. X. Han, Z. P. Wang, S. K. Duan, *Appl. Phys. Lett.*, **86**, 2051 (1999).
58. A. Kuramata, K. Horino, K. Domen, K. Shinohara, and T. Tanahashi, *Appl. Phys. Lett.*, **67**, 2521 (1995).
59. S. Nakamura, M. Senoh, S. Nagahama, N. Iwasa, T. Yamada, T. Matsushita, H. Kiyoku, and Y. Sugimoto, *Appl. Phys. Lett.*, **68**, 3269 (1996).
60. T. Kozawa, T. Kachi, H. Kano, H. Nagase, N. Koide, and K. Manabe, *J. Appl. Phys.*, **77**, 4389 (1995).

61. W. Rieger, T. Metzger, H. Angerer, R. Dimitrov, O. Ambacher, and M. Stutzmann, *Appl. Phys. Lett.*, **68**, 970 (1996).
62. F. Demangeot, J. Frandon, M. A. Renucci, O. Briot, B. Gil, and R. L. Aulombard, *Solid State Commun.*, **100**, 207 (1996).
63. In-H. Lee, In-H. Choi, C.-R. Lee, E.-J. Shin, D. Kim, S. K. Noh, S.-J. Son, K. Y. Lim, and H. J. Lee, *J. Appl. Phys.*, **83**, 5787 (1998).
64. R. J. Briggs and A. K. Ramdas, *Phys. Rev. B*, **13**, 5518 (1976).
65. M. Balkanski, R. F. Wallis, and E. Haro, *Phys. Rev. B*, **28**, 1928 (1983).
66. A. Link, K. Bitzer, W. Limmer, R. Sauer, C. Kirchner, V. Schwegler, M. Kamp, D. G. Ebling, and K. W. Benz, *J. Appl. Phys.*, **86**, 6256 (1999).
67. M. Stavola, in *Identification of Defects in Semiconductors*, edited by M. Stavola, Semiconductors and Semimetals (Academic Press, London, 1999) Vol. 51B, p. 153.
68. H. Amano, M. Kito, K. Hiramatsu, and I. Akasaki, *Jpn. J. Appl. Phys.*, 28, L2112 (1989).
69. S. Nakamura, N. Iwasa, M. Senoh, and T. Mukai, *Jpn. J. Appl. Phys.*, **31**, 1258 (1992).
70. W. Götz, N. Johnson, D. Bour, M. McCluskey, and E. Haller, *Appl. Phys. Lett.*, **69**, 3725 (1996).
71. H. Harima, T. Inoue, S. Nakashima, M. Ishida, and M. Taneya, *Appl. Phys. Lett.*, **75**, 1383 (1999).
72. H. Harima, T. Inoue, Y. Sone, S. Nakashima, M. Ishida, and M. Taneya, *Phys. Status Solidi B*, **216**, 789 (1999).
73. A. Kaschner, H. Siegle, G. Kaczmarczyk, M. Strassburg, A. Hoffmann, C. Thomsen, U. Birkle, S. Einfeldt, and D. Hommel, *Appl. Phys. Lett.*, **74**, 3281 (1999).
74. J. Neugebauer and C. G. Van de Walle, *Phys. Rev. Lett.*, **75**, 4452 (1995).
75. V. J. B. Torres, S. Cerg, and R. Jones, *MRS Internet J. Nitride Semicond. Res.*, **2**, 35 (1997).
76. Y. Okamoto, M. Saito, and A. Oshiyama, *Jpn. J. Appl. Phys.*, **35**, 807 (1996).
77. M. S. Brandt, J. W. Ager III, W. Götz, N. M. Johnson, J. S. Harris, Jr., R. J. Molnar, and T. D. Moustakas, *Phys. Rev. B*, **49**, 14758 (1994).
78. A. Kaschner, G. Kaczmarczyk, A. Hoffmann, C. Thomsen, U. Birkle, S. Einfeldt, and D. Hommel, *Phys. Status Solidi B*, **216**, 551 (1999).
79. S. J. Pearton, J. W. Corbett, and M. Stavola, in *Hydrogen in Crystalline Seminconductors*, Materials Science, edited by U. Gonser, A. Mooradian, R. M. Osgood, M. B. Panish, and H. Sakaki, (Springer-Verlag, Berlin, 1992) Vol. 23.
80. F. A. Reboredo and S. T. Pantelides, *Phys. Rev. Lett.*, **82**, 1887 (1999).
81. L. Sugiura, M. Suzuki, and J. Nishio, *Appl. Phys. Lett.*, **72**, 1748 (1998).
82. W. Limmer, W. Ritter, R. Sauer, B. Mensching, C. Liu, and B. Rauschenbach, *Appl. Phys. Lett.*, **72**, 2589 (1998).
83. C. H. Seager, S. M. Myers, G. A. Petersen, J. Han, and T. Headley, *J. Appl. Phys.*, **85**, 2568 (1999).
84. J. Q. Duan, B. R. Zhang, Y. X. Zhang, L. P. Wang, G. G. Qin, G. Y. Zhang, Y. Z. Tong, S. X. Jin, Z. J. Yang, X. Zhang, and Z. H. Xu, *Appl. Phys. Lett.*, **82**, 5745 (1997).
85. M. G. Weinstein, C. Y. Song, M. Stavola, S. J. Pearton, R. G. Wilson, R. J. Shul, K. P. Killeen, and M. J. Ludowise, *Appl. Phys. Lett.*, **72**, 1703 (1998).
86. C. G. Van de Walle, *Phys. Rev. B*, **56**, R10020 (1997).
87. M. Kuball, F. Demangeot, J. Frandon, M. A. Renucci, J. Massies, N. Grandjean, R. L. Aulombard, and O. Briot, *Appl. Phys. Lett.*, **73**, 960 (1998).
88. G.-C. Yi and B. W. Wessels, *Appl. Phys. Lett.*, **70**, 357 (1997).
89. M. O. Manasreh, J. M. Baranowski, K. Pakula, H. X. Jiang, and J. Lin, *Appl. Phys. Lett.*, **72**, 659 (1999).
90. J. M. Zhang, T. Ruf, M. Cardona, O. Ambacher, M. Stutzmann, J.-M. Wagner, and F. Bechstedt, *Phys. Rev. B*, **56**, 14399 (1997).
91. M. V. Klein, in *Light Scattering in Solids I*, edited by M. Cardona (Springer-Verlag, Berlin, 1975) p. 147.
92. G. Abstreiter, M. Cardona, and A. Pinczuk, in *Light Scattering in Solids*, edited by M. Cardona and G. Güntherrodt, Topics in Applied Physics (Springer-Verlag, Heidelberg, 1983) Vol. 54, p. 5.

93. S. J. Sheih and K. T. Tsen, D. K. Ferry, A. Botchkarev, B. Sverdlov, A. Salvador, and H. Morkoç, *Appl. Phys. Lett.*, **67**, 1757 (1995).
94. K. T. Tsen, R. P. Joshi, D. K. Ferry, A. Botchkarev, B. Sverdlov, A. Salvador, and H. Morkoç, *Appl. Phys. Lett.*, **68**, 2990 (1996).
95. K. T. Tsen, D. K. Ferry, A. Botchkarev, B. Sverdlov, A. Salvador, and H. Morkoç, *Appl. Phys. Lett.*, **71**, 1852 (1997).
96. K. T. Tsen, D. K. Ferry, A. Botchkarev, B. Sverdlov, A. Salvador, and H. Morkoç, *Appl. Phys. Lett.*, **72**, 2132 (1998).
97. A. Mooradian and G. B. Wright, *Phys. Rev. Lett.*, **16**, 999 (1966).
98. H. Harima, S. Nakashima, and T. Uemura, *J. Appl. Phys.*, **78**, 1996 (1995).
99. H. Harima, H. Sakashita, and S. Nakashima, *Mate. Sci. Forum,* **264–268**, 1363 (1998).
100. A. S. Barker, Jr. and M. Ilegems, *Phys. Rev. B*, **7**, 743 (1973).
101. C. Wetzel, W. Walukiewicz, E. E. Haller, J. Ager III, I. Grzegory, S. Porowski, and T. Suski, *Phys. Rev. B*, **53**, 1322 (1996).
102. F. A. Ponce, J. W. Steeds, C. D. Dyer, and G. D. Pitt, *Appl. Phys. Lett.*, **69**, 2650 (1996).
103. H. Siegle, A. Hoffmann, L. Eckey, C. Thomsen, J. Christen, F. Bertram, D. Schmidt, D. Rudloff, and K. Hiramatsu, *Appl. Phys. Lett.*, **71**, 2490 (1997).
104. F. Bertram, T. Riemann, J. Christen, A. Kaschner, A. Hoffmann, C. Thomsen, K. Hiramatsu, T. Shibata, and N. Sawaki, *Appl. Phys. Lett.*, **74**, 359 (1999).
105. M. Holtz, M. Seon, T. Prokofyeva, H. Temkin, R. Singh, F. P. Dabkowski, and T. D. Moustakas, *Appl. Phys. Lett.*, 75, **1757** (1999).
106. G. Irmer, V. V. Toporov, B. H. Bairamov, and J. Monecke, *Phys. Status Solidi B*, **119**, 595 (1983).
107. M. V. Klein, B. N. Ganguly, and P. J. Colwell, *Phys. Rev. B*, **6**, 2380 (1972).
108. W. L. Faust and C. H. Henry, *Phys. Rev. Lett.*, **17**, 1265 (1966).
109. F. Demangeot, J. Frandon, M. A. Renucci, N. Grandjean, B. B. Beaumont, J. Massies, and P. Gilbart, *Solid State Commun.*, **106**, 491 (1998).
110. A. Mooradian, in *Light Scattering Spectra of Solids*, edited by G. B. Wright (Springer-Verlag, Berlin, 1968) p. 285.
111. F. Demangeot, J. Frandon, M. A. Renucci, C. Meny, O. Briot, and R. L. Aulombard, *J. Appl. Phys.*, **82**, 1305 (1997).
112. H. Harima, T. Inoue, S. Nakashima, K. Furukawa, and M. Taneya, *Appl. Phys. Lett.*, **73**, 2000 (1998).
113. G. Popovici, G. Y. Xu, A. Botchkarev, W. Kim, H. Tang, A. Salvador, and H. Morkoç, *J. Appl. Phys.*, **82**, 4020 (1997).
114. M. Ramsteiner, O. Brandt, and K. H. Ploog, *Phys. Rev. B*, **58**, 1118 (1998).
115. F. Cerdeira, T. A. Fjeldly, and M. Cardona, *Phys. Rev. B*, **8**, 4734 (1973).
116. D. Olego and M. Cardona, *Phys. Rev. B*, **23**, 6592 (1981).
117. U. Fano, *Phys. Rev.*, **124**, 1866 (1961).
118. M. Ramsteiner, J. Menniger, O. Brandt, H. Yang, and K. H. Ploog, *Appl. Phys. Lett.*, **69**, 1276 (1996).
119. H. Siegle, I. Loa, P. Thurian, L. Eckey, A. Hoffmann, I. Broser, and C. Thomsen, *Appl. Phys. Lett.*, **70**, 909 (1997).
120. M. Ramsteiner, J. Menniger, O. Brandt, H. Yang, and K. H. Ploog, *Appl. Phys. Lett.*, **70**, 910 (1997).
121. H. Siegle, A. Kaschner, A. Hoffmann, I. Broser, and C. Thomsen, *Phys. Rev. B*, **58**, 13619 (1998).
122. P. Perlin, T. Suski, H. Teisseyre, M. Leszczynski, I. Grzegory, J. Jun, S. Porowski, P. Bogusawski, J. Bernholc, J. C. Chervin, A. Polian, and T. D. Moustakas, *Phys. Rev. Lett.*, **75**, 296 (1995).
123. C. Wetzel, T. Suski, J. W. Ager III, E. R. Weber, E. E. Haller, S. Fischer, B. K. Meyer, R. J. Molnar, and P. Perlin, *Phys. Rev. Lett.*, **78**, 3923 (1997).
124. See, e.g., A. Pinczuk and E. Burstein, in *Light Scattering in Solids I*, edited by M. Cardona, (Springer-Veralag, Berlin, 1975) p. 23.
125. D. Behr, J. Wagner, J. Schneider, H. Amano, and I. Akasaki, *Appl. Phys. Lett.*, **68**, 2404 (1996).
126. J. Wagner, A. Ramakrishnan, H. Obloh, and M. Maier, *Appl. Phys. Lett.*, **74**, 3863 (1999).

127. D. Behr, J. Wagner, A. Ramakrishnan, H. Obloh, and K.-H. Bachem, *Appl. Phys. Lett.*, **73**, 241 (1998).
128. N. Wieser, O. Ambacher, H.-P. Felsl, L. Görgens, and M. Stutzmann, *Appl. Phys. Lett.*, **74**, 3981 (1999).
129. L. H. Robins, A. J. Paul, C. A. Parker, J. C. Roberts, S. M. Bedair, E. L. Piner, and N. A. El-Masry, *MRS Internet J. Nitride Semicond. Res.*, **4S1**, G3.22 (1999).
130. E. Silveira, A. Tabata, J. R. Leite, R. Trentin, V. Lemos, T. Frey, D. J. As, D. Schikora, and K. Lischka, *Appl. Phys. Lett.*, **75**, 3602 (1999).
131. J. Gleize, F. Demangeot, F. Frandon, M. A. Renucci, M. Kuball, F. Widmann, and B. Daudin, *Phys. Status Solidi B*, **216**, 457 (1999).
132. See, e.g., J. Barbillat, in *Raman Microscopy*, edited by G. Turrell, and J. Corset (Academic Press, London, 1996) p. 175.
133. H. Siegle, P. Thurian, L. Eckey, A. Hoffmann, C. Thomsen, B. K. Meyer, H. Amano, I. Akasaki, T. Detchprohm, and K. Hiramatsu, *Appl. Phys. Lett.*, **68**, 1265 (1996).
134. H. Harima, T. Inoue, S.-I. Nakashima, H. Okumura, Y. Ishida, S. Yoshida, and H. Hamaguchi, *J. Cryst. Growth*, **189/190**, 435 (1998).
135. E. M. Goldys, T. Paskova, I. G. Ivanov, B. Arnaudov, and B. Monemar, *Appl. Phys. Lett.*, **73**, 3583 (1998).
136. M. Pophristic, F. H. Long, M. Schurman, J. Ramer, and I. T. Ferguson, *Appl. Phys. Lett.*, **74**, 3519 (1999).
137. D. C. Hamilton and A. L. McWhorter, in *Light Scattering in Solids III*, edited by G. B. Wright (Springer, New York, 1969).

CHAPTER 8

Raman Studies of Wurtzite GaN and Related Compounds

J. FRANDON, F. DEMANGEOT, and M.A. RENUCCI

Laboratoire de Physique des Solides (ESA 5477 CNRS), Université Paul Sabatier, 118 Route de Narbonne, 31062 Toulouse Cédex 04, France

8.1 INTRODUCTION

Raman spectroscopy has become widely used in the field of semiconductor physics for studying all types of solid-state excitations (phonons, one-electron excitations, plasmons, polaritons, spin-waves, etc.). This technique is based on the inelastic scattering of light by elementary excitations. In a first-order scattering process, one excitation with wavevector $\mathbf{q}$ and frequency ω is either created (Stokes process) or destroyed (anti-Stokes process). Considering Stokes scattering events, the most frequently encountered in experiments, conservation of momentum and energy requires both

$$\mathbf{q} = \mathbf{k}_i - \mathbf{k}_s$$

and

$$\omega = \omega_i - \omega_s,$$

where $\mathbf{k}_i$ and ω_i are, respectively, the wavevector *inside* the medium and the frequency of the incident photon, and $\mathbf{k}_s$ and ω_s are the corresponding physical quantities for the scattered photon. In typical experiments performed in the visible, only zone-center excitations (i.e., with negligible $\mathbf{q}$) are involved in first-order scattering, since $\lambda \gg a_0$ (where a_0 is the lattice constant); however, breakdown of the $\mathbf{q} \cong 0$ selection rule may arise when translational symmetry is lacking.

From a macroscopic point of view, light scattering originates from the modulation $\delta\overleftrightarrow{\chi}$ of the optical susceptibility induced by the excitation of the medium, which is linear in the excitation amplitude for first-order scattering. When lattice vibrations are concerned, for example, $\delta\overleftrightarrow{\chi}$ proceeds from the atomic displacements (**u**-mechanism) and possibly from the associated macroscopic electric field (**E**-, or electrooptic, mechanism) in polar materials. In doped semiconductors, charge-density fluctuations of the solid-state electron plasma provide an additional modulation mechanism (ρ-mechanism).

In the microscopic approach, the mechanism by which electromagnetic radiation ensures the energy transfer to the crystal is a three-step process mediated by the electrons. This mechanism implies:

(1) the destruction of an incident photon

(2) the creation of a phonon, in a Stokes event

(3) the creation of a scattered photon

Virtual electronic transitions take place at each step. Let us consider the particular process corresponding to the following sequence: step 1 is accompanied by the excitation of an electron–hole pair via the electron–radiation interaction H_{eR}; step 2 involves the transition of the electron or the hole to another state via the electron–phonon interaction H_{ep}; and step 3 occurs through the recombination of the pair via H_{eR}.

The scattering probability is obtained by a quantum mechanical third-order perturbation calculation. For the process just described, the probability is given by

$$P(\omega_s) = \frac{2\pi}{\hbar}\left|\sum_{\alpha,\beta}\frac{\langle 0|H_{eR}|\beta\rangle\langle\beta|H_{ep}|\alpha\rangle\langle\alpha|H_{eR}|0\rangle}{(h\omega_i - E_\alpha - i\eta_\alpha)\cdot(h\omega_s - E_\beta - i\eta_\beta)}\right|^2 \delta(\hbar\omega_i - \hbar\omega_s - \hbar\omega), \tag{8.1}$$

where $|\alpha\rangle$ and $|\beta\rangle$ are intermediate electronic states, with energy E_α and E_β, and damping constants η_α and η_β. The energy of the electronic ground state $|0\rangle$ is taken as reference.

When the energy of the incident or (and) scattered photon is tuned to an electronic transition energy, an enhancement of the Raman cross section may be expected from the preceding expression, which becomes all the more spectacular as the damping of the intermediate state involved in the process becomes smaller. This effect is known as resonant Raman scattering.

Electron–phonon interaction may take place through various processes. The most common, and efficient in any material, is the deformation potential interaction: the periodic crystal potential is directly modulated by the atomic displacements. Fröhlich interaction must be considered in the case of polar phonons in crystals with partially ionic bonding, as electrons may interact with the macroscopic electric field associated with the atomic displacements. Matrix elements of the Fröhlich interaction can be separated into q-independent interband terms, which involve states $|\alpha\rangle$ and $|\beta\rangle$ in different bands, and intraband terms, implying states $|\alpha\rangle$ and $|\beta\rangle$ in the same band, which contribute for $\mathbf{q} \neq 0$.

Table 8.1 Raman tensors for "allowed" scattering by phonons in wurtzite crystals.

$\begin{pmatrix} a & 0 & 0 \\ 0 & a & 0 \\ 0 & 0 & b \end{pmatrix}$	$\begin{pmatrix} 0 & 0 & c \\ 0 & 0 & 0 \\ c & 0 & 0 \end{pmatrix}$	$\begin{pmatrix} 0 & 0 & 0 \\ 0 & 0 & c \\ 0 & c & 0 \end{pmatrix}$
$A_1(z)$	$E_1(x)$	$E_1(y)$
$\begin{pmatrix} 0 & d & 0 \\ d & 0 & 0 \\ 0 & 0 & 0 \end{pmatrix}$	$\begin{pmatrix} d & 0 & 0 \\ 0 & -d & 0 \\ 0 & 0 & 0 \end{pmatrix}$	
E_2	E_2	

Constraints on the Raman cross section are imposed by the symmetry properties of the scattering medium. The intensity of scattered light by a phonon of given symmetry J in a crystal (i.e., belonging to the irreducible representation Γ_J of the crystallographic point group) is written as

$$I_s \propto |\mathbf{e_i} \cdot \overleftrightarrow{R}_J \cdot \mathbf{e_s}|^2 \tag{8.2}$$

where $\mathbf{e_i}$ and $\mathbf{e_s}$ are the polarization vectors of incident and scattered photons, and $\overleftrightarrow{R}_J$ is the Raman tensor that characterizes the phonon. Raman tensors for "allowed" scattering, involving the q-independent matrix elements of deformation potential and interband Fröhlich electron–phonon interactions, are given in Table 8.1 for the wurtzite structure. They are related to the $\delta \overleftrightarrow{\chi}$ changes associated with the **u**- and **E**-mechanisms in macroscopic theory. In near-resonant conditions, "forbidden" scattering associated with q-dependent matrix elements of the intraband Fröhlich interaction becomes dominant. The corresponding Raman tensors [1] are given in Table 8.2.

A typical Raman experiment requires a monochromatic laser source, a multistage spectrometer with high resolution and high stray-light rejection, and a sensitive light detector (today, multichannel CCDs are the most commonly used systems). For Raman investigation of nitride semiconductors, excitation may be achieved either with visible light in the transparency range of samples, or with UV lines for resonant scattering near the fundamental energy gap. Note that only a few discrete excitations are available in the UV range, emitted at 3.00, 3.05, 3.41, 3.53, 3.71, and 5.08 eV (frequency doubled) by the argon ion laser, or at 3.80 eV by the He–Cd laser. Small areas on the samples, about 1 μm^2, can be probed by Raman microscopy. This is particularly useful for checking selection rules in backscattering geometry

Table 8.2 Raman tensors for "forbidden" (Fröhlich) scattering by polar phonons in wurtzite crystals.

For phonon wavevector **q** parallel to the x-direction:

$$\begin{pmatrix} 0 & 0 & e' \\ 0 & 0 & 0 \\ e' & 0 & 0 \end{pmatrix} \qquad \begin{pmatrix} a' & 0 & 0 \\ 0 & a' & 0 \\ 0 & 0 & b' \end{pmatrix} \quad \begin{pmatrix} 0 & d' & 0 \\ d' & 0 & 0 \\ 0 & 0 & 0 \end{pmatrix}$$

$A_1(z)$ $E_1(x)$ $E_1(y)$

For phonon wavevector **q** parallel to the z-direction:

$$\begin{pmatrix} h' & 0 & 0 \\ 0 & h' & 0 \\ 0 & 0 & g' \end{pmatrix} \qquad \begin{pmatrix} 0 & 0 & f' \\ 0 & 0 & 0 \\ f' & 0 & 0 \end{pmatrix} \quad \begin{pmatrix} 0 & 0 & 0 \\ f' & 0 & g' \\ 0 & g' & 0 \end{pmatrix}$$

$A_1(z)$ $E_1(x)$ $E_1(y)$

with incident light normal to the edge of epitaxial layers, typically 1–2 μm thick.

8.2 LONG-WAVELENGTH PHONONS IN WURTZITE GaN, AlN AND InN

8.2.1 Raman-Active Phonons in GaN

Gallium nitride and related compounds crystallize, in their stable phase, in the wurtzite structure belonging to space group C_{6v}^4 ($P6_3mc$). The unit cell with its four inequivalent atoms is shown in Figure 8.1, which illustrates the relationship between wurtzite and zinc-blende structures. The stacking along the [0001] direction (**c**-axis) is of type ABAB... in the wurtzite structure, whereas the stacking along the [111] axis is of type ABCA... in the zinc-blende structure: the periodicity of the latter is twice the periodicity of the former in real space. The Brillouin zone of the wurtzite structure is given in Figure 8.2.

The mechanical representation based on the atomic motions has 12 dimensions. Its decomposition into irreducible representations of the C_{6v} point group gives three acoustical and nine optical phonons at the Brillouin zone center. For the optical modes the representation reduces to

$$\Gamma^{opt} = A_1 + E_1 + 2B_1 + 2E_2 \tag{8.3}$$

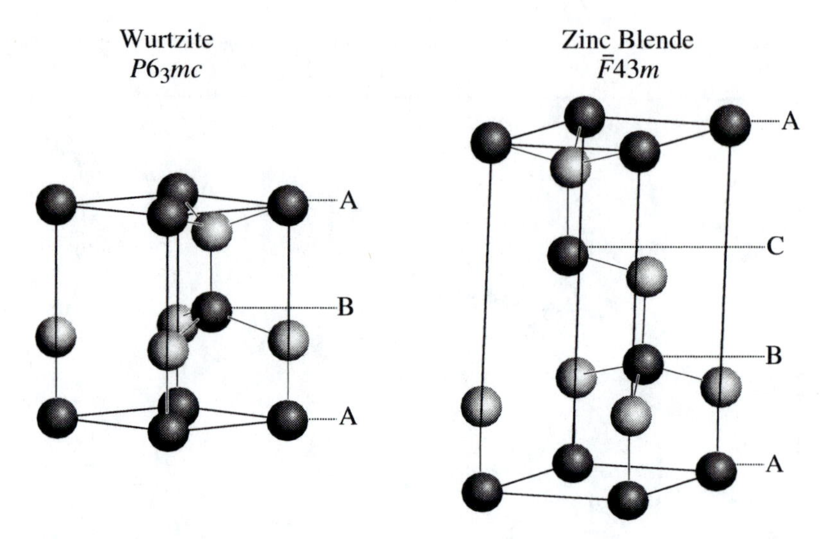

Figure 8.1 Unit cells of hexagonal and cubic GaN. Gallium and nitrogen atoms are represented in gray and black, respectively.

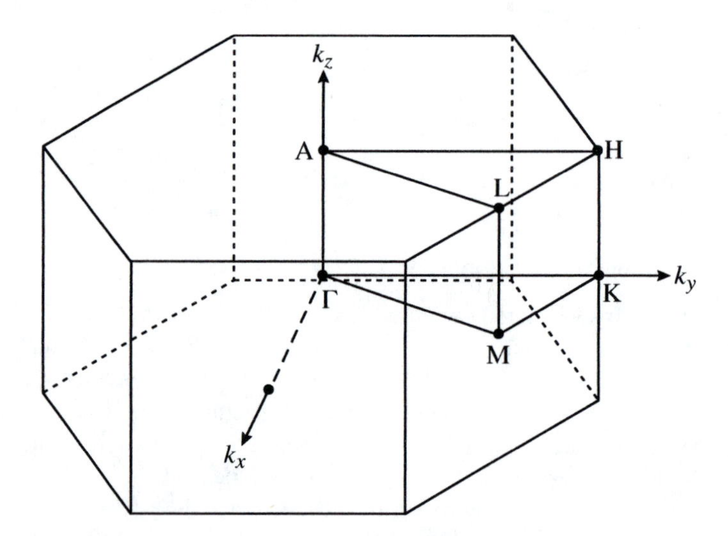

Figure 8.2 Brillouin zone of wurtzite GaN.

Thus the A_1 and E_1 symmetries are found once, whereas the B_1 and E_2 symmetries appear twice [2]. The B_1 modes are "silent", i.e., optically inactive. The nonpolar, doubly degenerate E_2 modes are Raman-active only. Due to the absence of an inversion center in the wurtzite structure, two modes are both IR and Raman active, the A_1 and the doubly degenerate E_1

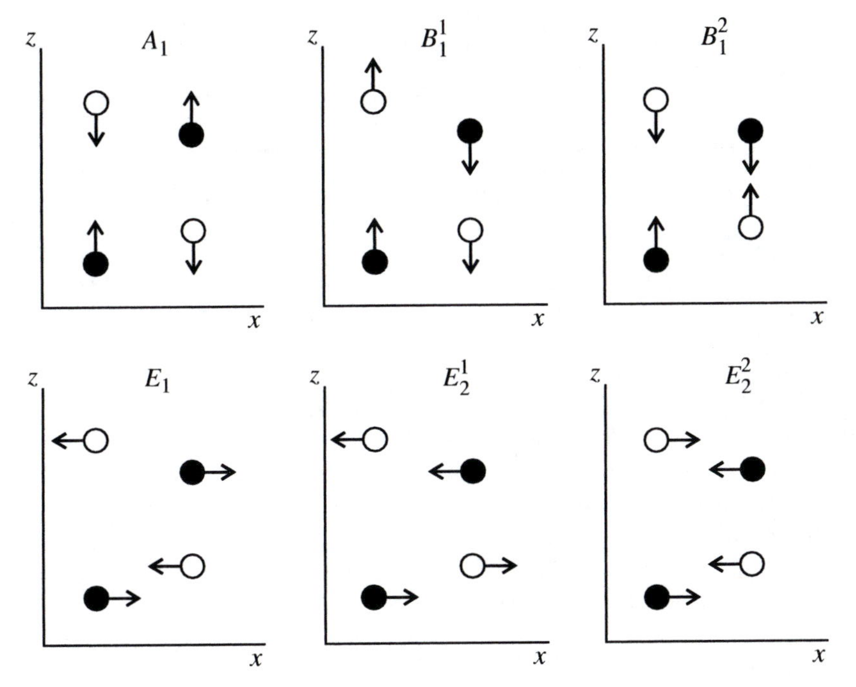

Figure 8.3 Atomic displacements corresponding to long-wavelength phonons in wurtzite GaN.

polar phonons. The associated atomic motions are shown in Figure 8.3. The A_1 and E_1 phonons correspond to displacements respectively parallel and normal to the **c**-axis, and their frequency splitting is directly related to the crystal anisotropy.

Owing to the relationship between both structures, the phonon dispersion in the direction $\Gamma - A$, i.e., along the **c**-axis of the Brillouin zone of wurtzite, may be obtained ideally by folding the phonon branches along the direction $\Gamma - L$ of the Brillouin zone of zinc blende: the zone-center B_1 and E_2 upper optical phonons in the wurtzite structure correspond to the LO(L) and TO(L) in the zinc blende structure.

The strong ionicity of the bond in GaN crystals results in an important frequency splitting of the transverse and longitudinal polar phonons due to the macroscopic electric field associated with the displacements; the corresponding long-range forces largely prevail over the short-range forces responsible for the anisotropy.

Far from resonant conditions, i.e., when the deformation potential and the electrooptic processes are dominant in the electron–phonon interaction, the selection rules for Raman scattering by optical phonons in wurtzite,

Table 8.3 Selection rules for "allowed" Raman scattering by phonons in wurtzite structure.

$z(xx)\bar{z}$	$z(xy)\bar{z}$	$x(zz)\bar{x}$	$x(yy)\bar{x}$	$x(yz)\bar{x}$	$x(yz)\bar{y}$
A_1(LO), E_2	E_2	A_1(TO)	A_1(TO), E_2	E_1(TO)	E_1(LO), E_1(TO)

Note: The symmetries of $\mathbf{q} = 0$ phonons are given for various configurations (x and y are any perpendicular directions in the plane normal to the **c**-axis).

found by using the Raman tensors of Table 8.1, are given in Table 8.3. The experimental geometry is characterized by Porto's notation $\mathbf{k_i}$ $(\mathbf{e_i}$ $\mathbf{e_s})$ $\mathbf{k_s}$, where the wavevector and the polarization of incident (scattered) light are $\mathbf{k_i}$ and $\mathbf{e_i}$ ($\mathbf{k_s}$ and $\mathbf{e_s}$).

GaN samples are usually thin epitaxial layers whose plane is normal to the **c**-axis; in this case, TO phonons cannot be discerned by Raman spectroscopy in the simplest experimental configuration (backscattering normal to the layer surface), as shown in Table 8.3. Micro-Raman measurements on the edge are thus needed for their observation. Note also that the E_1(LO) phonon is allowed only in a right-angle scattering geometry, with the **q** phonon wavevector normal to the **c**-axis.

It may be noted, however, that "forbidden" phonons can be activated by structural disorder, induced by defects or sample damage, for example. Moreover, the aforementioned selection rules are violated near resonance as the Fröhlich intraband electron–phonon mechanism becomes dominant; selection rules for the latter are governed by the Raman tensors given in Table 8.2. Note that "forbidden" A_1(LO) phonons are evidenced in backscattering along the **c**-axis for parallel polarizations of incident and scattered light, as in the case of "allowed" scattering: the distinction between both contributions is not straightforward in resonant conditions.

Manchon et al. [3] published the first Raman determination of all phonons but the A_1(LO), in bulk nonintentionally doped GaN needles. A few years later, Barker and Ilegems [4] performed IR reflectivity measurements on GaN thick layers; they deduced the frequencies of the E_1(TO) and A_1(LO) modes from a standard dielectric modelization of their spectra.

Today, most studied samples are epitaxial layers grown on various substrates (sapphire, 6H–SiC [5], $MgAl_2O_4$ [6]). Consequently, phonons from the substrate must be carefully identified in Raman spectra, in addition to the GaN phonons (for the most commonly used sapphire substrate, see Porto and Krishnan [7]).

Typical Raman spectra of GaN in various scattering geometries are shown in Figure 8.4. The most reliable values of Raman-active phonon frequencies are given in Table 8.4. They were recently obtained by Deguchi et al. [8] from high-quality GaN layers grown by the lateral epitaxial overgrowth technique, confirming previous determinations by Azuhata et al. [9].

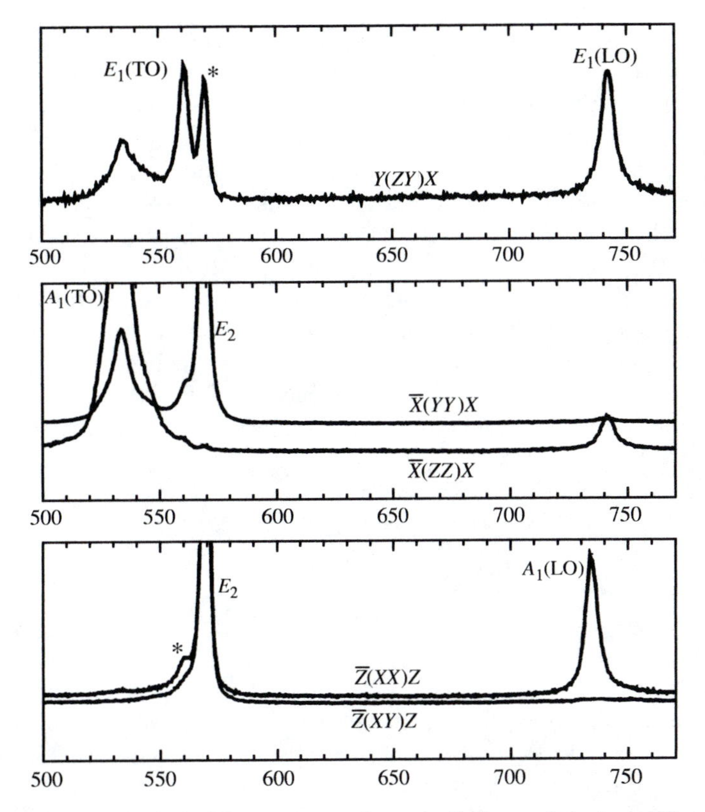

Figure 8.4 Typical polarized Raman spectra of wurtzite GaN, recorded using a 488-nm laser line. Peaks marked with an asterisk correspond to the leakage of signal coming from a slightly defective experimental geometry.

Table 8.4 Long-wavelenth phonon frequencies (cm^{-1}) for GaN, AlN, and InN.

	$E_2(low)$	$A_1(TO)$	$E_1(TO)$	$E_2(high)$	$A_1(LO)$	$E_1(LO)$
GaN	144[a]	533[a]	561[a]	569[a]	735[a]	743[a]
AlN	249[b]	614[b]	673[b]	662[b]	891[b]	912[b]
InN	87[c]	448[c]	476[c]	488[c]	586[c]	593[c]

[a]Ref. 8.
[b]Ref. 10.
[c]Ref. 12.

8.2.2 Raman-Active Phonons in AlN and InN

The phonon frequencies of wurtzite AlN were determined by McNeil et al. [10], who were the first to correctly assign all the peaks observed in various

Raman scattering geometries. The samples used for this study were an AlN single crystal and sintered polycrystals.

In contrast, phonons of InN have been investigated only recently. Using very thin ($< 0.2\,\mu$m) InN layers grown on (0001) sapphire by MOCVD at temperatures above 450 °C, Lee et al. [11] detected the E_2 and A_1(LO) phonons in Raman spectra recorded in backscattering geometry on the layer surface, whereas Davydov et al. [12] measured the frequencies of the E_1(TO), A_1(TO), E_2, A_1(LO), and E_1(LO) modes. However, the studied samples exhibited nonintentional n-doping with electron concentrations in the 1×10^{19}–$2 \times 10^{20}\,\text{cm}^{-3}$ range. Thus, the frequencies directly obtained from the spectra in the LO range more likely refer to coupled LO–phonon–plasmon modes than to uncoupled phonons.

The most reliable phonon frequencies for AlN and InN are also listed in Table 8.4.

8.2.3 Silent Phonons in GaN

The two "silent" long-wavelength B_1 modes in GaN, which can be evidenced in first-order scattering by disorder activation, may be responsible for the features observed near $300\,\text{cm}^{-1}$ and $670\,\text{cm}^{-1}$ in the Raman spectra of ion-implanted GaN, published by Limmer et al. [13]. The unambiguous signature of these modes at $300\,\text{cm}^{-1}$ and $667\,\text{cm}^{-1}$ were assigned very recently by Wieser et al. [14] using isotopically pure GaN epitaxial films (with two different isotopes at the nitrogen sites) damaged by ion implantation. When light ^{14}N atoms are replaced by heavy ^{15}N atoms in the lattice, a significant shift is observed in the high-frequency feature only, contrary to its low-frequency counterpart; the assignment is straightforward, as the low-frequency B_1 mode corresponds to motion of Ga atoms, and the upper-frequency mode involves that of N atoms, according to the calculation of Ref. 15. In addition, the observed frequency shift is quantitatively explained by the atomic masses of different isotopes. As a comparison, the frequencies of both B_1 modes obtained from first-principles calculations for GaN and AlN [15–17], and from a phenomenological model for InN [12], are listed in Table 8.5.

8.2.4 Angular Dispersion of Zone-Center Optical Phonons

Crystals with wurtzite structure exhibit anisotropic physical properties. They are uniaxial crystals in which two types of long-wavelength polar phonons may propagate: the ordinary phonon, characterized by a polarization normal to the **c** optical axis, and the extraordinary phonon, whose polarization is set in a plane containing this axis. In contrast to the former, the latter exhibits

Table 8.5 Silent B_1-mode frequencies (cm^{-1}) for GaN and AlN.

	Calculations						*Experiments*	
	Lower B_1			*Upper* B_1			*Lower* B_1	*Upper* B_1
GaN	335[a]	330[b]	337[c]	697[a]	677[b]	720[c]	300[d]	660[d]
							300[e] (^{14}N)	667[e] (^{14}N)
							300[e] (^{15}N)	646[e] (^{15}N)
AlN	534[a]	553[b]		703[a]	717[b]			
InN	220[f]			565[f]				

[a] Ref. 15.
[b] Ref. 16.
[c] Ref. 17.
[d] Ref. 13.
[e] Ref. 14.
[f] Ref. 12.

an angular dispersion when the angle θ between the **c**-axis and the phonon wavevector **q** *inside* the crystal varies.

Moreover, the extraordinary LO phonon with A_1 symmetry (polarization along the **c**-axis) for $\theta = 0$ exhibits E_1 symmetry (polarization in the basal plane) for $\theta = 90°$. For any θ between 0 and 90°, the LO mode is a "quasi-LO phonon" with mixed (A_1, E_1) symmetry. Conversely, the extraordinary TO phonon, which corresponds to E_1 and A_1 for $\theta = 0$ and 90°, respectively, behaves like a "quasi-TO phonon" with mixed symmetry between these θ values. The modes mostly keep their longitudinal or transverse character for variable θ, with the macroscopic long-range electric field, which dominates over the anisotropic short-range forces due to the strong ionicity of the Ga–N bond, mediating the coupling of different symmetries.

According to Loudon's treatment in the case of weak anisotropy, the variation of frequency with θ is given for the quasi-LO and -TO phonons by the following formulas [18]:

$$\omega_{QTO}^2 = \omega_{E_1(TO)}^2 \cos^2\theta + \omega_{A_1(TO)}^2 \sin^2\theta; \tag{8.4}$$

$$\omega_{QLO}^2 = \omega_{A_1(LO)}^2 \cos^2\theta + \omega_{E_1(LO)}^2 \sin^2\theta. \tag{8.5}$$

The angular dispersion of GaN and AlN has been investigated in the experimental works of Azuhata et al. [9], Bergman et al. [19], and Filippidis et al. [20]. The latter authors studied only the variation of the QTO phonon frequency for $\theta > 40°$, whereas both the QLO and QTO excitations were measured between 65° and 90° in Ref. 19. In any case, the phonon frequencies obey the aforementioned law, as illustrated in Figure 8.5 for GaN and AlN. These frequency variations are in agreement with the results of

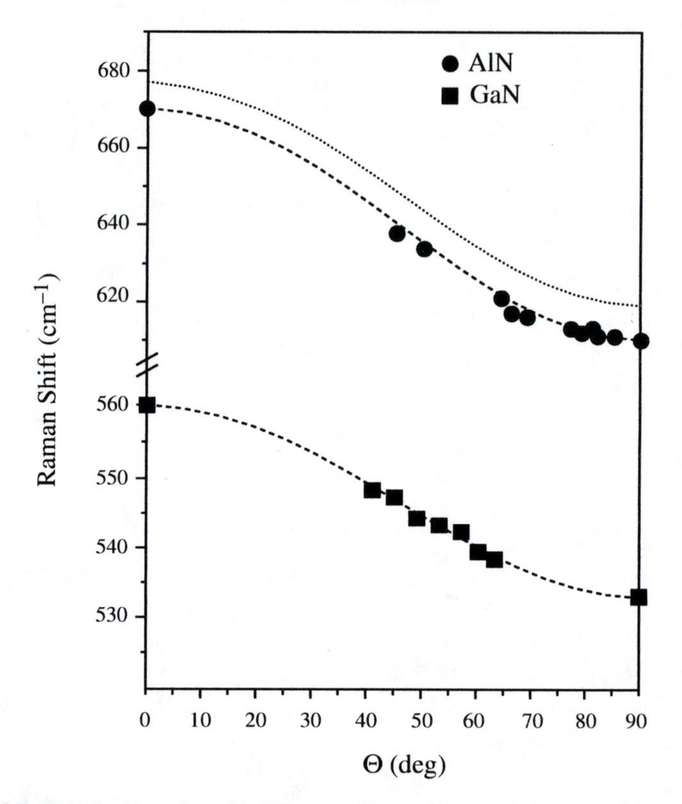

Figure 8.5 Angular dispersion of the TO extraordinary phonon measured in Ref. 20, for GaN (solid squares) and AlN (solid circles). The dashed lines correspond to the variation calculated according to Loudon [18].

calculations by Wei et al. [21], who used a phenomenological microscopic model of the zone-center optical phonons in GaN.

8.2.5 Phonon Lifetimes, Electron–Phonon Scattering Times

Phonon lifetimes at room temperature were indirectly determined by Bergman et al. [22] using a procedure that eliminates the spectral broadening of the Raman lines due to the finite slit widths of the spectrometer; the lifetime was deduced from the spectral width through the energy–time uncertainty relation. A lifetime of about 1 ps was found for the two TO phonons and for the high-frequency E_2 phonon in AlN and GaN. It was 10 times greater for the low-frequency E_2 mode.

By means of picosecond experiments on the A_1(LO) anti-Stokes Raman signal of GaN at low temperature, Tsen et al. [23] directly measured

5 ps for the lifetime of the LO phonons, which achieve the thermalization of hot electrons, and about 50 fs for the electron–phonon scattering time.

8.2.6 Structural Characterization of GaN Layers

Raman spectroscopy can yield very useful informations on some structural problems that may affect the layers. For example, a reorientation of GaN was detected by Siegle et al. [24] by means of micro-Raman measurements on the edge of very thick (400 μm) layers, as a function of the distance from the buffer interface: the **c**-axis of the crystallites experiences a rotation that has been correlated locally to the yellow photoluminescence (PL) band.

The same team [25] revealed the hexagonal minority phase in GaN layers with the zinc-blende structure, which eluded PL and X-ray diffraction measurements. The presence of the minority phase is not easy to establish, as the frequencies of optical phonons, TO and E_1(TO) on one hand, LO and A_1(LO) on the other hand, are very close to each other in the cubic and hexagonal phases; however, it is possible to obtain the signature of the minority phase in the case of weak tilt angles of the hexagonal [0001] axis with respect to the [001] axis of the cubic phase: the LO modes of both cubic and hexagonal GaN are allowed in the $z(y'y')\bar{z}$ configuration, whereas the former is forbidden in the $z(yy)\bar{z}$ configuration. Here y and y' stands, respectively, for the [010] and [110] axes of the cubic structure. Likewise, the TO mode of the majority phase, allowed in the $x(yy)\bar{x}$ configuration, can be distinguished from the E_1(TO) of the minority phase, which is forbidden in this case. Thus, cubic and hexagonal GaN present in the same sample can be separately characterized, using these experimental configurations.

8.3 SHORT-WAVELENGTH PHONONS AND PHONON DENSITY OF STATES

Invaluable information on phonon dispersion can be obtained from inelastic light scattering when the size of available samples precludes the use of a more direct technique such as neutron spectroscopy. Indeed, two-phonon Stokes processes involve the participation of excitations of opposite wavevectors, allowing observation of phonons of the whole Brillouin zone in second-order Raman scattering. Otherwise, the **q**-conservation rule can be lifted by the breaking of translational symmetry in disordered materials. Activation of ordinarily forbidden phonons at $\mathbf{q} \neq 0$ can thus be achieved in first-order scattering by defects in crystalline solids. As defect-induced Raman spectra mirror the phonon density of states (DOS), sharp features can be interpreted in terms of critical points lying along high-symmetry directions or at boundaries of the Brillouin zone.

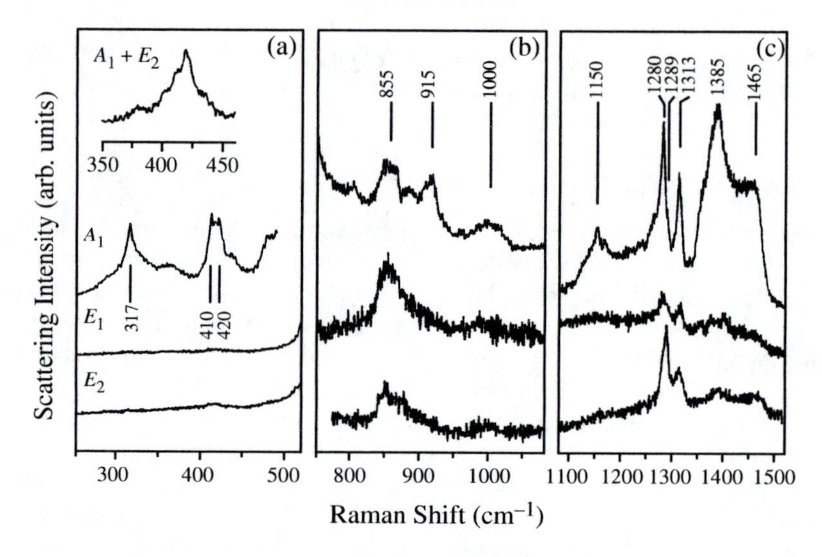

Figure 8.6 Polarized second-order Raman spectra of GaN, from Ref. 26.

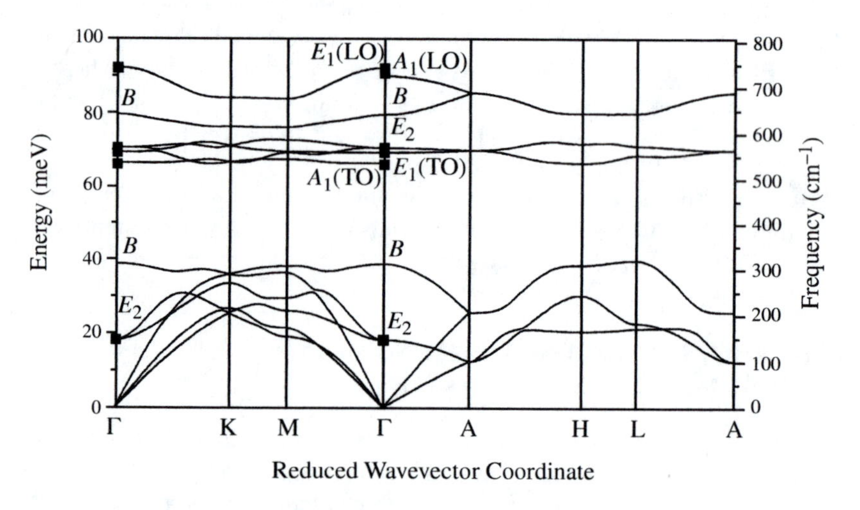

Figure 8.7 Phonon dispersion curves of GaN calculated in Ref. 26.

A thorough study of second-order Raman scattering was first performed by Siegle et al. [26] on hexagonal GaN. Various scattering geometries and polarization configurations were used to decompose the experimental spectra in the A_1, E_1, and E_2 irreducible components, as illustrated in Figure 8.6. An interpretation was made according to selection rules derived from group theory, and calculations of phonon dispersion curves. For the latter, which are displayed in Figure 8.7, Siegle et al. adapted to wurtzite-lattice compounds the modified valence-force model developed by Kane for diamond- and zinc-blende-structure semiconductors [27]. In their nine-parameter model, Siegle et al. considered short-range interactions between first and second neighbors, described by bond bending and bond stretching, plus the fifth-neighbor interaction introduced by McMurry et al. for cubic semiconductors [28]. Long-range Coulomb forces were treated within a rigid-ion model. Calculations were made using as starting values the Kane parameters of the isostructural ZnO and AlN compounds and then adjusting the model to data obtained from Raman experiments. Frequencies of the most intense second-order Raman features together with observed symmetries are listed in Table 8.6 also included are their assignments to definite processes implying phonons at special points of the Brillouin zone, showing that the strongest scattering appearing in the polarized A_1 spectrum is mainly due to overtone bands, as for other tetrahedral semiconductors [29].

More recently, Davydov et al. reported on second-order Raman scattering of GaN and AlN, as well as on defect-induced first-order scattering from intentionally damaged samples [30]. Detailed experimental information obtained at room and at low temperature was analyzed through group theory and lattice dynamical calculations. A seven-parameter model, based on pairwise interatomic potentials and rigid-ion Coulomb interactions, was

Table 8.6 Assignments, frequencies, and symmetries of strongest features in second-order Raman spectra of GaN.

Process	*Point*	*Frequency* (cm^{-1})	*Symmetry*
Overtone of acoustic phonons	H	317	A_1
Overtone of acoustic phonons	A, K	410	A_1
Overtone of acoustic phonons	M	420	A_1, E_2
Overtone of acoustic phonons	L	640	A_1
Acoustic–optical combination		855	A_1, E_2, E_1
Acoustic–optical combination		915	A_1
Acoustic–optical combination		1000	$A_1, (E_2)$
Overtone of optical phonons	K, H	1150	A_1
Combination of zone-boundary optical phonons		1280	$A_1, (E_1)$
Combination of zone-boundary optical phonons		1289	E_2
Combination of zone-boundary optical phonons		1313	$A_1, (E_1, E_2)$
Optical overtone of phonons	A, K	1385	A_1
Overtone of A_1(LO) phonons	Γ	1465	A_1
Overtone of E_1(LO) phonons	Γ	1495 (cutoff)	A_1, E_2

Source: From Siegle et al. [26].
Note: Parentheses indicate the possibility of additional symmetries.

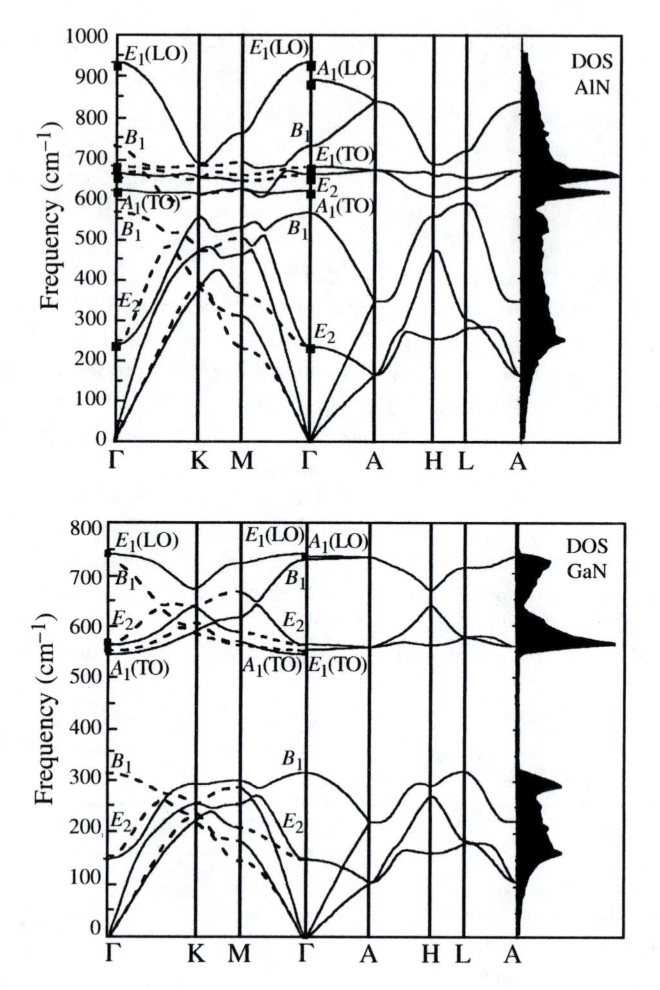

Figure 8.8 Phonon dispersion curves and phonon density of states of GaN and AlN, calculated in Ref. 30.

fitted to available data. The model includes the experimental values of elastic constants and Raman-active mode frequencies, and the estimate of the upper-B_1 silent-mode frequency from first-principles lattice dynamical calculations by Karch et al. [31]. The phonon dispersion and DOS calculated for GaN and AlN are shown in Figure 8.8.

Concerning GaN, the analysis is consistent on the whole with that of Siegle [26]. Davydov et al. underlined the flatness of the longitudinal optic branches in most of the Brillouin zone, which manifests itself in second-order scattering by a camelback-like structure between 1340 cm^{-1} and 1495 cm^{-1}, peaking at 1385 cm^{-1} and 1465 cm^{-1}. First-order scattering

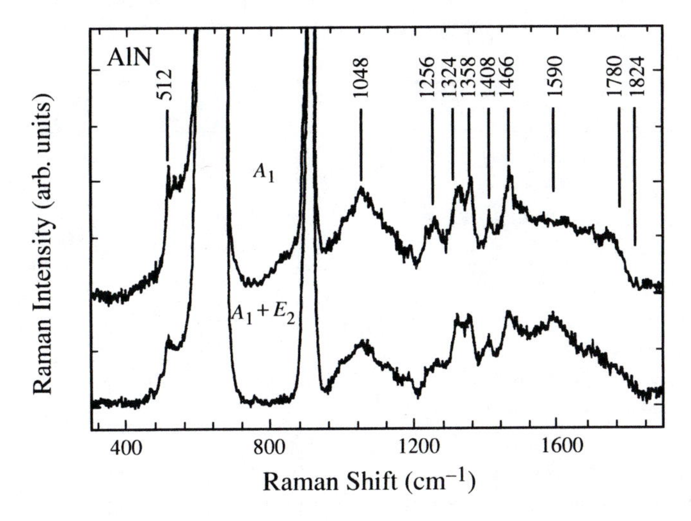

Figure 8.9 Polarized second-order Raman spectra of AlN, from Ref. 30.

from Er^+-implanted GaN samples reproduces this structure scaled by a factor of 2 and is in reasonable agreement with the calculated phonon DOS. The assignment of the highest-frequency peak to the upper phonon branch along $\Gamma - A$ is in disagreement with other phenomenological calculations, such as those developed by Azuhata et al. [32] to account for optical data on zone-center phonon frequencies measured by IR and Raman spectroscopies, or by Nipko and Loong [33] to explain the phonon DOS determined by time-of-flight neutron spectroscopy of powder samples. The assignment clearly proceeds from the high value of the upper-B_1-mode frequency used in Davydov's model as one of the input data.

As for AlN, polarized spectra recorded by Davydov in A_1 and $A_1 + E_2$ configurations are shown in Figure 8.9. The three independent components of the two-phonon Raman tensor could not be measured due to the rather weak intensity. Consequently, frequencies, symmetries, and assignments of the strongest second-order features given in Table 8.7 should be considered as tentative only. Experimental evidence for larger dispersion of the longitudinal optic branches, in comparison with GaN, is the lack of well-defined structures in the highest-frequency range of the spectra and corroborates the phonon dispersion calculations. The evidence agrees with the phonon DOS determined in the same report from defect-induced Raman scattering of Er^+-implanted samples, and by the neutron measurements of Nipko and Loong on AlN powders [34].

Recently, Raman data on phonons of the whole Brillouin zone were obtained for InN by Davydov et al. [12] and compared with calculations of the dispersion curves, shown in Figure 8.10, within the same empirical

Table 8.7 Assignments, frequencies, and symmetries of strongest features in second-order Raman spectra of AlN.

Process	*Point*	*Frequency* (cm^{-1})	*Symmetry*
Overtone of acoustic phonons		512	$A_1, (E_2)$
Overtone of acoustical phonons, acoustical-optical combination		1048	A_1, E_2
Overone of optical phonons	K or M	1256	$A_1, (E_2)$
Combination and overtone of optical phonons		1324	A_1
Overtone of zone-boundary optical phonons		1358	$A_1, (E_2)$
Combination of optical phonons		1408	A_1
Combination of optical phonons	M	1466	A_1, E_2
Overtone of optical phonons	K	1500	A_1
Overtone of optical phonons	M	1580	A_1, E_2
Overtone of E_1(LO) phonons	Γ	1824 (cutoff)	A_1, E_2

Source: From Davydov et al. [30].
Note: Parentheses indicate the possibility of additional symmetries.

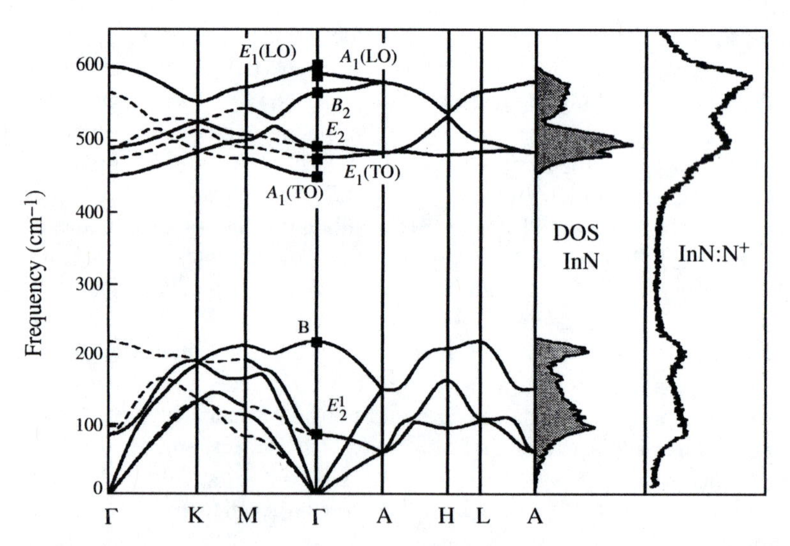

Figure 8.10 Phonon dispersion curves and phonon density of states of InN, calculated in Ref. 12.

model they used for GaN and AlN. As illustrated in the same figure, good agreement was found between the calculated phonon DOS and the disorder-induced Raman spectrum they recorded from N^+-implanted InN. Strong similarity with the phonon DOS of GaN at high energy reflects the major role played by nitrogen atoms in the dynamics of both lattices.

8.4 LONG-WAVELENGTH PHONONS IN TERNARY SOLID SOLUTIONS

8.4.1 $Al_xGa_{1-x}N$ Alloys

The first experimental study of phonons in $Al_xGa_{1-x}N$ alloys (for $x < 0.15$ only) was published by Hayashi et al. [35]. Then, the variation of phonon frequencies with x was investigated in the whole compositional range by Demangeot et al. [36] and by Cros et al. [37] on $Al_xGa_{1-x}N$ layers grown by MOVPE and plasma-enhanced MBE. In both cases, the vibrational modes in Raman spectra can be easily followed for $x < 0.5$; unfortunately, for Al contents in the 60–80% range, a partial relaxation of selection rules and a line broadening are observed in the Raman spectra, leading to questionable assignments of some of the experimental features in the correponding compositional range. According to Ref. 36, the best fit for the measured A_1(TO) and A_1(LO) frequencies is given (only for $x > 0.05$) by the following formulas:

$$\omega_{A_1(TO)} = 540 + 19.7x + 31.7x^2 + 22.2x^3; \qquad (8.6)$$

$$\omega_{A_1(LO)} = 746 + 169.5x + 11.7x^2 - 36.6x^3. \qquad (8.7)$$

The variation of phonon frequencies given in Ref. 36, which are very similar to the findings of Cros et al. [37], is shown in Figure 8.11. The experimental data suggest that the E_1(TO) and A_1(TO) modes, and more clearly the A_1(LO) and E_1(LO) modes, follow the so-called one-mode behavior: the frequency of the unique phonon varies continuously from the value measured for pure GaN to the one found in pure AlN. In contrast, in spite of the lack of unambiguous symmetry assignment for high Al contents, the high-frequency E_2 mode seems to exhibit a two-mode behavior.

Demangeot et al. [36] proposed for both LO phonons a dielectric modeling derived from the work of Hon and Faust [38], which takes into account the strong coupling between the GaN- and AlN-like oscillators via the macroscopic electric field. The frequency variation is satisfyingly reproduced by this calculation. Moreover, the apparent one-mode behavior of the LO phonon is interpreted as a true two-mode behavior, obscured by a quasi-complete intensity transfer from one type of oscillator (GaN) to the other (AlN), in most of the compositional range.

Very recently, Wieser el al. [14] found a two-mode behavior not only for the E_2 but also for the E_1(TO) mode. The latter point is supported by analysis of polarized Raman spectra of an $Al_{0.7}Ga_{0.3}N$ alloy, though the selection rules seem to be only approximately verified. A similar conclusion for the E_1(TO) phonon was drawn from IR reflectivity measurements by Wisniewski et al. [39]. Grille et al. [40] recently calculated phonons

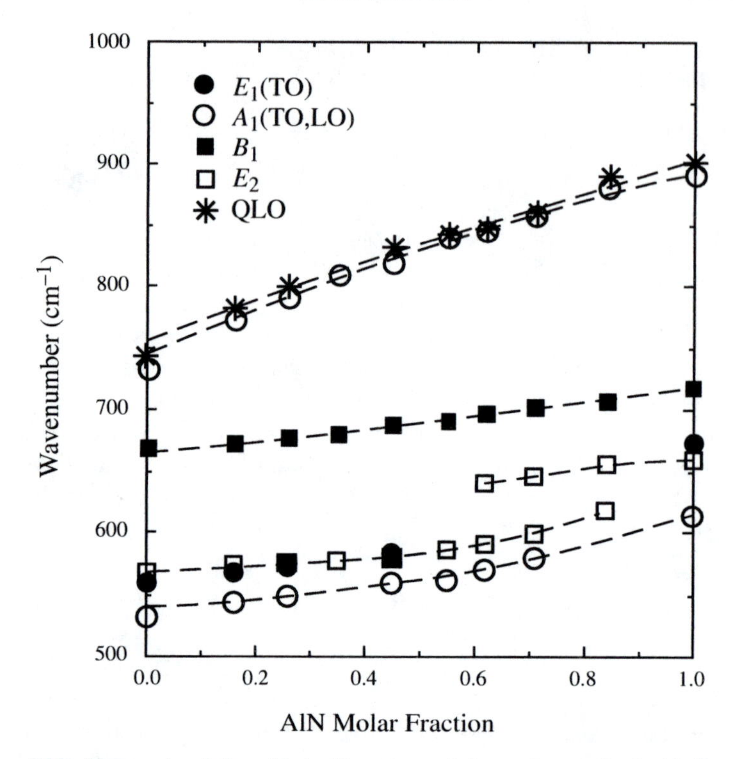

Figure 8.11 Measured variation with the Al content x of phonon frequencies for $Al_xGa_{1-x}N$ solid solutions, from Ref. 36.

in wurtzite $Al_xGa_{1-x}N$ alloys in the framework of the modified random-isodisplacement model. A two-mode behavior for both A_1(TO) and E_1(TO) modes was found, in contrast to the one-mode behavior obtained for the A_1(LO) and E_1(LO) phonons. Finally, the configurational disorder of MOCVD $Al_xGa_{1-x}N$ layers was studied by Bergman et al. [41]: the asymmetrical line broadening of the E_2 mode, which is maximum for aluminum content $x = 0.5$, was accounted for by the spatial correlation model, whereas no long-range order was evidenced in the solid solution.

8.4.2 $In_xGa_{1-x}N$ Alloys

The literature on Raman determination of vibrational modes in such alloys is very poor. Indeed, this solid solution is known to suffer phase segregation from an In content x of 0.1, leading to important compositional inhomogeneities. Only Behr et al. [42] have published the variation with x of the E_2 and A_1(LO) phonon frequencies in InGaN layers for an In content lower than 0.11. This variation suggests a one-mode behavior with significant bowing for these phonons.

8.5 PHONONS IN STRAINED WURTZITE MATERIALS

8.5.1 Strain in GaN Layers

Layers of nitride semiconductors with wurtzite structure are grown at high temperature on various substrates, among which the most commonly used are hexagonal (0001) sapphire (corundum) and 6H–SiC. In-plane stress in GaN films originates from growth on lattice-mismatched substrates and from a difference in the thermal expansion coefficients of layer and substrate. It is generally agreed that the only thermal mismatch stress (compressive for GaN on sapphire or tensile for GaN on SiC) persists after a few nanometers of growth; however, Edwards et al. [5] have reported weak compressive stresses due to lattice mismatch for thicknesses up to 0.5 μm in GaN layers grown on SiC.

The internal strain may easily be evidenced by means of Raman spectrometry. Indeed, the phonon frequencies are shifted from strain-free values by the stress experienced by the lattice. For example, the gradual relaxation of compressive stress in a very thick (220 μm) GaN layer grown on sapphire has been demonstrated by Siegle et al. [43] using measurements on the high-frequency E_2 phonon: the stress is completely relaxed at about 100 μm from the interface only, as shown in Figure 8.12. Thus, layers grown on sapphire, which are typically 1 μm thick, exhibit at room temperature a significant residual biaxial stress, which is always found to be compressive.

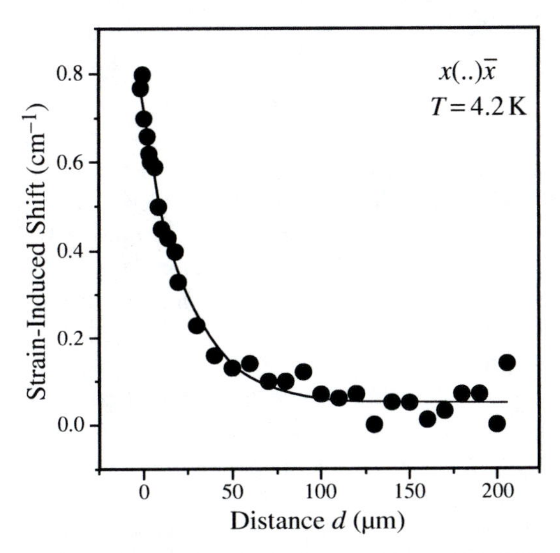

Figure 8.12 Strain-induced shift of the E_2 phonon in a 220-μm-thick GaN layer for various distances from the interface, measured in Ref. 43.

8.5.2 Phonon Deformation Potentials of Wurtzite GaN

The first determination of phonon deformation potentials of wurtzite crystals was performed by Briggs and Ramdas [44], who investigated the effect of an external uniaxial stress on the Raman spectrum of CdS. Demangeot et al. [45] applied a similar analysis to shifts in phonon frequencies under the built-in stress in epitaxial GaN layers grown on (0001) sapphire. Biaxial strain results from the in-plane stress $\sigma_{xx} = \sigma_{yy} = -\sigma_b$ ($\sigma_b > 0$ for compressive stress) and $\sigma_{zz} = 0$, where z is the growth axis of the layer. The effects of strain are considered in terms of linear deformation potential theory. In order to be consistent with the definition usually adopted for the hydrostatic pressure coefficient, we define the stress coefficient of phonon J as $K_J^B = dv'_J/d\sigma_b$, where $J = A_1, E_1, E_2$, as

$$K_J^B = -2(S_{11} + S_{12})a_J - 2S_{13}b_J, \tag{8.8}$$

where the S_{ij} are the elastic constants. The deformation potential constants a_J and b_J are the changes in frequency for a 100% strain along the z- and x-axes, respectively. One more relation is needed for the experimental determination of the $\{a_J, b_J\}$ set of constants. It may be obtained from the pressure coefficient of phonon J under hydrostatic pressure p, $K_J^H = dv_I/dp$, where $\sigma_{xx} = \sigma_{yy} = \sigma_{zz} = -p$ ($p > 0$ for compression). The relation is given by the following equation:

$$K_J^H = -2(S_{11} + S_{12} + S_{13})a_J - (2S_{13} + S_{33})b_J. \tag{8.9}$$

Gil et al. [46] performed a stress calibration on the same GaN samples in an independant experiment by analyzing the energy shift of the excitons in reflectivity spectra recorded in the UV range. In these samples, the internal strain was correlated to the III/V molar ratio used for the MOVPE growth of the layers [47]. Taking into account the stress calibration, the Raman measurements gave for the stress coefficients K_J^B the values 2.9 and 0.8 cm^{-1}/GPa for the E_2 and A_1(LO) phonons, respectively. Finally, the hydrostatic data needed for the determination of deformation potentials were taken from Ref. 48.

More recently, Davydov et al. [49] studied GaN layers grown on 6H–SiC after the deposition of various buffer layers and derived the deformation potentials for A_1(TO), E_1(TO), and both E_2 modes from a direct stress calibration based on X-ray measurements of lattice parameters. The stress coefficient of the high-frequency E_2 mode (2.7 cm^{-1}/GPa) is very close to the value given in Ref. 45. Note that the positive sign of $K_{E_2}^B$ corresponds to our definition of the pressure coefficient. The low-frequency E_2 mode exhibits a weaker stress coefficient with opposite sign. This

Table 8.8 Biaxial stress coefficients of long-wavelenth phonons for GaN (cm^{-1}/GPa).

$K^B_{E_2}$ (low)	$K^B_{A_1(TO)}$	$K^B_{E_1(TO)}$	$K^B_{E_2}$ (high)	$K^B_{A_1(LO)}$
-0.5^b	2.8^b	1.4^b	2.9^a, 2.7^b	0.8^a

[a]Ref. 45.
[b]Ref. 49.

Table 8.9 Phonon deformation potentials for GaN (cm^{-1}/unit strain).

J	E_2 (low)	$A_1(TO)$	$E_1(TO)$	E_2 (high)	$A_1(LO)$
a_J	$+115^b$	-630^b	-820^b	-818^a, -850^b	-685^a
b_J	-80^b	-1290^b	-680^b	-797^a, -920^b	-997^a

[a]Ref. 45.
[b]Ref. 49.

particular behavior can be understood if we recall that this zone-center mode in the wurtzite structure corresponds to the longitudinal acoustic phonon at L in the cubic structure, which is known to soften under pressure.

Values obtained for biaxial stress coefficients and for phonon deformation potentials are given in Tables 8.8 and 8.9.

These experimental data may be compared to calculated values. In the recent work by Wagner and Bechstedt [50], shifts under biaxial strain were determined for GaN and AlN phonons from first-principles calculations. The shift in frequency of the upper E_2 mode was estimated as $-1350\,cm^{-1}$ ($-1200\,cm^{-1}$) for GaN (AlN) per 100% biaxial extensive strain; the stress coefficient $K^B_{E_2} = 2.8\,cm^{-1}$/GPa was derived for GaN using the corresponding elastic constants given by Kim et al. [51] and Polian et al. [52]. The agreement with the experimental determinations was satisfactory.

An analysis of various experimental data by Kisielowski et al. [53] suggested that both biaxial and hydrostatic strains may take place in GaN layers. The former is due to lattice mismatch and to the difference in thermal expansion coefficients between substrate and layer, whereas the latter is due mainly to doping or to point defects (Ga or N substitutional atoms) induced by an inadequate growth rate of the epitaxial layers. Combining strain determination by DRX measurements and Raman experiments, these authors deduced a biaxial stress coefficient $K^B_{E_2} = 4.2\,cm^{-1}$/GPa from Raman spectra; however, they pointed out that the latter result, determined from the Young modulus and Poisson ratio (estimated in this work as 290 GPa and 0.23, respectively), depends strongly on the values used for the elastic constants.

Note that a lower value of $K^B_{E_2}$ can be derived using for the elastic constants the more recent data of Ref. 52 or 53.

The aforementioned measurements are not in agreement with the previous results of Kozawa et al. [54]; the latter authors found a larger stress coefficient for the E_2 phonon ($K^B_{E_2} = 6.2\,\mathrm{cm}^{-1}$/GPa). In this work, the strain of GaN layers of different thicknesses, deposited on sapphire, was estimated by measuring the wafer curvature using surface profilometry.

8.5.3 Hydrostatic Pressure Effects

For GaN, the main study on the phonon shift under hydrostatic pressure was carried out by Perlin et al. [48] on bulk GaN needles or platelets. They used diamond anvil cell to apply the pressure to the samples. The measured pressure coefficient was $K^H_{E_2} = 4.17\,\mathrm{cm}^{-1}$/GPa for the high-frequency E_2 phonon; similar frequency shifts were found for the other phonons, except for the low-frequency E_2, which exhibits a negative pressure coefficient. These data are listed in Table 8.10, together with values of the Grüneisen parameter. In addition, these authors reported on phonon line vanishing for hydrostatic pressure greater than 47 GPa. The evolution of the Raman spectra was correlated to a structural phase transformation evidenced in the X-ray absorption near-edge structure.

Siegle et al. [55] performed similar experimental determinations; the measured values of the Grüneisen parameter for wurtzite GaN are compared with those from Ref. 48 in Table 8.10.

For wurtzite AlN, the first study of phonons under hydrostatic pressure was performed by Sanjurjo et al. [56] (the zero-pressure phonon line measured at $659.3\,\mathrm{cm}^{-1}$, which actually corresponds to the E_2 mode, was assigned to the A_1(TO) mode). These authors found pressure coefficients rather close to the values found for GaN in Ref. 48. Then, Perlin et al. [57] measured the frequency shift of A_1(TO), E_1(LO), and E_2 phonons under hydrostatic pressure for bulk wurtzite AlN. Except for the E_1(LO) mode, the values of pressure coefficients are also close to those found for GaN [48], as shown in Table 8.11.

Table 8.10 Hydrostatic pressure coefficients (cm^{-1}/GPa) and Grüneisen parameters of long-wavelength phonons for GaN.

J	E_2 (low)	A_1(TO)	E_1(TO)	E_2 (high)	A_1(LO)
K^H_J	-0.25^a	$4.06^a, 4.0^b$	$3.68^a, 3.94^b$	$4.17^a, 4.24^b$	4.4^b
γ_J	-0.426^a	$1.184^c, 1.51^b$	$1.609^c, 1.41^b$	$1.798^a, 1.50^b$	1.20^b

[a] Ref. 48.
[b] Ref. 55.

Table 8.11 Hydrostatic pressure coefficients (cm^{-1}/GPa) and Grüneisen coefficients of long-wavelength phonons for AlN.

J	$A_1(TO)$	E_2 (high)	$E_1(LO)$
K_J^H	4.63	3.99	1.67
γ_J	1.58	1.26	0.38

Source: From Ref. 57.

8.6 DOPING OF GaN

8.6.1 LO Phonon–Plasmon Coupled Modes in Doped Semiconductors

Raman spectroscopy is known to be a powerful tool for probing free electrons or holes in a semiconductor, without any electrical contact on the sample. Indeed, the presence of the free-carrier gas may be detected in the spectra through coupling of the longitudinal optical (LO) phonon with plasma oscillation.

In solids, the collective oscillation (plasmon) of the free-electron or free-hole gas is a longitudinal excitation. For a vanishing wavevector, its frequency ω_P is written as

$$\omega_P = \sqrt{\frac{ne^2}{m^* \varepsilon_0 \varepsilon_\infty}}, \tag{8.10}$$

where n and m^* are the density and the effective mass of the free carriers. On the other hand, the damping constant γ_P of the collective oscillation, related to the mobility μ, is given by

$$\gamma_P = \frac{e}{\mu m^*}. \tag{8.11}$$

If the plasmon frequency is nearly close to the frequency of the LO phonon, both longitudinal excitations may couple together via their associated macroscopic electric field, and the coupled plasmon–LO phonon mode (CM) exhibits two components, with frequencies ω_+ and ω_- [29]. The latter, obtained by solving the coupled equations of ionic and electronic motion when damping is neglected are

$$\omega_\pm^2 = \tfrac{1}{2}(\omega_L^2 + \omega_P^2) \pm \sqrt{(\omega_L^2 + \omega_P^2)^2 - 4\omega_T^2 \omega_P^2} \tag{8.12}$$

where ω_L and ω_T are the frequencies of the LO and TO modes, respectively. In Figure 8.13, the variation of the CM frequencies as a function

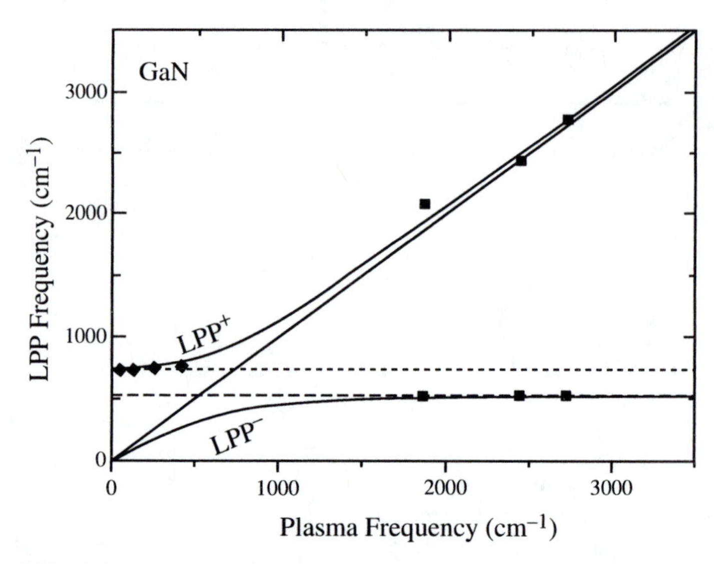

Figure 8.13 Calculated variation of frequencies for the coupled mode (CM) versus the doping level in GaN (when plasmon damping is neglected). LPP$^-$ (LPP$^+$) corresponds to the low- (high-) frequency component of the CM. Solid squares correspond to CM frequencies measured in Ref. 58.

of n is represented for GaN and compared with measured frequencies for various doping levels [58]. For weak carrier densities, the low-frequency, plasmonlike component of the CM is faintly seen as a weak feature in the Raman spectrum; the frequency ω_+ of the phononlike component is slightly upshifted from the uncoupled LO phonon frequency. For high doping levels, the high-frequency, plasmon like component of the CM usually appears as a very broad band shifted toward very high frequencies, whereas the frequency ω_- of the phonon like component shifts close to the TO frequency.

Scattering of light is achieved by the phonon component of the coupled modes via the deformation potential and electrooptic (**u** and **E**) mechanisms, and by their plasmon part via the electrooptic and charge density fluctuation (**E** and ρ) mechanisms. Within the dielectric model [38,59], the frequency spectra of fluctuating variables **u**, **E**, ρ are calculated from the generalized form of the fluctuation–dissipation theorem. For the (**u**, **E**) mechanism, the scattered intensity is given by

$$I_s \propto A(\omega) \cdot \mathrm{Im}\left[-\frac{1}{\varepsilon(\omega)}\right], \tag{8.13}$$

and for the ρ-mechanism, by

$$-I_s \propto q^2 B(\omega) \cdot \mathrm{Im}\left[-\frac{1}{\varepsilon(\omega)}\right], \tag{8.14}$$

where the dielectric function is expressed as

$$\varepsilon(\omega) = \varepsilon_\infty \left[1 + \frac{\omega_L^2 - \omega_T^2}{\omega_T^2 - \omega^2 - i\gamma\omega} - \frac{\omega_P^2}{\omega^2 - i\gamma_P} \right]. \tag{8.15}$$

In this approach, the free-carrier contribution is thus treated in the Drude model.

The explicit form of functions $A(\omega)$ and $B(\omega)$ may be found in the paper by Irmer [60]. It depends on the Faust–Henry coefficient C, which measures the relative strength of the deformation potential compared with the electrooptic coupling in the electron–phonon interaction [29].

In the preceding equations, $\mathbf{q}$ is the wavevector of the excitation, ω_T and ω_L are the transverse and longitudinal frequencies of the polar phonon, γ and γ_p are the damping constants of the phonon and plasmon, respectively, and ε_∞ takes into account the contribution of core electrons. In most cases, the $(\mathbf{u}, \mathbf{E})$ mechanism dominates for wide-bandgap materials excited with visible lasers [61].

In order to verify the validity of the dielectric model, it may be convenient to determine first the parameters ω_P and γ_p by means of independant experiments, like IR reflectivity or Hall measurements, and then to check the reliability of values obtained from fits of the Raman lineshape. Actually, independent measurements of n and μ by electrical or optical methods are generally used to obtain starting values of both fitting parameters ω_P and γ_p.

8.6.2 Dielectric Modeling of Coupled Modes for GaN

Two extreme situations concerning the values of ω_P with respect to the uncoupled phonon frequency ω_L are frequently encountered with doped GaN samples. Non-intentionally doped materials, particularly single crystals, are degenerate semiconductors with typical electron densities in the range 10^{19}–10^{20} cm^{-3}, for which $\omega_P \gg \omega_L$. In contrast, silicon- or magnesium-doped GaN exhibit doping levels reaching at the most 5×10^{18} cm^{-3}, and the reverse inequality $\omega_P \ll \omega_L$ holds true in that case. Rather broad features, significantly shifted from the well-known phonon frequencies and assigned to the CM's components, are expected in the Raman spectra. Practically, the phononlike component is very often the only one that can be resolved thanks to its higher scattering efficiency, lying close to the ω_T frequency for high doping levels, or near ω_L in the opposite case.

Moreover, the CMs must reflect the anisotropy of hexagonal GaN. Actually, they may be evidenced at different frequencies for axial and planar excitations, corresponding to A_1 and E_1 symmetries, respectively. Note that the selection rules of the CM are similar to those obeyed by the uncoupled LO phonon for GaN. Whatever the scattering mechanism is, the CM exhibiting the A_1 symmetry, for example, should be observed in backscattering along the **c**-axis for parallel polarizations of the incident and scattered light.

When the CM's profile is calculated within the dielectric model, the ω_P plasmon frequency is estimated from Eq. (8.10) using the high-frequency dielectric constant $\varepsilon_\infty = 5.35$ [4] and the carrier effective mass of GaN. For electrons, the value $m_e^* = 0.2m_0$ is widely used [62]. The same is not true for free holes: there is a significant dispersion in calculated or measured effective masses. The effective mass $m_h^* = 0.8m_0$ proposed by Pankove et al. [63] has been used for the CM modeling, in the case of p-type GaN. This value is very close to the average effective mass of holes as calculated recently by Suzuki et al. [64].

It should be pointed out that the Faust–Henry coefficient, needed to fit the CM, has not yet been unambiguously determined for GaN. In the case of the A_1 phonon, for example, it may be deduced from intensity measurements of scattering by the LO and TO modes [29] in Raman spectra recorded in the $z(yy)\bar{z}$ and $x(yy)\bar{x}$ configurations, respectively, in order to compare the same Raman tensor component. Using this method, Wetzel et al. [65] and Demangeot et al. [66] have proposed for C the values -5.2 and -3.8, respectively; however, several authors have tentatively introduced either negative values close to the Faust–Henry coefficient for GaAs (-0.5) [67], or weak positive values [68].

8.6.3 n-Type Intentional Doping

The most commonly used donor for intentional n doping of GaN is silicon. Free-electron density as high as $5 \times 10^{19}\,\mathrm{cm}^{-3}$ can be achieved. The first Raman study of the CM in Si-doped GaN, grown by MOCVD, was published by Kozawa et al. [69], who revealed only its high-frequency component, for electron densities up to $2.5 \times 10^{18}\,\mathrm{cm}^{-3}$. The electron density and mobility, deduced from the fit of Raman data with the dielectric modeling, were compared with the electrical measurements, and an overall agreement was found; however, these authors introduced very large phonon damping constants ($\gamma = 70\,\mathrm{cm}^{-1}$ at high doping levels), which seem to be rather unrealistic values of the parameter.

Another study was performed by Harima et al. [70] on MOCVD samples, for similar doping levels: both components of the CM were observed in micro-Raman spectra, and the dielectric fit required large values of the plasmon damping constant (up to $\gamma_P = 380\,\mathrm{cm}^{-1}$, for $\omega_P = 550\,\mathrm{cm}^{-1}$, corresponding to $n = 3.6 \times 10^{18}\,\mathrm{cm}^{-3}$). From the fit of the CM's low-frequency component, the authors inferred a significant contribution from the charge density fluctuation mechanism, at large carrier concentration, as shown in Figure 8.14. In addition, the CM lineshape analysis was used to

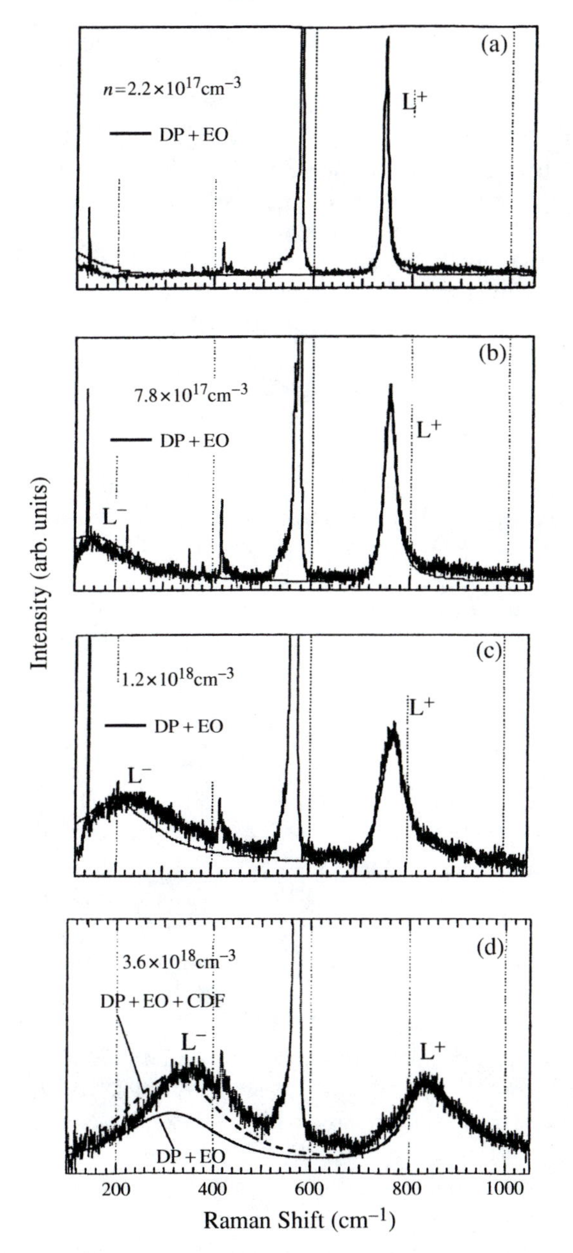

Figure 8.14 Raman spectra of n-type GaN for free-electron densities $2.2 \times 10^{17}\ cm^{-3}$ (a), $7.8 \times 10^{17}\ cm^{-3}$ (b), $1.2 \times 10^{18}\ cm^{-3}$ (c), $3.6 \times 10^{18}\ cm^{-3}$ (d), measured in Ref. 70. Smooth solid lines indicate theoretical fits by deformation potential and electrooptic mechanisms; dashed line in (d) includes the charge density fluctuation contribution.

reveal in epitaxial layers a gradient in carrier density along the direction of gas flow.

Clear evidence for carrier inhomogeneities has also been produced by Ponce et al. [71] using micro-Raman imaging. In this experiment the scattered light was passed through a dielectric filter (with a very narrow spectral pass band), allowing the distinction at a microscopic scale between the undoped GaN, whose signature is the uncoupled LO phonon at 734 cm^{-1}, and the n-type regions, characterized by the CM feature observed at 765 cm^{-1} in this sample. The maximum donor concentration was found on the edge between the facets of hexagonal GaN microcrystallites with typical 30-μm size.

Wetzel et al. [65] have reported on the effect of very slow cooling down and warming up of n-type GaN layers; the temperature-dependent variation of the carrier density is monitored by the evolution of the CM observed in the Raman spectrum. A significant retardation is found in both the "freeze-out" of the electrons and the ionization of the donors, suggesting some intermediate steps during the latter processes.

To summarize, the analysis of the CM in the Raman spectra of Si-doped GaN indicates values of electron densities in the 10^{17}–10^{19} cm^{-3} range with maximum mobilities of several hundred cm^{2}V^{-1}s^{-1}. In the present case, plasmons are never overdamped.

8.6.4 p-Type GaN

The doping of GaN with acceptors has been a difficult challenge: diffusion of magnesium followed by thermal annealing achieves p-type doping, but the hole concentration is lower than 1×10^{18} cm^{-3}. Indeed, a self-compensation mechanism takes place at high Mg concentrations, resulting in limitation of hole density and in severe decrease of hole mobility. Thus, the plasmon damping constant introduced in the dielectric modeling must be significantly larger for p-type GaN than in the case of n-type material.

The first Raman study of p-type doping of GaN was published by Popovici et al. [72], for a carrier concentration of up to 7×10^{17} cm^{-3} and a hole mobility of 17 cm^{2}V^{-1}s^{-1}. The frequency shift of the CM was detected in the experimental data and compared with the shift calculated using Eq. (8.2); the frequency shift of the nonpolar E_2 vibration mode was attributed to the extensive stress induced by the Mg incorporation at high concentrations in the GaN layers.

Demangeot et al. [73] used the standard dielectric model for a lineshape analysis of CMs in Raman spectra of p-type GaN layers grown either by MOVPE or by MBE; the samples had previously been characterized by Hall measurements, which yielded hole concentrations between 3×10^{17} cm^{-3} and 3×10^{18} cm^{-3}, with mobilities as low as 2–8 cm^{2}V^{-1}s^{-1}. The CM's high-frequency component is observed in Raman spectra, only slightly upshifted and broadened with respect to the LO phonon: both lineshape and location of the CM are thus very different from that of n-type layers. For p-type samples, the hole plasmon is overdamped ($\omega_P \ll \gamma_P$), due to lattice disorder induced by the Mg incorporation. Strictly speaking, the experimental feature is not a CM but rather corresponds to the LO phonon that is only slightly influenced by the nonresonant carrier gas. A significant difference has been observed between the CMs related to the E_1(LO) and the A_1(LO) phonons, but it has not been taken into account in the modeling, due to the lack of reliable data concerning the anisotropy of the hole effective mass.

Finally, for frequencies lower than 400 cm^{-1} in Raman spectra of p-type GaN layers, Harima et al. [74] observed a significant background related to Mg incorporation, which was attributed to inter-valence-band hole transitions. The intensity of the continuum, normalized to the E_2 phonon, increases linearly with the hole density, up to 10^{18} cm^{-3}.

8.6.5 Nonintentionally n-Doped GaN

Bulk GaN single crystals, grown under high-pressure and high-temperature conditions, always exhibit a high level of nonintentional n-type doping. The nature of the corresponding donor is controversial, but the nitrogen vacancy is considered to be a good candidate [75].

Perlin et al. [76] observed both components of the CM in the Raman spectra of GaN needles, with electron densities of up to 5×10^{19} cm^{-3}, corresponding to values of ω_P as high as 2700 cm^{-1}; they did not find in this case that the plasmon was overdamped. As displayed in Figure 8.15, their measurements established a clear correlation between the broad high-frequency component of the CM in the Raman spectra and the plasma resonance in the IR reflectivity spectrum.

Using both techniques, the same authors used micro-Raman measurements as evidence for an inhomogeneous n-type doping in bulk GaN, which can almost be suppressed by a subsequent surface polishing [76]: according to these observations, it was suggested that this behavior may be due to impurities, probably oxygen atoms, introduced into the lattice during growth under high pressure.

A drastic decrease of the free-electron density was observed in Ref. 77, for nonintentionally doped GaN submitted to a 32-GPa hydrostatic pressure, through the decoupling of the LO phonon and plasmon. Taking into account other experimental data such as IR absorption measurements, these authors suggest that a defect state, resonant with the conduction band under atmospheric pressure, moves down into the gap under high pressure, trapping free electrons. Calculations indicate that this defect may be the nitrogen vacancy.

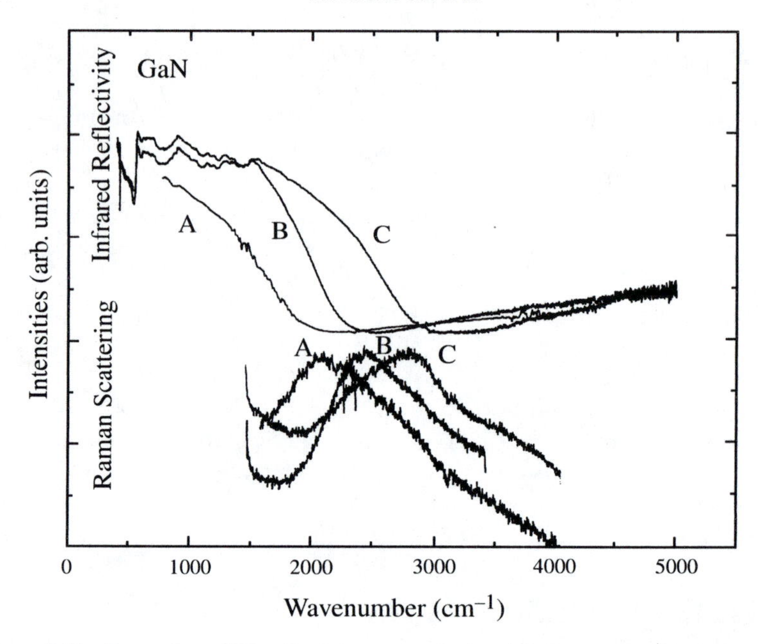

Figure 8.15 Comparison of IR reflectivity spectra (top) and Raman spectra (bottom) measured on three GaN single crystals, from Ref. 58. The free-electron concentrations are $3.9 \times 10^{19}\,cm^{-3}$, $5.1 \times 10^{19}\,cm^{-3}$ and $8.7 \times 10^{19}\,cm^{-3}$ for samples A, B, C respectively.

A similar study by Wetzel et al. [78], using Raman and IR measurements, showed that a high hydrostatic pressure (29 GPa) results in a 97% decrease of free-electron density in bulk platelets with a high doping level; this decrease has been assigned to the localization of carriers on an energy level lying 126 meV below the bottom of the GaN conduction band.

Actually, nonintentional doping can be seen not only on single crystals but also on GaN thin layers on sapphire substrates, for example when inadequate temperature conditions are chosen for the MOVPE growth [79]. The CM's low-frequency component observed in micro-Raman spectra of such layers has been fitted within the framework of the dielectric model, giving the signature of a moderately damped plasmon corresponding to a $10^{20}\,cm^{-3}$ electron density; these findings have been confirmed by IR reflectivity measurements [66]. Moreover, an additional broad feature peaking around $660\,cm^{-1}$ has been attributed to scattering by the **q**-dependent electron charge density fluctuation mechanism. This assignment was confirmed in a paper by Ramsteiner et al. [80], although this work was devoted to cubic GaN: calculations taking into account the contributions to the Raman spectra of Fröhlich and charge density fluctuation mechanisms showed that the latter contribution is responsible for broad scattering covering the LO–TO frequency range, observed far from resonant conditions.

8.7 RESONANT RAMAN SCATTERING IN GaN AND RELATED SEMICONDUCTORS

When the energy of the incident or scattered photon approaches that of electronic transitions, the intensity of Raman scattering increases rapidly. Indeed, the real part of one or two of the denominators in the expression of the scattering cross section vanishes [see Eq. (8.1)]. An important enhancement of the Raman signal by several orders of magnitude may thus be expected. The resonant behavior is particularly useful for obtaining a significant signature of thin layers or crystallites with very small volume.

Moreover, resonance provides information on the electron–phonon interaction in semiconductors. Actually, in resonant conditions, the dominant contribution to light scattering by polar phonons no longer proceeds from the deformation potential mechanism and electrooptic coupling, but from the intraband Fröhlich mechanism.

Observation of resonant Raman scattering by optical phonons in GaN below the absoption edge was first reported by Lemos et al. [81]. They showed that exciton effects have to be considered in the range they covered with an argon laser ($E_L < 2.7$ eV). Then, Behr et al. [82] published a resonance study of GaN in the violet and UV spectral range. Using the excitation energies $E_L = 3.00$ and 3.05 eV, the approach of resonant conditions was monitored by temperature, as the bandgap E_G decreased from 3.44 eV to 3.08 eV when the sample temperature was increased from 300 to 870 K. In the chosen $x(zz)\bar{x}$ configuration, the quasi-LO phonon, forbidden to scatter via the deformation potential or the electrooptic mechanisms, appears in near-resonant conditions via the Fröhlich mechanism. In order to get rid of the temperature-dependent phonon number, Raman spectra were corrected for statistical factors. In this experiment, separate resonances for the incoming and scattered photons could not be resolved. For a vanishing $\Delta E = E_L - E_G(T)$ detuning energy parameter, an enhancement of one and a half orders of magnitude was evidenced. Figure 8.16 gives the resonance profile, experimentally determined for the negative detuning parameter only. Resonance effects were also observed on multi-LO–phonon structures in the spectra. Using an excitation in the UV range, the same authors reported on multiphonon scattering up to the fifth order, for temperatures decreasing from 300 to 15 K [83]. The fine structure in the sidebands of the phonon features was discussed in terms of scattering of the exciton–polariton by an acoustical phonon followed by the emission of an LO phonon.

Resonant Raman scattering in $Ga_{1-x}Al_xN$ layers with variable Al contents ($0 < x < 0.85$) grown by MOVPE on sapphire was investigated [84]. The experiments were made in backscattering geometry on the layer surface under a fixed excitation at 244 nm, using a frequency-doubled argon laser. Under these conditions, the nonresonant E_2 mode could not be observed, and

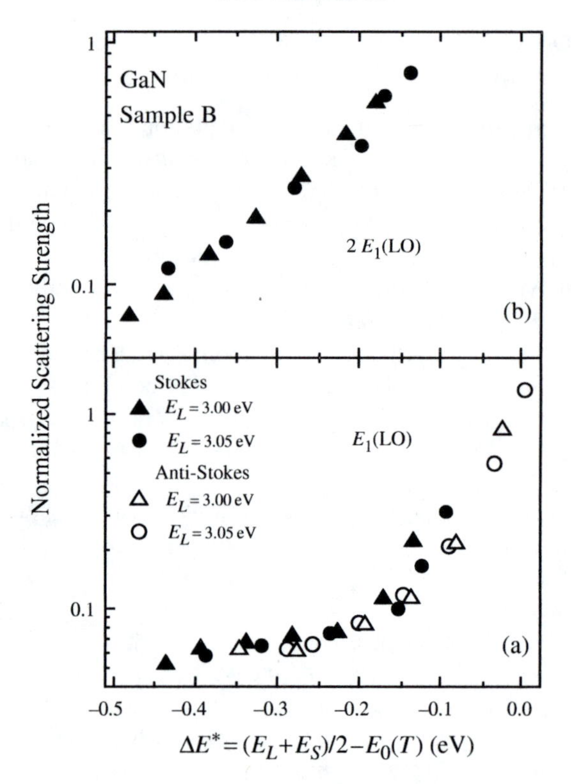

Figure 8.16 Normalized intensity of first- and second-order scattering by LO phonons in GaN versus the detuning energy parameter, for excitation energies below the bandgap, measured in Ref. 82.

only the A_1(LO) phonon and its multiphonon structure were seen in the spectra. The resonance profile was plotted versus the detuning energy parameter $\Delta E = E_L - E_G(x)$, as shown in Figure 8.17. An important enhancement of the LO phonon scattering, by about three orders of magnitude, was observed when the semiconductor bandgap was close to the excitation energy (5.08 eV), i.e., for the optimum Al molar ratio ($x = 0.72$) in the series.

In addition, the frequency of the LO phonon measured under UV excitation was found to differ significantly from its value obtained using visible light; see Figure 8.18. This unusual shift depends on the detuning parameter and changes sign around $x = 0.7$. It was attributed to selective enhancement of the Raman signal from locally GaN- or AlN-rich regions, achieved via Fröhlich interaction and quantitatively interpreted in terms of statistical fluctuations of the composition at the nanoscale of the resonant Raman probe.

Recent work by Behr et al. [85] on resonant Raman scattering in $In_xGa_{1-x}N$ thin layers with low In content ($x = 0.11$) also gives evidence for alloy compositional inhomogeneities. As in Ref. 82, the excitation is

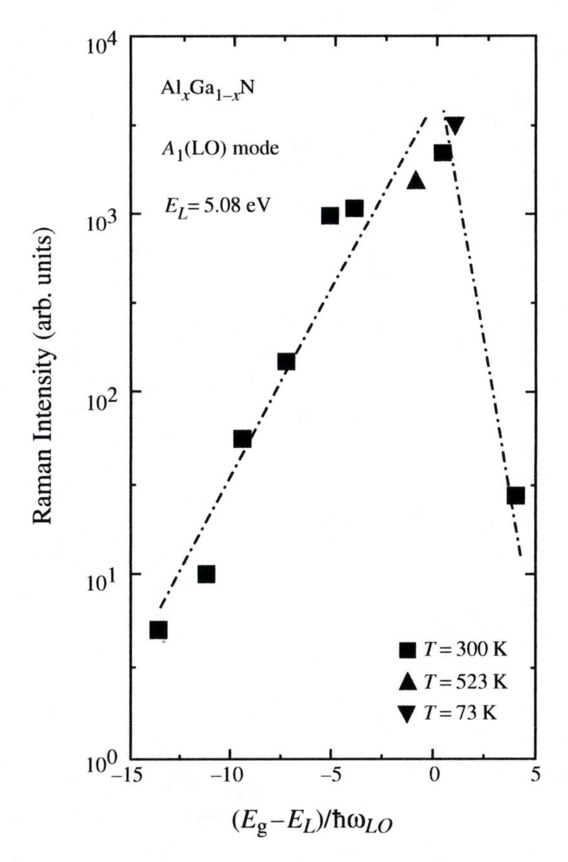

Figure 8.17 Resonance profile of $Al_xGa_{1-x}N$ alloys, from Ref. 84. The intensity of the A_1(LO) phonon is plotted versus the detuning energy parameter (normalized to the phonon energy), for a fixed excitation energy (5.08 eV).

made at $E_L = 3.00$ and 3.05 eV, and the sample is heated up to 870 K, in order to tune its bandgap energy $E_G(T)$ at the incident photon energies. The bandgap energy (3.15 eV) at room temperature is deduced from the resonance profile of the A_1(LO) phonon. An anomalous change in the relative frequency shift with temperature of the A_1(LO) phonon with respect to the nonpolar E_2 mode is explained by selective enhancement of the LO-phonon scattering due to local compositional inhomogeneities.

8.8 RAMAN STUDIES OF NANOSIZE STRUCTURES

$Al_xGa_{1-x}N$/GaN and $In_xGa_{1-x}N$/GaN heterostructures and superlattices made of layers with the wurtzite structure are used for the design of

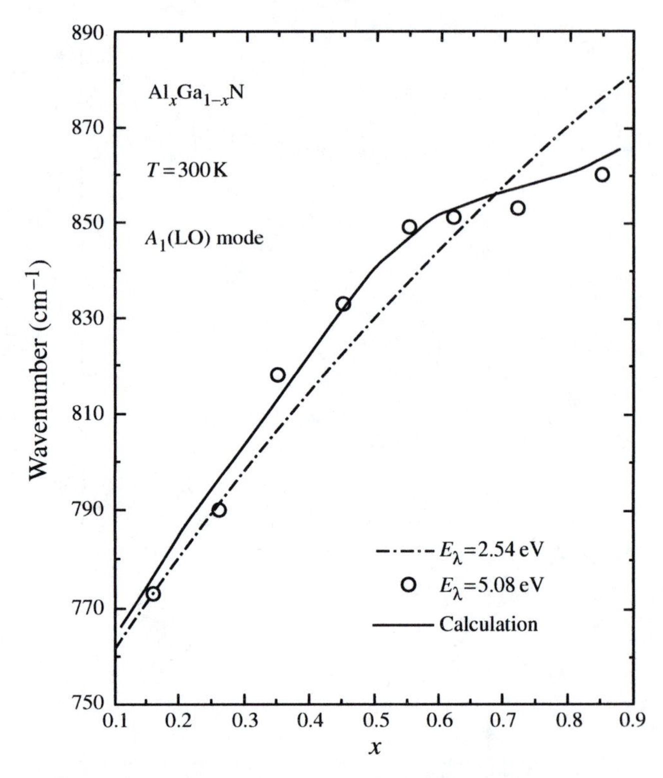

Figure 8.18 A_1(LO) phonon frequency in $Al_xGa_{1-x}N$ layers with various Al contents, measured using visible (488 nm) and UV (244 nm) excitations, from Ref. 84.

optoelectronic devices like LEDs and LDs in the UV or blue spectral range. However, until now, only a few papers have been devoted to the basic properties of these structures and to their study by Raman spectroscopy. Raman spectroscopy may conveniently be used to evaluate internal strains and confinement effects, and to assess the quality of the interfaces, for example. The resonant enhancement of LO-phonon scattering provides a selective and efficient probing of QWs or barriers with submicrometric characteristic sizes. From a more fundamental point of view, resonant Raman scattering allows one to identify intermediate electron–hole pair states in the scattering process and to get information on their degree of confinement in systems of reduced dimensionality.

Before the start of experimental work on heterostructures and superlattices (SL), microscopic calculations were performed on vibrational modes of short-period GaN/AlN or $GaN/Ga_{1-x}Al_xN$ SL for layers with the zinc-blende structure [86,87]. Then, the polar modes in structures made of hexagonal materials were studied within the framework of the dielectric continuum model by Komirenko et al. [88] in GaN/AlN single QWs, and by

Gleize et al. [89] in GaN/AlN SLs. In the latter case, proper confinement of A_1(LO) and E_1(TO) extraordinary phonons inside each type of SL layer was achieved only for zero in-plane wavevector. The Fuchs–Kliewer interface (IF) and quasi-confined (QC) modes with finite **q** were also discussed: the former are evanescent in both types of layers, whereas the latter are confined in one type of layer but can penetrate into the surrounding layers, where they decay exponentially. Due to the anisotropy of GaN and of AlN, a change of mode type from QC to IF occurs along the same dispersion branch; note that the frequency range of the modes strongly depends on the actual strain in the SL layer.

8.8.1 Strain and Confinement Effects in Quantum Wells and Superlattices

The first experimental study of artificial structures with single GaN QWs of various thicknesses (2 to 4 nm) embedded in wide $Ga_{0.75}Al_{0.15}N$ barriers was published by Behr et al. [90]. Calculations for the interband transition energies give values between 3.62 and 3.52 eV, depending on the well width. Near-resonant conditions were achieved for excitation at 3.54 eV. The A_1(LO) phonon from GaN, which is clearly seen in the Raman spectra shown in Figure 8.19, can be unambiguously related to the well, the only part made up of GaN in this structure grown on an AlN nucleation layer. In the case of the 2-nm-thick QW, the feature assigned to the GaN phonon is broadened and shifted toward high frequencies, probably due to intermixing effects at the well/barrier interface. The same authors also reported on a heterostructure containing a $Ga_{0.89}In_{0.11}N$ single QW between GaN layers [91]: the E_2 and A_1(LO) modes from the well were observed with an excitation at 3.0 eV, not far from resonant conditions (the bandgap energy of the $Ga_{0.89}In_{0.11}N$ alloy is around 3.16 eV).

Very recently, a similar study of Raman scattering was performed on a collection of four GaN single QWs with different thicknesses (4, 8, 12, and 16 ML), separated by 10-nm-thick $Al_{0.17}Ga_{0.83}N$ barriers [92]. Raman spectra, recorded using various laser lines in the UV range, are shown in Figure 8.20. Strong features, superimposed on PL broad peaks, are observed. Under the 3.53-eV excitation, scattering by the GaN A_1(LO) phonon is enhanced by resonance through the incoming channel in the 8-ML-thick QW, and through the outgoing channel in the 12-ML-thick QW. In contrast, when the energy of the incident photon (3.70 eV) is nearly matched to the energy bandgap of the barriers, preferential scattering by $Al_xGa_{1-x}N$ phonons is observed, suggesting a delocalized exciton as the first intermediate state of the process.

The first superlattice with wurtzite structure to be investigated was a 100-period GaN (6.3 nm)/AlN (5.1 nm) system [93]. Due to the significant

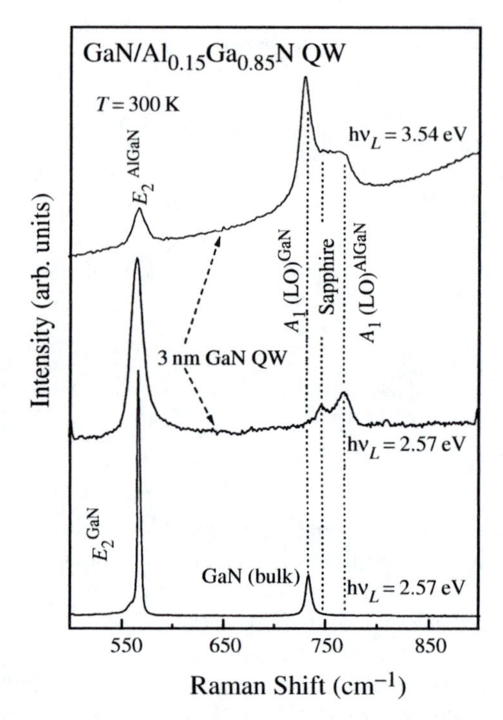

Figure 8.19 Resonant Raman spectra of GaN single QW structures embedded in $Al_{0.15}Ga_{0.85}N$ barriers, recorded with laser lines at 3.54 eV and 2.57 eV, from Ref. 90. Arrows indicate PL signal from the GaN QWs or from the buffer layer.

volume of the SL, micro-Raman spectra could be recorded under visible excitation, far from resonance. The E_1(TO), E_2, and A_1(LO) phonons confined in GaN layers were unambiguously identified through a polarization analysis displayed in Figure 8.21. As quantization effects could be ruled out in view of layer widths, frequency shifts were attributed to the internal biaxial stress and used to determine the corresponding strain (−1.3%) in the GaN layers of the SL. The latter is consistent with the value of the SL in-plane parameter, calculated from those of both constituents weighted by the relative thickness of layers in the structure. E_1(TO) and E_2 phonons from the AlN layers under tensile stress in the SL show up in the spectra. Additional measurements on a beveled edge of the sample gave evidence for a gradual strain relaxation in the first SL layer close to the interface with the buffer layer.

8.8.2 Electric Field Effects in Quantum Wells

Due to the anisotropy of the wurtzite structure, GaN and related compounds present interesting electric properties, such as spontaneous polarization and

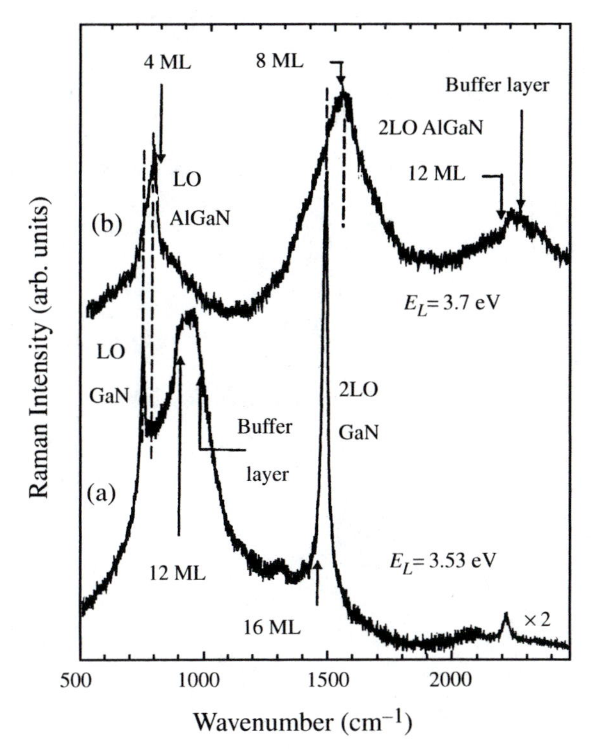

Figure 8.20 Resonant Raman spectra of a collection of four GaN QWs (4, 8, 12, 16 ML) in $Al_{0.17}Ga_{0.83}N$ barriers, excited at 3.53 eV and 3.70 eV, from Ref. 92. Arrows mark the PL peaks from QWs.

the piezoelectric effect, which is about 10 times stronger than in zinc-blende III–V semiconductors [94]. These effects, seen in structures with small characteristic sizes, such as QWs and dots, may be studied by Raman spectroscopy. The effective bandgap energy of QWs, which is increased by confinement effects, is reduced by the built-in electric field. The energy corresponding to resonant excitation may thus be significantly shifted and eventually lowered from the value found for the bulk material. In addition, scattering by polar phonons via the intraband Fröhlich electron–phonon mechanism may be significantly enhanced by the field-induced separation of electron and hole; however, for large spatial separation, the overlap between electron and hole envelope wavefunctions decreases, and the enhancement of the LO-phonon scattering cancels [95].

Field-induced enhancement of Raman scattering in GaN QWs was recently reported [92]. A comparison was made between two structures made of four identical GaN QWs, 16 monolayers thick, separated by $Al_{0.17}Ga_{0.83}N$ layers. These samples differed only by their barrier thickness (5 and 30 nm).

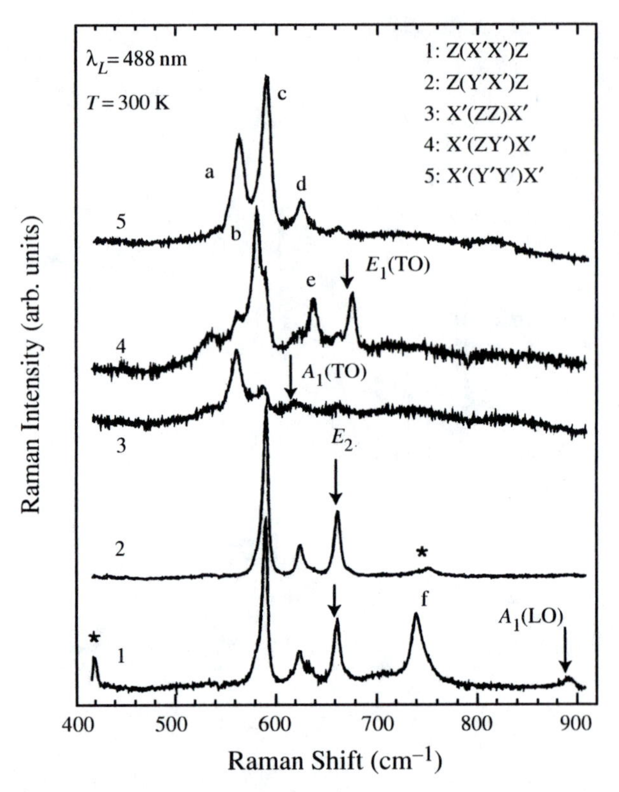

Figure 8.21 Polarized Raman spectra of a GaN (6.3 nm)–AlN (5.1 nm) strained superlattice (SL), recorded using an excitation at 2.54 eV, from Ref. 93. Peaks labeled b, c, f are assigned, respectively, to E_1(TO), E_2, A_1(LO) phonons confined in GaN layers of the SL; peaks labeled d and e are assigned to E_2 and E_1(TO) phonons confined in AlN layers of the SL. Arrows indicate phonons from the AlN buffer layer.

Raman spectra were recorded with an excitation at 3.70 eV, which was significantly higher than the fundamental transition in the QWs (3.37 eV) and close to the bandgap of the barriers (3.71 eV). For both GaN/$Al_{0.17}Ga_{0.83}N$ structures, the authors clearly demonstrated first-order scattering by the LO phonons of barriers and wells. The intensity of the latter was found stronger for the structure exhibiting wider barriers. These observations were interpreted as an effect of scattering enhancement by the internal electric field in the QWs, which increases with the barrier layer width [96].

Another work by Wagner et al. [97] was devoted to strained $In_{0.13}Ga_{0.87}N$ single QWs located between wide GaN barriers. The Raman spectra were excited at 3.0 eV, whereas the bandgap energy of the QWs was expected at 2.91 eV. For wells thinner than 12 nm, the PL signal is shifted to lower energies, due to strong piezoelectric effects, and both E_2 and A_1(LO) phonons from the $In_{0.13}Ga_{0.87}N$ layer are clearly seen. The LO phonon line exhibits an increasing intensity with the incident light power, for which the following explanation is given: at high-power excitation, the high density of photogenerated carriers screens the piezoelectric field inside the QW, which is responsible for the reduced exciton lifetime, and thus ensures full recovery of resonant enhancement.

8.8.3 Quantum Dots

Raman studies of structures containing quantum dots (QDs) with the wurtzite structure are just beginning. Obviously, strain and confinement effects, together with electric field effects, also take place in these type of samples.

To our knowledge, only the work of Gleize et al. [98] has thus far been published on that subject. It concerns periodically stacked GaN QDs grown

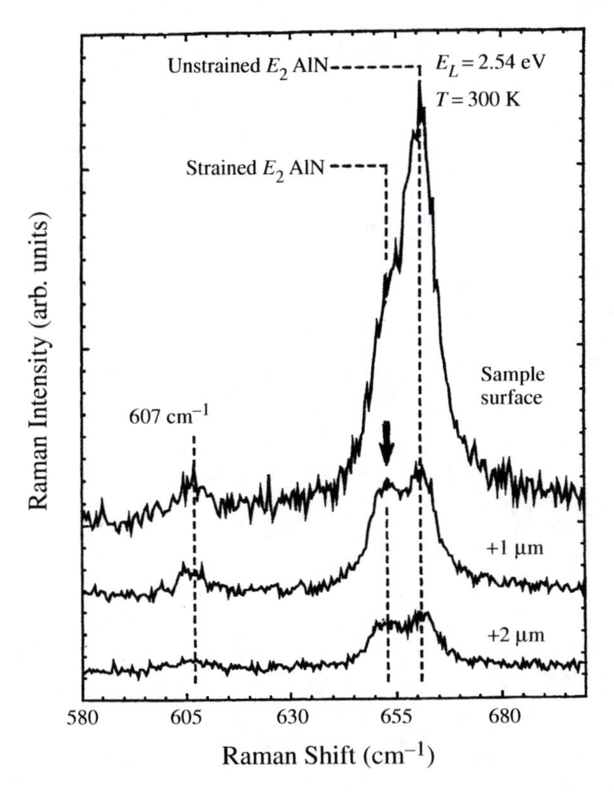

Figure 8.22 Micro-Raman spectra of a structure made of correlated GaN quantum dots (QDs), from Ref. 98. Stacked layers of QDs are embedded in AlN spacers and grown on a thick AlN buffer layer. Spectra were recorded under 488-nm excitation in a confocal geometry, for various focusing locations inside the structure. The intermediate spectrum gives the signature of strain in the spacers.

on very thin wetting layers, two monolayers thick, and embedded between AlN spacers. The QDs formed spontaneously in the Stranski–Krastanov growth mode of GaN on AlN. For adequate growth conditions and spacer widths, QDs in successive layers were strongly correlated along the growth direction. Two types of structures were studied, with respective QD heights of 2 and 4 nm, and corresponding spacer widths of 12.5 and 7.5 nm. In the sample exhibiting vertical ordering of bigger dots, strain-driven self-organization of GaN QDs was evidenced in Raman spectra recorded far from resonant conditions by the significant downshift in frequency of the E_2 phonon of AlN spacers from its strain-free value, as shown in Figure 8.22. On the other hand, the signature of GaN QDs was obtained from resonant scattering by the confined A_1(LO) phonons of GaN using a UV excitation.

Acknowledgments

The authors would like to thank J. Gleize for critical reading of the manuscript.

References

1. W. Richter, *Springer Tracts in Modern Physics*, **78**, 121–272 (1976).
2. C. A. Arguello, D. L. Rousseau, and S. P. S. Porto, *Phys. Rev.*, **181**, 1351 (1969).
3. D. D. Manchon Jr., A. S. Barker, P. J. Dean, and R. B. Zetterstrom, *Solid State Commun.*, **8**, 1227 (1970).
4. A. S. Barker Jr. and M. Ilegems, *Phys. Rev. B*, **7**, 743 (1973).
5. N. V. Edwards, M. D. Bremser, R. F. Davies, A. D. Batchelor, S. D. Yoo, C. F. Karan, and D. E. Aspnes, *Appl. Phys. Lett.*, **73**, 2808 (1998).
6. G. H. Liu, W. Zhang, H. X. Han, Z. P. Wang, and S. K. Duan, *J. Appl. Phys.*, **86**, 2051 (1999).
7. S. P. S. Porto and R. S. Krishnan, *Phys. Rev.*, **47**, 1009 (1967).
8. T. Deguchi, D. Ichiryu, K. Sekiguchi, T. Sota, R. Matsuo, T. Azuhata, M. Yamaguchi, T. Yagi, S. Chichibu, and S. Nakamura, *J. Appl. Phys.*, **86**, 1860 (1999).
9. T. Azuhata, T. Sota, K. Suzuki, and S. Nakamura, *J. Phys.: Condens. Matter*, **7**, L129 (1995).
10. L. E. McNeil, M. Grimsditch, and R. H. French, *J. Am. Ceram. Soc.*, **76**, 1132 (1993).
11. M. C. Lee, H. C. Lin, Y. C. Pan, C. K. Shu, W. H. Chen, and W. K. Chen, *Appl. Phys. Lett.*, **73**, 2606 (1998).
12. V. Yu. Davydov, V. V. Emtsev, I. N. Goncharuk, A. N. Smirnov, V. D. Petrikov, V. V. Mamutin, V. A. Vekshin, S. V. Smirnov, and T. Inushima, *Appl. Phys. Lett.*, **75**, 3297 (1999).
13. W. Limmer, W. Ritter, R. Sauer, B. Mensching, C. Liu, and B. Rauschenbach, *Appl. Phys. Lett.*, **72**, 2589 (1997).
14. N. Wieser, O. Ambacher, H. Angerer, R. Dimitrov, M. Stutzmann, B. Stritzker, and J. K. N. Lindler, *Proc. 3rd Int. Conf. Nitride Semicond.*, Montpellier, France, *Phys. Status Solidi B*, **216**, 807 (1999).
15. K. Miwa and A. Fukumoto, *Phys. Rev. B*, **48**, 7897 (1993).
16. I. Gorczycka, N. E. Christensen, E. L. Peltzer y Blanca, and C. O. Rodriguez, *Phys. Rev. B*, **51**, 11936 (1995).
17. K. Karch, J. M. Wagner, and F. Bechstedt, *Phys. Rev. B*, **57**, 7043 (1998).
18. R. Loudon, *Adv. Phys.*, **13**, 423 (1965).

19. L. Bergman, M. Dutta, C. Balkas, R. F. Davis, J. A. Christman, D. Alexson, and R. J. Nemanitch, *J. Appl. Phys.*, **85**, 3535 (1999).
20. L. Fillipidis, H. Siegle, A. Hoffmann, C. Thomsen, K. Karch, and F. Bechstedt, *Phys. Status Solidi B*, **198**, 621 (1996).
21. G. Wei, J. Zi, K. Zhang, and X. Xie, *J. Appl. Phys.*, **82**, 4693 (1997).
22. L. Bergman, D. Alexson, P. L. Murphy, R. J. Nemanich, M. Dutta, and M. A. Stroscio, *Phys. Rev. B*, **59**, 12977 (1999).
23. K. T. Tsen, D. K. Ferry, A. Botchkarev, B. Sverdlov, A. Salvador, and H. Morkoç, *Appl. Phys. Lett.*, **71**, 1852 (1997).
24. H. Siegle, P. Thurian, L. Eckey, A. Hoffmann, C. Thomsen, B. K. Meyer, H. Amano, I. Akasaki, T. Detchprom, and K. Hiramatsu, *Appl. Phys. Lett.*, **68**, 1265 (1996).
25. H. Siegle, L. Eckey, A. Hoffmann, C. Thomsen, B. K. Meyer, D. Schikora, M. Hankeln, and K. Lischka, *Solid State Commun.*, **96**, 943 (1995).
26. H. Siegle, G. Kaczmarczyk, L. Filippidis, A. P. Litvinchuk, A. Hoffmann, and C. Thomsen, *Phys. Rev. B*, **55**, 7000 (1997).
27. E. O. Kane, *Phys. Rev. B*, **31**, 7865 (1965).
28. H. L. McMurry, A. W. Solbrigh Jr., J. K. Boyter, and C. Noble, *J. Phys. Chem. Solids*, **28**, 2359 (1967).
29. M. Cardona, in *Light Scattering in Solids* II, edited by M. Cardona and G. Güntherodt (Springer-Verlag, 1982).
30. V. Y. Davydov, Y. E. Kitaev, I. N. Goncharuk, A. N. Smirnov, J. Graul, O. Semchinova, D. Uffmann, M. B. Smirnov, A. P. Mirgorodsky, and R. A. Evarestov, *Phys. Rev. B*, **58**, 12899 (1998).
31. K. Karch and F. Bechstedt, *Phys. Rev. B*, **56**, 7404 (1997); K. Karch, J. M. Wagner, and F. Bechstedt, *Phys. Rev. B*, **57**, 7043 (1998).
32. T. Azuhata, T. Matsunaga, K. Shimada, K. Yoshida, T. Sota, K. Suzuki, and S. Nakamura, *Physica B*, **219** & **220**, 493 (1996).
33. J. C. Nipko and C.-K. Loong, C. M. Balkas, and R. F. Davis, *Appl. Phys. Lett.*, **73**, 34 (1998).
34. J. C. Nipko and C.-K. Loong, *Phys. Rev. B*, **57**, 10550 (1998).
35. K. Hayashi, K. Itoh, N. Sawaki, and I. Akasaki, *Solid State Commun.*, **77**, 115 (1991).
36. F. Demangeot, J. Groenen, J. Frandon, M. A. Renucci, O. Briot, S. Clur, and R. L. Aulombard, *Proc. 2nd Eur. GaN Workshop, Internet J. Nitride Semicond. Res.*, **2**, 40 (1997); F. Demangeot, J. Groenen, J. Frandon, M. A. Renucci, O. Briot, S. Clur, and R. L. Aulombard, *Appl. Phys. Lett.*, **72**, 2674 (1998).
37. A. Cros, H. Angerer, R. Handschuh, O. Ambacher, and M. Stutzmann, *Solid State Commun.*, **104**, 35 (1997).
38. D. T. Hon and W. L. Faust, *J. Appl. Phys.*, **1**, 241 (1973).
39. P. Wisniewski, W. Knap, J. P. Malzac, J. Camassel, M. D. Bremser, R. F. Davis, and T. Suski, *Appl. Phys. Lett.*, **73**, 1760 (1998).
40. H. Grille, C. Schnittler and F. Bechstedt, *Phys. Rev. B.*, forthcoming.
41. L. Bergman, M. D. Bremser, W. G. Perry, R. F. Davis, M. Dutta, and R. J. Nemanitch, *Appl. Phys. Lett.*, **71**, 2157 (1997).
42. D. Behr, H. Obloh, J. Wagner, K. H. Bachem, and U. Kaufmann, *Proc. MRS Symp.*, **468**, 213 (1997).
43. H. Siegle, A. Hoffmann, L. Eckey, C. Thomsen, J. Christen, F. Bertram, D. Schmidt, D. Rudloff, and K. Hiramatsu, *Appl. Phys. Lett.*, **71**, 2490 (1997).
44. R. J. Briggs and A. K. Ramdas, *Phys. Rev. B*, **13**, 5518 (1976).
45. F. Demangeot, J. Frandon, M. A. Renucci, O. Briot, B. Gil, and R. L. Aulombard, *Solid State Commun.*, **100**, 207 (1996).
46. B. Gil, O. Briot, and R. L. Aulombard, *Phys. Rev.*, **52**, 17028 (1995).
47. O. Briot, J. P. Alexis, M. Tchounkeu, and R. L. Aulombard, *Mater. Sci. Eng., B*, **43**, 147 (1997).
48. P. Perlin, C. Jauberthie-Carillon, J. P. Itie, A. San Miguel, I. Grzegory, and A. Polian, *Phys. Rev. B*, **45**, 83 (1992).
49. V. Yu. Davydov, N. S. Averkiev, I. N. Goncharuk, D. K. Nelson, I. P. Nikitina, A. S. Polovnikov, A. N. Smirnov, and M. A. Jacobson, *J. Appl. Phys.*, **82**, 5097 (1997).

50. J. M. Wagner and F. Bechstedt, *Proc. 3rd Int. Conf. Nitride Semicond.*, Montpellier, France, *Phys. Status Solidi B*, **216**, 793 (1999).
51. K. Kim, W. R. L. Lambrecht, and B. Segall, *Phys. Rev. B*, **53**, 16310 (1996).
52. A. Polian, M. Grimsditch, and I. Grzegory, *J. Appl. Phys.*, **79**, 3343 (1996).
53. C. Kisielowski, J. Krüger, S. Ruvimov, T. Suski, J. W. Ager III, E. Jones, Z. Liliental-Weber, M. Rubin, and E. R. Weber, *Phys. Rev. B*, **54**, 17745 (1996).
54. T. Kozawa, T. Kachi, H. Kano, and H. Nagase, *J. Appl. Phys.*, **77**, 4389 (1995).
55. H. Siegle, A. R. Goni, C. Thomsen, C. Ulrich, K. Syassen, B. Schöttker, D. J. As, and D. Schikora, *MRS Symp. Proc.*, **468**, 225 (1997).
56. J. A. Sanjurjo, E. Lopez-Cruz, P. Vogel, and M. Cardona, *Phys. Rev. B*, **28**, 4579 (1983).
57. P. Perlin, A. Polian, and T. Suski, *Phys. Rev. B*, **47**, 2874 (1993).
58. P. Perlin, J. Camassel, W. Knap, T. Taliercio, J. C. Chervin, T. Suski, I. Grzegory, and S. Porowski, *Appl. Phys. Lett.*, **67**, 2524 (1995).
59. M. V. Klein, B. N. Ganguly, and P. T. Colwell, *Phys. Rev. B*, **6**, 2380 (1972).
60. G. Irmer, V. V. Toporov, B. H. Bairamov, and J. Monecke, *Phys. Status Solidi*, **119**, 595 (1983).
61. M. V. Klein, *Light Scattering in Solids*, edited by M. Cardona (Springer-Verlag, 1975) p. 160.
62. P. Perlin, E. Litwin-Staszewska, B. Suchanek, W. Knap, J. Camassel, T. Suski, R. Piotrzkowski, I. Grzegory, S. Porowski, E. Kaminska, and J. C. Chervin, *Appl. Phys. Lett.*, **68**, 1116 (1996).
63. J. Pankove, S. Bloom, and G. Harbeke, *RCA Rev.*, **36**, 163 (1975).
64. M. Suzuki, T. Uenuyama, and A. Yanase, *Phys. Rev. B*, **52**, 8132 (1995).
65. C. Wetzel, W. Walukiewicz, and J. W. Ager III, *MRS Symp.*, **449**, 567 (1997).
66. F. Demangeot, J. Frandon, M. A. Renucci, C. Meny, O. Briot, and R. L. Aulombard, *J. Appl. Phys.*, **82**, 1305 (1997).
67. D. Kirilov, H. Lee, and J. Harris Jr., *J. Appl. Phys.*, **80**, 4058 (1996).
68. H. Harima, S. Sakashita, T. Inoue, and S. Nakashima, *J. Cryst. Growth*, **189–190**, 672 (1998).
69. T. Kozawa, T. Kachi, H. Kano, M. Hashimoto, N. Koide, and K. Manabe, *J. Appl. Phys.*, **75**, 1098 (1994).
70. H. Harima, H. Sakashita, and S. I. Nakashima, Conference Stockholm (1997).
71. F. Ponce, J. W. Steeds, C. D. Dyer, and G. D. Pitt, *Appl. Phys. Lett.*, **69**, 2650 (1996).
72. G. Popovici, G. Y. Xu, A. Botchakarev, W. Kim, H. Tang, A. Salvador, H. Morkoç, R. Strange, and J. O. White, *J. Appl. Phys.*, **82**, 4020 (1997).
73. F. Demangeot, J. Frandon, M. A. Renucci, N. Grandjean, B. Beaumont, J. Massies, and P. Gibart, *Solid State Commun.*, **106**, 491 (1998).
74. H. Harima, T. Inoue, S. Nakashima, K. Furukawa, and M. Taneya, *Appl. Phys. Lett.*, **73**, 2000 (1998).
75. D. C. Look, D. C. Reynolds, J. W. Hemsky, J. R. Sizelove, R. L. Jones, and R. J. Molnar, *Phys. Rev. Lett.*, **79**, 2273 (1997).
76. P. Perlin, T. Suski, A. Polian, J. C. Chervin, A. Litwin-Staszewska, I. Grzegory, S. Porowski, and J. W. Erickson, *MRS Symp.*, **449**, 519 (1997).
77. P. Perlin, T. Suski, H. Teisseyre, M. Leszczynski, I. Grzegory, J. Jun, S. Porowski, P. Bogulawski, J. Bernholc, J. C. Chervin, A. Polian, and T. D. Moustakas, *Phys. Rev. Lett.*, **75**, 296 (1995).
78. C. Wetzel, W. Walukievicz, I. F. Haller, and J. Ager III, *Phys. Rev. B*, **53**, 1322 (1996).
79. F. Demangeot, J. Frandon, and M. A. Renucci, *Proc. E-MRS Spring Meeting*, Strasbourg, France, *Mater. Sci. Eng., B*, **43**, 3, 246 (1997).
80. M. Ramsteiner, O. Brandt, and K. H. Ploog, *Phys. Rev. B*, **58**, 1118 (1998).
81. V. Lemos, C. A. Arguello, and R. C. C. Leite, *Solid State Commun.*, **11**, 1351 (1971).
82. D. Behr, J. Wagner, J. Schneider, H. Amano, and I. Akasaki, *Appl. Phys. Lett.*, **68**, 2404 (1996).
83. D. Behr, J. Wagner, R. Niebuhr, C. Merz, K. H. Bachem, H. Amano, and I. Akasaki, *Proc. 23rd Int. Conf. on Physics of Semiconductors*, Berlin, edited by M. Scheffler and R. Zimmermann (World Scientific, 1996).

84. F. Demangeot, J. Frandon, M. A. Renucci, H. Sands, D. Batchelder, S. Ruffenach-Clur, and O. Briot, *Solid State Commun.*, **109**, 519 (1999).
85. D. Behr, J. Wagner, A. Ramakrishnan, H. Obloh, and K. H. Bachem, *Appl. Phys. Lett.*, **73**, 241 (1998).
86. G. Wei, J. Zi, K. Zhang, and X. Xie, *J. Appl. Phys.*, **82**, 622 (1997).
87. H. Grille et F. Bechstedt, *J. Raman Spectrosc.*, **27**, 201 (1996).
88. S. M. Komirenko, K. W. Kim, M. A. Stroscio, and M. Dutta, *Phys. Rev. B*, **59**, 5013 (1999).
89. J. Gleize, M. A. Renucci, J. Frandon, and F. Demangeot, *Phys. Rev. B*, **60**, 15985 (1999).
90. D. Behr, R. Niebuhr, J. Wagner, K. H. Bachem, and U. Kaufmann, *Appl. Phys. Lett.*, **70**, 363 (1997).
91. D. Behr, R. Niebuhr, H. Obloh, J. Wagner, K. H. Bachem, and U. Kaufmann, *MRS Proc.*, **468**, 213 (1997).
92. F. Demangeot, J. Gleize, J. Frandon, M. A. Renucci, M. Kuball, N. Grandjean, and J. Massies, *Proc. 3rd Int. Conf. Nitride Semicond.*, Montpellier, France, *Phys. Status Solidi B*, **216**, 799 (1999).
93. J. Gleize, F. Demangeot, J. Frandon, M. A. Renucci, F. Widmann, and B. Daudin, *Appl. Phys. Lett.*, **74**, 703 (1998).
94. F. Bernardini, V. Fiorentini, and D. Vanderbilt, *Phys. Rev. B*, **56**, R10024 (1997).
95. A. J. Shields, C. Trllero-Giner, M. Cardona, H. T. Grahn, K. Ploog, V. A. Haisler, D. A. Tenne, N. T. Moshegov, and A. I. Toporov, *Phys. Rev. B*, **46**, 6990 (1992).
96. M. Leroux, N. Grandjean, M. Laügt, J. Massies, B. Gil, P. Lefebvre, and P. Bigenwald, *Phys. Rev. B*, **58**, R13371 (1998).
97. J. Wagner, A. Ramakrishnan, H. Obloh, and M. Maier, *Appl. Phys. Lett.*, **74**, 3863 (1999).
98. J. Gleize, F. Demangeot, J. Frandon, M. A. Renucci, M. Kuball, F. Widmann, and B. Daudin, *Proc. 3rd Int. Conf. Nitride Semicond.*, Montpellier, France, *Phys. Status Solidi B*, **216**, 457 (1999).

CHAPTER 9

Light Emission from Rare Earth–Doped GaN

J.M. ZAVADA[1], U. HÖMMERICH[2] and A.J. STECKL[3]

[1]*US ARL—European Research Office, London, NW1 5TH, UK*
[2]*Hampton University, Department of Physics, Research Center for Optical Physics, Hampton, VA 23668*
[3]*Nanoelectronics Laboratory, University of Cincinnati, Cincinnati, OH 45221-0030*

9.1 INTRODUCTION TO RARE EARTH ELEMENTS

The rare earth (RE) or lanthanide elements denote the metallic elements having atomic numbers 57 to 71. These elements, which are listed in Table 9.1, have very similar chemical properties [1]. They are known as *rare earths* due to the difficulty in extracting them from materials known as *earths*,

Table 9.1 Basic properties of the lanthanide rare earth elements.

Atomic number	*Element*	*RE^{3+} Electron configuration*	*RE^{3+} Ground state*	*RE^{3+} Ionic radius (Å)*
57	Lanthanum	$4f^0 5s^2 5p^6$	$^1S_0$	1.15
58	Cerium	$4f^1 5s^2 5p^6$	$^2F_{5/2}$	1.02
59	Praseodymium	$4f^2 5s^2 5p^6$	$^3H_4$	1.00
60	Neodymium	$4f^3 5s^2 5p^6$	$^4I_{9/2}$	0.99
61	Promethium	$4f^4 5s^2 5p^6$	$^5I_4$	0.98
62	Samarium	$4f^5 5s^2 5p^6$	$^6H_{5/2}$	0.97
63	Europium	$4f^6 5s^2 5p^6$	$^7F_0$	0.97
64	Gadolinium	$4f^7 5s^2 5p^6$	$^8S_{7/2}$	0.97
65	Terbium	$4f^8 5s^2 5p^6$	$^7F_6$	1.00
66	Dysprosium	$4f^9 5s^2 5p^6$	$^6H_{15/2}$	0.99
67	Holmium	$4f^{10} 5s^2 5p^6$	$^5I_8$	0.97
68	Erbium	$4f^{11} 5s^2 5p^6$	$^4I_{15/2}$	0.96
69	Thulium	$4f^{12} 5s^2 5p^6$	$^3H_6$	0.95
70	Ytterbium	$4f^{13} 5s^2 5p^6$	$^2F_{7/2}$	0.94
71	Lutetium	$4f^{14} 5s^2 5p^6$	$^1S_0$	0.93

Source: Ref. 1.

e.g., lime and alumina. Actually, rare earth elements constitute about one-fourth of the known metals, and several are more abundant than gold or silver. The electronic structure of each trivalent rare earth (RE^{3+}) element consists of a partially filled inner 4f subshell and completely filled outer $5s^2$ and $5p^6$ subshells. With increasing nuclear charge, electrons enter into the underlying 4f subshell rather than the external 5d subshell, as shown in Table 9.1. The ground state of each RE^{3+} is represented as ^mL_J, where L is the total orbital angular momentum, J is the total angular momentum, and m is the multiplicity of terms in the atomic configuration. Since the filled $5s^2$ and $5p^6$ subshells screen the 4f electrons, the rare earth elements have very similar chemical properties. They are highly reactive with many other elements and readily form a variety of organic complexes.

The screening of the partially filled 4f subshell by the outer closed $5s^2$ and $5p^6$ subshells also gives rise to sharp absorption/emission spectra approximately independent of the host material [2]. The intrasubshell transitions of the 4f electrons lead to narrow spectral peaks in the ultraviolet, visible, and near-infrared regions. In particular, the trivalent Er ion (Er^{3+}) exhibits well-defined emission lines in the IR, near 1.54 μm, and in the visible (green), near 0.54 and 0.56 μm. The Pr^{3+} ion yields emission lines in the IR, near 1.3 μm and in the visible (red), near 0.65 μm. In the free RE^{3+} state, these electric dipole transitions are forbidden to first order, because of parity conservation; however, when a rare earth ion is embedded in a crystal or host, the local electric fields produce a Stark splitting of the energy levels, and certain electric dipole transitions are permitted. Transitions of 4f electrons between these split levels lead to the observed absorption/emission spectra. In Figure 9.1 are shown some of the major energy-level transitions for different RE^{3+} ions in GaN that have technological importance for optoelectronics.

The optical properties of rare earth ions in insulating materials have been extensively studied for applications in solid-state lasers and optical fiber amplifiers. Solid-state lasers, such as the Nd:YAG system, which is based on the 4f intrasubshell transitions of Nd^{3+} ions, exhibit a very stable lasing wavelength and minimum temperature dependence [3,4]. Because of these characteristics, these lasers have found widespread applications in laboratory and military systems.

Long-distance optical fiber communications systems require amplifiers, operating at the primary wavelengths (1.54 μm and 1.30 μm), to boost the transmitted signal. Initially, this amplification was done at repeater stations in which the optical signal was converted into an electrical signal, amplified, and then optically regenerated. In recent years, erbium-doped fiber amplifiers (EDFA) have been used to perform this amplification while keeping the signal in the optical domain. The intrasubshell transitions of 4f electrons in the Er^{3+} ions lead to an emission peak centered at 1.54 μm, which is the region of minimum loss in silica fibers. These optical amplifiers are the

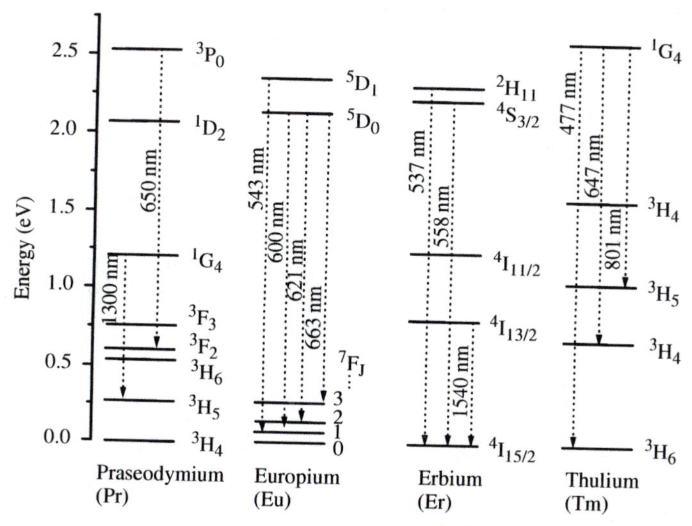

Figure 9.1 Simplified energy level diagram for selected rare earth ions: Pr^{3+}, Eu^{3+}, Er^{3+}, and Tm^{3+}. Relevant optical transitions discussed in the text are indicated by arrows.

cornerstone of wavelength division multiplexing (WDM) systems operating at 1.54 μm [5–7]. Major improvements in link distance and data rates, and reduced needs for signal regeneration have been demonstrated; however, due to the demand for greater transmission data rates, there has been widespread research and development to increase the bandwidth of EDFAs to increase the number of optical channels.

Nevertheless, there are disadvantages to the use of EDFAs in optical fiber communication systems. The EDFAs typically have a low optical gain and require elaborate optical pumping systems [8,9]. If Er atoms could be introduced into an appropriate semiconductor host and effectively activated by electrical means, then charge carriers could be used to induce the 4f intrasubshell optical transitions. Optical amplification at 1.54 μm might thereby be achieved using compact, rugged semiconductor components.

Investigations of the optical properties of RE-doped III–V semiconductors have been conducted for more than a decade. Beginning with the work of Ennen et al., the luminescence of rare earth ions in various III–V compound semiconductors has received considerable attention [10]. Studies of rare earth ions in a variety of different semiconductors have been conducted [11–13]. Due to the importance of the 1.54-μm region for optical communications, Er has been the main rare earth element to be investigated in these semiconductors; however, the intensity of the light emission appears to be strongly dependent upon the sample temperature and the energy bandgap of the host material. Wide-bandgap semiconductors, such as the III–V nitride semiconductors, and SiC, appear to be the best materials for device applications. Such RE-doped wide-gap semiconductors offer the prospect of temperature-stable optical amplifiers and light-emitting devices operating at wavelengths from the visible to the IR.

9.2 RARE EARTH DOPING OF GaN THIN FILMS

Two main methods have been used for doping GaN films with RE atoms: postgrowth ion implantation and in situ doping during epitaxial growth. Each of the methods presents certain advantages as well as difficulties. Although there have been efforts to incorporate Er atoms into GaN materials during bulk growth, this work is in its early stages of development.

9.2.1 Ion Implantation

Ion implantation has been widely used [14] in processing integrated electronic circuits and optoelectronic devices. Because this method is a nonequilibrium process, it is not limited by solubility constraints or by surface chemistry and can be used at ambient or elevated temperatures. A high concentration of dopant atoms can be produced in localized regions of the semiconductor film. Consequently, a number of research groups have used ion implantation to dope GaN and other III–V semiconductors with RE atoms. However, ion implantation of RE atoms leads to considerable damage to the GaN crystal, and postimplantation annealing is required to achieve RE^{3+} luminescence.

Wilson et al. reported [15] the first observations of optical emission from Er ions incorporated in III–V nitride semiconductors. In particular, they implanted GaN films grown on GaAs and on sapphire substrates, and AlN films grown on sapphire substrates. These films were coimplanted with Er^+ and O^+ ions and then for 20 min annealed at ~650–700 °C. The Er densities were confirmed [16] using secondary ion mass spectrometry (SIMS) analysis that was quantified using Er-implanted standards. Atomic depth profiles of Er in the GaN epilayers were obtained [17], as shown in Figure 9.2. The SIMS measurements were made using O_2 ion bombardment and positive secondary ion detection for the Er atoms. SIMS analysis confirmed that a maximum Er density on the order of 10^{19} cm^{-3} range was achieved.

Questions remain concerning the locations of the implanted RE atoms in the GaN crystal lattice and the atomic complexes that are formed with

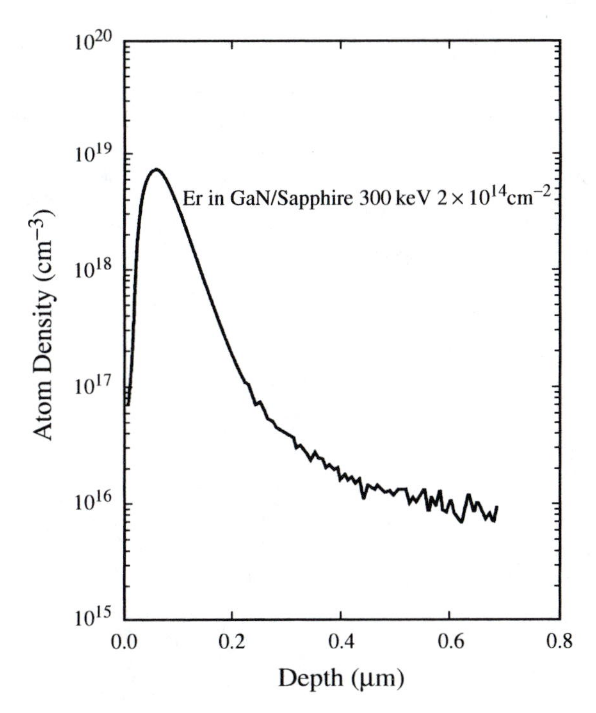

Figure 9.2 Atomic depth profile for Er ions implanted into a GaN film grown on a sapphire substrate. Implantation was at an energy of 300 keV to a fluence of 2×10^{14} cm^{-2}.

annealing. After ion implantation there are probably a number of different sites that are available to the RE atoms. Following annealing, the RE atoms tend to occupy substitutional sites. Rutherford backscattering (RBS) analysis has been useful in obtaining a better understanding of the actual locations.

Alves et al. examined [18] Er-implanted GaN films using RBS. A He ion beam was used to probe the crystallinity of the material and to detect atoms in interstitial sites. One GaN film was implanted with Er ions to a fluence of 5×10^{14} cm^{-2}. Another GaN film was implanted with both Er and O ions, each to a fluence of 5×10^{14} cm^{-2}. After furnace annealing for 30 min at 600 °C, the RBS data showed that the lattice damage was considerably reduced with a minimum yield along the c-axis of about 12% in the case of the Er + O implanted sample, and 20% for the Er-only implanted sample. In both cases the Er dip overlapped the host dip completely, indicating complete substitutionality of the Er atoms, as shown in Figure 9.3. Additional scans along the ⟨1011⟩ axis confirmed that the Er atoms occupy sites on the Ga sublattice. These results indicated that the strong Er luminescence reported by Wilson et al. [15] occurred when the Er ions were on the Ga sublattice sites.

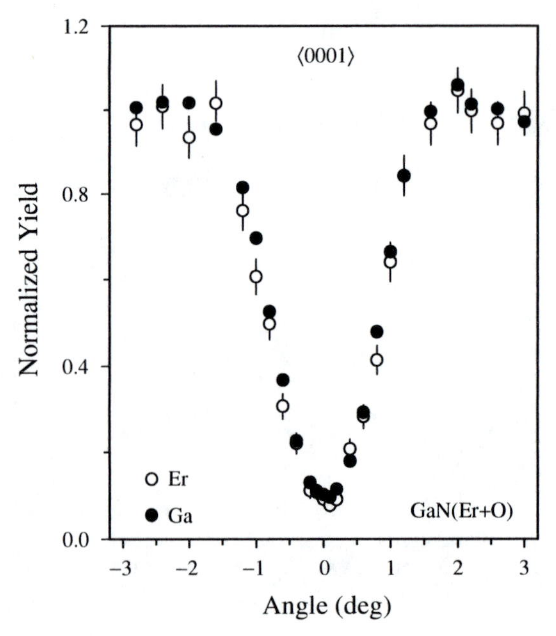

Figure 9.3 Normalized RBS yield along the ⟨0001⟩ axis for a GaN film after Er + O coimplantation and annealing.

9.2.2 Doping During Epitaxial Growth of GaN Films

Several types of epitaxial growth have successfully been used for doping GaN films with RE atoms including gas-source metal–organic molecular beam epitaxy (MOMBE), solid-source molecular beam epitaxy (SSMBE), and metal–organic vapor-phase epitaxy (MOVPE). These techniques have made possible precise control of epilayer composition, thickness, and doping profile. Concentrations of RE atoms on the order of 10^{19}–10^{20} cm^{-3} have been realized in these films.

MacKenzie et al. [19] were the first to succeed in growing GaN films doped with Er atoms. The GaN films were grown on either Si or sapphire substrates at ~700 °C using a MOMBE system with a solid elemental Er source and triethylgallium (TEGa) for the Ga flux. Reactive nitrogen for the growth of the GaN films was provided by an SVT rf plasma source operating at 400 W of forward power and 5 sccm of N_2. A shuttered effusion oven charged with 4 N Er was used for doping, with cell temperatures varying from 900 to 1450 °C. The films were single crystalline hexagonal and layer thicknesses were 0.5–1.0 μm. Cell temperatures greater than 1000 °C were required for the Er source in order to achieve significant Er concentrations in the epilayers.

SIMS analysis of the Er-doped GaN layers indicated that Er densities in the 10^{19}–10^{20} cm^{-3} range were achieved. Different Er cell temperatures were used in this experiment. The lowest cell temperature, ~900 °C, resulted in an Er density of $\sim 3 \times 10^{19}$ cm^{-3}. The highest cell temperature, ~1000 °C, yielded a density of $\sim 6 \times 10^{19}$ cm^{-3}. The total thickness of the epilayer was approximately 0.6 μm. SIMS measurements also showed that high concentrations of O were present in the GaN films.

Overberg et al. studied [20] the surface morphology of the GaN epilayers grown by MOMBE as a function of Er flux. The Er-doped GaN films showed improved surface smoothness, with rms roughness values ranging from 18.1 to 2.0 nm, as the Er cell temperature was increased from 1250 to 1450 °C. The surface morphology was characterized by atomic force microscopy (AFM) and by scanning electron microscopy (SEM). Several SEM micrographs of the GaN surface as a function of Er cell temperature are shown in Figure 9.4. The highest cell temperature, 1450 °C (lower right), resulted in the smoothest surface.

Steckl et al. at the University of Cincinnati investigated [21–24] the in situ incorporation of Er from a solid source into GaN films during MBE growth on sapphire and Si substrates utilizing solid sources to supply the Ga and RE fluxes. An SVTA rf plasma source was used to generate atomic nitrogen. Typical GaN growth rates were 0.8–1.0 μm/hr at growth temperatures of ~750 °C. Er concentration in the GaN film was primarily controlled by the Er cell temperature, with growth temperature

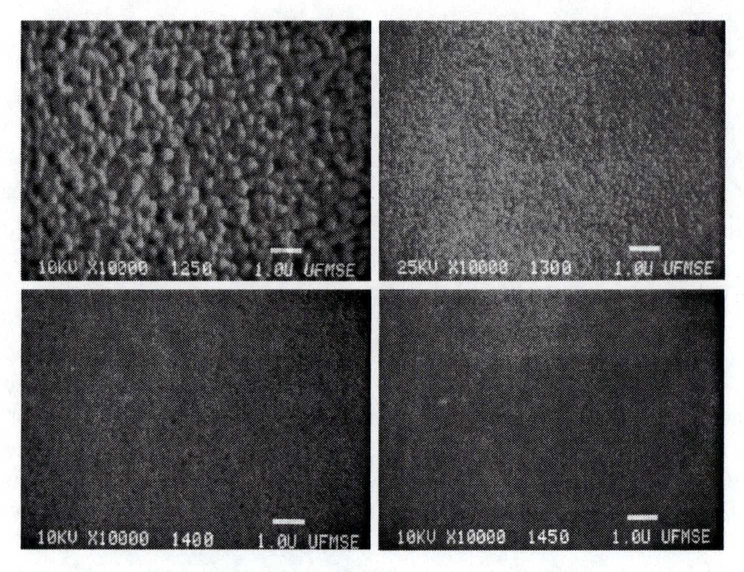

Figure 9.4 Scanning electron micrographs of the GaN surface with a progression of Er cell temperatures: 1250 °C (upper left), 1300 °C (upper right), 1400 °C (lower left), and 1450 °C (lower right).

and Ga flux also playing an important role. Experiments with GaN:Er on sapphire showed that the incorporated Er concentration tracks very well with the Er cell temperature. No evidence of reaching the solid solublity limit was observed for concentrations up to $3 \times 10^{20}\,\mathrm{cm}^{-3}$. For GaN growth on Si, even higher Er concentrations were observed. For example, in Figure 9.5 is shown a representative SIMS depth profile of an Er-doped GaN film grown on Si (111). For an Er cell temperature of 1100 °C, a very uniform Er concentration of $\sim 6.5 \times 10^{20}\,\mathrm{cm}^{-3}$ was measured, corresponding to ~0.7 at.% throughout the GaN film. The maximum concentration achieved under the conditions reported was ~5 at.%.

Using RBS techniques, Lorenz et al. [25] examined Er location in the in situ–doped GaN films grown on Si substrates. The RBS yield along the ⟨1000⟩ axis showed that the Er dip overlapped the host dip almost completely, indicating nearly 100% substitutionality of the Er atoms, as shown in Figure 9.6. Additional scans along the ⟨1011⟩ axis confirmed *that the Er* atoms occupy sites on the Ga sublattice. These results indicate *that the Er* ions are in the trivalent state and compete with the Ga ions for the same sublattice sites.

The research group at the University of Cincinnati also succeeded in doping GaN films with other RE elements including Pr, Eu, and Tm [26–28], using solid sources for the Ga and RE fluxes. An rf plasma source

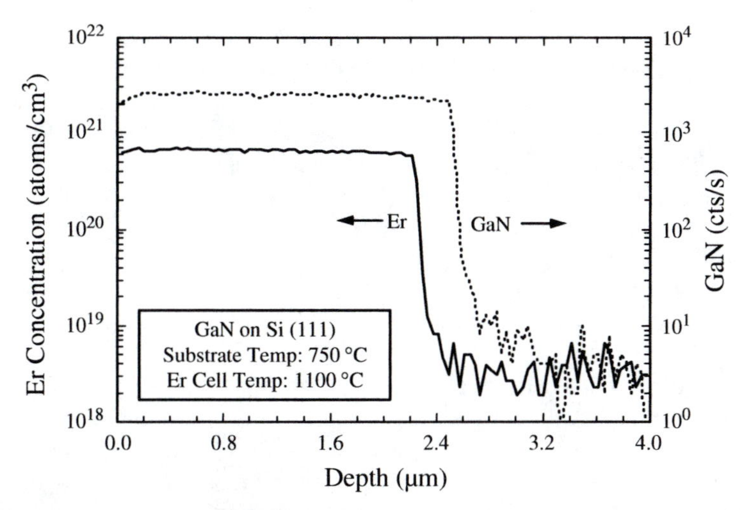

Figure 9.5 Er concentration depth profile obtained [27] by SIMS from in situ–doped GaN film grown by MBE on Si (111). The Er cell temperature during growth was 750 °C. A thin (~0.25 μm) undoped GaN initialization was grown first, followed by a ~2.2-μm Er-doped GaN layer.

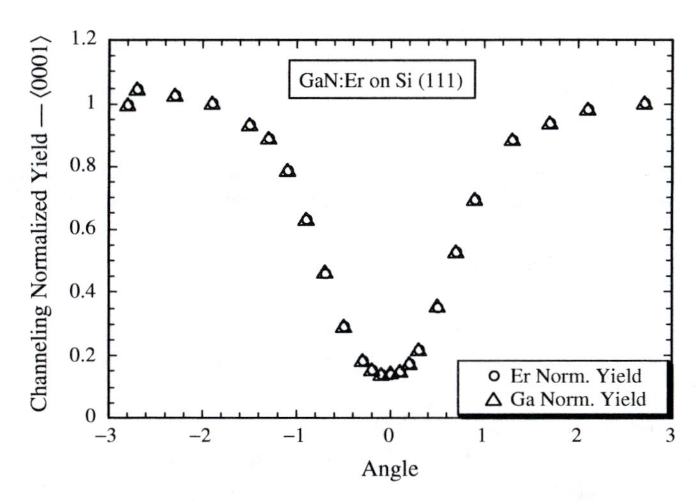

Figure 9.6 Normalized RBS yield along the ⟨1000⟩ axis for a GaN film doped with Er during MBE growth.

was used to generate atomic nitrogen. Subsequently, Morishima et al. [29] reported on the in situ doping of GaN films with Eu using gas-source MBE. The films were grown on sapphire substrates using uncracked ammonia (6 N) as the nitrogen source and metallic sources for Ga (6 N) and Eu (3 N). The growth temperature was varied from 500 to 700 °C. In another development, Hara et al. [30] reported the in situ doping of GaN films with Tb using a MOVPE system. The GaN layers were grown on sapphire substrates in an atmospheric-pressure, vertical reactor using TMGa and NH_3 as the source materials. The Tb source was tris-(dipivaloylmethanato)-terbium ($Tb(DPM)_3$), which is a solid compound with a vapor pressure of 9×10^{-2} Torr at 150 °C. This source was vaporized in a stainless-steel container kept at 150 °C and delivered to the reactor by the H_2 carrier gas.

9.3 OPTICAL EXCITATION OF RE^{3+} IONS

Photoluminescence (PL) spectroscopy has been the main optical technique used to characterize the emission of RE ions in semiconductor materials. This technique involves optical excitation of the RE ions and measurement of the emission as a function of intensity and energy. In a semiconductor, optical excitation at above bandgap energy, leads to creation of electron–hole pairs. Some of the electron–hole pairs may transfer energy to RE ions, leading to excitation of the 4f electrons. Alternatively, the RE ions may be excited directly by the optical pump radiation, provided that the energy of the radiation is resonant with one of the higher-energy states of the RE ion. Most of the PL measurements of RE-doped semiconductors have involved use of a laser operating at a wavelength above that of the bandgap of the host semiconductor. This technique corresponds to an indirect excitation of the RE^{3+} ions by electron–hole pairs. A few experiments have been carried out in which the RE^{3+} ions have been excited directly by the optical pump radiation [31].

In general, the optical emission of Er^{3+} ions in semiconductors has not been as efficient as in dielectric materials. It may be that ionic bonds found in dielectrics are better for forming the required Er^{3+} energy levels than the covalent bonds present in most III–V semiconductors. Co-doping the host semiconductor with impurity elements, such as O or F, seems to enhance the optical emission. This may be due to the formation of ligands between the impurity atoms and the Er atoms and conversion of the local bonds into a more ionic state. Favennec et al. [32] and Michel et al. [33] observed a significant enhancement of the Er^{3+} emission in Si due to the presence of O impurities. In a similar series of experiments, Colon et al. [34] showed that the addition of O enhances the Er^{3+} emission from Er-doped AlGaAs films.

Favennec et al. reported [35] a strong dependence of the emission intensity of Er^{3+} ions on the bandgap of the host semiconductor and on the sample temperature. Several different semiconductors were implanted with Er^{+} ions, and the emission intensity was measured. It was found that the intensity decreased at higher sample temperatures. This thermal quenching of the emission intensity was more severe for the smaller-bandgap materials. The wide-bandgap II–VI compounds, such as ZnTe and CdTe, exhibited the least temperature dependence. Neuhalfen and Wessels observed [36] a similar temperature dependence of the intensity of Er^{3+} emission for different compositions of $In_{1-x}Ga_xP$ epilayers grown by MOVPE. Based on their data, they proposed a model to explain the thermal quenching as a function of bandgap energy. In their model the Er ions act as radiative recombination centers. With increased sample temperatures, the electrons become delocalized from these centers into the conduction band and recombine via nonradiative channels. This results in a quenching of the PL emission at higher temperatures, which is stronger for smaller-bandgap semiconductors.

Another explanation for the decrease of thermal quenching with wider-band gap semiconductor materials is based on the energy backtransfer model as proposed by Takahei and Taguchi [37,38] for Yb-doped InP and other RE-doped semiconductors. Based on this model, Er doping introduces an electron trap close to the conduction band. The Er-related trap captures an electron, and this captured electron subsequently attracts a hole by Coulomb force, resulting in the formation of a bound exciton localized on the RE center. The bound exciton recombines and transfers its energy through a multiphonon process to the Er^{3+} 4f-shell electrons. Backtransfer is the reverse process, in which an excited Er^{3+} ion transfers an excited carrier back to the trap level followed by a thermalization into the conduction band. Subsequently, the carriers recombine radiatively or nonradiatively through the bandedge. Therefore, energy backtransfer provides an additional nonradiative decay channel for excited Er^{3+} ions, which reduces the Er^{3+} PL intensity with increasing temperature. The backtransfer process is less likely to occur in wide-gap semiconductors because of the large energy mismatch between the bound exciton and the intra-4f Er^{3+} transition.

Since the wider-bandgap semiconductors lead to less thermal quenching of Er^{3+} emission, the III–V nitride alloys appear to be especially promising host materials for RE doping. These alloys have a bandgap ranging from 1.9 eV for InN to 3.4 eV for GaN and 6.2 eV for AlN; however, due to the lack of a lattice-matched substrate, high-quality III–V nitride crystals have been unavailable for these investigations. The currently available epilayers contain a high density of dislocation defects and various impurity elements. Nevertheless, very encouraging results have been obtained with RE-doped III–V nitrides.

Photoluminescence excitation (PLE) spectroscopy is a powerful tool for identifying the absorption bands of various emitting centers [2]. In PLE the luminescence intensity of an optical emitting center, in this case the RE^{3+} ion, is measured as a function of excitation wavelength. Strong emission at a particular excitation wavelength indicates that the emitting center also absorbs strongly at that wavelength. Consequently, the spectral location and shape of the absorption bands of the emitting center can be accurately determined.

9.3.1 Infrared Luminescence

Er^{3+} optical emission[15] from GaN films coimplanted with Er^+ and O^+ ions was observed using an Ar^+ laser at a wavelength of 457.9 nm for optical excitation. Although this wavelength does not correspond to above-bandgap radiation, it was conjectured that some of the higher-energy manifolds of the Er^{3+} ion may be excited. Strong IR luminescence was measured even at room temperature. As shown in Figure 9.7, IR spectra were measured at 6, 77, and 300 K. The spectra were centered at 1.54 μm and displayed many distinct lines, indicative of the allowed transitions between the $^4I_{13/2}$ and the $^4I_{15/2}$ manifolds of the Er^{3+} system. In addition, the integrated Er^{3+} luminescence intensity at room temperature was nearly 50% of that at 77 K. The data provided further support that wide-gap materials, such as GaN, tend to suppress the thermal quenching of Er^{3+} luminescence.

Torvik et al. performed a systematic study of GaN films coimplanted with Er^+ and O^+ ions and then annealed [39]. The films were grown by chemical vapor deposition (CVD) on R-plane sapphire. Different combinations of fluences for the implants and annealing temperatures were examined. After each annealing stage, the samples were examined for Er^{3+} luminescence using a laser at a wavelength of 980 nm, which corresponded to a resonant

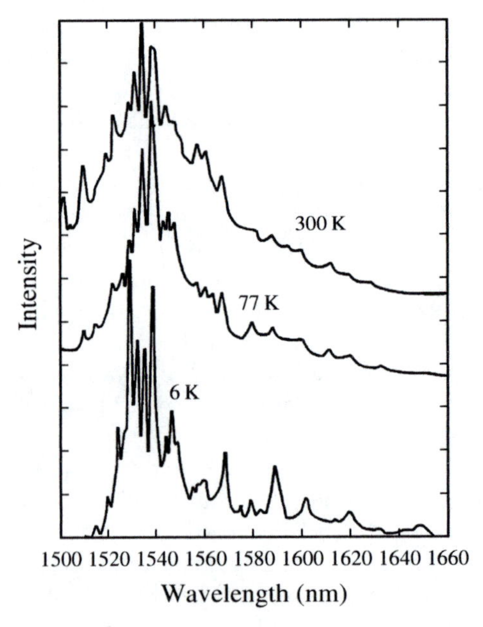

Figure 9.7 PL spectra of Er^{3+} ions implanted into a GaN/sapphire host as a function of measurement temperature. The sample film was coimplanted with O^+ ions and annealed at 700 °C.

excitation of the Er^{3+} ions by optical pumping to the $^4I_{13/2}$ manifold. In their experiments the samples with the highest Er fluence ($\sim 10^{15}$ Er^+ cm^{-2}) yielded the strongest PL intensity. In a subsequent work [40], Torvik et al. measured the luminescence of Er-doped GaN films using a laser at a wavelength of 360 nm, which represented an indirect excitation of the Er^{3+} ions by electron–hole pairs being created by above-bandgap radiation; however, the resulting emission at 1.54 μm was not as intense as that under resonant excitation.

Silkowski et al. [41] examined the luminescence of both Er and Nd ions implanted into GaN films that were grown by MOCVD on sapphire substrates. The ion-implanted films were annealed and then excited using an Ar^+ laser operating at 514.5 nm. No coimplantation with O was performed. Since the films were prepared by MOCVD, it is possible that high levels of O were already present. The Er-implanted samples showed the $^4I_{13/2} \rightarrow {}^4I_{15/2}$ transitions around 1.54 μm and the $^4I_{11/2} \rightarrow {}^4I_{15/2}$ transitions near 1.0 μm. Based on the data, it appeared that multiple Er^{3+} radiative centers were present in the sample.

Hömmerich et al. reported [42] on the luminescence properties of Er-doped GaN films grown by MOMBE and by SSMBE on Si substrates. Both types of samples emitted characteristic 1.54-μm PL resulting from the intra-4f Er^{3+} transition $^4I_{13/2} \rightarrow {}^4I_{15/2}$. Under below-gap excitation the samples exhibited very similar 1.54-μm PL features and intensities, as shown in Figure 9.8. The spectra were taken under identical experimental conditions. The pump power was kept constant at ~0.64 W/cm^2. These spectra do not exhibit the multiple emission lines of the Er-implanted sample shown in

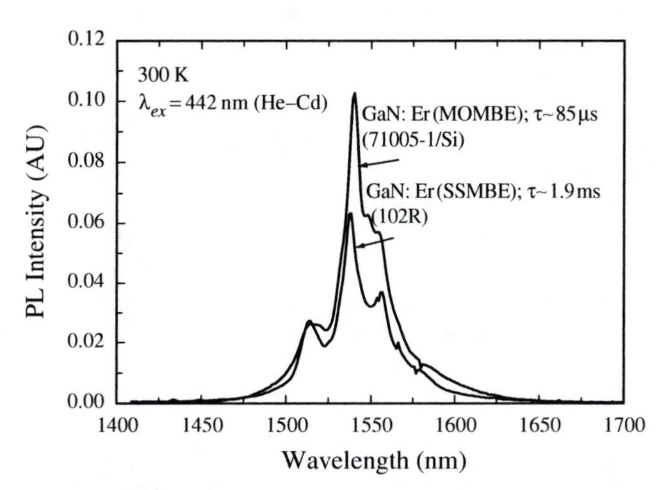

Figure 9.8 PL spectra at 1.54 μm of GaN:Er (MOMBE) and GaN:Er (SSMBE) at room temperature. The samples were excited with the 442-nm (below-gap) line of a He–Cd laser.

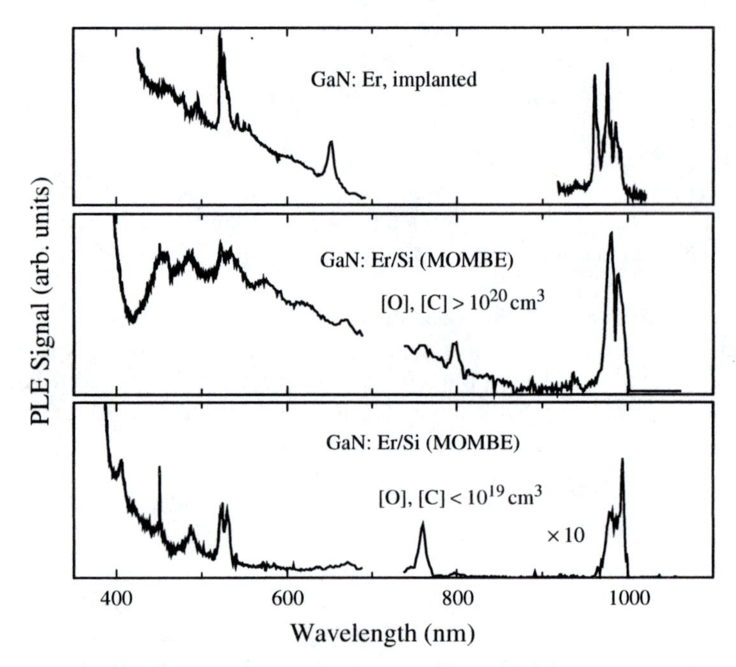

Figure 9.9 PLE spectra of Er-doped GaN films: (a) Er-implanted GaN; (b) and (c) Er-doped GaN samples prepared by MOMBE with varying oxygen and carbon content. The incorporation of high oxygen and carbon backgrounds (b) leads to a broad Er^{3+} defect-excitation band covering the entire visible region. Incorporation of low levels of oxygen and carbon (c) leads to a reduced Er^{3+} defect-excitation band.

Figure 9.7 due to the higher-resolution measurement made in the previous study; however, there is good agreement in the overall shape of the spectra.

Using a pulsed laser system, Thaik et al. [43] measured the PLE spectrum for the Er-implanted GaN film that had been studied by Wilson et al. [15]. For these measurements, the tunable laser consisted of a broadband optical parametric oscillator (OPO) system pumped by a Q-switched Nd^{3+}: YAG laser. The PLE signal was recorded at 300 K as the ratio between the PL intensity detected at 1.535 μm and the intensity of excitation laser. The room-temperature PLE spectrum, shown in Figure 9.9a, covered the wavelength range 425–1020 nm. There were several sharp lines (at 495 nm, 525 nm, and 651 nm) superimposed on a broad background absorption signal, from ~425 to 680 nm. The sharp lines appear to be due to a direct optical excitation of the Er^{3+} ions involving an intra-4f transition, e.g., ${}^4I_{15/2} \rightarrow {}^2H_{11/2}$. The broad absorption background is suggestive of a carrier-mediated process involving defects in the GaN host. Eventhough the sample had been annealed at 650 °C, there was still probably considerable residual implantation damage in the film [44]. There are also absorption sharp lines around 980 nm corresponding to the intra-4f transitions ${}^4I_{15/2} \rightarrow {}^4I_{13/2}$.

Hömmerich et al. [45] also measured the PLE spectrum for the Er-doped GaN films prepared by MOMBE with varying oxygen and carbon content. Triethylgallium (TEGa), dimethylethylamine alane (DMEAA), and thermally evaporated 8N Ga metal provided the group III fluxes. A shuttered effusion oven with 4N Er was used for solid-source doping. Reactive nitrogen species were provided by an SVT rf plasma source. Due to the incorporation of carbon and oxygen from residual ether in TEGa, the C and O backgrounds observed in TEGa-derived GaN were $\sim 10^{21}\,\text{cm}^{-3}$ and $\sim 10^{20}\,\text{cm}^{-3}$, respectively, as determined by SIMS measurements. GaN grown using thermally evaporated 8N pure Ga as the group III source was found [46] to have oxygen and carbon backgrounds of less than $10^{19}\,\text{cm}^{-3}$. The PLE spectrum of GaN:Er/Si (TEGa), shown in Figure 9.9b, reveals that the incorporation of high O and C backgrounds leads to a broad absorption band extending over the entire visible region (~400–800 nm). The PLE spectrum of GaN:Er (Ga), shown in Figure 9.9c, indicates that low C and O backgrounds lead to a reduced absorption band. Several absorption sharp lines arising from direct intra-4f Er transitions (e.g., ~453 and ~525 nm) can easily be detected. The data provide evidence that the background absorption is strongly dependent upon the sample preparation and the defects in the film.

The PLE data give a partial explanation of why it was possible to excite the Er-doped GaN films with below-bandgap optical radiation. The broad, below-bandgap absorption bands in these films provide an effective excitation mechanism of the Er^{3+} ions. Eventhough the wavelength of the optical pump was below-bandgap and did not correspond to an Er^{3+} intra-4f transition, the broad background absorption could couple optical energy into the Er^{3+} ions. Therefore, the Er^{3+} ions in these films could be excited either through resonant optical pumping into one of the 4f manifolds or through a carrier-mediated process involving defects in the host crystal.

Using PLE spectroscopy, the group at the University of Illinois also studied [47] the site location of Er^{3+} ions in implanted GaN films. The GaN epilayers were grown on sapphire substrates using MOCVD and implanted with Er^{+} and O^{+} ions. A Hg lamp was used as the excitation source, and PLE spectra were measured at 6 K over the range 2.3–3.4 eV. Several broad absorption bands, similar to the data in Figure 9.9 were observed. These bands could be used to excite distinct Er^{3+} PL spectra. At least three different Er^{3+} sites were identified in this manner. Excitation of two of the bands apparently involved optical absorption by defects and impurities with subsequent energy transfer to Er^{3+} centers. Excitation of the third band appeared to involve a bound exciton at an Er-related trap center.

In subsequent work, Kim et al. extended [48] their studies to above-bandgap excitation of Er^{3+} ions in implanted GaN films. As in the earlier study, the GaN epilayers were grown on sapphire substrates using MOCVD

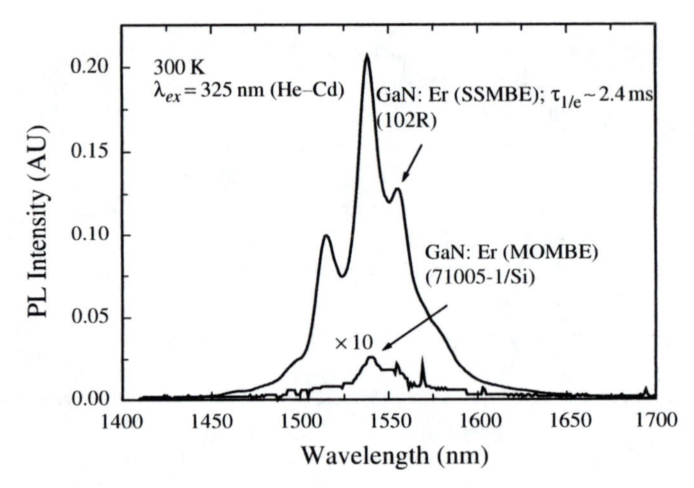

Figure 9.10 PL spectra at 1.54 μm of GaN:Er (MOMBE) and GaN:Er (SSMBE) at room temperature. The samples were excited with the 325-nm (above-gap) line of a He–Cd laser.

and implanted with Er^+ and O^+ ions. The PL spectra from the Er^{3+} ions were measured at 6 K using a variety of optical sources at different wavelengths. Four distinct Er^{3+} site-selective PL spectra, using above-bandgap excitation, were identified. These spectra were associated with trap-mediated excitation of the Er^{3+} centers. Above-bandgap excitation, involving a He–Cd laser at 325 nm, led to excitation of two of the trap-mediated emission bands. The other two bands were not strongly excited in this process. The data indicated that trap-mediated processes dominate the above-bandgap excitation of the Er^{3+} centers. Reduced thermal quenching of Er^{3+} emission in GaN may also be due to these defect centers.

With above-gap excitation (λ_{ex} = 325 nm) the MOMBE and the SSMBE samples exhibited very different 1.54-μm PL features and intensities, as shown in Figure 9.10. The spectra were taken under identical experimental conditions, and the pump power was kept constant at ~0.64 W/cm^2. The most striking feature is the large difference in luminescence intensity observed for the different samples under above-gap excitation. The SSMBE sample exhibited a strong 1.54-μm PL that was nearly 80 times more intense than that observed from GaN:Er prepared by MOMBE. The weak above-gap PL observed from the MOMBE sample can be explained [49] by a significantly reduced Er^{3+} excitation efficiency compared with below-gap excitation. Visible luminescence studies revealed that for MOMBE samples the GaN bandedge provides an efficient radiative combination channel, reducing the excitation efficiency of intra-4f Er transitions.

In a semiconductor, the lifetime of electron–hole pairs is primarily determined by the band structure and the defect density of the material. Typically,

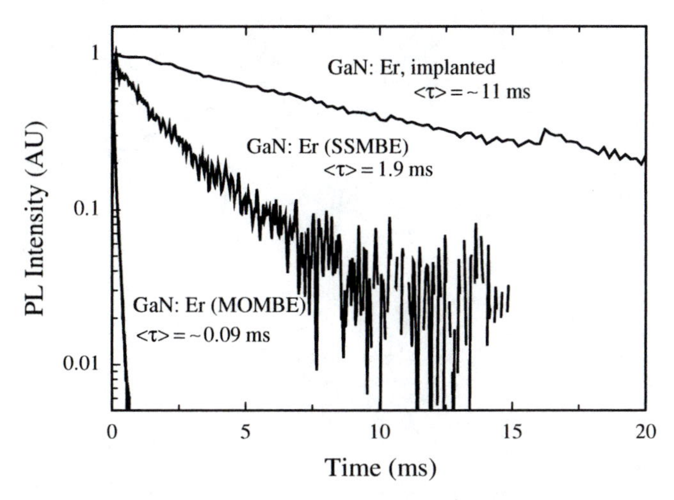

Figure 9.11 PL decay transients at 1.54 μm of Er-implanted GaN; GaN:Er (SSMBE), and GaN:Er (MBE) at 300 K with below-gap excitation.

in direct-bandgap materials, this lifetime is on the order of a few nanoseconds. Generally, the intra-4f transitions of Er^{3+} ions in a semiconductor exhibit a much longer lifetime, on the order of a few milliseconds. Using time-resolved PL techniques, Hömmerich et al. measured [42] the luminescence transients of the different GaN:Er samples. As shown in Figure 9.11, the decay curves were nonexponential, which suggests the existence of multiple Er sites. The existence of multiple Er sites in GaN has previously been reported [50]. To describe the lifetime decay, an average lifetime $\langle\tau\rangle = \int tI(t)\,dt / \int I(t)\,dt$ was used, where $I(t)$ is the luminescence transient, and t is the time. The time decay of 1.54-μm emission of Er^{3+} ions in the SSMBE sample was ~1.9 ms. This result is in agreement with prior data [51] by Klein and Pomrenke concerning the time decay of 1.54-μm emission of Er^{3+} ions in several different semiconductors. They showed that the emission decay in GaAs, GaP, InP, and Si was on the order of 1 ms; however, the time decay of 1.54-μm emission of Er^{3+} ions in the MOMBE sample was measured to be only ~85 μs. This very fast time decay of the Er^{3+} emission may be due a higher level of defects, or impurities, in the MOMBE sample. On the other hand, the time decay of 1.54-μm emission of Er^{3+} ions in the Er-implanted sample was ~11 ms, which is longer than reported elsewhere.

Thaik et al. performed [43] a series of experiments to determine the thermal quenching of the 1.54-μm luminescence of an Er-implanted GaN film. The Er^{3+} 1.54-μm emission was measured over a range of temperatures from 13 to 550 K. The sample was excited with both above-bandgap radiation, using a He–Cd laser operating at 325 nm, and below-bandgap radiation

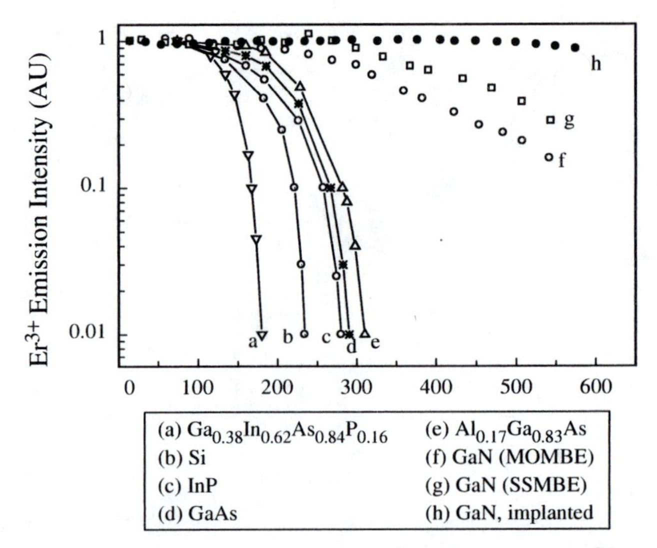

Figure 9.12 Comparison of the temperature dependence of the integrated Er^{3+} 1.54-μm PL for various Er-doped semiconductors.

from an Ar laser operating at 488 nm. There were minor changes in the PL spectra between 300 and 550 K. The FWHM of the PL spectrum at 300 K was ~80 nm, suggesting inhomogeneous broadening of the emission. This broadening indicated that the Er^{3+} ions occupy a wide range of sites, with slightly different atomic configurations, in the GaN host. The integrated PL intensity of the luminescence was found to be nearly constant over the entire range of measurement temperatures; see Figure 9.12h. Relative to its value at 15 K, the integrated PL intensity at 550 K decreased by only about 10%. This remarkable temperature stability of the Er^{3+} luminescence is the best reported data from any Er-doped semiconductor, including Er-doped SiCi [52].

When below-bandgap excitation was used there were significant changes in the PL spectrum depending on the excitation wavelength. This observation indicated that different subsets of Er^{3+} ions were excited with distinct PL and thermal quenching properties. The FWHM of the PL spectrum at 300 K was ~50 nm. There was also a large change in the integrated PL intensity over the range of measurement temperatures. Relative to its value at 15 K, the integrated PL intensity at 550 K decreased by about 50%.

Hömmerich et al. also investigated [42] the thermal quenching properties of Er-doped GaN films grown by MOMBE and by SSMBE. Compared with Si:Er, GaAs:Er, and AlGaAs:Er, both of the in situ–doped GaN:Er samples exhibited very stable 1.54-μm PL up to temperatures as high as 550 K with above-gap excitation; see Figure 9.12. However, neither of the in situ–doped

GaN:Er samples showed the temperature stability that was found in the Er-implanted sample. The data indicate that in addition to the bandgap of the semiconductor host, the local environment of the Er^{3+} ions determines the overall temperature dependence of the 1.54-μm emission.

9.3.2 Visible Luminescence

Steckl et al. were the first to observe [21] visible emission from higher excited 4f levels of RE ions incorporated into GaN. The initial experiments centered on GaN:Er films that were prepared by SSMBE. With above-bandgap excitation, these films exhibited intense green PL from the $^2H_{11/2}$ and $^4S_{3/2}$ excited levels to the $^4I_{15/2}$ ground state; see Figure 9.1. The emission was visible under ambient conditions to the naked eye at room temperature. In addition, these films exhibited well-defined IR emission, as was shown in Figure 9.10. In contrast, none of the GaN:Er films that had been prepared by MOMBE exhibited green PL emission. The visible luminescence spectra of the three different GaN:Er samples are shown in Figure 9.13. Following optical excitation at 325 nm, the GaN:Er (SSMBE) exhibited a weak band-edge PL at ~369 nm (3.36 eV) and two "green" lines located at 537 nm (2.309 eV) and 558 nm (2.222 eV). The GaN:Er (MOMBE) sample showed strong band-edge PL located at ~381 nm (3.25 eV); however, no indication

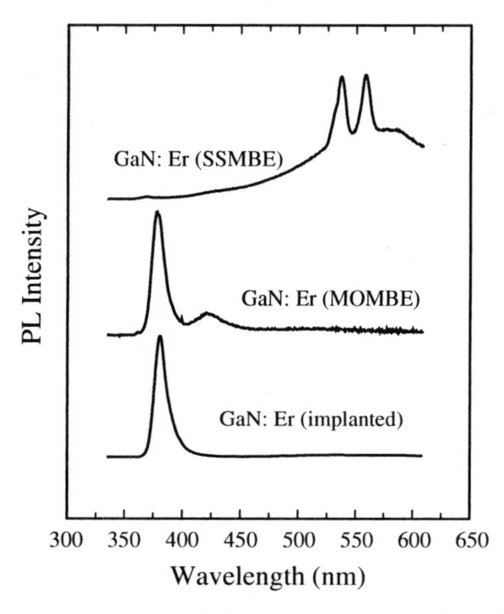

Figure 9.13 Visible PL spectra from GaN:Er (SSMBE), GaN:Er (MOMBE), and Er-implanted GaN at 300 K. The lines at 537 nm and 558 nm that are observed in the GaN:Er (SSMBE) sample are assigned to intra-4f Er^{3+} transitions. All spectra were taken using above-gap excitation at 325 nm.

of green Er^{3+} luminescence was found. Apparently, for GaN:Er (MOMBE) samples, the band-edge provides an efficient radiative combination channel, which reduces the excitation efficiency for visible and IR Er^{3+} transitions. A similar situation exists in the Er-implanted GaN films. Eventhough the RBS data (Figures 9.2 and 9.6), indicate that the Er ions are mainly on Ga-sublattice sites for both GaN:Er (SSMBE) and Er-implanted GaN samples, the PL properties are quite different. Chen et al. reported [53] both IR and green emission from GaN:Er films prepared by a sputtering technique; however, the green emission was superimposed on a broad yellow band found in defect-laden GaN. The local complexes that are formed with the Er ions seem to dominate the optical characteristics in these GaN:Er films.

Visible PL emission has also been observed in GaN films doped with other RE elements, including Pr, Eu, Tb, and Tm [54,55]. Red emission has been found in GaN films doped with either Pr or Eu. Blue emission has been found in GaN films doped with Tm, and green emission for those doped with Tb. It is noteworthy that visible emission from Dy-, Er-, Tm-, Sm-, and Ho-implanted GaN has also been obtained through cathodoluminescence [56,57].

In Figure 9.14 are shown the visible PL spectra, measured at room temperature, of a Pr-doped GaN film grown by SSMBE. The GaN:Pr film was grown on (111) Si substrates, and solid sources were used to supply the Ga and Pr fluxes. The spectra contain a narrow red emission located at ~650 nm (1.90 eV) corresponding to transitions from the $^3P_0$ to the $^3F_2$ excited levels,

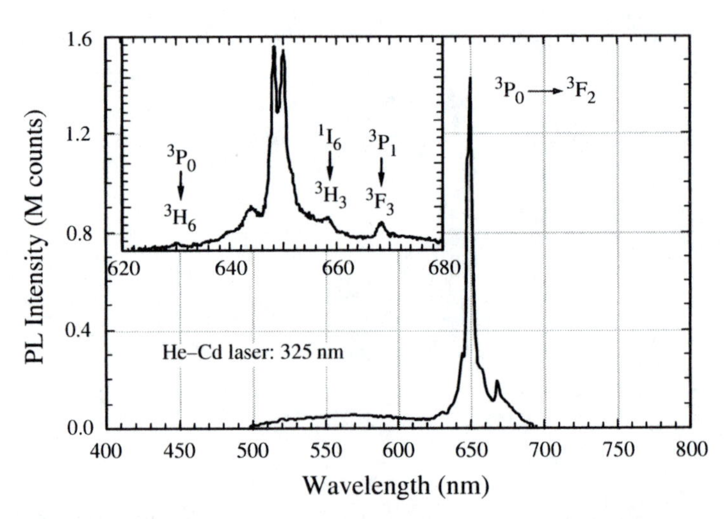

Figure 9.14 Visible PL spectra, measured at 300 K, for Pr-doped GaN films grown by SSMBE. The inset contains high-resolution spectra of the red emission taken with above-bandgap (He–Cd laser at 325 nm) and below-bandgap (Ar^+ laser at 488 nm) optical excitation.

see Figure 9.1. High-resolution spectra of this emission (see the inset to Figure 9.14) reveal that there are two emission lines, one located at ~648 nm and the other at ~650 nm. The two emission lines are probably due to Stark splitting of the excited Pr^{3+} levels. Furthermore, these two emission lines are found in PL measurements taken with both above-bandgap (325 nm) and below-bandgap (488 nm) optical excitation, however, as shown in the inset to Figure 9.14, above-bandgap excitation results in a higher intensity of the red emission. As was the case with GaN:Er films, below-bandgap photons appear to excite the Pr^{3+} ions by means of a defect-mediated mechanism.

Steckl et al. also prepared Pr-doped GaN films using focused-ion-beam implantation (FIB). Since FIB technology is a maskless and resistless "direct-write" process, it can be applied with great versatility [58] to the fabrication of photonic devices. FIB micro- and nanofabrication (with ion beam diameters ranging from less than 100 nm to a few micrometers) can be utilized to reduce the complexity required of conventional photonic fabrication technology. Using Pr ion sources, they produced an implanted pattern covering a 141-μm × 141-μm square area [59]. The implantation energy and dose were 290 keV and 4.7×10^{14} atoms/cm^2, respectively. The sample was annealed at 1050 °C for one hour in Ar after FIB implantation. Under UV excitation from the He–Cd laser, the implanted region emitted red light at 650 nm similar to that shown in Figure 9.14.

Zavada et al. studied the use of conventional ion-implantation techniques to incorporate Pr into GaN films [60]. The Pr-implanted GaN epilayers were examined using PL spectroscopy, and strong emission lines near 650 nm were observed. The dependence of the Pr-related emission intensity on postimplantation annealing temperature, sample temperature, excitation intensity, and oxygen coimplantation was systematically measured. The GaN films used for Pr ion-implantation were grown by using MOCVD on *c*-plane sapphire substrates. Pr was ion-implanted into the GaN epilayers at a dose of 5.7×10^{13}/cm^2 and an energy of 300 keV. SIMS was used to quantify the actual concentration of Pr after implantation. Some of the GaN epilayers were implanted with both Pr and O at similar doses. After implantation, the samples were annealed for 10 min under nitrogen at temperatures ranging from 750 °C to 1050 °C. For the PL measurements an optical source consisting of 290-nm laser pulses with 10-ps width and 9.5-MHz repetition rate was used. Figure 9.15 shows PL spectra, measured at 10 and 300 K over the wavelength range 640–680 nm, for a Pr-implanted GaN sample annealed at 1050 °C. At least five distinct transitions can be seen in the spectra, and the room-temperature linewidths are clearly much smaller than expected for typical band- or impurity-related transitions of a semiconductor. Furthermore, the lines show virtually no temperature-related wavelength shift, as one might expect for band- or impurity-related transitions in semiconductors. The PL emission from the Pr-doped samples exhibited a strong dependence

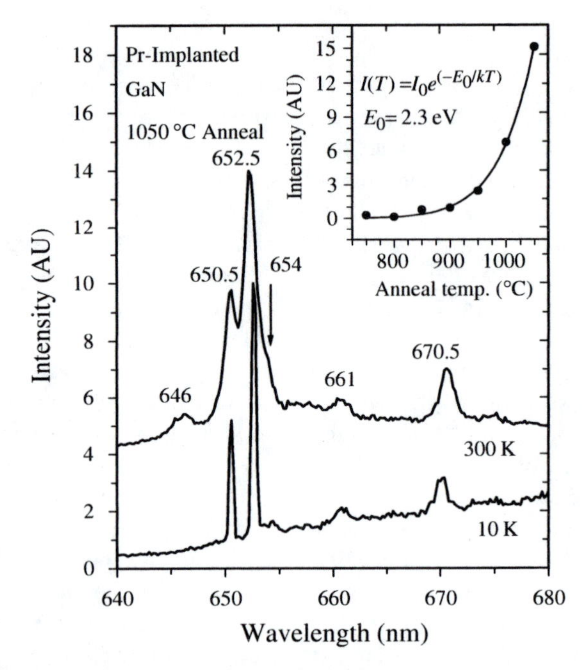

Figure 9.15 Visible PL spectra measured at 10 K and 300 K from a Pr-implanted GaN sample annealed at 1050 °C. The spectra were shifted vertically for a clear presentation. The inset shows the dependence of the integrated room-temperature PL emission, between 647.5 and 655 nm, on sample annealing temperature.

upon sample annealing temperature. In the inset to Figure 9.15 is shown the integrated PL intensity as a function of annealing temperature for the most intense pair of transitions at 650.5 and 652.5 nm. The PL intensity was found to increase exponentially with annealing temperature up to a temperature of 1050 °C, after which it decreased sharply. A least-squares fit of exponential form implies a thermal activation energy of 2.3 eV for the Pr-implanted GaN. The study by Chao and Steckl showed [59] a similar increase in the Pr-related PL efficiency with increased annealing time at a fixed temperature of 950 °C. The GaN band-edge PL near 3.4 eV at room temperature (not shown) was also studied for the Pr-implanted samples annealed at various temperatures; however, very little band-edge PL could be detected, and the PL intensity did not follow any obvious trend with annealing temperature.

Jadwisienczak et al. also studied Pr-implanted GaN films [61]. They examined the cathodoluminescence (CL) as well as the PL properties of these films. They observed richly structured spectra over a wide spectral range, 400–1000 nm, which is representative of the transition energies of Pr^{3+} ions. The intensity of the PL and CL emission remained strong from 11 K to 330 K.

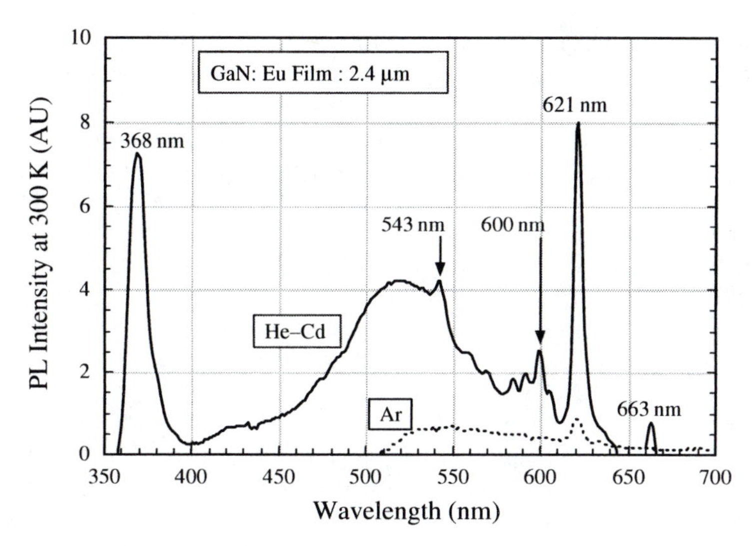

Figure 9.16 Visible and UV PL spectra, measured at 300 K, for Eu-doped GaN films grown by SSMBE on Si at 800 °C with the Eu cell temperature at 400 °C. Above-bandgap (He–Cd laser, solid line) and below-bandgap (Ar laser, dotted line) optical sources were used.

Another red emission line was achieved by Heikenfeld et al. by doping GaN films with Eu during SSMBE growth [27]. Solid sources were used for the Ga (7N purity) and Eu (3N purity) fluxes. Since Eu is the most reactive of the RE elements, extra precautions had to be maintained in the handling of this material. The Eu cell temperature was between 350 and 450 °C. The resulting GaN:Eu films had a thickness of about 2.4 μm. PL characterization was done using above-bandgap (He–Cd laser at 325 nm) and below-bandgap (Ar laser at 488 nm) optical excitation. Above-bandgap excitation at ~5 mW power led to a red emission with a dominant peak at 621 nm; see Figure 9.16. Several other visible lines were observed along with a broad yellow/green band. The line at 621 nm was attributed to a transition from the $^5D_0$ to the $^7F_2$ excited level; see Figure 9.1. The other lines appear to be due to transition from the $^5D_0$ to other ^7F_j excited levels. The above-bandgap excitation also produced a narrow emission at 368 nm corresponding to the band-edge of GaN. This result is in contrast with the data shown in Figure 9.13, in which the GaN:Er prepared by SSMBE did not exhibit significant band-edge emission. It seems that for GaN:Eu (SSMBE) samples, the band-edge does not provide an efficient radiative combination channel for reducing the excitation efficiency in the visible. Below-bandgap excitation of the GaN:Eu samples, at ~25 mW power, resulted in a much weaker visible emission, with only a single line located at 620 nm. Defect-mediated excitation of Eu^{3+} ions by below-bandgap photons does not seem to be very efficient for these samples.

Morishima et al. also observed red PL emission from GaN films doped with Eu [29]. The films were grown on sapphire or Si substrates using gas-source MBE. The optimum growth temperature was $\sim 700\,°C$. As in the work by Heikenfeld et al. [27] the authors found a dominant peak at 622 nm in the above-bandgap PL measurements, corresponding to a transition from the $^5D_0$ to the $^7F_2$ excited levels, and several other weaker emission lines in this region; however, they did not observe any band-edge emission from the GaN host. Furthermore, the intensity of the Eu^{3+} red emission was stronger for thin films grown on sapphire substrates than on Si substrates. This may be due to the polycrystalline nature of the GaN:Eu films grown on Si. Using PLE techniques the authors concluded that the Eu^{3+} red emission was due to excitation of GaN.

Hara et al. observed green PL emission from GaN films doped with Tb [30]. The GaN layers were grown on sapphire substrates in an atmospheric-pressure MOVPE system. Using a He–Cd laser source (325 nm) the authors recorded PL spectra at 24 and 290 K. The principal emission was at 541 nm, corresponding to a transition from the $^5D_4$ to the $^7F_5$ excited levels of Tb^{3+}. Although narrow emission lines were observed at 24 K, the room-temperature PL spectrum was dominated by a broad yellow band emission centered at ~560 nm.

The third primary color was produced by doping GaN films with Tm. Steckl et al. observed intense blue emission at room temperature from GaN:Tm films grown by SSMBE [28]. Solid sources were used for the Ga (7N purity) and Tm (3N purity) fluxes. The main emission peak was at 477 nm, corresponding to the $^1G_4 \rightarrow {}^4H_6$ transition of the Tm^{3+} ion. Additional emission lines were observed at 647 and 801 nm, corresponding to the $^1G_4 \rightarrow {}^3H_4$ and $^1G_4 \rightarrow {}^3H_5$ transitions; see Figure 9.1.

The preceding data indicate that it is possible to form a full-color display based on a single GaN film doped with RE elements. If a GaN film could be doped with two or more RE elements, and if each RE ion would emit simultaneously, then the human eye would perceive an additive mixture of the wavelengths. Various colors may be produced by adjusting the relative concentrations of the RE dopants and hence the intensities of the corresponding wavelengths. By using different RE elements (such as Eu, Er, and Tm) most of the colors in the visible spectrum may be realized.

9.4 ELECTRICAL EXCITATION

It has been estimated that the electrical excitation cross section for the $^4I_{15/2} \rightarrow {}^4I_{11/2}$ transition of Er^{3+} ions in a semiconductor is about 10^5 greater than the optical excitation cross section [62]. Consequently, considerable research activity has centered on producing electroluminescent semiconductor devices. Initially, this research was focused on Er-doped Si devices, later on Er-doped narrow-gap III–V devices, and more recently, on Er-doped wide-gap III–V devices. The results on RE-doped GaN based devices have been very encouraging.

9.4.1 Infrared Electroluminescence

Pankove and Torvik were the first to demonstrate [62,63] room-temperature operation of an electroluminescent device (ELD) emitting at 1.54 μm based on an Er-doped GaN semiconductor film. The device consisted of a metal/ i-GaN/n-GaN (m-i-n) structure grown by CVD on an R-plane sapphire substrate. The active region of the device, the i-GaN layer, was coimplanted with Er^+ and O^+ ions and annealed as in their prior studies [39]. The topography of the i-GaN top surface consisted of sharp ridges and valleys due to the growth conditions. This surface topography permitted high-field injection of electrons from the metal under reverse bias leading to strong electroluminescence (EL) at 1.54 μm. When the device was operated under forward bias, no Er^{3+}-related luminescence was observed. The excitation of Er^{3+} centers was apparently due to impact of energetic electrons. PL emission from the Er-implanted layer was obtained using optical pumping with a diode laser at 980 nm. The EL and PL spectra were very similar; however, at room temperature, the EL intensity was about three times greater than that of the PL. Nevertheless, the external quantum efficiency of the ELD, measured at −126 V and 318 μA, was very low, on the order of $\sim 10^{-7}$.

The research group at the University of Cincinnati produced a variety of ELDs using in situ RE-doped GaN films grown by SSMBE [55]. Initial results were obtained with Er-doped GaN films grown on Si [64]. EL measurements were performed on GaN:Er/Si devices with indium–tin (90%–10%) oxide (ITO) rectifying contacts. The electrical and optical properties of the ITO contact can be tailored for high transmission in the visible and/or IR regions and for low sheet resistance by adjusting the film thickness and annealing conditions [65]. Typical properties of ITO films were a thickness of ~100–500-nm, a sheet resistance of 5–100 Ω/square, and a transmission of ~70–95% in the visible and ~5–90% in the near-IR (at 1–2 μm). Er-related emission spectra from an ITO/GaN:Er ELD are shown in Figure 9.17. The near-IR emission spectrum exhibits three well-defined peaks between 1.51 and 1.56 μm, which correspond quite well to the PL spectrum shown in Figure 9.10. In addition there is a prominent peak at ~1 μm, which is probably associated with the $^4I_{11/2} \rightarrow {}^4I_{15/2}$ transition of Er^{3+} ions. The IR spectra were measured at a constant bias of 250 mW over a temperature range from 250 to 500 K. The emission intensity at 400 K was nearly as strong as at 250 K with little change in the spectrum. Even at 500 K significant emission from the ELD was observed.

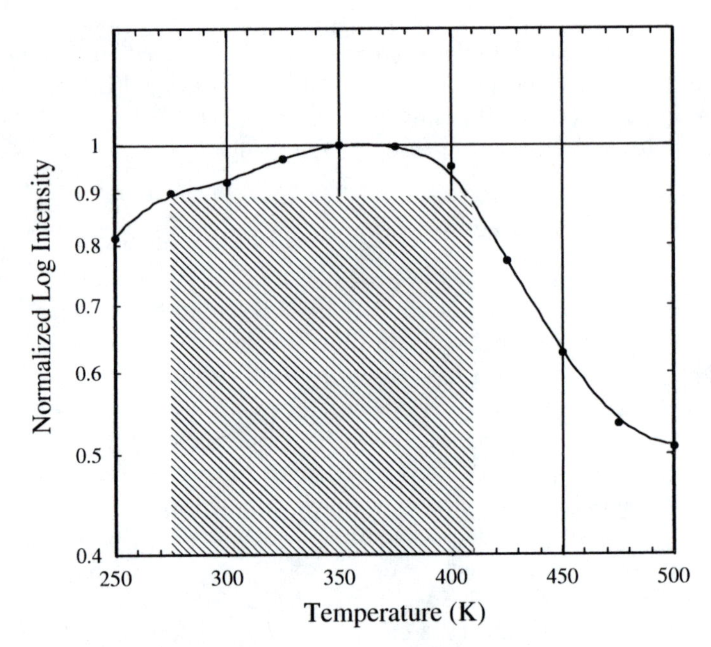

Figure 9.17 Temperature dependence of 1.5-μm electroluminescence from an Er-doped GaN device. The shaded region indicates the useful operating temperature range, from 275 to 410 K, for which the signal intensity changes by a maximum of ±5%.

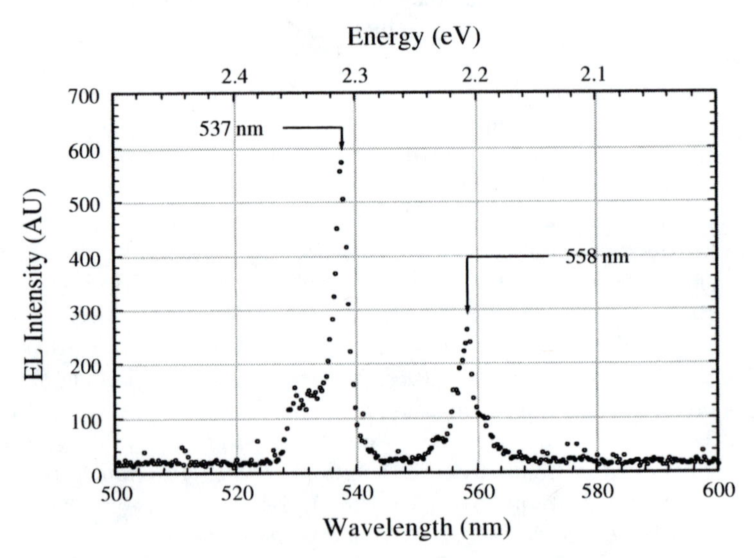

Figure 9.18 Green electroluminescence, taken at 300 K, for an ELD based on a GaN:Er film grown by SSMBE. The spectrum was measured at a constant bias of 4.4 mA.

9.4.2 Visible Electroluminescence

Steckl et al. [66] were the first to observe visible EL emission from an ITO/GaN:Er ELD; see Figure 9.18. The GaN:Er films were grown by SSMBE using solid sources for the Ga and the Er. The EL emission was in the green, with prominent peaks at wavelengths of 537 and 558 nm, corresponding to radiative transitions from the $^2H_{11/2}$ and $^4S_{3/2}$ excited states to the $^4I_{15/2}$ ground level of the Er^{3+} ion, as shown in Figure 9.1. The visible EL spectrum was in good agreement with the PL spectrum shown in Figure 9.13. The FWHM of each of the green lines was ~2.5–3.0 nm. This is much narrower than that from commercially available GaN-based green LEDs, which have a FWHM ranging from 45 to 90 nm depending on the structure and center wavelength [8,67].

Red emission from a GaN ELD was achieved by in situ RE-doping of GaN films with either Pr or Eu. SSMBE growth was used at the University of Cincinnati to produce ITO/GaN:RE devices. The EL spectrum from the ITO/GaN:Pr ELD had a prominent peak at ~650 nm, whereas the EL spectrum from the ITO/GaN:Eu device had a prominent peak at ~623 nm, see Figure 9.19. Morishima et al. [29] used gas-source MBE growth with a metallic source for Eu to produce MIS diodes on n-type Si(111). The EL spectrum from the GaN:Eu device, operated under a dc bias of ~5 V, had a prominent peak at ~622 nm. In each of these experiments the EL spectrum coincided quite well with the PL spectrum.

Steckl et al. achieved blue EL by doping GaN with Tm [28]. Strong blue emission, visible at room temperature, was observed from an ITO/GaN:Tm

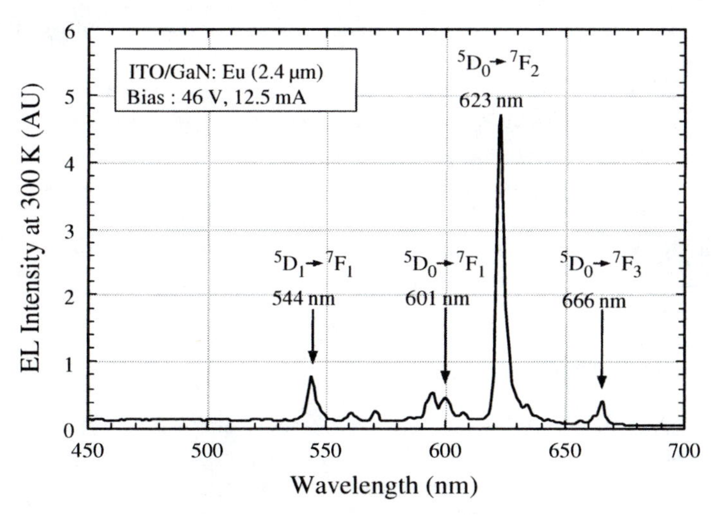

Figure 9.19 Red electroluminescence, taken at 300 K, for an ELD based on a GaN:Eu film grown by SSMBE. The spectrum was measured at a dc bias of 46 V and a current of 12.5 mA.

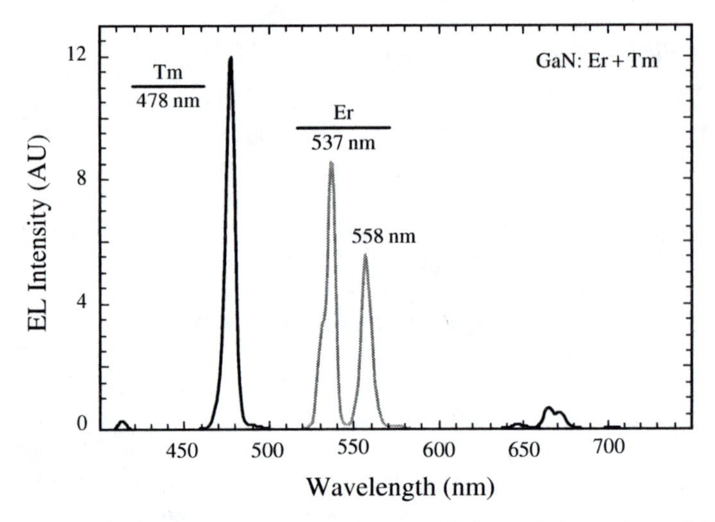

Figure 9.20 Visible electroluminescence, taken at 300 K, for an ELD based on a GaN film co-doped with Er and Tm. Bias current was 2.72 mA, and the resulting overall color was aqua.

ELD. The main emission peak was at 478 nm, and the EL spectrum coincided with the PL spectrum. In addition, Steckl et al. were able to co-dope [54,55] a GaN film with both Tm and Er. The visible EL spectrum from an ITO/GaN ELD codoped during SSMBE growth with Er and Tm is shown in Figure 9.20. Both the Tm blue emission at 478 nm and the Er green emission at 537–558 nm are present. This blue/green combination results in the overall color of aqua or turquoise.

In Figure 9.21 is a photograph of the visible emissions from different ITO/GaN:RE ELDs. The Er-doped device, shown in the upper left-hand corner, gives a green emission, and the Eu-doped device, shown in the upper right-hand corner, yields a red emission. The Tm-doped device, shown in the lower right-hand corner, gives a blue emission, and the Er+Tm–doped device, shown in the lower left-hand corner, produces the aqua emission. These results indicate a large potential impact of RE-doped GaN ELDs on display technology. Nearly every hue within the chromaticity diagram can be achieved using a combination of the basic colors shown in Figure 9.21.

9.5 CONCLUSIONS

Rare earth doping of GaN has recently attracted interest for optoelectronic applications ranging from optical communications to flat-panel display materials. Significant progress has been made during the last few years in incorporating rare earth ions into GaN and achieving light emission from the visible to the infrared. Prototype electroluminescent devices based

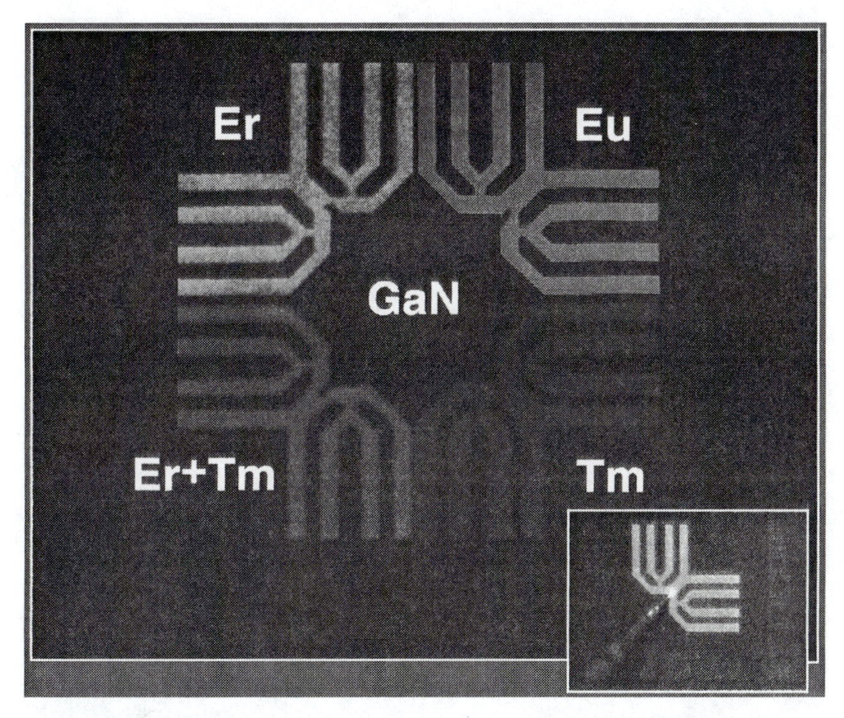

Figure 9.21 Visible electroluminescence, taken at 300 K, for different ELDs based on RE-doped GaN films. The three primary colors, red, blue, and green, are produced by the REs Eu, Er, and Tm, respectively. The aqua color is from a co-doping with Er and Tm.

on RE-doped GaN have been fabricated at wavelengths including 477 nm (blue), ~550 nm (green), ~620 and ~650 nm (red), ~800 nm (IR), 1 μm (IR), and 1.54 μm (IR). In addition, mixed colors using multiple RE doping of GaN have been demonstrated. Interestingly, many of these devices were fabricated on Si substrates, making integration with existing Si technology possible. Compared with semiconductor alloying of GaN/InN/AlN, RE-doped GaN ELDs are relatively simple, monolithic devices. Eventhough current RE-based GaN ELDs are still limited by a low efficiency, a significant increase in the overall efficiency of this novel class of ELDs is expected after optimization of materials and device structures.

References

1. S. Hufner, *Optical Spectra of Transparent Rare Earth Compounds* (Academic Press, 1978).
2. B. Henderson and G. F. Imbusch, *Optical Spectroscopy of Inorganic Solids* (Clarendon Press, Oxford, 1989).
3. A. A. Kaminski, *Crystalline Lasers: Physical Properties and Operating Schemes* (CRC Press, Boca Raton, FL, 1996).
4. W. Koechner, *Solid State Laser Engineering*, 3rd ed. (Springer-Verlag, 1992).
5. S. Saito, T. Imai, and T. Ito, *J. Lightwave Technol.*, **9**, 161 (1991).

6. G. R. Walker, N. G. Walker, R. C. Steele, M. J. Creaner, and M. C. Brain, *J. Lightwave Technol.*, **9**, 182 (1991).
7. L. F. Mollenauer, S. G. Evangelides, and H. A. Haus, *J. Lightwave Technol.*, **9**, 194 (1991).
8. S. B. Poole, D. N. Payne, R. J. Mears, M. E. Ferrman, and R. I. Lannings, *J. Lightwave Technol.*, **4**, 870 (1986).
9. E. Desurvire, J. R. Simpson and P. C. Becker, *Opt. Lett.*, **12**, 888 (1987).
10. H. Ennen, J. Schneider, G. Pomrenke, and A. Axmann, *Appl. Phys. Lett.*, **43** (10): 943–945 (1983).
11. See *Rare Earth Doped Semiconductors* I, *Mater. Res. Soc. Proc.*, **301**, edited by G. S. Pomerenke, P. B. Klein, and D. W. Langer (1993).
12. See *Rare Earth Doped Semiconductors* II, *Mater. Res. Soc. Proc.*, **422**, edited by S. Coffa, A. Polman, and R. N. Schwartz (1996).
13. J. M. Zavada and D. Zhang, *Solid-State Electron.*, **38**, 1285 (1995).
14. S. J. Pearton, *Mater. Sci. Rep.*, **4**, 313 (1990).
15. R. G. Wilson, R. N. Schwartz, C. R. Abernathy, S. J. Pearton, N. Newman, M. Rubin, T. Fu, and J. M. Zavada, *Appl. Phys. Lett.*, **65** (8): 992–994 (1994).
16. R. G. Wilson, F. A. Stevie, and C. M. Magee, *Secondary Ion Mass Spectrometry: A Practical Guide for Depth Profiling and Bulk Impurity Analysis* (Wiley, New York, 1989).
17. J. M. Zavada, R. G. Wilson, R. N. Schwartz, J. D. MacKenzie, C. R. Abernathy, S. J. Pearton, X. Wu, and U. Hömmerich, *Mater. Res. Soc. Symp. Proc.*, **422**, 193–197 (1996).
18. E. Alves, M. F. DaSilva, J. C. Soares, J. Bartels, R. Vianden, C. R. Abernathy, and S. J. Pearton, *MRS Internet J. Nitride Semicond. Res.*, **4S1**, G11.2 (1999).
19. J. D. MacKenzie, C. R. Abernathy, S. J. Pearton, U. Hömmerich, J. T. Seo, R. G. Wilson, J. M. Zavada, *Appl. Phys. Lett.*, **72**, 2710 (1998).
20. M. E. Overberg, J. Brand, J. D. MacKenzie, C. R. Abernathy, S. J. Pearton, and J. M. Zavada, *Electrochem. Soc. Proc.*, **99–4**, 195 (1999).
21. A. J. Steckl and R. Birkhahn, *Appl. Phys. Lett.*, **73**, 1702 (1998).
22. R. Birkhahn and A. J. Steckl, *Appl. Phys. Lett.*, **73**, 2143 (1998).
23. R. H. Birkhahn, R. Hudgins, D. S. Lee, A. J. Steckl, R. J. Molnar, A. Saleh, and J. M. Zavada, *J. Vac. Sci. Technol.*, **D17** (3), 1195 (1999).
24. R. H. Birkhahn, R. Hidgins, D. S. Lee, A. J. Steckl, R. G. Wilson, A. Saleh, and J. M. Zavada, *MRS Internet J. Nitride Semicond. Res.*, **421**, G3.80 (1999).
25. K. Lorenz, R. Vianden, R. Hirkhahn, A. J. Steckl, M. F. da Silva, J. C. Soares, E. Alves, *Elsevier Sci., B.V.*, **7513**, 1–6 (1999).
26. R. H. Birkhahn, M. Garter, and A. J. Steckl, *Appl. Phys. Lett.*, **74**, 2161 (1999).
27. J. Heikenfeld, M. Garter, D. S. Lee, R. Birkhahn, and A. J. Steckl, *Appl. Phys. Lett.*, **75**, 1189 (1999).
28. A. J. Steckl, M. Garter, D. S. Lee, J. Heikenfeld, and R. Birkhahn, *Appl. Phys. Lett.*, **75**, 2184 (1999).
29. S. Morishima, T. Maruyama, M. Tanaka, Y. Masumoto, and K. Akimoto, *Phys. Status Solidi A*, **176**, 113 (1999).
30. K. Hara, N. Ohtake, and K. Ishu, *Phys. Status Solidi B*, **216**, 625 (1999).
31. R. A. Hogg, K. Takahei, A. Taguchi, and Y. Horikoshi, *Mater. Res. Soc. Symp. Proc.*, **422**, 167–172 (1996).
32. P. N. Favennec, H. L'Haridon, D. Moutonnet, M. Salvi, and M. Gauneau, *Jpn. J. Appl. Phys.*, **29** (4), L524–526 (1990).
33. J. Michel, J. L. Benton, R. F. Ferrante, D. C. Jacobson, D. J. Eaglesham, E. A. Fitzgerald, Y. Xie, J. M. Poate, and L. C. Kimerling, *J. Appl. Phys.*, **70** (5), 2672–2678 (1991).
34. J. E. Colon, D. W. Elsaesser, Y. K. Yeo, R. L. Hengehold, and G. S. Pomrenke, *Mater. Res. Soc. Symp. Proc.*, **301**, 169–174 (1993).
35. P. N. Favennec, H. L'Haridon, M. Salvi, D. Moutonnet, and Y. L. Guillou, *Electron. Lett.*, **25** (11), 718–719 (1989).
36. A. J. Neuhalfen and B. W. Wessels, *Appl. Phys. Lett.*, **60** (21): 2657–2659 (1992).
37. K. Takahei and A. Taguchi, *J. Appl. Phys.*, **74**, 1979 (1993).
38. A. Taguchi, K. Takahei, and Y. Horikoshi, *J. Appl. Phys.*, **76** (11), 7288–7295 (1994).
39. J. T. Torvik, R. J. Feuerstein, C. H. Qui, M. W. Leksono, F. Namavar, and J. I. Pankove, *Mater. Res. Soc. Symp. Proc.*, **422**, 199–204 (1996).

40. J. T. Torvik, C. H. Qui, R. J. Feuerstein, J. I. Pankove, and F. Namavar, *J. Appl. Phys.*, **81** 6343 (1997).
41. E. Silkowski, Y. K. Yeo, R. L. Hengehold, B. Goldenberg, and G. S. Pomerenke, *Mater. Res. Soc. Symp. Proc.*, **422**, 69–74 (1996).
42. U. Hömmerich, J. T. Seo, J. D. MacKenzie, C. R. Abernathy, R. Birkhahn, A. J. Steckl, and J. M. Zavada, MRS Fall 1999 meeting, paper W11.65.
43. M. Thaik, U. Hömmerich, R. N. Schwartz, R. G. Wilson, and J. M. Zavada, *Appl. Phys. Lett.*, **71**, 2641–2643 (1997).
44. J. C. Zolper, H. H. Tan, J. S. Williams, J. Zou, D. J. H. Cockayne, S. J. Pearton, M. Hagerott-Crawford, R. J. Karlicek, *Appl. Phys. Lett.*, **70**, 2729–2731 (1997).
45. U. Hömmerich, J. T. Seo, Myo Thaik, J. D. MacKenzie, C. R. Abernathy, S. J. Pearton, R. G. Wilson, and J. M. Zavada, *Internet J. Nitride Semicond. Res.*, **4S1**, G11.6 (1999).
46. J. D. MacKenzie, C. R. Abernathy, S. J. Pearton, U. Hömmerich, J. T. Seo, R. G. Wilson, and J. M. Zavada, *Appl. Phys. Lett.*, **72**, 2710 (1998).
47. S. Kim, S. J. Rhee, D. A. Turnbull, E. E. Reuter, X. Li, J. J. Coleman, and S. G. Bishop, *Appl. Phys. Lett.*, **71**, 231–233 (1997).
48. S. Kim, S. J. Rhee, D. A. Turnbull, X. Li, J. J. Coleman, and S. G. Bishop, and P. B. Klein, *Appl. Phys. Lett.*, **71**, 2662–2664 (1997).
49. U. Hömmerich, J. T. Seo, M. Thaik, C. R. Abernathy, J. D. MacKenzie, and J. M. Zavada, *J. Alloys Compd.*, in press.
50. S. Kim, S. J. Rhee, D. A. Turnbull, E. E. Reuter, X. Li, J. J. Coleman, and S. G. Bishop, *Appl. Phys. Lett.*, **71**, 231 (1997).
51. P. B. Klein and G. S. Pomrenke, *Electron. Lett.*, **24** (24), 1502–1503 (1988).
52. W. J. Choyke, R. P. Devaty, L. I. Clemen, M. Yoganathan, G. Pensl, and Ch. Hassler, *Appl. Phys.*, **65** (13), 1668–1671 (1994).
53. H. Chen, K. Gurumurugan, M. E. Kordesch, MRS Fall 99 meeting, paper W3.16.
54. A. J. Steckl and J. M. Zavada, *MRS Bulletin*, **24** (9), 33–38, 1999.
55. A. J. Steckl, J. Heikenfeld, M. Garter, R. Birkhahn, and D. S. Lee, *Compd. Semicond.*, **6** (1), 48 (2000).
56. H. J. Lozykowski, W. M. Jadwisienczak, and I. M. Brown, *Appl. Phys. Lett.*, **74**, 1129 (1999).
57. H. J. Lozykowski, W. M. Jadwisienczak, and I. M. Brown, *Solid State Commun.*, **110**, 253 (1999).
58. A. J. Steckl, *Proc. Advanced Workshop on Frontiers in Electronics*, IEEE Cat. # 97TH8292, 47, (Jan. 1997).
59. L. C. Chao and A. J. Steckl, *Appl. Phys. Lett.*, **74**, 2364 (1999).
60. J. M. Zavada, R. A. Mair, C. J. Ellis, J. Y. Lin, H. X. Jiang, R. G. Wilson, P. A. Grudowski, R. D. Dupuis, *Appl. Phys. Lett.*, **75**, 790 (1999).
61. W. M. Jadwisienczak, H. J. Lozykowski, MRS Fall 1999 meeting, paper W11.62.
62. J. T. Torvik, R. J. Feuerstein, J. I. Pankove, C. H. Qui, and F. Namavar, *Appl. Phys.*, **69** (14), 2098–2100 (1996).
63. C. H. Qui, M. W. Leksono, J. I. Pankove, J. T. Torvik, R. J. Feuerstein, and F. Namavar, *Appl. Phys.*, **66**, 562–563 (1995).
64. M. Garter, J. Scofield, R. Birkhahn, and A. J. Steckl, *Appl. Phys. Lett.*, **74**, 182 (1999).
65. M. Garter, R. Birkhahn, A. J. Steckl, and J. Scofield, *MRS Internet J. Nitride Semicond. Res.*, **4S1**, G11.3 (1999).
66. A. J. Steckl, M. Garter, R. Birkhahn, and J. Scofield, *Appl. Phys. Lett.*, **73**, 2450 (1999).
67. S. Nakamura, M. Senoh, N. Iwasa, and S. Nagahama, *Jpn. J. Appl. Phys.*, **34**, L797 (1995).

Index